T0192647

Geometry from a Differentiable Viewpoint

The development of geometry from Euclid to Euler to Lobachevskiĭ, Bolyai, Gauss, and Riemann is a story that is often broken into parts – axiomatic geometry, non-Euclidean geometry, and differential geometry. This poses a problem for undergraduates: Which part is geometry? What is the big picture to which these parts belong?

In this introduction to differential geometry, the parts are united with all of their interrelations, motivated by the history of the parallel postulate. Beginning with the ancient sources, the author first explores synthetic methods in Euclidean and non-Euclidean geometry and then introduces differential geometry in its classical formulation, leading to the modern formulation on manifolds such as space-time. The presentation is enlivened by historical diversions such as Hugyens's clock and the mathematics of cartography. The intertwined approaches will help undergraduates understand the role of elementary ideas in the more general, differential setting.

This thoroughly revised second edition includes numerous new exercises, together with newly prepared solutions to selected exercises.

JOHN MCCLEARY is professor of mathematics at Vassar College on the Elizabeth Stillman Williams Chair. His research interests lie at the boundary between geometry and topology, especially where algebraic topology plays a role. His research papers have appeared in journals such as *Inventiones Mathematicae* and the *American Journal of Mathematics*, and he has also written expository papers in the *American Mathematical Monthly*. He is interested in the history of mathematics, particularly the history of geometry in the nineteenth century and of topology in the twentieth century. He is the author of *A User's Guide to Spectral Sequences* and *A First Course in Topology: Continuity and Dimension*, and he has edited proceedings in topology and in history, as well as a volume of the collected works of John Milnor.

Geometry
from a Differentiable Viewpoint

Second Edition

JOHN McCLEARY
Vassar College

CAMBRIDGE
UNIVERSITY PRESS

University Printing House, Cambridge CB2 8BS, United Kingdom

One Liberty Plaza, 20th Floor, New York, NY 10006, USA

477 Williamstown Road, Port Melbourne, VIC 3207, Australia

314-321, 3rd Floor, Plot 3, Splendor Forum, Jasola District Centre, New Delhi - 110025, India

79 Anson Road, #06-04/06, Singapore 079906

Cambridge University Press is part of the University of Cambridge.

It furthers the University's mission by disseminating knowledge in the pursuit of education, learning and research at the highest international levels of excellence.

www.cambridge.org
Information on this title: www.cambridge.org/9780521133111

First published 1995
Second edition published 2013

A catalogue record for this publication is available from the British Library

Library of Congress Cataloging in Publication data
McCleary, John, 1952–
Geometry from a differentiable viewpoint / John McCleary. – 2nd ed.
p. cm.
Includes bibliographical references and indexes.
ISBN 978-0-521-11607-7 (hardback) – ISBN 978-0-521-13311-1 (pbk.)
1. Geometry, Differential. I. Title.
QA641.M38 2012
516.3′6–dc23 2012017159

ISBN 978-0-521-11607-7 Hardback
ISBN 978-0-521-13311-1 Paperback

To my sisters

Mary Ann, Denise, Rose

Contents

Preface to the second edition

Giving an author a chance to rewrite a text is a mixed blessing. The temptation to rewrite every sentence is strong, as is the temptation to throw everything out and start over. When I asked colleagues who had taught from the book what I might change, I was surprised in one case to hear that I should change nothing. That presented a challenge to preparing a second edition.

Several folks were kind enough to point out errors in the first edition that I have fixed. Many thanks to you. Among these errors was a mishandling of congruences that has led to additional material in this edition. The notion of congruence leads to the important theory of transformation groups and to Klein's Erlangen Program. I have taken some of the opportunities to apply arguments using transformations, exposing another of the *pillars of geometry* (Stillwell 2005) to the reader.

Succumbing to the second temptation, I have reordered the material significantly, resulting in one fewer chapter and a better story line. Chapter 4 is a more or less self-contained exposition of non-Euclidean geometry, and it now parallels Chapter 14 better. I have added material including Euclid's geometry of space, further results on cycloids, another map projection, Clairaut's relation, and reflections in the Beltrami disk. I have also added new exercises. Due to a loss of files for the first edition, all of the pictures have been redrawn and improved. In many places I have made small changes that I hope improve the clarity of my telling of this amazing and rich story.

Introduction

ΑΓΕΩΜΕΤΡΗΤΟΣ ΜΗΔΕΙΣ ΕΙΣΙΤΩ

ABOVE THE ENTRANCE TO PLATO'S ACADEMY

One of the many roles of history is to tell a story. The history of the Parallel Postulate is a great story. It spans more than two millennia, stars an impressive cast of characters, and contains some of the most beautiful results in all of mathematics. My immodest goal for this book is to tell this story.

Another role of history is to focus our attention. We can then see a thread of unity through a parade of events, people, and ideas. My more modest goal is to provide a focus with which to view the standard tools of elementary differential geometry, and discover how their history emerges out of Geometry writ large, and how they developed into the modern, global edifice of today.

In recent years, to offer a course in differential geometry to undergraduates has become a luxury. When such a course exists, its students often arrive with a modern introduction to analysis, but without having seen geometry since high school. In the United States high school geometry is generally elementary Euclidean geometry based on Hilbert's axiom scheme. Such an approach is a welcome introduction to the rigors of axiomatic thinking, but the beauty of Euclidean geometry can get lost in the carefully wrought two-column proof. If mentioned at all, the marvels of non-Euclidean geometry are relegated to a footnote, enrichment material, or a "cultural" essay. This situation is also the case in most current introductions to differential geometry. The modern subject turns on problems that have emerged from the new foundations that are far removed from the ancient roots of geometry. When we teach the new and cut off the past, students are left to find their own ways to a meaning of the word *geometry* in differential geometry, or failing that, to identify their activity as something different and unconnected.

This book is an attempt to carry the reader from the familiar Euclid to the state of development of differential geometry at the beginning of the twentieth century. One narrow thread that runs through this large historical period is the search for a proof of Euclid's Postulate V, the Parallel Postulate, and the eventual emergence of a new and non-Euclidean geometry. In the course of spinning this tale, another theme enters – the identification of properties of a surface that are intrinsic, that is, independent of the manner in which the surface is embedded in space. This idea, introduced by Gauss, provides the analytic key to properties that are really geometric, opening new realms to explore.

The book is written in sonata-allegro form. Part A opens with a prelude—a small and orienting dose of spherical geometry, whose generalizations provide important guideposts in the development of non-Euclidean geometry. One of the main themes of the sonata is played out in Chapters 2 and 3, which focus on Book I of Euclid's *The Elements*, one of the most important works of Western culture, and the later criticism of Euclid's theory of parallels. The other main themes are found in Chapter 4, which contains an account of synthetic non-Euclidean geometry as introduced and developed by Lobachevskiĭ, Bolyai, and Gauss. I have tried to follow the history in Part A basing my account on the masterfully written books of Gray (1979) and Rosenfeld (1998).

What remains unresolved at the end of Part A is the existence of a concrete representation of non-Euclidean geometry, that is, a rigorous model. An analogous situation is given by the ontological status of complex numbers in the time before Argand and Gauss. The utility of $\sqrt{-1}$ in algebraic settings does not present a model in which such a number exists. Identifying the plane with the complex numbers, and $\sqrt{-1}$ with rotation through a right angle, gives concrete representation of the desired structure. The work of Lobachevskiĭ, Bolyai, and Gauss poses the need for a model of some sophistication. With the introduction of analytic ideas, formulas like the Lobachevskiĭ–Bolyai Theorem (Theorem 4.29) reveal the basic role that analysis can play in geometry. The portrait of the non-Euclidean plane through its trigonometry so perfectly parallels the trigonometry of the sphere that, once developed, led the founders of non-Euclidean geometry to trust in its existence.

Part B begins with curves, a success story based on the introduction of appropriate coordinates and measures such as curvature and torsion. There is a brief interlude in Chapter 5 where the story of involutes, evolutes, and Hugyens's clock is told. Though it does not bear on the Parallel Postulate, Hugyens's work is paradigmatic for differential geometry; questions of an applied nature (cartography, motion, gravity, optics) press the geometer to find new ways to think about basic notions.

Chapter 7 presents the basic theory of surfaces in space. In another interlude, Chapter 7bis presents map projections, a particular application of the definitions and apparatus associated to a surface, in this case, the sphere. Chapters 8 and 9 develop the analogue of curvature of curves for a surface. This curvature, Gaussian curvature, is shown to be independent of the manner in which the surface lies in space, that is, it is an intrinsic feature of the surface. Gauss (1828) found this property to be remarkable (*egregium*) because it identifies a new point of view:

...we see that two essential different relations must be distinguished, namely, on the one hand, those that presuppose a definite form of the surface in space; on the other hand, those that are independent of the various forms which the surface may assume.

Guided by the intrinsic, in Chapter 10 we introduce geodesics, that is, "lines" on a surface. We compute the integral of Gaussian curvature in Chapter 11, which leads to the Gauss–Bonnet Theorem and its global consequences. In Chapter 12 we finally arrive at an analytic recipe for a model of the non-Euclidean plane – it is a complete, simply connected surface of constant negative Gaussian curvature. Hilbert's Theorem shows us that our investigations have reached an impasse; there are no such surfaces in space. And so a more general notion of surface is needed.

The final part of the sonata is a recapitulation of themes from Part A. Led by insights of Riemann, we introduce the key idea of an abstract surface, a generalization of the surfaces in space. In Chapter 14 a theorem of Beltrami leads to a model of non-Euclidean geometry as an abstract surface, and further development leads to the other well-known models by Poincaré. After reprising the non-Euclidean geometry of Lobachevskiĭ, Bolyai, and Gauss in these models, I end with a coda based on the theme of the intrinsic. Riemann's visionary lecture of 1854 is discussed along with the structures motivated by his ideas, which include the modern idea of an n-dimensional manifold, Riemannian and Lorentz metrics, vector fields, tensor fields, Riemann–Gauss curvature, covariant differentiation, and Levi–Civita parallelism. Chapter 15 is followed by a translation of Riemann's *Habiliationsvortrag*: "On the hypotheses which lie at the foundation of geometry." My translation is based on Michael Spivak's found in Spivak (1970, Vol. 2). My thanks to him for permission to use it. I have tweaked it a little to restore Riemann's rhetorical structure and to clean up the language a bit.

Exercises follow each chapter. Those marked with a dagger have a solution in the final appendix. My solutions are based on work of Jason Cantarella, Sean Hart, and Rich Langford. Any remaining errors are mine, however.

The idea of presenting a strict chronology of ideas throughout the book would have limited the choice of topics, and so I have chosen some evident anachronisms in the pursuit of clarity and unity of story. I have also chosen to restrict my attention to functions that are smooth, though this restriction is not required to prove most of the theorems. The interested reader should try to find the most general result by identifying the appropriate degree of differentiability needed for each construction. This concession is to uniformity and simplicity in the hopes that only the most geometric details remain.

How to use this book

This book began as a semester-long course at Vassar College, first taught this way in 1982 (my thanks to Becky Austen, Mike Horner, and Abhay Puri for making it a good experience). Since then I have added sections, details, and digressions that make it impossible to cover the entire book in a semester. In order to use the book in a thirteen-week semester, I recommend the following choices:

Chapters 1 and 4 (Chapters 2 and 3 as a reading assignment)
Chapter 5 through the Fundamental Theorem
Chapter 6 (Appendix as a reading assignment)
Chapters 7, 8, and 9 (Chapter 7bis as a reading assignment)
Chapter 10 up to the statement of the Hopf–Rinow Theorem
Chapter 11, up to Jacobi's Theorem
Chapter 12 (skip the proof of Hilbert's Theorem)
Chapter 13
Chapter 14 with a review of stereographic projection

If students have had a multivariable calculus course in which parametric surfaces are covered well, then Chapters 1, 7, and 7bis make a nice unit on the sphere and can be done first. If the focus is on differential geometry, make Chapters 2, 3, and 4 a reading assignment and begin at Chapter 5. Then tie Chapters 4 and 14 together at the appropriate time. Getting to Chapter 15 serves students who want to get ready to study General Relativity. It is also possible to do a course on non-Euclidean geometry by presenting Chapters 1 through 4, then Chapters 7bis and 14, cherry-picking the results of the previous chapters as needed to fill gaps.

The general prerequisites for the book are a good knowledge of multivariable calculus, some elementary linear algebra including determinants and inner products, and a little advanced calculus or real analysis through compactness. A nodding acquaintance with differential equations is nice, but not required. A student with strong courses in multivariable calculus and linear algebra can take on faith results from classical analysis, such as convergence criteria and the extreme-value principle, and comfortably read the text. To read further into all of the nooks and crannies of the book, an acquaintance with point set topology is recommended.

Acknowledgments

Many folks have offered their encouragement, time, and advice during the preparation of the first edition, and its subsequent rewrite. Most thanks are due to my grammatical and mathematical conscience, Jason Cantarella, who combed the first edition for errors of expression and exposition. A small part of his efforts were supported by a grant from the Ford Foundation and by research positions at Vassar College. I hope that I have followed up on all his suggestions for improvements in the second edition. Thanks to MaryJo Santagate for typing up the handwritten notes from my course. Thanks to Diane Winkler, Joe Chipps, Griffin Reiner-Roth, and Tian An Wong for help with getting a text file together to edit for the second edition. Anthony Graves-McCleary lent his expert eye to improving the diagrams in the second edition. David Ellis had the courage to try the unedited text when he taught differential geometry. His advice was helpful, as was the advice of Elizabeth Denne who taught between editions. My thanks to Bill Massey, Larry Smith, Jeremy Shor, Harvey Flad, Tom Banchoff, David Rowe, Ruth Gornet, Erwin Kreyszig, John Stillwell, David Cox, Janet Talvacchia, and anonymous editors for sharing knowledge, materials, and guidance. Thanks for the help with translations to Rob Brown and Eliot Schreiber. Special thanks to Thomas Meyer and Robert Schmunk who shared their expertise in geodesy and cartography. Thanks to John Jones for introducing me to the publisher. I thank Bindu Vinod for expert shepherding of the second edition from files to book. For all her patience and encouragement, special thanks to Lauren Cowles of Cambridge University Press. Greg Schreiber copy-edited the first edition, providing improvements beyond this author's expectations.

The copy editor of the second edition likewise provided sage advise to improve my prose. Special thanks to Jeremy Gray whose books, articles, and conversations have provided much inspiration and encouragement in this project. Finally, thanks to my family for support of another writing project.

PART A

Prelude and themes

Synthetic methods and results

1

Spherical geometry

If the river carried away any portion of a man's lot he appeared before the king, and related what had happened; upon which the king sent persons to examine and determine by measurement the exact extent of the loss .... From this practice, I think, Geometry first came to Egypt, whence it passed to Greece.

HERODOTUS, *The Histories*

The earliest recorded traces of geometry among the ancient Babylonian and Egyptian cultures place its origin in the practical problems of the construction of buildings (temples and tombs), and the administration of taxes on the land (Katz, 2008). Such problems were mastered by the scribes, an educated elite in these cultures. The word geometry is of Greek origin, γεωμετρία, to "measure the earth." Geometric ideas were collected, transformed by rigorous reasoning, and eventually developed by EUCLID (*ca.* 300 B.C.E.) in his great work *The Elements*, which begins with the geometry of the plane, the abstract field of the farmer.

Another source of ancient geometric ideas is astronomy. The motions of the heavens determined the calendar and hence times for planting and for religious observances. The geometry at play in astronomy is *spherical geometry*, the study of the relations between figures on an idealized celestial sphere. Before discussing Euclid's work and its later generalizations, let us take a stroll in the garden of spherical geometry where many of the ideas that will later concern us arise naturally.

The *sphere* of radius $R > 0$ is the set of points in space the distance R from a given point O, the *center* of the sphere. Introducing rectangular coordinates on space $\mathbb{R}^3$, we choose the center to be the origin $(0,0,0)$. With the familiar Pythagorean distance between points in $\mathbb{R}^3$, the sphere is the algebraic set

$$S_R = \{(x,y,z) \in \mathbb{R}^3 \mid x^2 + y^2 + z^2 = R^2\}.$$

Basic plane geometry is concerned with points and lines, with their incidence relations and congruences of figures. To study such notions on a sphere, we first choose what we mean by the words "congruence," "line," and "line segment." A congruence of a sphere is formally a motion of the sphere that does not change the distance relations between points on it. There are two motions that are familiar to anyone who has held a globe, namely, a *rotation* around an axis through the center of the sphere, and a *reflection* across a plane through the center of the sphere. In both cases the sphere goes over to itself and distance relations between points are preserved. In fact, we prove later that any congruence of the sphere is a product of these basic congruences.

For lines and line segments, we distinguish a particular class of curves on the sphere.

Definition 1.1. *A* **great circle** *on a sphere is the set of all points on the sphere that also lie on a plane that passes through the center of the sphere (for example, on the Earth as a sphere, the Equator or an arc of constant longitude).*

Ancient geometers understood that great circles share many formal properties with lines in the plane, making them a natural choice for lines on a sphere:

(1) Given two points on the sphere that are not *antipodal* (P and Q are antipodal if the line in space joining P to Q passes through the center of the sphere), there is a unique great circle joining that pair of points (THEODOSIUS end of the 2nd century B.C.E.): To construct this great circle, take the plane determined by the pair of points and the center of the sphere and form the intersection of that plane with the sphere. The analogue of a line segment on the sphere is a *great circle segment* defined to be a portion of a great circle between two points on it.
(2) In the plane, if we reflect points across a fixed line, then the line itself is fixed by the reflection. If we reflect the sphere across the plane that determines a great circle, then this great circle is also fixed by the reflection.
(3) Finally, any pair of great circles meet in a pair of antipodal points. The great circles can then be related by a rotation of the sphere around the diameter determined by their intersection. Since rotations around a line through the origin preserve all the measurable quantities on the sphere, such as length, area, angle, and so on, we find, as in the plane, all great circles are geometrically identical.

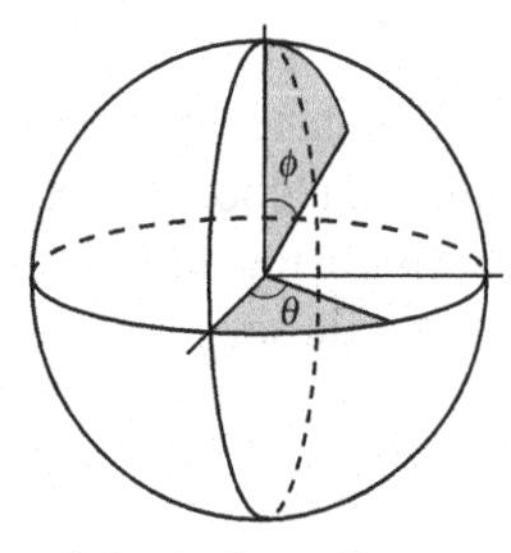

Spherical coordinates.

To track the positions of heavenly objects and to calculate the positions of the sun and moon, an analytic expression for points on the sphere is desirable. In terms of rectangular coordinates, the sphere S_R may be given as the set of points $(x,y,z) \in \mathbb{R}^3$ satisfying $x^2+y^2+z^2=R^2$. However, these coordinates were not used by the first astronomers. Measuring from the intersection of the ecliptic with the Equator, the angle θ corresponds to longitude, and measuring down from the axis of rotation of the Earth gives ϕ, the colatitude and (θ,ϕ) are the *spherical coordinates* on S_R. They translate to rectangular coordinates by the representation:

$$S_R = \{(R\cos\theta\sin\phi, R\sin\theta\sin\phi, R\cos\phi) \mid 0 \le \theta \le 2\pi, 0 \le \phi \le \pi\}.$$

Using this representation we can see why a great circle shares the property of lines of being the shortest path between two points. A curve on the sphere may be represented as a function $\alpha\colon [a,b] \to S_R$ given by $\alpha(t) = (\theta(t),\phi(t))$ in spherical coordinates, or by

$$\alpha(t) = (R\cos(\theta(t))\sin(\phi(t)), R\sin(\theta(t))\sin(\phi(t)), R\cos(\phi(t)))$$

in rectangular coordinates. As $\alpha(t)$ is a curve in $\mathbb{R}^3$, we can apply a little multivariable calculus to compute the *length* of α, denoted $l(\alpha)$:

$$l(\alpha) = \int_a^b \sqrt{\alpha'(t)\cdot\alpha'(t)}\,dt.$$

If P and Q are points in S_R and $\alpha\colon [a,b] \to S_R$ is a curve joining $P = \alpha(a)$ to $Q = \alpha(b)$, then we can compare $l(\alpha)$ with the length of the great circle segment joining P to Q. By rotating the sphere we can take $P = (1,0,0)$ and Q to lie along the $\theta = 0$ meridian. The great circle joining P to Q may be coordinatized by $\beta(t) = (0,t)$ in spherical and $\beta(t) = (R\sin(t), 0, R\cos(t))$ in rectangular coordinates for $0 \le t \le \theta_0$. Computing the derivative we get $\beta'(t) = (R\cos(t), 0, -R\sin(t))$ and $\beta'(t)\cdot\beta'(t) = R^2$. Hence $l(\beta) = \displaystyle\int_0^{\theta_0} R\,dt = R\theta_0$.

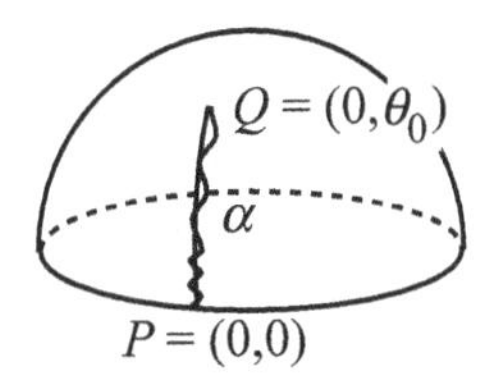

Suppose $\alpha(t)$ varies from the great circle by zigzagging horizontally while moving vertically like the great circle. Then in spherical coordinates we can write $\alpha(t) = (\eta(t), t)$ for a function $\eta\colon [0,\theta_0] \to \mathbb{R}$ with $\eta(0) = 0$ and $\eta(\theta_0) = 0$. This gives rectangular coordinates and derivative

$$\begin{aligned}
\alpha(t) &= (R\cos(\eta(t))\sin(t), R\sin(\eta(t))\sin(t), R\cos(t)) \\
\alpha'(t) &= (-R\sin(\eta(t))\eta'(t)\sin(t), R\cos(\eta(t))\eta'(t)\sin(t), 0) \\
&\quad + (R\cos(\eta(t))\cos(t), R\sin(\eta(t))\cos(t), -R\sin(t)) = \mathbf{u} + \mathbf{v}.
\end{aligned}$$

Then, since $\mathbf{u}\cdot\mathbf{v} = 0$,

$$\alpha'(t)\cdot\alpha'(t) = \mathbf{u}\cdot\mathbf{u} + 2\mathbf{u}\cdot\mathbf{v} + \mathbf{v}\cdot\mathbf{v} = R^2\sin^2(t)(\eta'(t))^2 + R^2,$$

it follows that

$$l(\alpha) = \int_0^{\theta_0} \sqrt{R^2\sin^2(t)(\eta'(t))^2 + R^2}\,dt \ge \int_0^{\theta_0} R\,dt = R\theta_0 = l(\beta).$$

Thus the great circle segment is less than or equal in length to the length of the curve α of the type considered. It is a small step to all curves joining P to Q (taken later), and so great circles satisfy another property of lines, that is, following ARCHIMEDES (287–212 B.C.E.), a line is a curve that is the shortest path joining any two points that lie on it.

Trigonometry is concerned with the relations between lengths of sides and angles of triangles. The *length* of a path between two points along a great circle is easy to define; it is the measure in radians of the angle made by the radii at each point multiplied by the radius of the sphere. This length is unchanged when we rotate the sphere around some axis or reflect the sphere across a great circle.

In his influential work on trigonometry (Euler, 1753), L. EULER (1707–87) introduced the convention of naming the interior angles of a triangle in upper case letters for the vertices, and the lengths of the opposite sides in the corresponding lower case letters. In this notation we prove a fundamental relation among the sides of a right spherical triangle.

Theorem 1.2 (Spherical Pythagorean Theorem). *If $\triangle ABC$ is a right triangle on a sphere of radius R with right angle at the vertex C, then*

$$\cos\frac{c}{R} = \cos\frac{a}{R}\cdot\cos\frac{b}{R}.$$

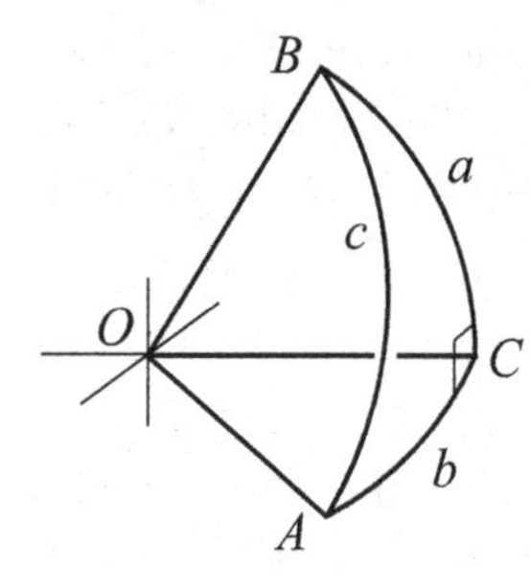

PROOF: By rotating the sphere we can arrange that the point C has coordinates $(0,R,0)$ and that the point A lies in the xy-plane. The point B then has spherical coordinates $((\pi/2)-(a/R),\pi/2)$. This follows because the central angle subtending a great circle segment of length a is a/R. With these choices we have

$$A=(R\sin\frac{b}{R},R\cos\frac{b}{R},0)\quad B=(0,R\cos\frac{a}{R},R\sin\frac{a}{R})$$
$$C=(0,R,0).$$

The central angle subtending AB is c/R. From the elementary properties of the dot product on $\mathbb{R}^3$ we compute the cosine of the angle between the vectors A and B:

$$\cos\frac{c}{R}=\frac{A\cdot B}{\|A\|\cdot\|B\|}=\frac{R^2\cos(a/R)\cos(b/R)}{R^2}=\cos\frac{a}{R}\cos\frac{b}{R}. \qquad \blacksquare$$

This result is called the Pythagorean Theorem because it relates the hypotenuse of a right spherical triangle to its sides. To see the connection with the classical Pythagorean Theorem (see Chapter 2), recall the Taylor series for the cosine (at $x=0$):

$$\cos x = 1-\frac{x^2}{2}+\frac{x^4}{4}-\frac{x^6}{6}+\cdots.$$

The Spherical Pythagorean Theorem gives the equation

$$\begin{aligned} 1-\frac{c^2}{2R^2}+\cdots &= \left(1-\frac{a^2}{2R^2}+\cdots\right)\left(1-\frac{b^2}{2R^2}+\cdots\right)\\ &= 1-\frac{a^2}{2R^2}-\frac{b^2}{2R^2}+\frac{a^2b^2}{4R^4}+\cdots. \end{aligned}$$

On both sides subtract 1 and multiply by $-2R^2$ to obtain

$$c^2+\frac{\text{stuff}}{R^2}=a^2+b^2+\frac{\text{other stuff}}{R^2}.$$

The terms "stuff" and "other stuff" converge to finite values and so, if we let R go to infinity, we deduce the classical Pythagorean Theorem. Since the Earth is a sphere of such immense radius compared to everyday phenomena, small right triangles would seem to obey the classical Pythagorean Theorem.

With great circles as lines, the angle between two intersecting great circles is defined to be the *dihedral angle* between the two planes that determine the great circles. The dihedral angle is the angle made by intersecting lines, one in each plane, which are perpendicular to the line of intersection of the planes. This angle is also the angle formed by the intersection of these planes with the plane tangent to the sphere at the vertex of the angle.

To see the abundance of congruences, suppose we are given angles at two points P and Q on the sphere of the same magnitude. Form the great circle joining P and Q: Rotating around the axis through the center and perpendicular to the plane that determines this great circle, we can move Q to P. Take the line through P and the center of the sphere and rotate the transported angle. Either the angle lies over the given angle at P, or its reflection across the plane of one of the sides of the angle at P lies over the given angle. Thus we have enough congruences to compare angles of the same measure anywhere on the sphere.

With this definition of angle we can measure the interior angles of a triangle of great circle segments on the sphere. In the proof of the Spherical Pythagorean Theorem we used the dot product on $\mathbb{R}^3$ and the embedding of the sphere in $\mathbb{R}^3$. We go further with this idea and prove another of the classical formulas of spherical trigonometry that relates the sides and interior angles of a spherical triangle.

Theorem 1.3 (Spherical Sine Theorem). *Let $\triangle ABC$ be a spherical triangle on a sphere of radius R. Let a, b, and c denote the lengths of the sides, and let $\angle A$, $\angle B$, and $\angle C$ denote the interior angles at each vertex. Then*

$$\frac{\sin(a/R)}{\sin(\angle A)} = \frac{\sin(b/R)}{\sin(\angle B)} = \frac{\sin(c/R)}{\sin(\angle C)}.$$

PROOF: We first treat the case of a right triangle. We restrict our attention to triangles lying entirely within a quarter of a hemisphere. Suitable modifications of the proof can be made for larger triangles.

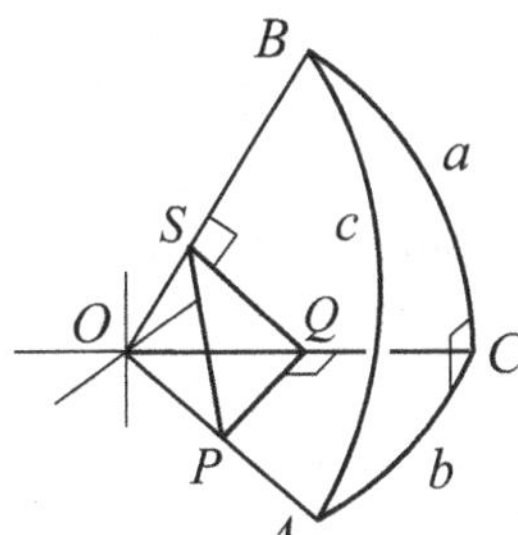

Let $\triangle ABC$ be a right triangle with right angle at C. Choose a point P on the radius OA. In the plane of OAC let Q be on OC with QP perpendicular to OC. Let S be on OB with QS perpendicular to OB. This leads to right triangles $\triangle OQP$ and $\triangle OSQ$ and hence the relations

$$OP^2 = OQ^2 + QP^2, \quad OQ^2 = OS^2 + QS^2.$$

We add one more relation that follows because $\angle C$ is a right angle. This means that the planes of OAC and OBC are perpendicular.

(A thorough discussion of the geometry of lines and planes in space is found in Chapter 4.) From there we take the proposition:

When two planes T_1 and T_2 are perpendicular and a line $\overleftrightarrow{PQ} = \ell$ in T_1 is perpendicular to the line of intersection $m = T_1 \cap T_2$, then any line n in T_2 passing through the point of intersection of the lines $Q = \ell \cap m$ is perpendicular to ℓ.

With this observation (Lemma 4.17) we know that QS is perpendicular to PQ and ΔPQS is a right triangle. We get another relation: $PS^2 = PQ^2 + QS^2$.

Putting together the relations among the lengths, we obtain

$$OP^2 = OQ^2 + PQ^2 = OS^2 + QS^2 + PQ^2 = OS^2 + PS^2.$$

Therefore ΔOSP is a right triangle with right angle at S.

The central angles subtended by a, b, and c at the center of the sphere are given (in radians) by a/R, b/R, and c/R. With all our right triangles we can compute

$$\sin\frac{b}{R} = \sin(\angle POQ) = \frac{PQ}{OP} = \frac{PQ}{PS}\cdot\frac{PS}{OP} = \sin\angle B \sin\frac{c}{R},$$

and so $\sin(c/R) = (\sin(b/R))(\sin\angle B)$. Similarly $\sin(c/R) = (\sin(a/R))(\sin\angle A)$, and so

$$\frac{\sin(a/R)}{\sin\angle A} = \frac{\sin(b/R)}{\sin\angle B}.$$

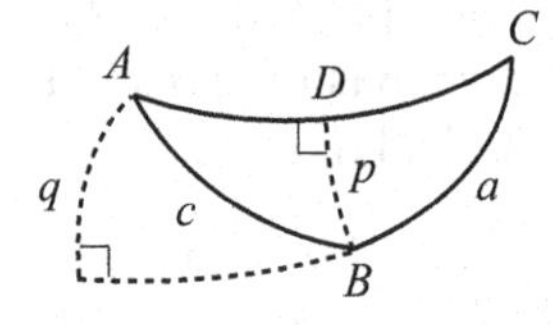

For an arbitrary triangle, we can construct the spherical analogue of an altitude to reduce the relation for two sides to the case of a right triangle. For example, in the adjoining figure we insert the altitudes from A and B. By the right triangle case we see

$$\sin(a/R)\sin\angle C = \sin(p/R) = \sin(c/R)\sin\angle A.$$

From the other altitude we find

$$\sin(b/R)\sin\angle C = \sin(q/R) = \sin(c/R)\sin(\pi - \angle B) = \sin(c/R)\sin\angle B.$$

Thus $\dfrac{\sin(a/R)}{\sin(\angle A)} = \dfrac{\sin(c/R)}{\sin(\angle C)} = \dfrac{\sin(b/R)}{\sin(\angle B)}$, and the theorem is proved. ■

The three interior angles and three sides of a spherical triangle are six related pieces of data. The problem of *solving the triangle* is to determine the missing three pieces of data from a given three. The problem arose in astronomy and geodesy. The solution was worked out by ancient astronomers. (See Rosenfeld and van Brummelen for more details.)

On the sphere we can form triangles whose interior angles sum to greater than π: For example, take the triangle that bounds the positive octant of the sphere S_R in $\mathbb{R}^3$. In fact, *every* triangle of great circle segments has interior angle sum greater than π. To see this, we study the area of a triangle on the sphere. It was the Flemish mathematician ALBERT GIRARD (1592–1632) who first published the relation between area and interior angle sum.

Area is a subtle concept that is best treated via integration (see Chapter 7). However, the properties we need to define the area of a polygonal region on the sphere are few and so we may take them as given:

(1) The sphere of radius R has area $4\pi R^2$.
(2) The area of a union of nonoverlapping regions is the sum of their areas.
(3) The area of congruent regions are equal.
(4) A **lune** is one of the regions enclosed by two great circles from one point of intersection to its antipode. The ratio of the area of a lune to the area of the whole sphere is the same as the ratio of the angle determined by the lune to 2π.

Proposition 1.4 (Girard's Theorem). *On the sphere of radius R, a triangle $\triangle ABC$ with interior angles α, β, and γ has area given by*

$$\text{area}(\triangle ABC) = R^2(\alpha + \beta + \gamma - \pi).$$

PROOF (Euler 1781): Two great circles cross in a pair of antipodal points and determine two antipodal lunes. Let θ denote the dihedral angle between the great circles that determine a lune. Assumption 4 implies

$$\text{area of the lune} = \frac{\theta}{2\pi} \cdot 4\pi R^2 = 2\theta R^2.$$

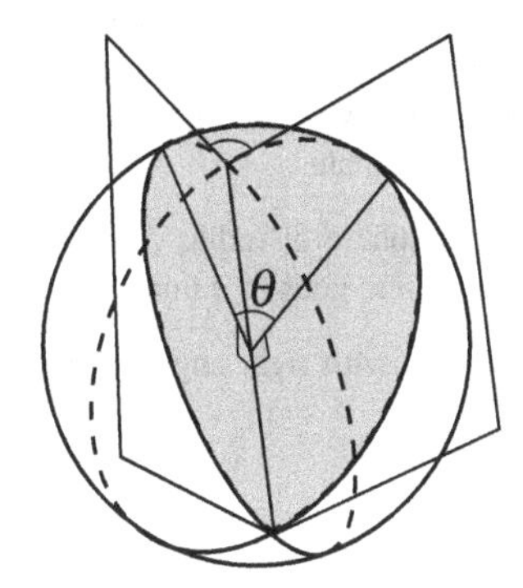

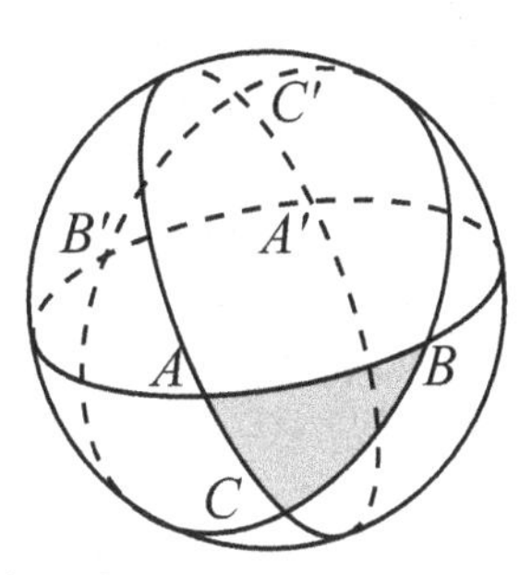

A lune and a spherical triangle.

We assume our triangle $\triangle ABC$ lies in one hemisphere; if not, subdivide it into smaller triangles and argue on each piece. A triangle is the intersection of three lunes. Extending the three lunes that determine $\triangle ABC$ to the rest of the sphere gives rise to antipodal lunes and an antipodal copy of the triangle, $\triangle A'B'C'$. Taking $\triangle ABC$ along with the three lunes that determine it, we cover half the sphere but count the area of $\triangle ABC$ three times. This gives the equation

$$2\pi R^2 = 2\alpha R^2 + 2\beta R^2 + 2\gamma R^2 - 3\,\text{area}(\triangle ABC) + \text{area}(\triangle ABC),$$

where α, β, and γ are the interior angles made by the lunes. Then

$$\text{area}(\triangle ABC) = R^2(\alpha + \beta + \gamma - \pi). \qquad \blacksquare$$

We call the value $\alpha + \beta + \gamma - \pi$ the **angle excess** of the spherical triangle. Because every triangle on the sphere has nonzero area, every triangle has an interior angle sum greater than π. If the radius of the sphere is very large and the triangle very small in area, as a triangle of human dimensions on this planet would be, there is only negligible angle excess.

The lessons to be learned from this short visit to a geometry different from the geometry of our school days set the stage for the rest of the book. The sphere has some striking geometric properties that differ significantly from those displayed by the plane. For example, angle sum and area are intimately related on the sphere. The Pythagorean Theorem is shared by the sphere and the plane in an analytic fashion by viewing the plane as a sphere of infinite radius. Other analytic tools, such as trigonometry, coordinates, and calculus, as well as tools from the study of symmetry–rotations and reflections–opened up the geometry of this surface to us. All of these ideas return to guide us in later chapters.

Exercises

1.1 Prove that two great circles bisect one another.

1.2 Prove that any circle on S^2 is the intersection of some plane in $\mathbb{R}^3$ with the sphere.

1.3 The sphere of radius 1 can be coordinatized as the set of points as the set of points $(1, \psi, \theta)$ in spherical coordinates, with $0 \leq \psi \leq 2\pi$, and $0 \leq \theta \leq \pi$. In this coordinate system, determine the distance along a great circle between two arbitrary points on the sphere as a function of their coordinates. Compare with rectangular coordinates.

1.4† Show that the circumference of a circle of radius ρ on a sphere of radius R is given by $L = 2\pi R \sin(\rho/R)$. What happens when the radius of the sphere goes to infinity?

1.5† Suppose $\triangle ABC$ is a right triangle on the sphere of radius R with right angle at vertex C. Prove the formula

$$\cos A = \sin B \cos(a/R).$$

1.6 Show that the dihedral angle between two planes is independent of the choice of point of intersection by the pair of perpendicular lines with the line shared by the planes.

1.7 If three planes Π_1, Π_2, and Π_3 meet in a point P and Π_2 is perpendicular to both Π_1 and Π_3, show that $\Pi_1 \cap \Pi_2$ and $\Pi_2 \cap \Pi_3$ give a pair of perpendicular lines.

1.8 A *pole* of a great circle is one of the endpoints of a diameter of the sphere perpendicular to the plane of the great circle. For example, the North and South Poles are poles to the Equator. Prove that through a given point, not a pole of a given great circle, there is a unique great circle through the given point and perpendicular to the given great circle.

1.9 Show that an isosceles triangle on the sphere has base angles congruent (Proposition I.2 of Menelaus's *Spherics*; (van Brummelen 2009)).

Solutions to Exercises marked with a dagger appear in the appendix, pp. 325–326.

1.10[†] Suppose $\triangle ABC$ is a spherical triangle on the unit sphere and A', B', and C' are poles of sides BC, AC, and AB, respectively, where we apply the right-hand rule to choose the poles (namely, OA, OB, OC' form the columns of a matrix with positive determinant). Show that $\angle A = \pi - a'$, $\angle B = \pi - b'$ and $\angle C = \pi - c'$.

1.11[†] Points on the terrestrial sphere are often coordinatized by latitude λ and longitude ϕ. These coordinates are related to spherical coordinates (θ, ψ) by $\lambda = \pi/2 - \theta$ and $\phi = \psi$ mod 2π but taking a representative $-\pi \leq \phi \leq \pi$ instead of $0 \leq \psi \leq 2\pi$. Given two points M, N with latitude and longitude given by (λ_1, ϕ_1), (λ_2, ϕ_2), determine (1) the distance from M to N; (2) the compass direction from M to N. (Part 2 solves the problem of locating the *Qibla* for Muslim daily prayer. The coordinates for N in this case are $N = (21.42252°N, 39.82621°E)$.)

1.12 Prove the law of cosines for sides in spherical trigonometry, given by

$$\cos(a/R) = \cos(b/R)\cos(c/R) + \sin(b/R)\sin(c/R)\cos\angle A.$$

Prove the law of cosines for angles is given by

$$\cos\angle A = -\cos\angle B\cos\angle C + \sin\angle B\sin\angle C\cos(a/R).$$

(*Hint*: Try the law of cosines for sides applied to the polar triangle.)

1.13 Prove that a right spherical triangle can be *solved* if any two of its data, not the right angle, are known; that is, any other angle or side can be found from the initial data. More generally, on a spherical triangle, if any three data are known, all the rest follow. Prove some cases of this assertion.

1.14 Show that there are no similar triangles on the sphere that are not congruent. This result is the congruence criterion for spherical triangles, Angle-Angle-Angle.

2

Euclid

There has never been, and till we see it we never shall believe that there can be, a system of geometry worthy of the name which has any material departures ... from the plan laid down by Euclid.

A. DEMORGAN (OCTOBER 1848)

Geometry, like arithmetic, requires for its logical development only a small number of simple, fundamental principles. These fundamental principles are called the axioms of geometry. The choice of the axioms and the investigation of their relations to one another is a problem ... tantamount to the logical analysis of our intuition of space.

DAVID HILBERT (1899)

The prehistory of geometry was a practical matter. It consisted of facts and rules that could be applied to determine the positions of the Sun and stars or to measure land areas. Merchants brought these ideas from Egypt and Babylonia to ancient Greece where the prevailing ideas of systematic thought changed geometry from a tool to a deductive discipline of the mind. Centuries of contemplation and careful reorganization culminated in Euclid's *The Elements* (Euclid *ca.* 300 B.C.E), whose thirteen surviving books summarized the mathematics of his day and influenced all subsequent generations.

The Elements proceeds by the *axiomatic method*; definitions and axioms are presented first, then propositions are shown to follow from these assumptions and from each other through logical deduction. When later mathematicians sought a model for the rigorous development of mathematical ideas, they turned to Euclid.

We begin with some definitions from *The Elements*:

(1) A point is that which has no part.
(2) A line is a breadthless length.
(3) The extremities of a line are points.
(4) A straight line is a line that lies evenly with the points on itself.
(5) A surface is that which has length and breadth only.
(6) The extremities of a surface are lines.
(7) A plane surface is a surface that lies evenly with straight lines on itself.

Euclid intends to define the familiar. The definition of point, however, holds little meaning: T.L. HEATH (1861–1940), the editor of *The Thirteen Books of Euclid's Elements (1926)*, explains the definition as a device to avoid circularity. The modern view, developed and exemplified by DAVID HILBERT (1862–1943), takes such terms as *point*, *line*, and *plane* as undefined. To define a geometric structure, we begin with a set S (space) of points and two classes of subsets: $\mathcal{L}$, the lines, and $\mathcal{P}$, the planes. A point P, that is, an element of S, *lies on* a line l in $\mathcal{L}$ if $P \in l$, and a line l *lies in* a plane Π in $\mathcal{P}$ if $l \subset \Pi$. Hilbert gave axioms to determine the relations among these notions in his *Grundlagen der Geometrie* (Hilbert 1901).

Definition 2.1. *A set S together with collections of subsets $\mathcal{L}$, of lines, and $\mathcal{P}$, of planes, is an* **incidence geometry** *if the following axioms of incidence hold:*

(1) *For any P, Q in S, there is an l in $\mathcal{L}$ with $P \in l$ and $Q \in l$. If $P \neq Q$, then l is unique. We denote the line determined by P and Q by $\overleftrightarrow{PQ}$.*
(2) *If l is in $\mathcal{L}$, then there are points P, Q in S, with $P \in l$, $Q \in l$, and $P \neq Q$.*
(3) *For any plane $\Pi \in \mathcal{P}$, there are distinct points P, Q, and R in Π so that $R \notin \overleftrightarrow{PQ}$.*

These axioms were chosen by Hilbert to capture the notion of incidence for points and lines in Book I of *The Elements*. Euclid sets out his assumptions about these objects in the five postulates and five common notions. A list of the propositions of Book I (Propositions I.1–I.48) follows this chapter in the appendix.

Postulate I. *To draw a straight line from any point to any point.*

Notice the uniqueness clause in Hilbert's Axiom 1; Euclid tacitly assumes uniqueness, and he applies it in the proofs of certain propositions, for example, Proposition I.4.

Hilbert gave further axioms for points, lines, and planes in space. The modern axiomatic approach proceeds without interpretation of the basic terms and focuses on the relations between the basic objects. This open structure allows for different realizations of the axioms. For example, certain finite sets with subsets and relations are incidence geometries and their properties are used to study problems in combinatorics and algebra (Buekenhout 1994).

In Postulate I, Euclid described a tool for drawing the ideal straight edge. In Postulate II, Euclid extends the properties of this tool.

Postulate II. *To produce a finite straight line continuously in a straight line.*

In order to give a modern version of Postulate II, we introduce two further primitive notions: Given three distinct points P, Q, and R lying on a line l, we want to be able to say when Q **lies between** P **and** R. This relation is denoted $B(P,Q,R)$ and defined as a ternary relation $B \subset S \times S \times S$ satisfying certain axioms (to be given later) that render abstract the notion of betweenness. A **line segment** $PQ \subset l$ is defined as the subset:

$$PQ = \{R \in l \mid R = P \text{ or } R = Q \text{ or } B(P,R,Q)\}.$$

The other primitive notion is that of *congruence of line segments*, denoted $PQ \cong AB$. The axioms for betweenness imply that the endpoints determine a line segment as a subset of a line, so we frame congruence of line segments as a four-place relation $E(P,Q;A,B)$ satisfying certain axioms. We write $PQ \cong AB$ whenever the relation $E(P,Q;A,B)$ holds.

If we take these primitive notions as given, then the modern version of Postulate II becomes:

Given two line segments PQ and AB there is a point R on the line through P and Q such that $QR \cong AB$ and Q is between P and R.

The notions of betweenness and congruence used in Postulate II can be based on a **distance function**, $d\colon S \times S \to \mathbb{R}$, on the set S that makes S a **metric space**. A distance function determines a metric space if the following conditions are satisfied:

(1) For all P and Q in S, $d(P,Q) = d(Q,P)$.
(2) For all P and Q in S, $d(P,Q) \geq 0$ and $d(P,Q) = 0$ if and only if $P = Q$.
(3) For all P, Q, and R in S, $d(P,Q) + d(Q,R) \geq d(P,R)$ (*the triangle inequality*).

In the presence of a metric space structure on S, betweenness can be defined as the ternary relation:

$$B(P,Q,R) \text{ if and only if } d(P,Q) + d(Q,R) = d(P,R).$$

Congruence of line segments can be defined by $PQ \cong AB$ if and only if $d(P,Q) = d(A,B)$. An incidence geometry along with a distance function is called a **metric geometry** (see Millman, Parker (1981) or Mac Lane (1959) for an alternate foundation of geometry based on metric postulates).

Alternatively, if we have a notion of betweenness and congruence for an incidence geometry $(S, \mathcal{L}, \mathcal{P})$, then we can ask for a **measure of length**. This is a real-valued function ϕ defined on the set of line segments in S which satisfies the properties:

(1) For all line segments PQ, $P \neq Q$, we have $\phi(PQ) > 0$.
(2) If $PQ \cong P'Q'$, then $\phi(PQ) = \phi(P'Q')$.
(3) If $B(P,Q,R)$, then $\phi(PQ) + \phi(QR) = \phi(PR)$.

This notion is equivalent to a metric space structure. One can prove that any two measures of length are proportional and, given a particular line segment AB, there is a measure of length ϕ_0 with $\phi_0(AB) = 1$. The chosen segment AB is the *unit of the measure* ϕ_0. Restricting ϕ_0 to a line $l \in \mathcal{L}$, together with a pair of points O, $O' \in l$, gives a function $\phi_0|_l\colon l \to \mathbb{R}$ that determines a **ruler** or **coordinate system** on l. The point O is the origin of the coordinate system and the **ray** :

$$\overrightarrow{OO'} = \{Q \in S \mid B(O,Q,O') \text{ or } B(O,O',Q)\}$$

determines the positive part of the line. In later chapters we will assume the existence of rulers on lines in order to introduce functions defined on lengths, that is, on congruence classes of line segments. For a proof of the existence of measures of length, the interested reader may consult the book of Borsuk and Szmielew (1960, especially Chapter III, §§9–10).

Postulates I and II describe the ideal straight edge (not a calibrated ruler). The Greeks allowed another ideal tool for geometric constructions, the compass.

Postulate III. *To describe a circle with any center and distance.*

In modern parlance, Postulate III reads:

In a plane Π, given points O and A, there is a circle $\mathcal{C}$ with center O and radius OA, defined as the set $\mathcal{C} = \{Q$ such that $OQ \cong OA\}$.

This postulate permits the construction of a circle with a compass that is idealized to have any finite radius. Without further description we must take Euclid's compass to be collapsible, that is, one cannot simply move it from place to place. This hindrance is overcome by Proposition I.2 of *The Elements* in which a clever construction permits one to move a compass without collapsing the arms.

The next postulate deals with angles. We first recall a few more of Euclid's definitions:

(8) A *plane angle* is the inclination to one another of two lines in the plane that meet one another and do not lie in a straight line.
(9) When the lines containing the angle are straight lines, the angle is called *rectilineal.*
(10) When a straight line set up on a straight line makes adjacent angles equal to one another, each of the equal angles is *right*, and the straight line standing on the other is called a *perpendicular* to that on which it stands.
(11) An *obtuse angle* is an angle greater than a right angle.
(12) An *acute angle* is an angle less than a right angle.

Postulate IV. *All right angles are congruent.*

In order to make the definition of right angle and Postulate IV precise, we need a notion of *congruence of angles*. Congruence of figures may be thought of as a motion in space of one figure that superimposes it on another figure such that corresponding points and line segments coincide. Euclid used superposition in the proofs of congruence criteria, as in Propositions I.4, I.8, and I.26. BERTRAND RUSSELL (1872–1970), in his 1902 article for the *Encyclopedia Britannica Supplement*, points out that "(a)ctual superposition ... is not required; all that is required is the transference of our attention from one figure to another." We take a **congruence**, a "transference of attention," to mean a mapping $\phi\colon S \to S$, defined on the underlying set S of an incidence geometry and satisfying:

(1) ϕ is a one-to-one correspondence.
(2) ϕ takes lines to lines and planes to planes. Furthermore, if $(S, \mathcal{L}, \mathcal{P}, d)$ is a metric geometry, then $d(\phi(P), \phi(Q)) = d(P, Q)$.

In Hilbert (1902), congruence of angles is a primitive notion satisfying certain axioms from which Postulate IV is deduced as a theorem.

From the modern viewpoint, Euclid's Postulate IV is a statement about the existence of enough congruences to compare every pair of right angles. For example, to superimpose adjacent right angles, we can reflect one right angle through space across a line onto its neighbor. Thus, Euclid assumes the existence of a congruence that is reflection across a line in his definition of right angle. Superposition of figures may be effected by rotations and translations of the plane. Such motions have the "homogeneity of space" as a consequence. The motion of a figure determined by two line segments at right angles onto another that preserves angles and distances implies that close to the vertices of the angles, the planes containing the right angles look the same.

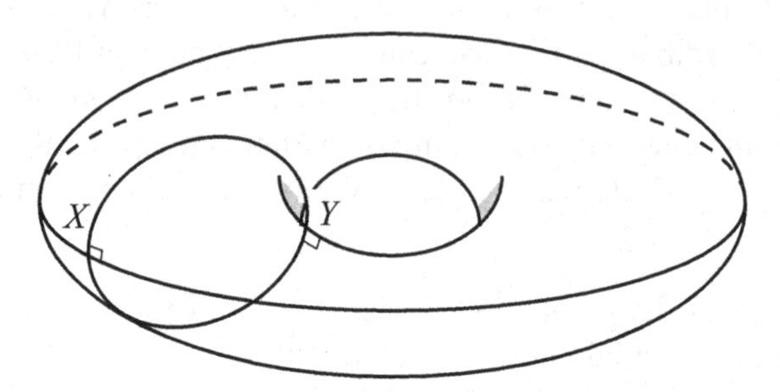

On a more general surface it may not be the case that right angles are congruent in the sense of superposition. Consider the case of a torus (an inner tube): The meridians (vertical slices) and the innermost and outermost horizontal circles on the torus may be taken for lines. There is no "motion" of the torus to itself, however, such that the right angle at the point Y of intersection of a given meridian with the innermost circle and the one at the point X of intersection of the given meridian with the outermost circle are superimposed (if there were, the Gaussian curvature on the torus would agree at X and Y; see chapter 9). Thus these right angles are not congruent. Homogeneity of general surfaces will be studied in chapter 12.

From the definitions, common notions, and Postulates I through IV, Euclid derived the first 28 propositions of Book I. Let us consider a few of these propositions and discuss his methods and assumptions further.

Proposition I.1. *Given a finite straight line, one can construct an equilateral triangle with the given line segment as base.*

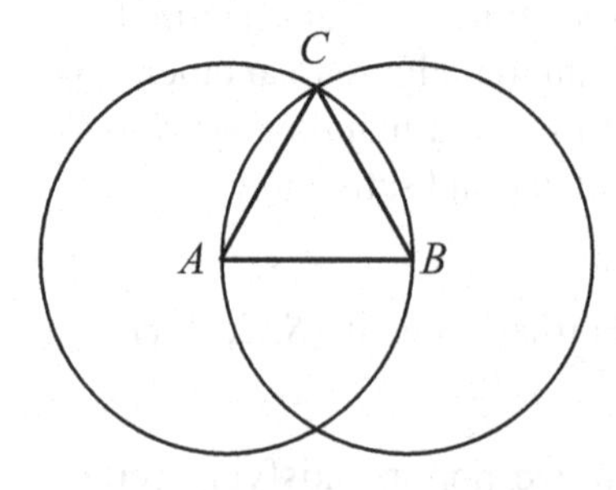

PROOF: Let AB be the given line segment. By Postulate III there is a circle with center A and radius AB, and a circle with center B and radius BA. Since AB is a radius for both circles, they meet at a point C. By Postulate I, we can form the line segments AC and BC and triangle $\triangle ABC$. The line segment BC is a radius and so it is congruent to $BA \cong AB$; AC is also a radius and is congruent to AB. Thus $\triangle ABC$ is an equilateral triangle. ■

REMARKS.

(1) Notice how the axiomatic method is applied here; only postulates and definitions are used.
(2) Euclid assumes that the constructed circles must meet at a point C. Criticism of this assumption can be found before the fifth century C.E. (Proclus). The problem is overcome in Hilbert's scheme (Hilbert 1902) by requiring the *axiom of continuity*. This axiom and a related statement, *Pasch's axiom*, guarantee the existence of points of intersections for many figures in Euclid.

Axiom of Continuity (Archimedes). *Given line segments AB and CD, there are points $A_1, A_2, \ldots, A_n$ lying on $\overleftrightarrow{AB}$ such that $CD \cong AA_1 \cong A_1A_2 \cong \cdots \cong A_{n-1}A_n$ and B lies between A and A_n.*

Pasch's Axiom. *If a line in a plane enters a triangle in that plane through at most one vertex of the triangle, then the line exits the triangle at a point on one of the sides not already cut by the given line.*

To illustrate why these axioms are necessary, suppose the plane is simply the set $\mathbb{Q} \times \mathbb{Q}$ of all ordered pairs with each entry a rational number. If we take $A = (0,0)$ and $B = (1,0)$ in Proposition I.1, then the point $C = (1/2, \sqrt{3}/2)$ in the proof is missing from the plane. The Axiom of Continuity provides the basis for the *completeness* of the real number line, the property of $\mathbb{R}$ that all Cauchy sequences of real numbers converge to a real number. With this property, the planes of Euclid are like $\mathbb{R}^2$ and completeness implies the existence of the desired points.

REMARK: The existence of rulers on lines requires the Axiom of Continuity. By using the fact that we can bisect a line segment (Proposition I.10) and Postulate II, we can apply the Axiom of Continuity to coordinatize a line by making points correspond to dyadic expansions of real numbers. The notion of Dedekind cuts may also be used to coordinatize lines; see (Borsuk–Szmielew 1960).

One of the highlights of Book I is the following useful theorem.

Propositon I.16 (Exterior Angles Theorem). *In any triangle, if one of the sides is extended, the exterior angle formed is greater than either of the opposite interior angles.*

PROOF: Consider this extraordinary figure. Given $\triangle ABC$, extend BC along to D. We want to show $\angle BAC$ is less than $\angle ACD$. That is, there is an angle contained in $\angle ACD$ that is congruent to $\angle BAC$. Let M be the midpoint of AC (which can be constructed as in Proposition I.10) and join B to M. Extend BM to a point E so that $BM \cong ME$ (Postulate II).

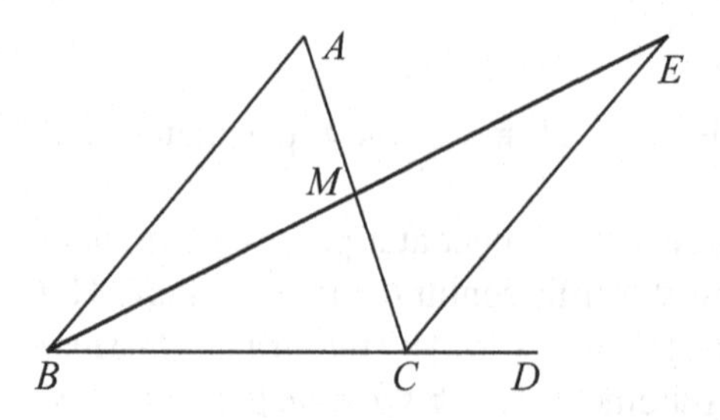

It follows that $CM \cong MA$, $BM \cong ME$, and $\angle AMB \cong \angle CME$; these are *vertical angles* and hence congruent by Proposition I.15. By the congruence criterion Side–Angle–Side (Proposition I.4), $\triangle AMB \cong \triangle CME$, and, in particular, $\angle BAM \cong \angle ECM$. Since E is on the ray $\overrightarrow{BM}$ and D is on the ray $\overrightarrow{BC}$, $\angle ECM$ is inside $\angle MCD$, and so $\angle BAM \cong \angle BAC$ is less than $\angle MCD = \angle ACD$. Thus, $\angle BAC$ is less than $\angle ACD$. To argue that $\angle ACD$ exceeds $\angle ABC$, construct the analogous diagram extending AC. ■

Though we are stepping out of Euclid for a moment, we derive an important theorem for the theory of parallels. The assumptions are in place for the proof of this result, and the method is analogous to the proof of Proposition I.16.

Theorem 2.2 (Saccheri–Legendre Theorem). *The sum of the angles interior to a triangle is less than or equal to two right angles.*

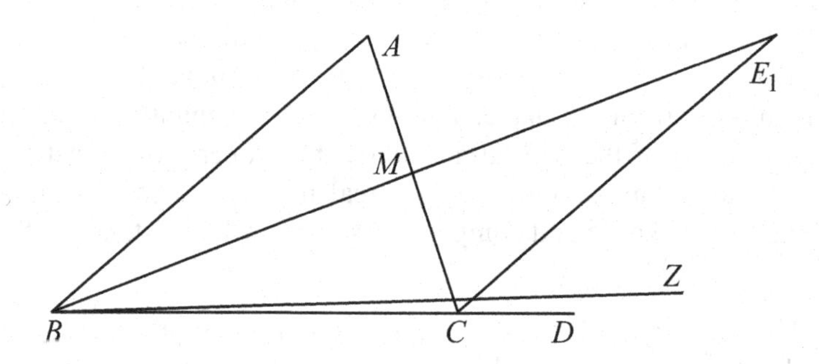

PROOF: Suppose we have a triangle $\triangle ABC$ and suppose the sum of the interior angles is two right angles plus an angle congruent to $\angle CBZ$. Suppose $\angle CBA$ is the least of the three interior angles. As in the proof of Proposition I.16, let M be the midpoint of AC and construct the ray $\overrightarrow{BM}$. Mark E_1 on $\overrightarrow{BM}$ so that $BM \cong ME_1$, and so $\triangle AMB \cong \triangle CME_1$. Notice that $\angle CBM + \angle ABM = \angle ABC$, so one of these angles is less than or equal to $\frac{1}{2}\angle ABC$, where the meaning of $\frac{1}{2}\angle BAC$ is clear. Hence, either $\angle CBE_1$ or $\angle CE_1M$ is less than or equal to $\frac{1}{2}\angle ABC$. Observe that:

$$\begin{aligned}\angle ABC + \angle CAB + \angle BCA &= \angle CBM + \angle ABM + \angle MCE_1 + \angle BCM \\ &= \angle CBE_1 + \angle CE_1B + \angle BCE_1.\end{aligned}$$

Thus, $\triangle BCE_1$ has the same angle sum as $\triangle ABC$ but has one angle less than half the smallest angle of $\triangle ABC$. If we iterate this procedure n times, we obtain a triangle in which the smallest angle is less than $(1/2^n)\angle ABC$. The Axiom of Continuity may be applied to angles, and so repeated halving produces an angle less than $\angle CBZ$. The triangle constructed at this stage still has an angle sum of two right angles plus $\angle CBZ$. Thus, the two remaining angles in the triangle must sum to more than two right angles, which contradicts Euclid's Postulate I.17: *The sum of any two interior angles in a triangle is less than two right angles.* ■

From this result it is clear that the first four postulates of Euclid cannot be satisfied by spherical geometry, since triangles on the sphere have angle sums greater than two right angles. One of the points of departure for the sphere is Postulate II; it is not possible to define a natural notion of "between" on a great circle, and so the modern version of the postulate cannot be realized. The uniqueness assertion of Postulate I also fails on the sphere; when two antipodal points are chosen, there are infinitely many great circles joining such points.

Euclid's theory of parallels

Book I of *The Elements* breaks naturally into two parts at Proposition I.28. It is here that Euclid begins to construct a theory of parallels and area in order to prove the Pythagorean Theorem (I.47). Euclid defines parallel lines as follows:

(23) *Parallel straight lines* are straight lines that, being in the same plane and being produced indefinitely in both directions, do not meet one another in either direction.

The first important result about parallels in *The Elements* is the following proposition.

Proposition I.27 (Alternate Interior Angles Theorem). *If a straight line falling on two straight lines makes the alternate interior angles congruent to one another, the straight lines will be parallel to one another.*

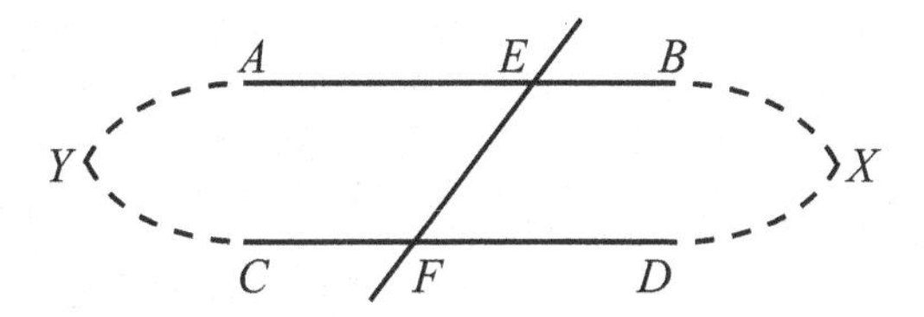

PROOF: Let the line $\overleftrightarrow{EF}$ cross lines $\overleftrightarrow{AB}$ and $\overleftrightarrow{CD}$ so that $\angle BEF$ is congruent to $\angle CFE$. Suppose $\overleftrightarrow{AB}$ meets $\overleftrightarrow{CD}$ on the side of B and D at X. Consider $\triangle EFX$. Because $\angle CFE$ is an exterior angle, it is greater than $\angle FXE$ and $\angle XEF$

(Proposition I.16). But $\angle XEF = \angle BEF \cong \angle CFE$. This is a contradiction. We can argue similarly that $\overleftrightarrow{AB}$ and $\overleftrightarrow{CD}$ do not meet on the side of A and C. ∎

The converse of Proposition I.27 is Proposition I.29 and it is the first case in *the Elements* in which Euclid calls upon the last of his assumptions.

Postulate V. *That, if a straight line falling on two straight lines makes the interior angles on the same side sum to less than two right angles, the two straight lines, if produced indefinitely, meet on that side on which the angles sum to less than two right angles.*

It is immediate that Postulate V is different in nature than the other four. It is simply more difficult to state than the others and it uses loose words such as "indefinitely." Postulate V was criticized by later generations of mathematicians who accused Euclid of introducing an "unproved theorem" to simplify proofs. This criticism is the source of centuries of interesting geometry and it is the focus of much of the rest of the book.

Euclid develops the theory of parallels in order to construct figures whose areas can be compared. He introduces a different use of the word "equals" (in Proposition I.35) to mean "of equal area." Along the way he shows that the interior angle sum of a triangle is two right angles, that corresponding angles and segments of a parallelogram are congruent, and that parallelograms may be constructed with area the same as the area of a given triangle. After showing that squares may be constructed on a given base, Euclid reaches the climax of Book I:

Proposition I.47 (Pythagorean Theorem). *In a right-angled triangle, the square on the side subtending the right angle is equal in area to the squares on the sides containing the right angle.*

PROOF: Refer to the figure on the next page for notation. Given the right triangle $\triangle ABC$ with the right angle at A, construct the squares on each of the sides (Proposition I.46). Using the fact that the sum of the angles in a square is four right angles and that they are all congruent, we see that $\overleftrightarrow{CA}$ is the line $\overleftrightarrow{AG}$ and $\overleftrightarrow{AB}$ is the line $\overleftrightarrow{AH}$. Since $\angle DCB + \angle BCA \cong \angle KCA + \angle BCA$ and sides $DC \cong CB$ and $AC \cong CK$, by Side-Angle-Side (Proposition I.4), we have $\triangle CKB \cong \triangle CDA$. If AM is an altitude from A for $\triangle ABC$, that is, AM is perpendicular to BC, consider the rectangles $\square CDNM$ and $\square CKHA$. By Proposition I.41, $\square CKHA$ is twice $\triangle CKB$ in area and $\square CDNM$ is twice $\triangle CDA$, since $\triangle CKB \cong \triangle CDA$, $\square CKHA$ and $\square CDNM$ have the same area. Similarly $\square ABFG$ has the same area as $\square MNEB$ and the result follows by putting $\square CDNM$ and $\square MNEB$ together to form the square $\square CDEB$. ∎

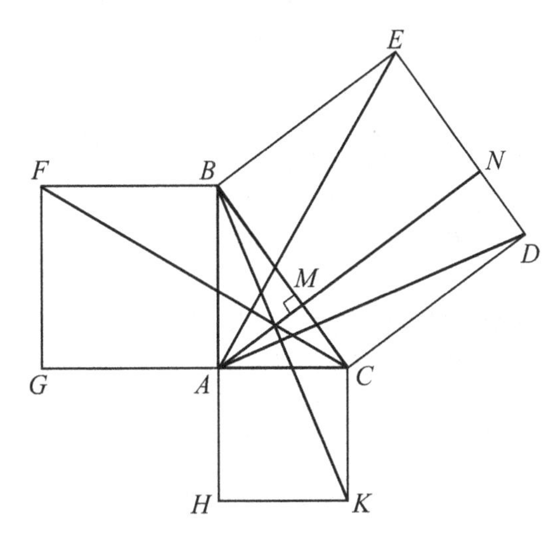

Reading the propositions of Book I that follow Proposition I.28, one is struck by the role that parallels play in the determination of relations among the areas of elementary figures. This use of areas is in part due to the lack of symbolic algebra in Greek mathematics. The square of a number was interpreted as the area of a square with sides of the given number. There are many other proofs of the Pythagorean Theorem, some of which are based on simple algebraic formulas (Euclid, trans. Heath 1956, pp. 350–368).

We list here the rest of Hilbert's system for describing planar geometry. This scheme provides the modern means to make Euclid's work precise, as well as giving an example of contrasting axiomatic systems.

Fix a set S with classes of subsets $\mathcal{L}$ of lines and $\mathcal{P}$ of planes. Assume that $(S,\mathcal{L},\mathcal{P})$ is an incidence geometry. Let us assume that we have relations $B \subset S \times S \times S$ and $E \subset S \times S \times S \times S$, the properties of which encode our ideas of betweenness and congruence. If a triple of points (P,Q,R) lies in B, we say that $B(P,Q,R)$ holds, or that Q lies between P and R. If a four-tuple of points lies in E, we say that $E(P,Q;A,B)$ holds, and we denote this by $PQ \cong AB$.

Axioms of order.

(1) *If P, Q, R lie on a line l, and $B(P,Q,R)$ holds, then $B(R,Q,P)$ holds.*
(2) *If P and R lie on a line l, then there are points Q and S on l such that $B(P,Q,R)$ and $B(P,R,S)$ hold.*
(3) *If P, Q, R lie on a line l, then either $B(P,Q,R)$ or $B(Q,R,P)$ or $B(R,P,Q)$ holds.*
(4) *Any four points P, Q, R, S on a line l can always be so arranged that $B(P,Q,R)$ and $B(P,Q,S)$ hold, as well as $B(P,R,S)$ and $B(Q,R,S)$.*
(5) *Pasch's Axiom.*

Axioms of congruence.

(1) *$PQ \cong PQ$ always holds. If P and Q lie on a line l and P' lies on l', then there is a point Q' on l' on a given side of P' such that $PQ \cong P'Q'$.*
(2) *If $PQ \cong AB$ and $PQ \cong CD$, then $AB \cong CD$.*
(3) *If P, Q, R lie on a line l, P', Q', R' lie on a line l', $B(P,Q,R)$ and $B(P',Q',R')$ hold, and $PQ \cong P'Q'$ and $QR \cong Q'R'$, then $PR \cong P'R'$.*
(4) *$\angle POQ \cong \angle POQ$ and $\angle POQ \cong \angle QOP$ always hold. In a plane we have an angle $\angle POQ$ and in another plane a ray $\overrightarrow{O'P'}$ along with a choice of half-plane determined by the line $\overleftrightarrow{O'P'}$. Then there is a unique ray $\overrightarrow{O'Q'}$ in the chosen half-plane such that $\angle POQ \cong \angle P'O'Q'$.*
(5) *If $\angle POQ \cong \angle P'O'Q'$ and $\angle POQ \cong \angle P''O''Q''$, then $\angle P'O'Q' \cong \angle P''O''Q''$.*
(6) *Given two triangles $\triangle ABC$ and $\triangle A'B'C'$, suppose $AB \cong A'B'$, $AC \cong A'C'$, and $\angle BAC \cong \angle B'A'C'$, then $\angle ABC \cong \angle A'B'C'$ and $\angle ACB \cong \angle A'C'B'$ (Side–Angle–Side).*

The first few exercises treat the relationship between these axioms and Euclid's work. Hilbert gave five groups of axioms—incidence, order, continuity, congruence, and parallels—in his scheme to put geometry on firm foundations. He proved that all of the axioms are independent and so all are required. In the next chapter parallel lines will be discussed further.

Exercises

2.1 From the axioms for an incidence geometry, show that two lines l and l', $l \neq l'$, either share a point in common or no points in common.

2.2 Using only the notions of incidence and order, prove that a line lying in a plane divides the plane into two distinct regions called half-planes.

2.3 From the axioms of congruence, prove $PQ \cong QP$.

2.4† From the congruence axioms prove Proposition I.4, that is, if given $\triangle ABC$ and $\triangle A'B'C'$ such that $AB \cong A'B'$, $AC \cong A'C'$, and $\angle BAC \cong \angle B'A'C'$, then $BC \cong B'C'$. Prove the other congruence criteria: Angle–Angle–Side, Angle–Side–Angle (Proposition I.26), and Side–Side–Side (Proposition I.8) from the axioms.

2.5 (a) Given congruent angles $\angle POQ \cong \angle P'O'Q'$ and a ray $\overrightarrow{OX}$ lying within $\angle POQ$, prove that there is a ray $\overrightarrow{O'X'}$ lying in $\angle P'O'Q'$ such that $\angle POX \cong \angle P'O'X'$ and $\angle XOQ \cong \angle X'O'Q'$.
(b) Given $\angle POQ \cong \angle P'O'Q'$ and $\angle QOR \cong \angle Q'O'R'$, then, if Q lies in $\angle POR$ and Q' lies in $\angle P'O'R'$, prove that $\angle POR \cong \angle P'O'R'$.
(c) From (a) and (b), prove Euclid's Postulate IV.

Solutions to Exercises marked with a dagger appear in the appendix, pp. 326–327.

2.6 Consider the sets $S = \{A, B, C, D, E, F, G\}$ and $\mathcal{L} = \{ABC, CDE, EFA, AGD, BGE, CGF, BDF\}$ with the obvious relation $\in$. Show that these choices form an incidence geometry. (Try to draw a picture of this finite geometry. It is known as the *Fano plane*. It encodes the multiplicative relations among the unit Cayley numbers that are not equal to 1.) *Prove or disprove*: In an incidence geometry, each line has at least three points on it.

2.7 Let S denote the set of all lines through the origin in $\mathbb{R}^3$. Take a plane through the origin as a line. We define $\mathcal{L}$ as the set of planes through the origin in $\mathbb{R}^3$ with $\in$ meaning $l \subset \Pi$. Show that this set S, the set of lines $\mathcal{L}$, and $\in$ so interpreted give an incidence geometry (with one plane, all of $\mathbb{R}^3$).

2.8 From Postulates I through IV prove Propositions I.9, I.10, I.11, and I.18 in the appendix to the chapter. You may assume any propositions previous to the one you are proving.

2.9 Assuming Postulates I through IV and Proposition I.29, prove Postulate V.

2.10† Prove that the segment joining the midpoints of two sides of a triangle has length less than or equal to half the third side. For the proof, did you need Postulate V?

2.11 Suppose that two right triangles have congruent hypotenuses and one corresponding side congruent. Show that the triangles are congruent. Does this hold for general triangles?

2.12 Prove that the internal bisectors of the angles of a triangle meet in a point. For the proof did you need Postulate V? (*Hint*: Take two bisectors that meet in a point. Join the third vertex to this point and consider the segments perpendicular to the sides through this point. Now apply the preceding exercise.)

2.13† In a circle, form a triangle with one side a diameter and the third a vertex on the circle. Show that this is a right triangle. Does your proof require Postulate V?

2.14† Assume the existence of a unit length and, using straight edge and compass, give constructions for the sum and product of the lengths of two line segments, the reciprocal of the length of a line segment, and the square root of the length of a line segment. (*Hint*: Assume Postulate V and the relations between sides of similar triangles.)

Appendix
The Elements: Book I

Common Notions

(1) Things that are equal to the same thing are also equal to each other.
(2) If equals are added to equals, then the wholes are equal.
(3) If equals are subtracted from equals, then the remainders are equal.
(4) Things that coincide with one another are equal to one another.
(5) The whole is greater than the parts.

Propositions

Proposition I.1. Given a line segment, one can construct an equilateral triangle with the given line segment as base.

Proposition I.2. Given a point and a line segment, one can construct a line segment with the given point as endpoint and congruent to the given line segment.

Proposition I.3. Given two unequal line segments, one can construct on the greater segment a line segment congruent to the lesser.

Proposition I.4 (*Side–Angle–Side*). If two triangles have the two sides congruent to two sides, respectively, and the angles enclosed by the two congruent sides congruent, then the triangles are congruent.

Proposition I.5 (*Pons Asinorum*). In an isosceles triangle, the angles at the base are congruent. If one of the congruent sides is extended to form a new triangle and the other side is extended by the same length, then the angles under the base are congruent.

Proposition I.6. If in a triangle two angles are congruent, then the sides opposite the congruent angles are congruent.

Proposition I.7. Given a triangle, there cannot be another triangle constructed on the same base to a different point that is congruent to the first triangle with the sides in the same order.

Proposition I.8 (*Side–Side–Side*). If two triangles have corresponding sides congruent, then they also have angles enclosed by corresponding sides congruent.

Proposition I.9. With straight edge and compass one can bisect a given angle.

Proposition I.10. One can bisect a given line segment.

Proposition I.11. One can construct a line perpendicular to a given line at a given point on the line.

Proposition I.12. To a given line and a point not on the line, one can construct a line through the point perpendicular to the given line.

Proposition I.13. If a line meets another line, on one side they form angles which sum to two right angles.

Proposition I.14. If two lines through a point on a given line form adjacent angles with that line and sum to two right angles, then the two lines are the same line.

Proposition I.15. If two lines meet, they make congruent vertical angles.

Proposition I.16 (*The Exterior Angles Theorem*). In any triangle, if one of the sides is extended, the exterior angle formed is greater than either of the opposite interior angles.

Proposition I.17. In any triangle two angles taken together are less than two right angles.

Proposition I.18. In any triangle the greater side is opposite the greater angle.

Proposition I.19. In any triangle the greater angle is opposite the greater side.

Proposition I.20 (*The Triangle Inequality*). In any triangle, two sides taken together are greater than the remaining one.

Proposition I.21. If a triangle is constructed inside a given triangle on its base, then the two sides of the new triangle are less than the corresponding sides of the given triangle and they form a greater angle.

Proposition I.22. Given three line segments so that any two taken together are greater than the third, there is a triangle with sides congruent to the given line segments.

Proposition I.23. At a point on a given line, one can construct an angle congruent to a given angle.

Proposition I.24. If two triangles have two sides congruent, then the greater included angle is opposite the greater base.

Proposition I.25. If two triangles have two sides congruent, then the greater base is opposite the greater included angle.

Proposition I.26 (*Angle-Side-Angle, Angle-Angle-Side*). If two triangles have two angles and one side congruent, then the triangles are congruent.

Proposition I.27 (*The Alternate Interior Angles Theorem*). If a line falling on two straight lines makes the alternate interior angles congruent to one another, the straight lines are parallel to one another.

Proposition I.28. If a line falling on two straight lines makes the exterior angle congruent to the interior and opposite angle on the same side, the straight lines are parallel to one another.

Proposition I.29. A line cutting two parallel lines makes alternate interior angles congruent, the exterior angle congruent to the interior and opposite angle, and the interior angles on one side congruent to two right angles.

Proposition I.30. Lines parallel to a given line are parallel to each other.

Proposition I.31. Through a given point not on a given line one can construct a line parallel to the given line.

Proposition I.32. In any triangle, if one of the sides is extended, the exterior angle is congruent to the two opposite interior angles taken together, and the three interior angles sum to two right angles.

Proposition I.33. The line segments joining corresponding endpoints of congruent and parallel line segments are themselves congruent and parallel.

Proposition I.34. In a parallelogram, the opposite sides and angles are congruent and the diameter bisects the area.

Proposition I.35. Parallelograms on the same base and in the same parallels have equal areas.

Proposition I.36. Parallelograms on congruent bases and in the same parallels have equal areas.

Proposition I.37. Triangles that are on the same base and in the same parallels have equal areas.

Proposition I.38. Triangles that are on congruent bases and in the same parallels have equal areas.

Proposition I.39. Triangles on the same base and of equal areas are on the same parallels.

Proposition I.40. Triangles on congruent bases on the same side of a given line and of equal areas are also in the same parallels.

Proposition I.41. If a parallelogram has the same base with a triangle and they are in the same parallels, the parallelogram has twice the area of the triangle.

Proposition I.42. In a given angle, one can construct a parallelogram of area equal to the area of a given triangle.

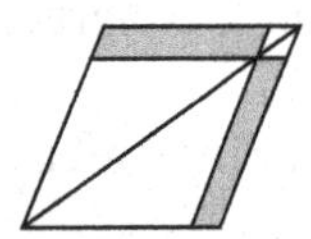

Proposition I.43. In any parallelogram, the complements of the parallelogram about the diameter are equal in area.

Proposition I.44. On a given line segment in a given angle, one can construct a parallelogram of area equal to a given triangle.

Proposition I.45. One can construct in a given angle a parallelogram of area equal to a given quadrilateral.

Proposition I.46. One can construct a square on a given line segment.

Proposition I.47 (*Pythagorean Theorem*). In a right-angled triangle, the square on the hypotenuse is equal in area to the sum of the squares on the sides.

Proposition I.48. In a triangle, if the square on one of the sides is equal in area to the sum of the squares on the remaining two sides of the triangle, then the angle contained by the two sides is a right angle.

3

The theory of parallels

This ought even to be struck out of the Postulates altogether, for it is a theorem ... the converse of it is actually proved by Euclid himself as a theorem ... It is clear then from this that we should seek a proof of the present theorem, and that it is alien to the special character of postulates.

PROCLUS (410–85 C.E.)

To be sure, it might be possible that non-intersecting lines diverge from each other. We know that such a thing is absurd, not by virtue of rigorous inferences or clear concepts of straight and crooked lines, but rather throught experience and the judgement of our eyes.

G. S. KLÜGEL (1763)

Some of the most reliable information about Euclid and early Greek geometry is based on the commentaries of Proclus, the leader of the Academy in Athens in the fifth century of the common era, whose objections to Postulate V are stated in the epigram. To its author and early readers, *The Elements* provided an idealized description of physical space. From this viewpoint it is natural to understand the objections to Postulate V. The phrase "if produced indefinitely" strains the intuition based on constructions with compass and straight edge. Furthermore, Euclid avoided using Postulate V in the proofs of the first twenty-eight propositions of Book I. It is first called upon in the proof of Proposition I.29, which is the converse of Propositions I.27 and I.28. Several of the previous propositions are converses of their neighbors with proofs that simply observe the contradiction to the earlier statement were the converse false (see, for example, Propositions I.5 and I.6, I.13 and I.14, I.18 and I.19). Proposition I.29 does not yield to this logic. *Why introduce such an unnatural statement to prove it?*

To eliminate this "blemish" on Euclid's great work, subsequent generations heeded Proclus's call and either sought a proof of Postulate V from the other assumptions, or tried to replace it with a more self-evident assumption. Some of the titles of their efforts indicate their intentions, for example, *Commentaries on the difficulties in the premises of Euclid's book* ('UMAR KHAYYĀM (1048–1131), of Rubā'iyat fame); *Treatise that heals doubts raised by parallel lines* (NAṢIR AL-DĪN AL-ṬŪSĪ (1201–74)); and later, *Euclid vindicated of every flaw* (GIROLAMO SACCHERI (1667–1733)). In most of these efforts some assumption is made that is equivalent to Postulate V; later generations, seeking to avoid this fate, corrected the work of their predecessors only to commit *petitio principii* (begging the question, that is, assuming what you want to prove) themselves.

Their efforts fall into certain categories that reveal the properties of parallels. Later in the chapter we consider negations of Postulate V. We derive properties of the subsequent *non-Euclidean* geometry from the equivalent statements.

Uniqueness of parallels

In a book on parallels that is said to have been written by CLAUDIUS PTOLEMY (100–178), Proposition I.29 is proved without the benefit of Postulate V. Ptolemy is known for his great work *Almagest* on astronomy and for his contributions to geography and spherical geometry. His proof was reported in Proclus (1970).

Proposition I.29. *A line cutting two parallel lines makes alternate angles congruent, the exterior angle congruent to the interior and opposite angle, and the sum of the interior angles on the same side congruent to two right angles.*

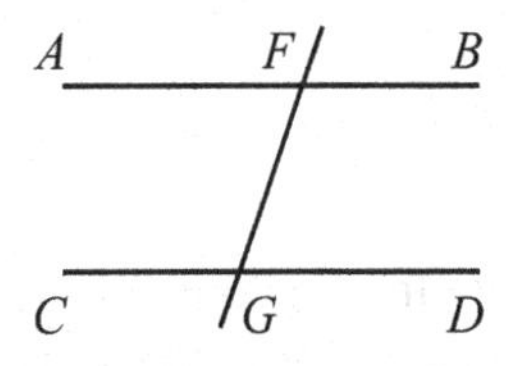

PROOF (Ptolemy): A straight line that cuts parallels must make the sum of the interior angles on the same side equal to, greater than, or less than two right angles. "Greater than" is not allowed because, following the figure, if $\angle BFG$ and $\angle DGF$ sum to more than two right angles, so also must $\angle AFG$ and $\angle CGF$ because $\overleftrightarrow{AF}$ and $\overleftrightarrow{CG}$ are no more parallel than $\overleftrightarrow{FB}$ and $\overleftrightarrow{GD}$. "Less than" is argued similarly. ■

Proclus identified the source of Ptolemy's *petitio principii*, which has since become known as *Playfair's Axiom* from a 1795 English edition of Euclid *Elements of Geometry* (1795) by JOHN PLAYFAIR (1748–1819).

Playfair's Axiom. *Through a point not on a line, there is one and only one line through the point parallel to the given line.*

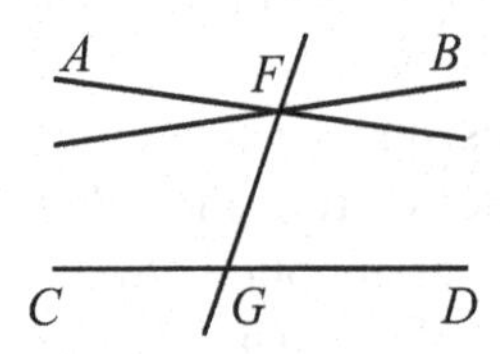

In the figure we see how two parallels through the point F, if they existed, would ruin Ptolemy's argument. In fact, Playfair's Axiom is equivalent to Postulate V, and so Ptolemy had assumed Postulate V in his "proof."

Theorem 3.1. *Playfair's Axiom is equivalent to Postulate V.*

PROOF: In one direction, suppose a pair of lines $\overleftrightarrow{AB}$ and $\overleftrightarrow{CD}$ is cut by another line so that the interior angles on one side sum to less than two right angles. Construct line $\overleftrightarrow{LM}$ through F so that $\angle GFM$ is congruent to $\angle FEB$, which is possible by Proposition I.23. By the alternate interior angles theorem (Proposition I.27), $\overleftrightarrow{LM}$

is parallel to $\overleftrightarrow{AB}$. By Playfair's Axiom, $\overleftrightarrow{CD}$ is not parallel, and furthermore, D is inside $\angle EFM$. Therefore, $\overleftrightarrow{CD}$ meets $\overleftrightarrow{AB}$ on the side of B and M.

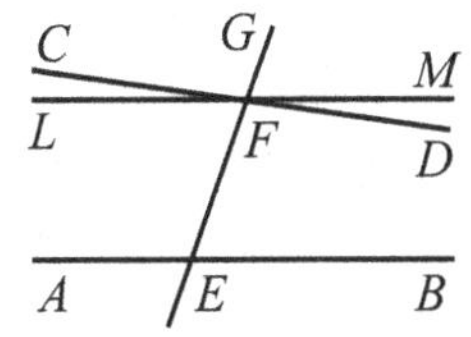

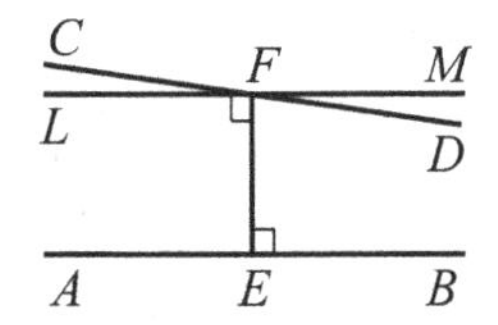

In the other direction, if we are given a point F not on a line $\overleftrightarrow{AB}$, construct (Proposition I.12) the line $\overleftrightarrow{LM}$ so that EF is perpendicular to $\overleftrightarrow{LM}$ and EF is perpendicular to $\overleftrightarrow{AB}$. By Proposition I.27, $\overleftrightarrow{LM}$ is parallel to $\overleftrightarrow{AB}$. Suppose $\overleftrightarrow{CD}$ is any line through F. If $\overleftrightarrow{CD} \neq \overleftrightarrow{LM}$, then $\overleftrightarrow{CD}$ makes an angle on one side of $\overleftrightarrow{EF}$ that is less than a right angle. By Postulate V, $\overleftrightarrow{CD}$ meets $\overleftrightarrow{AB}$. Hence, $\overleftrightarrow{LM}$ is the unique parallel. ■

Equidistance and boundedness of parallels

Proclus himself offered a "proof" of Postulate V, in which he asserts:

If from one point two straight lines forming an angle be produced indefinitely, the distance between the said lines, produced indefinitely, will exceed any finite magnitude.

Proclus's statement was known as the "Philosopher's Principle" to medieval Islamic geometers (Rosenfeld). The philosopher is ARISTOTLE (384–322 B.C.E.) who wrote on parallels as a source of logical errors. We first prove Proclus's assertion.

Proposition 3.2. *If $\triangle ACM$ is a right triangle with right angle at C, and B is the midpoint of AM, then if BD is the perpendicular to AC at D, BD is not greater than half of CM.*

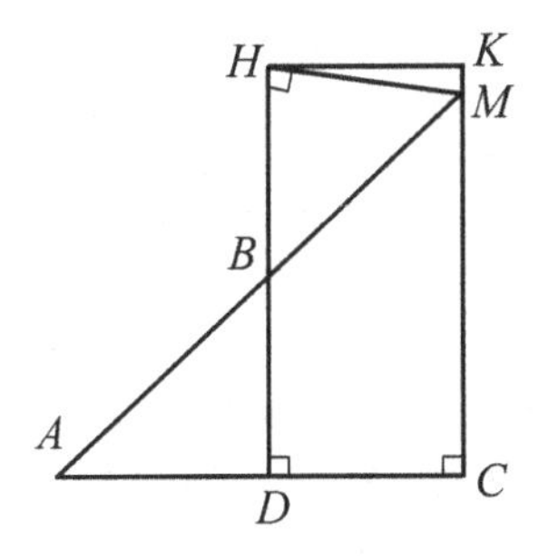

PROOF (Saccheri 1733, Proposition XX): Extend DB to DH with $BH \cong BD$. If BD is greater than half of CM, then DH is greater than CM. Extend CM to $CK \cong DH$. Consider the triangles $\triangle ADB$ and $\triangle MBH$. By the congruence criterion Side–Angle–Side, $\triangle ADB \cong \triangle MBH$ and so $DA \cong HM$. Furthermore $\angle BHM$ is a right angle. Hence $\angle DHK$ is greater than a right angle. The reader can prove (or look at the proof of Proposition 3.10) that since $DH \cong CK$, we have $\angle CKH$ is congruent to $\angle DHK$. Then the quadrilateral $DCKH$ has interior angle sum greater than four right angles, which is not possible by the Saccheri–Legendre Theorem (Theorem 2.2). Hence twice BD is less than or equal to CM. ■

Corollary 3.3. *Given an angle $\angle MAC$, if AM and AC are extended indefinitely, then the distance between them will exceed any given finite length.*

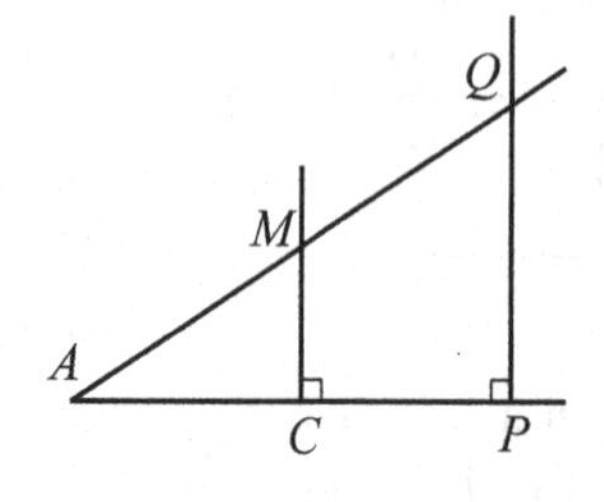

PROOF: The distance from a point to a line is the length of the perpendicular to the line through the point. If we double AM to AQ and construct the perpendicular to $\overleftrightarrow{AC}$ from Q to P, then $PQ \geq 2CM$. By iterated doubling and the Axiom of Continuity, we can construct a perpendicular from $\overrightarrow{AM}$ to $\overleftrightarrow{AC}$ that exceeds any given length. ■

From this property Proclus observes that if any straight line cuts one of two parallels, it will also cut the other. The argument is as follows: Let $\overleftrightarrow{AB}$ be parallel to $\overleftrightarrow{CD}$ and $\overleftrightarrow{FG}$ cut $\overleftrightarrow{AB}$ at F. If $\overrightarrow{FG}$ and $\overrightarrow{FB}$ are produced indefinitely, the distance between them will exceed any magnitude, including the interval between the parallels. Therefore, when $\overrightarrow{FB}$ and $\overrightarrow{FG}$ are at a distance greater than the gap between $\overleftrightarrow{AB}$ and $\overleftrightarrow{CD}$, $\overrightarrow{FG}$ must cut $\overleftrightarrow{CD}$.

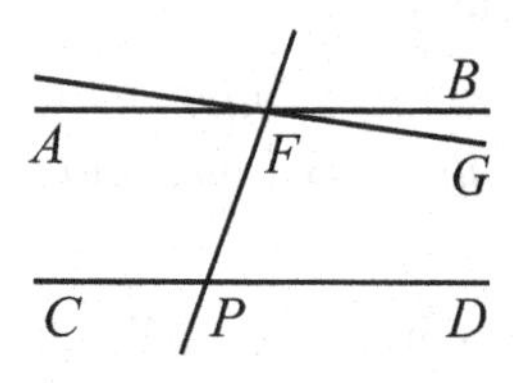

Postulate V can be proved as follows: If $\angle GFP$ and $\angle DPF$ sum to less than two right angles, construct the line $\overleftrightarrow{FB}$ such that $\angle BFP$ and $\angle DPF$ sum to two right angles. Proposition I.28 implies that $\overleftrightarrow{FB}$ is parallel to $\overleftrightarrow{CD}$ and so $\overrightarrow{FG}$ cuts a parallel at the point F. Thus $\overrightarrow{FG}$ meets $\overleftrightarrow{CD}$ on the side of the lesser angles.

The unspoken assumption made by Proclus is the following statement:

If two parallel lines that have a common perpendicular are produced indefinitely, then the distance from a point of one line to the other line remains bounded.

The assumptions that parallel lines are a fixed distance apart or that they are a bounded distance apart find their way into many of the attempts to prove Postulate V. Equidistance may have figured in Archimedes's definition of parallels (Rosenfeld 1988, p. 41) and it was the basis for proofs of Postulate V by POSIDONIUS (*ca.* 135–50 B.C.E.) and by IBN SĪNĀ (980–1037). Boundedness of parallels figures in Proclus's attempt as well as a more subtle proof by `Umar Khayyām. In the sixth century, the Byzantine commentator SIMPLICIUS revealed a proof of the fifth postulate due to AGHĀNIS (fifth century C.E.) who asserts that equidistant straight lines exist. This assumption is different from Proclus's and can be stated:

The set of points equidistant from a given straight line is also a straight line.

The Jesuit priest CHRISTOPHER CLAVIUS (1537–1612) asserted that this statement needed proof and he offered that it is false of other species of curves (such as a hyperbola and a line parallel but close to its asymptote). In fact, these ideas are equivalent to Postulate V.

Theorem 3.4. *That the set of points equidistant from a given line forms a line is equivalent to Postulate V.*

PROOF: From Postulates I through V, one can prove Euclid's Proposition I.34, which asserts that, in a parallelogram, the opposite sides and angles are congruent to each other. Construct lines $\overleftrightarrow{PQ}$ and $\overleftrightarrow{RS}$ perpendicular to $\overleftrightarrow{PR}$. Also construct (Proposition I.12) $\overleftrightarrow{ST}$ through S and perpendicular $\overleftrightarrow{RS}$ (the unique parallel). The line $\overleftrightarrow{ST}$ meets $\overleftrightarrow{PQ}$ at Q to form a parallelogram. Thus $PQ \cong RS$; that is, when we construct any pair of perpendiculars from $\overleftrightarrow{ST}$ to $\overleftrightarrow{PR}$, they are congruent by Proposition I.34.

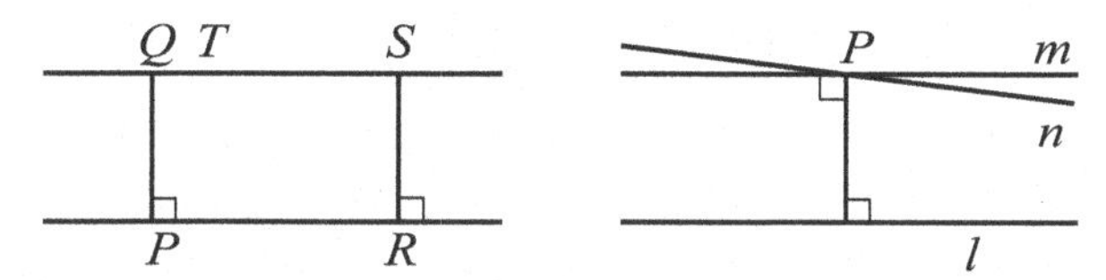

In the other direction, it suffices to prove Playfair's Axiom from our assumption. Suppose m is the line of points equidistant from l through P, and n is any other line through P. Since m and n form an angle at P we can argue, as Proclus did, that the lines m and n diverge. Since m remains a fixed distance from l, n must meet l and thus n is not parallel to l and m is the unique parallel to l through P. ■

Notice that the assertion that the locus of points equidistant from a given line forms a line fails on a sphere where great circles are lines.

On the angle sum of a triangle

Much of the literature from the golden age of Greece reached us through the translation and preservation of texts by the people of medieval Islam. The distinguished geometers of this culture also tried to improve Euclid by proving Postulate V or replacing it with a more reasonable assumption. The key to many of their attempts is the following assumption:

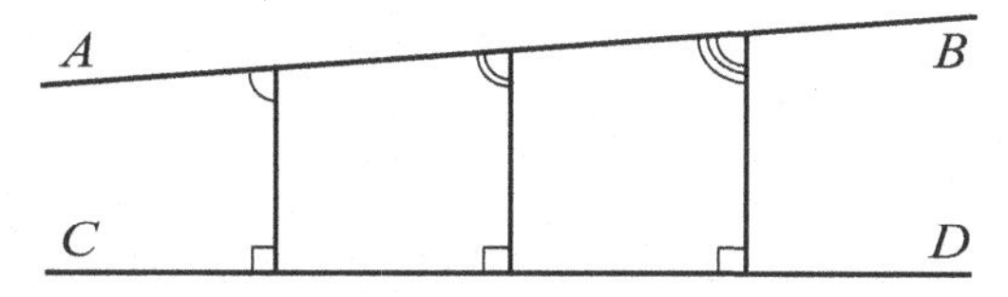

If $\overleftrightarrow{AB}$ and $\overleftrightarrow{CD}$ are two straight lines such that successive perpendiculars from $\overleftrightarrow{AB}$ to $\overleftrightarrow{CD}$ always make with $\overleftrightarrow{AB}$ unequal angles that are always acute on the side toward A and obtuse on the side toward B, then the lines approach continually nearer in the direction of the acute angles and diverge in the direction of the

obtuse angles, and the perpendiculars will diminish or increase in the respective directions. Furthermore, the converse holds.

This assumption is found in one form or another in the work of THĀBIT IBN QURRA (836–901), `Umar Khayyām, and Naṣir al-Dīn al-Ṭūsī (Rosenfeld 1988). The most important consequence of this assumption is the following:

Lemma 3.5. *Under the previous assumption, if AC and BD are at right angles to AB and $AC \cong BD$ and CD is joined, then $\angle ACD$ and $\angle BDC$ are right angles.*

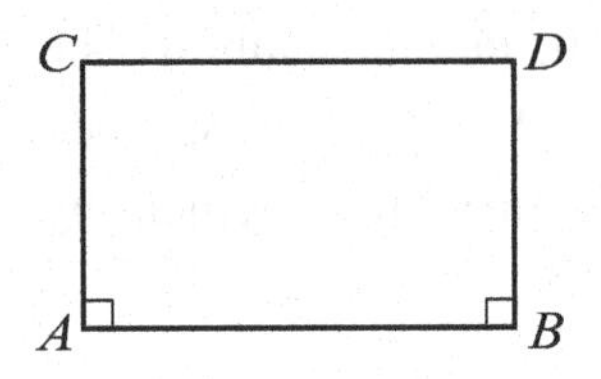

PROOF: Suppose not. Then $\angle ACD$ is either acute or obtuse. In either case, by the assumption, AC could not be congruent to BD because the sides AC and BD are diminishing or increasing in the appropriate direction. ■

The lemma can be applied to show that there are triangles for which the three interior angles sum to two right angles: Construct the diagonal of the quadrilateral of Lemma 3.5 to form two triangles and apply the Saccheri–Legendre Theorem. We obtain two triangles, each with an angle sum of two right angles. This fact is the key to a derivation of Postulate V in many of the Islamic efforts.

Theorem 3.6. *That the sum of the angles interior to a triangle is two right angles is equivalent to Postulate V.*

PROOF: We assume Postulate V in the form of Playfair's Axiom and Proposition I.29. Given $\triangle ABC$, let $\overleftrightarrow{ECD}$ be the unique parallel to $\overleftrightarrow{AB}$ through C. By Proposition I.29, $\angle ACE \cong \angle CAB$ and $\angle BCD \cong \angle CBA$. Thus, $\angle CAB + \angle CBA + \angle ACB = \angle ECA + \angle ACB + \angle BCD =$ two right angles.

In the other direction, we are given lines $\overleftrightarrow{AB}$, $\overleftrightarrow{CD}$ cut by a line $\overleftrightarrow{AC}$ perpendicular to $\overleftrightarrow{CD}$ and for which $\angle CAB$ is acute. Let G_1 be on $\overleftrightarrow{AB}$ and construct G_1H_1 perpendicular to $\overleftrightarrow{AC}$ at H_1 with the point H_1 on the side of $\overleftrightarrow{AC}$ toward C. If AH_1 is congruent to or exceeds AC, then $\overleftrightarrow{CD}$ meets $\overleftrightarrow{AB}$ by Pasch's Axiom.

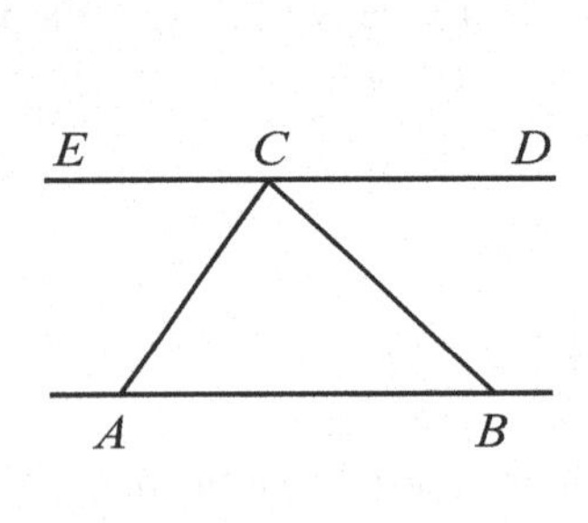

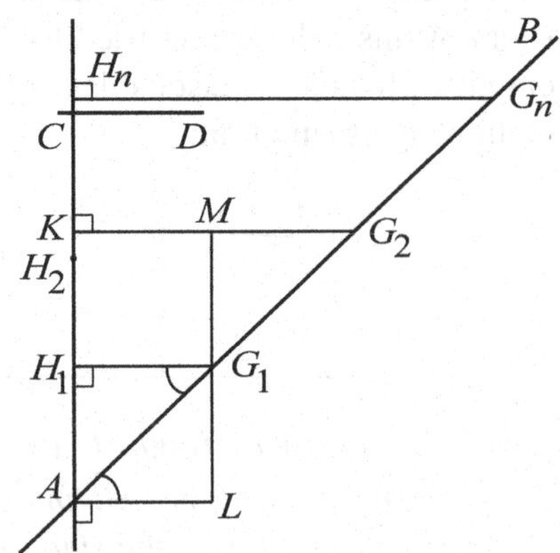

Suppose H_1 falls between A and C. Construct points $H_2, H_3, \ldots, H_n$ on $\overleftrightarrow{AC}$ with $AH_1 \cong H_1H_2 \cong \cdots \cong H_{n-1}H_n$ and AH_n greater than AC. This is possible by the Axiom of Continuity. Construct points $G_2, G_3, \ldots, G_n$ along $\overleftrightarrow{AB}$ with $AG_1 \cong G_1G_2 \cong \cdots \cong G_{n-1}G_n$.

Let G_2K be the perpendicular to $\overleftrightarrow{AC}$ from G_2 at K. We claim that $K = H_2$. Let AL be perpendicular to AC with $AL \cong G_1H_1$. Join LG_1. Since angles $\angle AG_1H_1$ and $\angle G_1AL$ are complementary to $\angle G_1AH_1$, they are congruent. (Here we have used the fact that the angle sum is two right angles.) By Side–Angle–Side, $\triangle G_1AH_1$ and $\triangle AG_1L$ are congruent and so $AH_1 \cong LG_1$ and $\angle ALG_1$ is a right angle.

Let M on G_2K be such that $KM \cong H_1G_1$. By the same argument $\angle KMG_1$ is congruent to a right angle and $G_1M \cong H_1K$. Since L, G_1, and M are now seen to be collinear, $\angle LG_1A \cong \angle G_2G_1M$ as vertical angles. By Angle–Angle–Side, $\triangle G_1MG_2 \cong \triangle G_1LA$ and so $H_1K \cong G_1M \cong LG_1 \cong AH_1 \cong H_1H_2$. Therefore $K = H_2$.

By iterating the argument, we prove that G_kH_k, and in particular, G_nH_n, is perpendicular to $\overleftrightarrow{AC}$. Thus $\overleftrightarrow{CD}$ is inside $\triangle AG_nH_n$ and $\overleftrightarrow{CD}$ is parallel to $\overleftrightarrow{G_nH_n}$. Pasch's Axiom implies $\overleftrightarrow{CD}$ leaves $\triangle AG_nH_n$ through side AG_n, that is, $\overleftrightarrow{CD}$ meets $\overleftrightarrow{AB}$. ■

This proof is credited by John Wallis (1676–1703) and by Saccheri to Naṣir al-Dīn al-Ṭūsī, but Rosenfeld (1988) argues that it was due to a later geometer, perhaps al-Ṭūsī's son.

To establish the full strength of Theorem 3.6, we prove the following theorem.

Theorem 3.7 (Three Musketeers Theorem). *If there exists one triangle with interior angle sum equal to two right angles, then every triangle has angle sum equal to two right angles.*

Proof: If $\triangle ABC$ has interior angle sum equal to two right angles, and from B we construct an interior altitude BT, then we can consider $\triangle BTC$ and $\triangle BTA$. Both are right triangles and $\angle BTC + \angle TCB + \angle CBT + \angle BTA + \angle TAB + \angle ABT =$ four right angles. By the Saccheri–Legendre Theorem this is possible only if:

$$\angle BTC + \angle TCB + \angle CBT = \angle BTA + \angle TAB + \angle ABT = 2 \text{ right angles.}$$

Therefore, there is a right triangle with interior angle sum equal to two right angles. We next show that every right triangle has angle sum equal to two right angles.

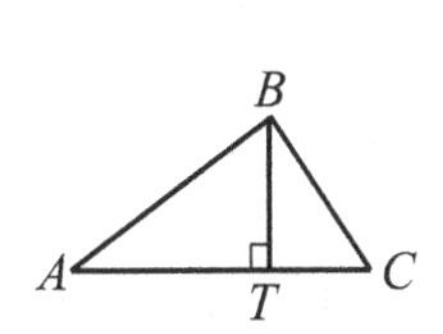

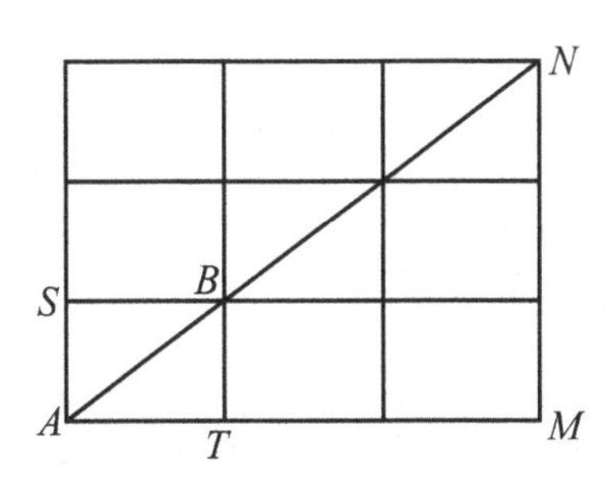

Construct the point S, as in the figure, so that S is not on the same side of the line $\overleftrightarrow{AB}$ as T and $AS \cong BT$, $BS \cong AT$. Then $\triangle ABT \cong \triangle BAS$ and $ATBS$ is a rectangle. If $\triangle AMN$ is obtained by marking off n copies of AT along $\overleftrightarrow{AT}$ and n copies of AB along $\overleftrightarrow{AB}$, by filling in with rectangles we get that $\triangle AMN$ has angle sum equal to two right angles.

Let $\triangle PQR$ be any right triangle and construct $\triangle A'QT'$ congruent to $\triangle ANM$ so that $A'Q > PQ$ and $T'Q > RQ$. The triangle $\triangle A'QT'$ has angle sum equal to two right angles and so angle sum $\triangle A'QT'$ = angle sum $\triangle A'PT'$ + angle sum $\triangle PT'Q$ − two right angles (from $\angle A'PT' + \angle T'PQ$). The Saccheri–Legendre Theorem once again forces the equation:

$$\text{Angle sum } \triangle A'PT' = \text{Angle sum } \triangle PT'Q = \text{Two right angles.}$$

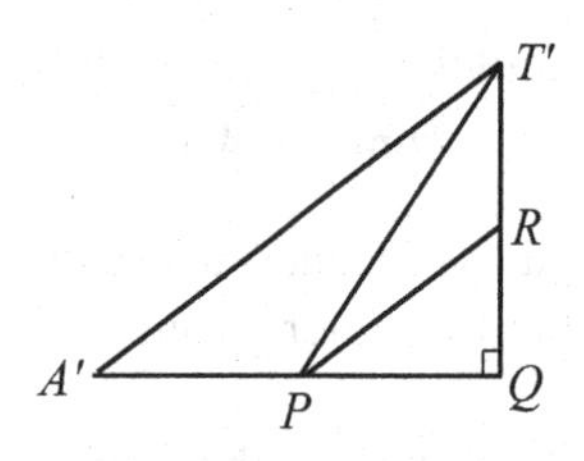

Similarly the angle sum of $\triangle PT'Q$ equals the angle sum of $\triangle PQR$ plus the angle sum of $\triangle PRT'$ minus two right angles. By the same argument, the angle sum of $\triangle PQR$ is two right angles.

Since every right triangle has interior angle sum equal to two right angles and an arbitrary triangle can be divided into two right triangles by an interior altitude, the theorem is proved. ■

This theorem gives a simple criterion for the failure of Postulate V: If *any* triangle has angle sum strictly less than two right angles, then all do. Also, if one wants to prove Postulate V from Postulates I through IV, then it suffices to produce one triangle with angle sum equal to two right angles. Saccheri tried in vain to prove this (Saccheri, 1733).

Similarity of triangles

Another noteworthy attempt to prove the fifth postulate is due to Wallis. To prove Postulate V he assumed Postulates I through IV and another assumption he judged to be more self-evident:

Wallis's Postulate. *To every triangle, there exists a similar triangle of arbitrary magnitude. That is, given a triangle $\triangle ABC$ and any line segment DE, there is a point F so that $\triangle ABC$ is similar to $\triangle DEF$, that is, $\angle ABC \cong \angle DEF$, $\angle BCA \cong \angle EFD$, and $\angle CAB \cong \angle FDE$.*

Wallis knew that he had assumed Postulate V in his postulate, which he considered more natural. Of course, if Postulates I through IV implied Wallis's postulate, then he would "remove the blemish" in Euclid's *The Elements*. Postulate III provides similar circles of arbitrary radius and Wallis's postulate is the analogous statement for triangles, and hence for polygons.

Theorem 3.8. *Wallis's postulate is equivalent to Postulate V.*

PROOF: In one direction, suppose we have lines $\overleftrightarrow{AB}$ and $\overleftrightarrow{CD}$ cut by $\overleftrightarrow{AC}$ so that $\angle BAC$ and $\angle DCA$ sum to less than two right angles. We "slide" $\overleftrightarrow{AB}$ along $\overleftrightarrow{AC}$ (*parallel transport*?) to the point C as follows: Choose points B_1 on $\overleftrightarrow{AB}$ and A_1 on $\overleftrightarrow{AC}$ as in the figure. To any point P on AC construct the triangle $\triangle APQ$ similar to $\triangle A_1AB_1$ by Wallis's postulate. The line $\overleftrightarrow{PQ}$ is parallel to $\overleftrightarrow{AB}$ by the alternate angles theorem (Proposition I.27).

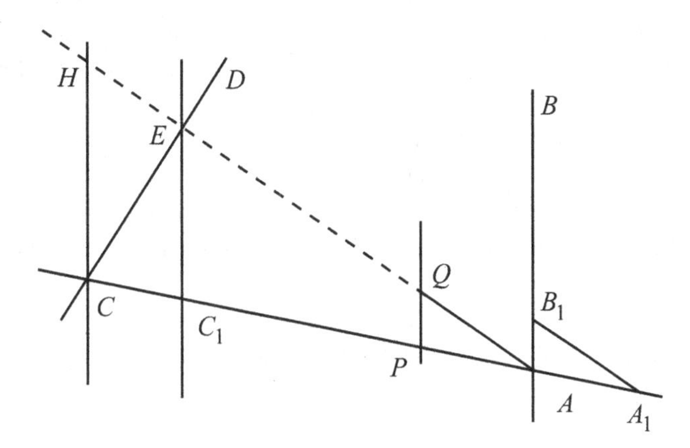

At C, the line $\overleftrightarrow{CH}$ so constructed forms angles $\angle ACH$ and $\angle CAB$ that sum to two right angles. Thus, the line $\overleftrightarrow{CD}$ enters the angle $\angle ACH$. By continuity we know that somewhere in the slide, the transport of $\overleftrightarrow{AB}$ must cross $\overleftrightarrow{CD}$, say as C_1E where E is on $\overleftrightarrow{CD}$. Construct the triangle with base AC similar to $\triangle C_1CE$. This new triangle has sides that are segments along $\overleftrightarrow{AB}$ and $\overleftrightarrow{CD}$ and so these lines meet.

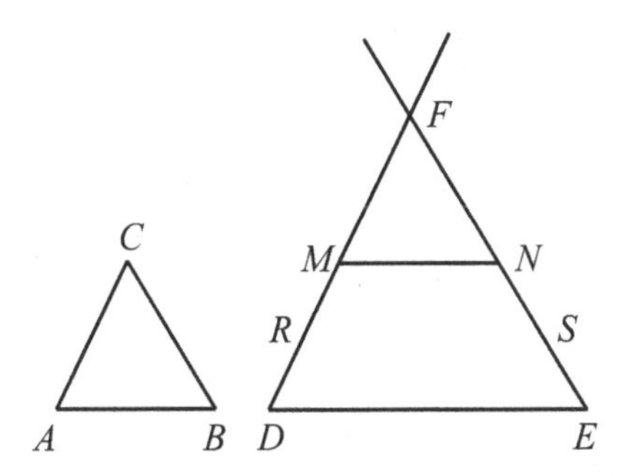

In the other direction we present a proof due to Saccheri (1733, Scholion III). Suppose we are given $\triangle ABC$ and segment DE. By Euclid's Proposition I.23, construct $\overrightarrow{DR}$ and $\overrightarrow{ES}$ so that $\angle EDR \cong \angle BAC$ and $\angle DES \cong \angle ABC$. By Proposition I.17 these base angles sum to less than two right angles and so by Postulate V, $\overrightarrow{DR}$ meets $\overrightarrow{ES}$ at a point F. It suffices to show $\angle DFE \cong \angle ACB$. Let M be on DF so that $MF \cong AC$. By Playfair's Axiom there is only one line through M parallel to DE. By Postulate V it meets EF at a point N. Since $\overleftrightarrow{MN}$ is parallel to $\overleftrightarrow{DE}$ and Postulate V holds, $\angle MNF \cong \angle DEF$. By Angle–Angle–Side

$\triangle MNF$ is congruent to $\triangle ABC$ and so $\angle ACB \cong \angle NFM \cong \angle DFE$. Therefore $\triangle ABC$ is similar to $\triangle DEF$. ■

This theorem is startling in its consequences. Suppose Postulate V were false. Then triangles have an interior angle sum less than two right angles. *No* similar triangles on different sides exist, and Angle–Angle–Angle becomes a congruence criterion for triangles. As J. H. LAMBERT (1728–77) observed, Angle–Angle–Angle allows one to have an absolute unit of length. The need for a standard meter or foot is eliminated by solving the problem:

Construct an equilateral triangle with a given interior angle.

If we agree on $\pi/4$ as the given angle, then the side of an equilateral triangle with such an angle, which could be constructed anywhere, could be a standard of length, and a geometric construction would replace the need for a Bureau of Standards. To compare this phenomenon with more familiar notions, we use an absolute unit, "1/360 of a circle" (the degree) to measure angles. Similarly, given a line segment, we could construct an equilateral triangle on it by Proposition I.1. The amount that the interior angle sum of this triangle differs from π could be fed into a function that gives out the length of a side, since the defect from π for an equilateral triangle completely characterizes it by Angle–Angle–Angle.

The complete lack of similarity has a major drawback. No one could carry a reliable street map, for it would have to be the size of the city. Also architectural plans would need to be the same size as the building.

Many other equivalents to Postulate V can be given. In the following theorem we mention some others and summarize our previous efforts:

Theorem 3.9. *Postulate V is equivalent to each of the following:*

(1) *Playfair's Axiom.*
(2) *The set of points on one side and equidistant from a given line forms a line.*
(3) *The interior angles of a triangle sum to two right angles.*
(4) *Wallis's Postulate.*
(5) *The converse of the Alternate Interior Angles Theorem* (Proposition I.29).
(6) *The perpendicular bisectors of the sides of a triangle are concurrent.*
(7) *There exists a point equidistant from any three noncollinear points.*
(8) *If C is on a circle with diameter AB and $A \neq C \neq B$, then $\angle ACB$ is a right angle.*
(9) *There exists a rectangle.*
(10) *There exists an acute angle such that every point in the interior of the angle is on a line intersecting both rays of the angle away from the vertex.*
(11) *Any pair of parallel lines has a common perpendicular.*
(12) *The Pythagorean Theorem.*

The work of Saccheri

Earlier geometers sought to prove that Postulates I, II, III, and IV imply V. One approach is a *reductio ad absurdum* argument: Assume Postulates I, II, III, and IV and the *negation* of Postulate V and then reason to a contradiction. But how do we make precise the failure of Postulate V? An efficient answer was introduced by Saccheri.

Girolamo Saccheri was a Jesuit priest and professor at the University of Pavia. His *Euclides ab omni naevo vindicatus* (1733) marks a triumph of logic in the pursuit of a proof of Postulate V. It also contains the beginnings of the study of non-Euclidean geometry, disguised by a flaw in Saccheri's work.

The general device that he introduced is called the Saccheri quadrilateral. In fact, this figure had been introduced much earlier in the work of `Umar Khayyām. It also played a role in a work by GIORDANO VITALE (1633–1711) who gave another *petitio principii* proof of Postulate V in his book *Euclides restituto* (1680).

Let AB be a line segment, the *base*, and suppose AD and BC are line segments with $AD \cong BC$, and AD and BC both perpendicular to AB. Join the points C and D to form the **Saccheri quadrilateral** $ABCD$. We develop the properties of these figures and consider $\angle ADC$ and $\angle BCD$.

The references to proposition numbers, scholia, and such in this section are to the original statements in Saccheri.

Proposition 3.10 (Proposition I). *In the Saccheri quadrilateral $ABCD$ on the base AB, $\angle ADC \cong \angle BCD$.*

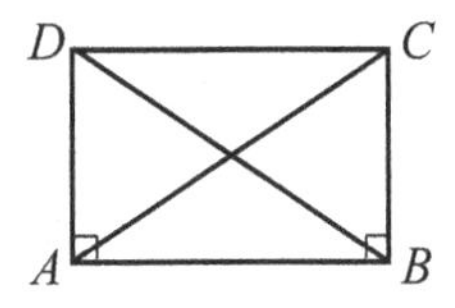

PROOF: Consider the line segments AC and BD. Since $AD \cong BC$ and $AB \cong BA$, by the congruence criterion Side–Angle–Side, $\triangle ACB$ is congruent to $\triangle BDA$, and so $AC \cong BD$. By Side–Side–Side, $\triangle BCD \cong \triangle ADC$ and the proposition follows. ■

Opposite sides of a Saccheri quadrilateral enjoy common perpendiculars.

Proposition 3.11 (Proposition II). *Let $ABCD$ be a Saccheri quadrilateral on the base AB, and let M and N be the midpoints of AB and CD, respectively. Then MN is perpendicular to AB and MN is perpendicular to CD.*

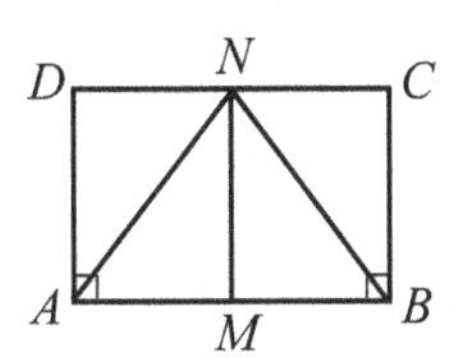

PROOF: Join AN and BN. By Side–Angle–Side, $\triangle ADN \cong \triangle BCN$. By Side–Side–Side, we find $\triangle AMN \cong \triangle BMN$ and MN is perpendicular to AB. By joining D to M and C to M the same argument applies to show that MN is perpendicular to CD. ■

Saccheri considers *all* possible cases of such a quadrilateral.

$\angle ADC$ is a right angle: **HRA**, the hypothesis of the right angle.
$\angle ADC$ is an obtuse angle: **HOA**, the hypothesis of the obtuse angle.
$\angle ADC$ is an acute angle: **HAA**, the hypothesis of the acute angle.

HRA is equivalent to Postulate V because a rectangle exists. To negate Postulate V we can take **HOA** or **HAA** to hold. Furthermore, as in the proof of Theorem 3.7, if one of **HRA**, **HOA**, or **HAA** holds for one quadrilateral, it holds for all.

Theorem 3.12 (Proposition III). *Let ABCD be a Saccheri quadrilateral on the base AB. Under the assumption* **HRA**, **HOA**, *or* **HAA**, *we have $AB = CD$, $AB > CD$, or $AB < CD$, respectively, and the interior angle sum of a triangle is $=$, $>$, or $<$ two right angles, respectively.*

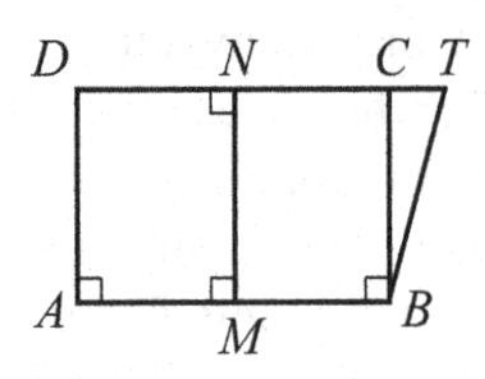

PROOF: Let M and N denote the midpoints of AB and CD, respectively. We consider the assumption **HAA** since **HRA** is known and the argument for **HOA** is similar. Suppose $CD < AB$, then $CN < BM$ and so extend CD to T with $NT \cong MB$. Then $\angle MBT >$ a right angle. But, in the Saccheri quadrilateral $NMBT$, $\angle BTN \cong \angle TBM$ by Proposition 3.10. However, $\angle BTN$ is opposite the exterior angle $\angle BCN$ for $\triangle BCT$. By **HAA**, $\angle BCN$ is acute. Since $\angle BTN \cong \angle TBM >$ a right angle, we get a contradiction to the exterior angles theorem (Euclid Proposition I.16). Thus $AB < CD$.

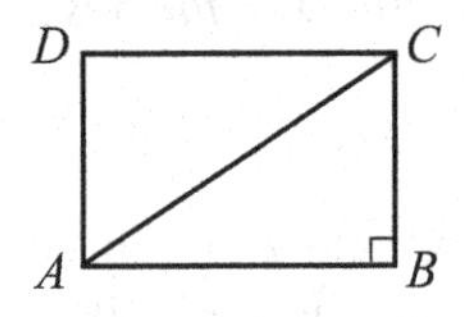

Suppose we have $\triangle ABC$, a right triangle with right angle at B. Construct AD perpendicular to AB with $AD \cong BC$. Assuming **HAA**, $CD > AB$. Thus $\angle DAC > \angle ACB$ since the greater side subtends the greater angle (Euclid Proposition I.25). But $\angle CAB + \angle DAC =$ a right angle. Therefore, the angle sum of $\triangle ABC$ is less than two right angles. ■

To apply *reductio ad absurdum*, Saccheri tried to prove that **HOA** and **HAA** lead to contradictions. **HOA** contradicts the Saccheri–Legendre Theorem (Theorem 2.2) and so **HOA** is disallowed. Saccheri then began *"a lengthy battle against* **HAA** *which alone opposes the truth of the axiom"* (Saccheri 1733, p. 13). He proved some remarkable theorems along the way.

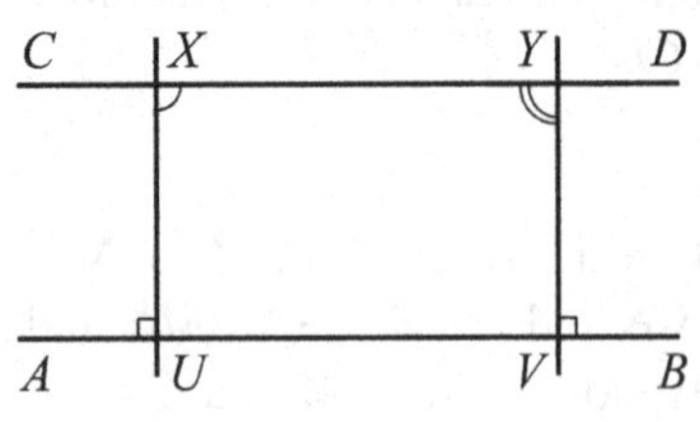

Consider two parallel lines $\overleftrightarrow{AB}$ and $\overleftrightarrow{CD}$, and construct lines $\overleftrightarrow{XU}$ and $\overleftrightarrow{YV}$ perpendicular to $\overleftrightarrow{AB}$ from X, Y on $\overleftrightarrow{CD}$. Consider $\angle UXY$ and $\angle VYX$. If both angles are acute, then, by continuity, $\overleftrightarrow{AB}$ and $\overleftrightarrow{CD}$ have a common perpendicular between the segments UX and VY. If one of $\angle UXY$ or $\angle VYX$ is a right angle, then $\overleftrightarrow{CD}$ and $\overleftrightarrow{AB}$ already share a common perpendicular. Finally, suppose $\angle UXY$ is acute and $\angle VYX$ is obtuse. If we move V away from U along $\overleftrightarrow{AB}$, then $\angle VYX$ may change to a right angle (giving a common perpendicular) or it may remain obtuse.

Theorem 3.13 (Proposition XXIII). *In the preceding diagram, and assuming* **HAA**, *if $\overleftrightarrow{AB}$ is parallel to $\overleftrightarrow{CD}$ and $\angle VYX$ remains obtuse as V moves away from U along $\overleftrightarrow{AB}$, then $\overleftrightarrow{CD}$ is asymptotic to $\overleftrightarrow{AB}$.*

PROOF: Suppose for some $V'Y'$ perpendicular to $\overleftrightarrow{AB}$ further from U than V, we have $V'Y' > VY$. Let W be on $V'Y'$ with $V'W \cong VY$. By Proposition 3.10 $\angle VYW \cong \angle V'WY$ and by assumption $\angle VYW$ is acute.

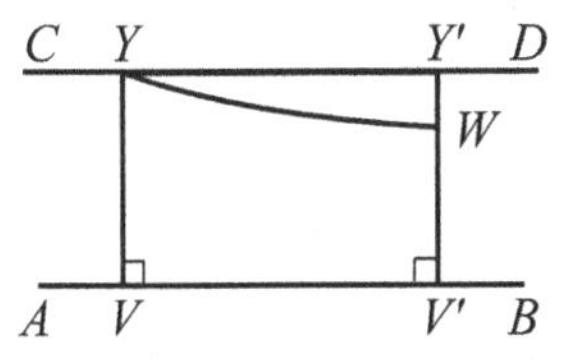

However, $\angle V'Y'Y$ is obtuse and the exterior angles theorem implies $\angle V'WY > \angle V'Y'Y$, which is absurd. Since $V'Y' \cong VY$ implies that there is a common perpendicular to $\overleftrightarrow{AB}$ and $\overleftrightarrow{CD}$ (Theorem 3.11), contrary to our assumption, then we see that $VY > V'Y'$.

It remains to show that $V'Y'$ can be made smaller than any fixed line segment. For this we use the fact that, under **HAA**, all triangles, and hence all quadrilaterals, have an *angle defect*. By Theorem 3.12 the sum of the interior angles in a triangle is less than two right angles and so the sum of the interior angles of a quadrilateral is less than four right angles. Denote the angle defect of a quadrilateral $KLMN$ by:

$$\delta(KLMN) = 2\pi - (\angle KLM + \angle LMN + \angle MNK + \angle NKL).$$

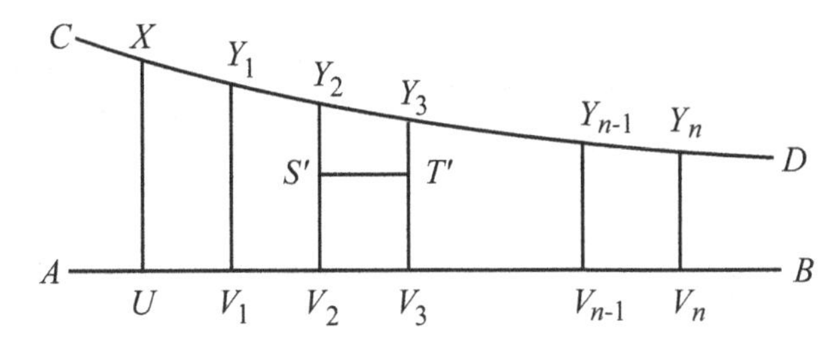

Suppose there is a line segment ST, such that $YY > ST$ for every choice of V to the right of U. Along the ray $\overrightarrow{AB}$ mark the points $V_1, V_2, \ldots, V_n$ with $UV_1 \cong V_1V_2 \cong \cdots \cong V_{n-1}V_n \cong ST$. Construct the line segments $V_1Y_1, V_2Y_2, \ldots, V_nY_n$ perpendicular to $\overleftrightarrow{AB}$ with $Y_1, Y_2, \ldots Y_n$ on $\overleftrightarrow{CD}$. Consider the angle defect of the quadrilateral UV_nY_nX. The basic property of angle defect is that it sums like area (Exercise 3.2) and so:

$$\delta(UV_nY_nX) = \delta(UV_1Y_1X) + \cdots + \delta(V_{n-1}V_nY_nY_{n-1}).$$

Since the vertical segments are perpendicular to $\overleftrightarrow{AB}$ we see that:

$$0 < \delta(UV_nY_nX) < \pi.$$

Each of the constituent quadrilaterals has angle defect greater than 0, and so there must be some $V_kV_{k+1}Y_{k+1}Y_k$ with $\delta(V_kV_{k+1}Y_{k+1}Y_k) \leq \pi/n$. Since $V_kY_k > ST$, we can always construct a quadrilateral $V_kS'T'V_{k+1}$ with $V_kS' \cong T'V_{k+1} \cong V_kV_{k+1} \cong ST$ and every such quadrilateral is congruent to any other. Let $\delta(V_kS'T'V_{k+1}) = \rho > 0$. However, we can choose n large enough so that $\pi/n < \rho$. Since

$\delta(V_kV_{k+1}Y_{k+1}Y_k) = \delta(V_kS'T'V_{k+1}) + \delta(S'Y_kY_{k+1}T')$, we get $\delta(S'Y_kY_{k+1}T') < 0$, a contradiction. Thus $\overrightarrow{CD}$ is asymptotic to $\overrightarrow{AB}$. ∎

In the figure notice that $\angle V_nY_nC < \angle V_{n-1}Y_{n-1}C$. It cannot be equal since this implies **HRA** and it cannot be greater because $\angle V_nY_nC + \angle V_{n-1}Y_{n-1}D$ would be greater than two right angles, which would violate **HAA**. Saccheri shows that asymptotic lines as in Theorem 3.13 exist under **HAA**.

Theorem 3.14 (Proposition XXXII). *Given a point P not on a line $\overleftrightarrow{AB}$ there are three classes of lines through P:*

(1) *lines meeting $\overleftrightarrow{AB}$,*
(2) *lines with a common perpendicular to $\overleftrightarrow{AB}$, and*
(3) *lines without a common perpendicular to $\overleftrightarrow{AB}$ and hence asymptotic to $\overleftrightarrow{AB}$.*

PROOF: Let AP be perpendicular to the line $\overleftrightarrow{AB}$. We can construct the line $\overleftrightarrow{PD}$ perpendicular to AP by Euclid's Proposition I.11. Suppose $\overrightarrow{PD'}$ is a line through P with common perpendicular $B'D'$ to $\overleftrightarrow{AB}$, and suppose $\overrightarrow{PD''}$ is a line so that:

$$\angle APD' < \angle APD'' < \text{a right angle.}$$

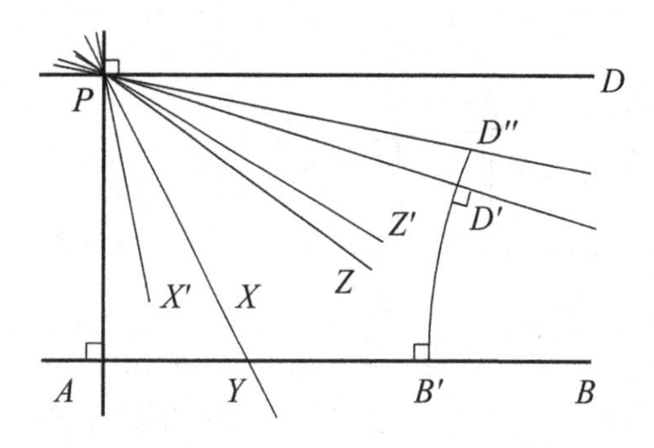

By the Exterior Angles Theorem, $\angle PD''D'$ is acute and $\angle APD''$ is acute; by continuity of the change of angle, $\overrightarrow{PD''}$ must have a common perpendicular with $\overleftrightarrow{AB}$ somewhere between A and B'.

Suppose $\overrightarrow{PX}$ is a line with $\overrightarrow{PX}$ meeting $\overleftrightarrow{AB}$ at Y. If $\overrightarrow{PX'}$ is another line with $0 < \angle APX' < \angle APX$, then $\overrightarrow{PX'}$ is trapped inside $\triangle APY$ and, by Pasch's Axiom, exits the triangle. It must exit through $\overrightarrow{AY}$.

This accounts for lines of types 1 and 2. Let $\overrightarrow{PZ}$ be a line through P that does not meet $\overrightarrow{AB}$ so that if $\overrightarrow{PW}$ is another line and $\angle APW$ is less than $\angle APZ$, $\overrightarrow{PW}$ meets $\overleftrightarrow{AB}$. That is, $\overrightarrow{PZ}$ is the upper limit of lines through P meeting $\overleftrightarrow{AB}$. Let $\overrightarrow{PZ'}$ be the line so that if $\overrightarrow{PV}$ is any line with $\angle APV$ acute and $\angle APV > \angle APZ'$, then $\overrightarrow{PV}$ has a common perpendicular with $\overleftrightarrow{AB}$. Thus, $\overrightarrow{PZ'}$ is the lower limit of lines with a common perpendicular.

Notice $\overrightarrow{PZ'}$ has no common perpendicular, for if it did one could find $\overrightarrow{PV}$ with $\angle APV < \angle APZ'$ and satisfying the criteria for $\overrightarrow{PZ'}$.

Claim: $\overrightarrow{PZ} = \overrightarrow{PZ'}$.

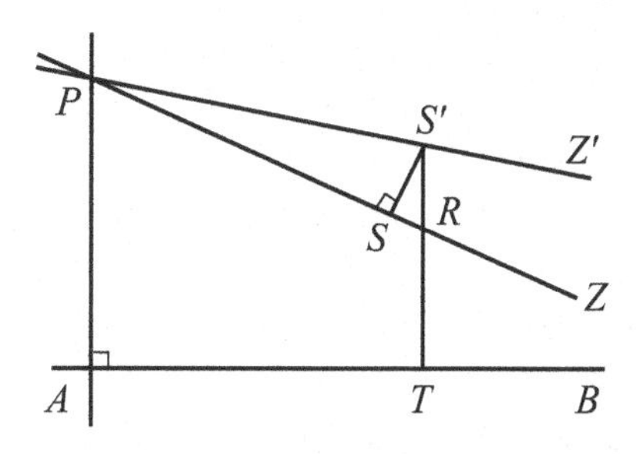

Suppose $\overrightarrow{PZ} \neq \overrightarrow{PZ'}$. Since $\angle ZPZ'$ is acute, the distance between $\overrightarrow{PZ}$ and $\overrightarrow{PZ'}$ can be made as large as desired (Corollary 3.3). Choose S' on $\overrightarrow{PZ'}$ and S on $\overrightarrow{PZ}$ so that $SS' > AP$. Construct the perpendicular from S' to $\overleftrightarrow{AB}$ at T. Since $\overrightarrow{PZ'}$ is asymptotic to $\overleftrightarrow{AB}$, $AP > S'T$. The distance between a line and a point is given by the perpendicular. Since S' is on the opposite side of $\overrightarrow{PZ}$ from T, $S'T$ meets $\overrightarrow{PZ}$ at a point R and $S'T > S'R \geq SS'$. Therefore, $AP > S'T > SS' > AP$, a contradiction. Thus, $\overrightarrow{PZ} = \overrightarrow{PZ'}$. ∎

Saccheri then concluded that *"the hypothesis of the acute angle is absolutely false, because it is repugnant to the nature of straight lines"* (Proposition XXXIII). He reasoned as follows: The lines $\overrightarrow{PZ}$ and $\overleftrightarrow{AB}$ are both "perpendicular" to the "line at infinity" at the same point (being asymptotic). Two lines cannot be perpendicular from the same point to a given line without being the same line and so the line $\overrightarrow{PZ}$ cannot exist. After a *tour de force* of flawless logic, he attributed properties at infinity that are only true in finite ranges. The flaw in his work was observed by Lambert (1786).

From Saccheri's work we have a wealth of theorems, true for the system of geometry determined by the first four postulates of Euclid with **HAA**:

(1) Angle sums of triangles are less than two right angles.
(2) There exist parallel lines without a common perpendicular that are asymptotic.
(3) Lines through a point not on a line fall into three classes (as in Theorem 3.14).

His conviction that Euclidean geometry was the only true geometry clouded his vision to a new world he had labored to reveal. This new world waited for almost another one hundred years for others to lay claim to it.

Exercises

3.1 Prove the assertions of the last theorem that Postulate V is equivalent to: (a) the existence of a rectangle (a quadrilateral with four interior right angles); (b) if C is on a circle with diameter AB and C does not lie on AB, then $\angle ACB$ is a right angle; and (c) any pair of parallel lines has a common perpendicular.

3.2 Suppose all triangles have an *angle defect*, that is, $\pi -$ the interior angle sum of any triangle is a positive number. Let $\delta(\triangle ABC) = \pi - \angle ABC - \angle CAB - \angle BCA$ denote the angle defect. Suppose that a triangle is subdivided into two nonoverlapping triangles (for example, by an interior altitude); then show that the angle defect of the large triangle is the sum of the angle defects of the constituent parts. Extend the notion of angle defect to quadrilaterals and polygons in general.

Solutions to Exercises marked with a dagger appear in the appendix, pp. 327–328.

3.3 On the sphere with great circles as lines we have seen several results that show that Postulates I through V cannot hold. Show how the various equivalents of Postulate V fail on the sphere and interpret their failure as a fact about the geometry of the sphere.

3.4 Suppose Postulate V. Show that there is a point equidistant from three given noncollinear points. How does Postulate V play a role?

3.5[†] Criticize the following paraphrase of a "proof" of Postulate V due to Legendre (1804) and determine which of the equivalents of Postulate V he assumed:

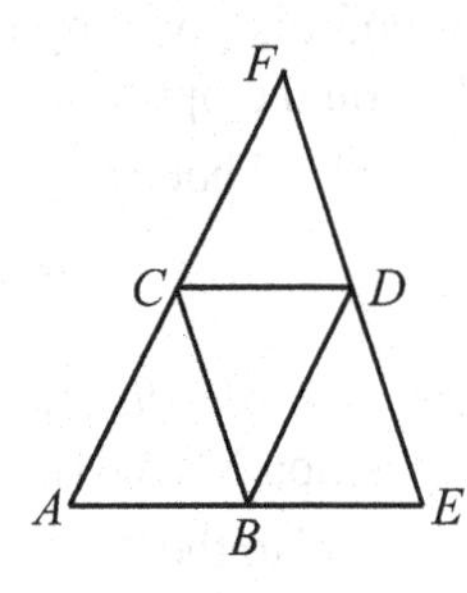

"Given $\triangle ABC$, construct on BC the triangle $\triangle BCD$ congruent to $\triangle ABC$ with $\angle DBC \cong \angle BCA$, $\angle DCB \cong \angle CBA$. Then draw through D any line which cuts $\overleftrightarrow{AC}$ and $\overleftrightarrow{AB}$. If the angle sum of $\triangle ABC$ is $\pi - \delta$, then the angle sum of $\triangle AEF$ is less than or equal to $\pi - 2\delta$. Repeating this process n times gives a triangle with angle sum less than or equal to $\pi - 2^n\delta$. If $\delta > 0$, then n can be chosen large enough that $2^n\delta > \pi$; this leads to a contradiction. Thus the angle sum of the triangle must be π." (See also the discussion in the book of Laubenbacher and Pengelley (p. 26–29, 1999).)

3.6 Prove that if Wallis's postulate holds, then given any polygon $A_1A_2\cdots A_n$ and a segment BC, there is a polygon $B_1B_2\cdots B_n$ with $B_1 = B$ and $B_2 = C$ and $B_1B_2\cdots B_n$ similar to $A_1A_2\cdots A_n$.

3.7[†] A *Lambert quadrilateral* is one in which three of the interior angles are right angles. Show that the fourth angle is a right angle if and only if **HRA** holds. If **HAA** holds, show that the sides adjacent to the acute angle are greater than their respective opposite sides.

3.8 Suppose $ABCD$ is a quadrangle and all four angles are right angles. Show that $AB \cong CD$. (*Hint*: If $AB < CD$, then there is a point D' on CD with $CD' \cong AB$ forming a Saccheri quadrilateral $ABCD'$.)

3.9 Prove a theorem first presented by `Umar Khayyām: If $ABCD$ is a Saccheri quadrilateral and three points of a line CD are equidistant from a line AB then all points on CD are equidistant.

3.10[†] Prove that if l and m are nonintersecting lines with a common perpendicular, then, under **HAA**, l and m diverge.

3.11 Complete the proof of part (b) Theorem 3.14 in Theorem 3.7 where you consider a ray $\overrightarrow{AC}$ that does not meet the line $\overleftrightarrow{PQ}$.

3.12[†] Fix a line $\overleftrightarrow{AX}$ and a perpendicular segment OA. Suppose l is a line through O, and M is a point on l in the direction of X. Suppose that the perpendicular to $\overleftrightarrow{AX}$ through M meets $\overleftrightarrow{AX}$ at P. Show that the length of PM is a continuous function of the length of the segment OM. (*Hint*: Use Saccheri quadrilaterals.)

4

Non-Euclidean geometry

I have created a new universe from nothing. All that I have hitherto sent you compares to this only as a house of cards to a castle.

J. BOLYAI (1823)

The non-Euclidean geometry throughout holds nothing contradictory.

C. F. GAUSS (1831)

In geometry I find certain imperfections which I hold to be the reason why this science, apart from transition into analytics, can as yet make no advance from that state in which it has come to us from Euclid.

N. I. LOBACHEVSKIĬ (1840)

Between Euclid's time and 1829, the year Lobachevskiĭ's *On the Principles of Geometry* appeared, most of the critics of *The Elements* were concerned with the "purification" of Euclid's work from its perceived imperfections. So strong was the conviction that Postulate V depended on Postulates I through IV that some of these critics did not see in their work the basis for a new geometry. An important contribution appeared in 1763 with the Göttingen doctoral thesis of G. S. KLÜGEL (1739–1812), written under the direction of ABRAHAM KAESTNER (1719–1800), in which Klügel examined 30 attempts to prove Postulate V, finding all of them guilty of some form of *petitio principii*. In his introduction he states:

> *If all the attempts are given thorough consideration, it becomes clear that Euclid correctly counted among the axioms a proposition that cannot be proven in a proper manner with any others.*

With Klügel, the first doubt about the program of "vindicating Euclid of every flaw" is voiced and it became possible to examine alternatives critically and logically. The development of analysis in the eighteenth century plays another role that was unavailable to earlier researchers. The use of limits and the properties of functions provides another framework in which to develop geometry. The sureness of analytical ideas lent credibility to the geometric ideas based upon them. The pioneers of non-Euclidean geometry saw the analytic features of the new geometry as a consistent picture of a new world.

The work of Gauss

We skip over the interesting work of the late eighteenth century, slighting Legendre, Lambert, Taurinus, Schweikart, F. Bolyai, and others to get to

CARL-FRIEDRICH GAUSS (1777–1855). (See Bonola (1955), Rosenfeld (1987), or Gray (2007) for more details.) His contributions are found in two brief memoranda (Gauss, 1870, VIII, pp. 202–8), letters, and unpublished notes. The following is an excerpt from a letter that expresses his feelings on the investigations:

> *The assumptions that the sum of three angles is less than* 180° *leads to a curious geometry, quite different from ours, but thoroughly consistent . . .*
>
> (Gauss to Taurinus, November 8, 1824)

This sentiment distinguishes his work from his predecessors. It denies the absolute nature of Euclid's geometry. Gauss's unwillingness to publish his work left the public introduction of this "curious geometry" to JÁNOS BOLYAI (1802–60) and NICOLAĬ I. LOBACHEVSKIĬ (1792–1850).

We begin with a *different* definition of parallel:

Definition 4.1 (Gauss 1831). *Given a line* $\overleftrightarrow{AB}$ *and a point P not on* $\overleftrightarrow{AB}$*, let AP be the perpendicular from P to* $\overleftrightarrow{AB}$*. A line* $\overleftrightarrow{PQ}$ *is* **parallel to** $\overleftrightarrow{AB}$ **through** *P if for any line* $\overleftrightarrow{PS}$ *with S lying in BAPQ and* $0 < \angle APS < \angle APQ$*, it follows that* $\overleftrightarrow{PS}$ *meets* $\overleftrightarrow{AB}$*.*

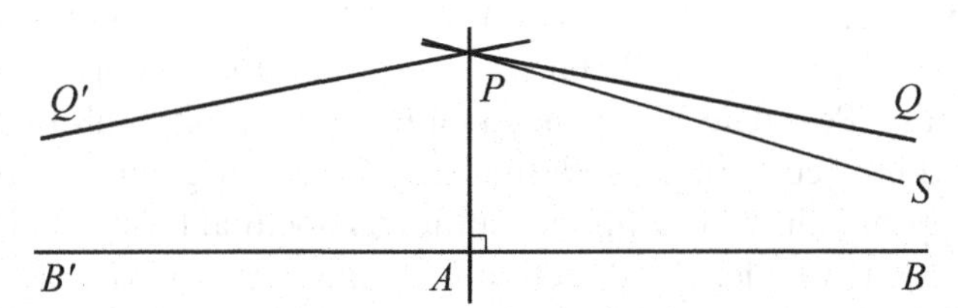

If we consider the set of lines through the point P, the parallel $\overleftrightarrow{PQ}$ is the "first" line in the set of lines through P, sweeping toward B from PA, not meeting $\overleftrightarrow{AB}$. Notice that without Postulate V we do not know if this first line is also the last (Playfair's Axiom). It also depends on which direction we sweep the lines. To include this distinction we speak of parallel lines *in a direction* (here $\overrightarrow{AB}$). Those lines through P that do not intersect $\overleftrightarrow{AB}$ will be called **nonintersecting** lines to distinguish them from these parallels.

Theorem 4.2. *In a given direction, "being parallel" is an equivalence relation.*

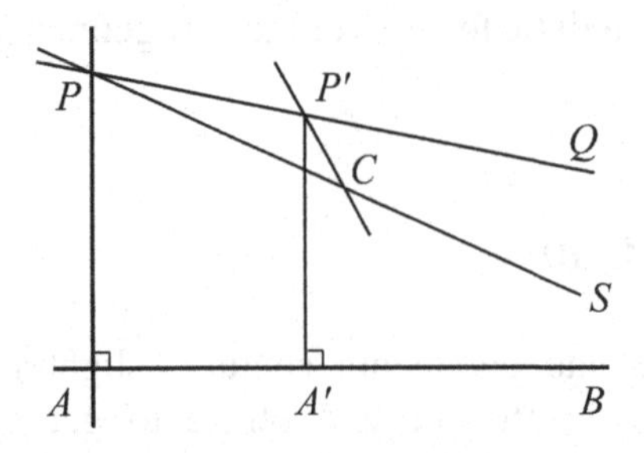

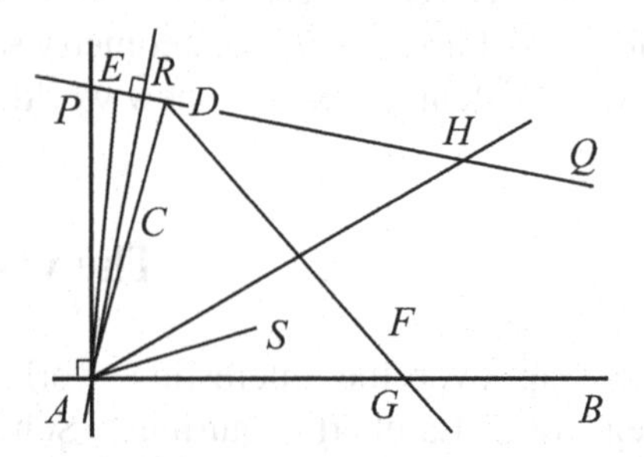

PROOF (Gauss, 1870, VIII, pp. 202–5):

(a) We first show that if $\overleftrightarrow{PQ}$ is parallel to $\overleftrightarrow{AB}$ in the direction of $\overrightarrow{AB}$, then for any point P' between P and Q, $\overleftrightarrow{P'Q}$ is parallel to $\overleftrightarrow{AB}$, that is, parallelism is well defined for lines in a direction. Let $P'A'$ be perpendicular to $\overleftrightarrow{AB}$. Consider any line through P', $\overleftrightarrow{P'C}$, such that C is inside $BAP'Q$. Join P to C. By assumption $\overleftrightarrow{PC}$ meets $\overleftrightarrow{AB}$. But $\overleftrightarrow{P'C}$ has entered a triangle and, by Pasch's axiom, must exit through $\overleftrightarrow{AB}$. Thus $\overleftrightarrow{P'Q}$ is parallel to $\overleftrightarrow{AB}$ in the direction $\overrightarrow{AB}$.

(b) Next we show that if $\overleftrightarrow{PQ}$ is parallel to $\overleftrightarrow{AB}$, then $\overleftrightarrow{AB}$ is parallel to $\overleftrightarrow{PQ}$. Let R be on $\overleftrightarrow{PQ}$ so that AR is perpendicular to $\overleftrightarrow{PQ}$. Let $\overleftrightarrow{AS}$ be any line through A with S lying in $QPAB$. Choose $\overrightarrow{AC}$ so that $\angle RAC \cong \frac{1}{2}\angle SAB$; then either $\overrightarrow{AC}$ meets $\overleftrightarrow{PQ}$ or it doesn't. Suppose it does at point D. Let E be such that $RE \cong RD$ and $E \neq D$. Then $\angle EAD \cong \angle SAB$. Choose $\overrightarrow{DF}$ with $\angle ADF \cong \angle AED$. The line $\overleftrightarrow{PQ}$ is parallel to $\overleftrightarrow{AB}$ and so $\overrightarrow{DF}$ meets $\overleftrightarrow{AB}$ at, say, G. Let H be on $\overleftrightarrow{PQ}$ so that $EH \cong DG$. Then, by Side–Angle–Side, $\triangle AEH$ is congruent to $\triangle ADG$. Thus, $\angle EAH \cong \angle DAG$. But $\angle EAH = \angle EAD + \angle DAH = \angle SAB + \angle DAH$. Then $\angle SAB = \angle DAG - \angle DAH = \angle HAG$ and $\overrightarrow{AH}$ coincides with $\overrightarrow{AS}$.

We leave it to the reader to prove the case when $\overrightarrow{AC}$ does not meet $\overleftrightarrow{PQ}$.

(c) Finally we show that if $\overleftrightarrow{AB}$ is parallel to $\overleftrightarrow{PQ}$ and $\overleftrightarrow{PQ}$ is parallel to $\overleftrightarrow{UV}$, then $\overleftrightarrow{AB}$ is parallel to $\overleftrightarrow{UV}$.

Case 1: $\overleftrightarrow{PQ}$ is between $\overleftrightarrow{AB}$ and $\overleftrightarrow{UV}$. Suppose $\overrightarrow{AS}$ is a line toward B with S inside $BAUV$. Since $\overleftrightarrow{AB}$ is parallel to $\overleftrightarrow{PQ}$, $\overrightarrow{AS}$ meets $\overleftrightarrow{PQ}$ at a point, say M. Let T be on $\overleftrightarrow{AS}$ on the other side of $\overleftrightarrow{PQ}$. Since $\overleftrightarrow{PQ}$ is parallel to $\overleftrightarrow{UV}$, $\overrightarrow{MT}$ meets $\overleftrightarrow{UV}$ and so $\overrightarrow{AS}$ meets $\overleftrightarrow{UV}$.

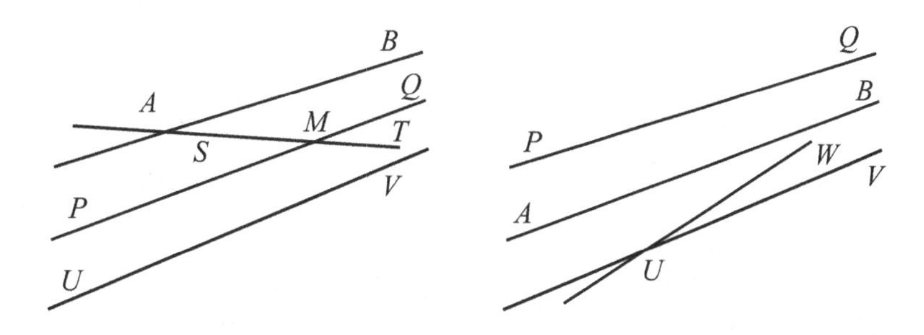

Case 2: $\overleftrightarrow{AB}$ is between $\overleftrightarrow{PQ}$ and $\overleftrightarrow{UV}$. If $\overleftrightarrow{AB}$ is not parallel to $\overleftrightarrow{UV}$, there is a line $\overleftrightarrow{UW}$ with $\overleftrightarrow{AB}$ parallel to $\overleftrightarrow{UW}$ in the direction $\overrightarrow{AB}$. If W is inside $UVBA$, since $\overleftrightarrow{PQ}$ is parallel to $\overleftrightarrow{UV}$, then $\overrightarrow{UW}$ meets $\overleftrightarrow{PQ}$, a contradiction. Also, if W is outside $UVBA$, since $\overrightarrow{UW}$ is parallel to $\overleftrightarrow{AB}$ and $\overleftrightarrow{AB}$ is parallel to $\overleftrightarrow{PQ}$, by Case 1 we have that $\overleftrightarrow{UW}$ is parallel to $\overleftrightarrow{PQ}$, and so $\overleftrightarrow{UV}$ meets $\overleftrightarrow{PQ}$, a contradiction. ∎

Gauss's notion of parallel leads to a function of a perpendicular line segment to a given line:

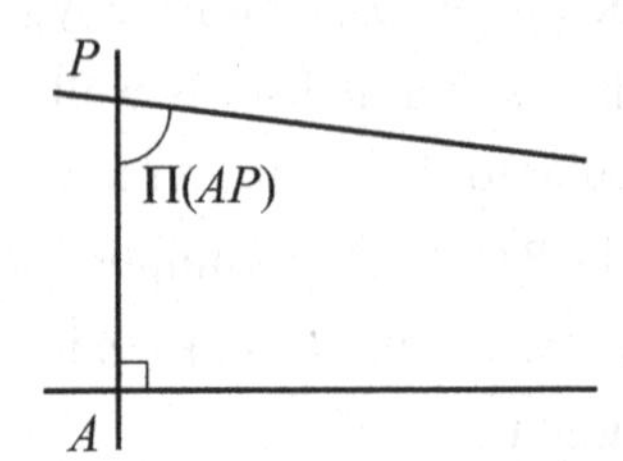

Definition 4.3. *Let P be a point not on a line ℓ and PA perpendicular to ℓ with A on ℓ. The* **angle of parallelism** $\Pi(AP)$ *is the angle between AP and a parallel to ℓ through P.*

Under the assumption **HRA**, $\Pi(AP)$ equals a right angle for any AP; **HAA** implies that $\Pi(AP)$ is strictly less than a right angle. We develop some elementary properties of $\Pi(AP)$ first.

(1) $\Pi(AP)$ depends only on the length of AP. That is, if $AP \cong CR$, then $\Pi(AP) = \Pi(CR)$.
To see this, suppose that $\Pi(AP) < \Pi(CR)$.

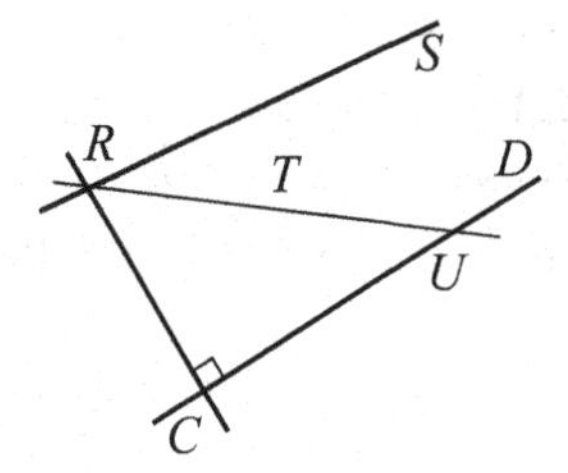

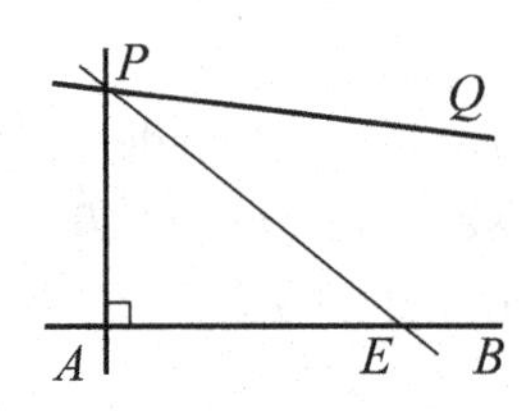

Then there is a line $\overrightarrow{RT}$ so that $\angle CRT \cong \angle APQ$ and $\overrightarrow{RT}$ meets $\overleftrightarrow{CD}$ at a point U. Let E be on $\overleftrightarrow{AB}$ with $AE \cong CU$ and consider the triangle $\triangle APE$. Since $AP \cong CR$ and $\angle PAE \cong \angle RCU$, by Side–Angle–Side, $\triangle APE \cong \triangle CRU$. Then $\angle APE \cong \angle CRU$, which equals the angle of parallelism, and the lines $\overrightarrow{PE}$ and $\overrightarrow{PQ}$ coincide, which is impossible. The case for $\Pi(AP) > \Pi(CR)$ is symmetric.

Suppose we have chosen a unit length and this choice determines a measure of length on line segments. Property 1 of the angle of parallelism allows us to define a function $\Pi : [0,\infty) \to [0,\pi/2]$ given by $\Pi(x) = \Pi(AP)$, where AP is any line segment of length x in the measure.

(2) If **HAA** holds and $AP > AR$, then $\Pi(AP) < \Pi(AR)$, that is, the angle of parallelism is a monotone decreasing function of length.

Suppose that $\Pi(AP) = \Pi(AR)$. Let M denote the midpoint of PR. Construct MQ' perpendicular to $\overleftrightarrow{PQ}$ and extend it to MS'. Since $\Pi(AP) = \Pi(AR)$, $\angle MPQ' \cong \angle MRS'$. We also have $MR \cong MP$ and $\angle PMQ' \cong \angle RMS'$, so $\triangle PQ'M \cong \angle RS'M$. It follows that MS' is perpendicular to $\overleftrightarrow{RS}$. This implies that $\overleftrightarrow{PQ}$ and $\overleftrightarrow{RS}$ share a common perpendicular. Since $\overleftrightarrow{PQ}$ is parallel to $\overleftrightarrow{RS}$, under **HAA** this cannot happen by Theorem 3.14.

Suppose $\Pi(AP) > \Pi(AR)$, then there is a line through P, say $\overrightarrow{PT}$ with $\angle RPT \cong \angle ARS$. Then $\angle APT < \angle APQ$ and $\overrightarrow{PT}$ meets $\overleftrightarrow{AB}$, and hence $\overleftrightarrow{RS}$. By the Alternate Interior Angles Theorem (Proposition I.27), however, $\overleftrightarrow{PT}$ does not meet $\overleftrightarrow{RS}$.

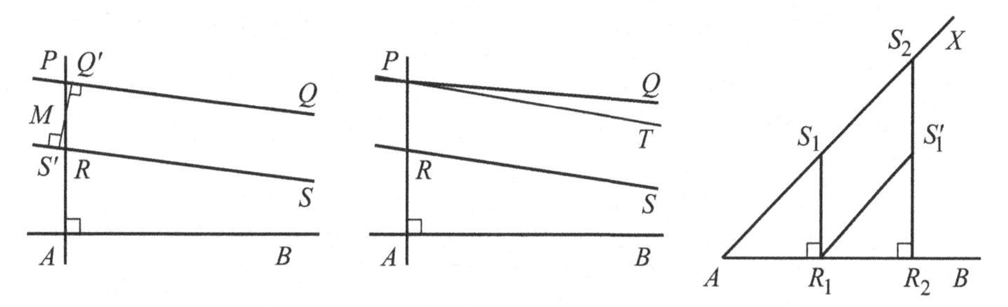

(3) In fact, any acute angle is the angle of parallelism for some length. To see this, fix the angle $\angle BAX$. Suppose R_1S_1 is perpendicular to $\overleftrightarrow{AB}$ at R_1 and S_1 lies on $\overrightarrow{AX}$. The triangle $\triangle AR_1S_1$ has some angle defect, say $\delta > 0$. Mark R_2 along $\overleftrightarrow{AB}$ with $R_1R_2 \cong AR_1$. If the perpendicular to $\overleftrightarrow{AB}$ from R_2 meets $\overrightarrow{AX}$ at S_2, then $R_2S_2 > R_1S_1$ and we can construct the point S_1' along R_2S_2 such that $R_2S_1' \cong R_1S_1$. It follows that $\triangle AR_2S_2$ contains two copies of $\triangle AR_1S_1$ and so the angle defect of $\triangle AR_2S_2$ is at least 2δ (see Exercise 3.2). If every perpendicular to $\overleftrightarrow{AB}$ within $\angle BAX$ meets $\overrightarrow{AX}$, then we can iterate this procedure to construct a triangle with angle defect greater than $n\delta$ for any n and so greater than two right angles, which is impossible. Hence, some perpendicular to $\overleftrightarrow{AB}$ does not meet AX and $\angle BAX$ is the angle of parallelism for some length.

The hyperbolic plane

Without the millstone of proving Postulate V around their necks, Gauss, J. Bolyai, and Lobachevskiĭ developed the properties of the plane with the new definition of parallelism and **HAA**. Auxiliary constructions and curves led to new tools and results, especially when developed in space.

We introduce properties and objects that do not depend on the choice of **HAA** or **HRA**. We call such a property or object **absolute**. The following collections of lines are absolute. A **pencil of lines** is a set of lines satisfying one of three properties: There is the pencil of all lines through a given point P, denoted $\mathbf{P}_P$; the pencil of all lines parallel to a given line l, denoted $\mathbf{P}_l$; and the pencil of all lines perpendicular to a given line, m, denoted $\mathbf{P}_m^{\perp}$.

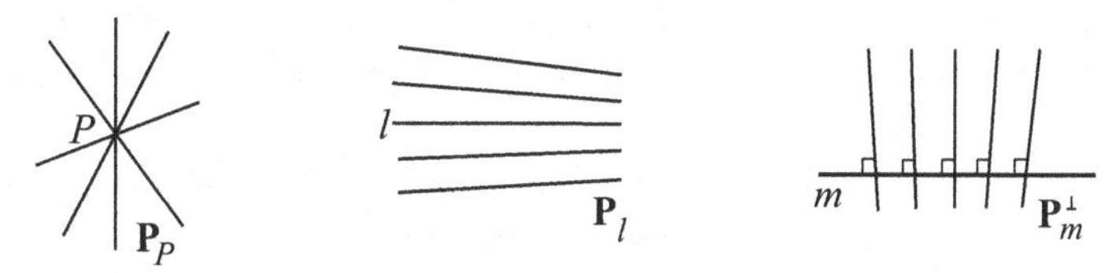

The pencil of lines $\mathbf{P}_l$ parallel to a given line l can be thought of as a limit of pencils of lines $\mathbf{P}_X$ through a point X, as X travels along l to "infinity." Notice that every point in the plane lies on a unique line in the pencils of parallels or perpendiculars, and every point other than the given point P lies on a unique line in $\mathbf{P}_P$.

A choice of three distinct points determines each sort of pencil of lines.

Theorem 4.4. *Given distinct points X, Y, and Z, the lines l, m, and n that are the perpendicular bisectors of the segments XY, YZ, and XZ, respectively, lie in a pencil of lines.*

PROOF (Shirikov 1964): Consider the triangle ΔXYZ. If l and m meet, then they meet at the center of a circle circumscribed around the triangle. Furthermore, the line n passes through the same point of intersection. We leave it to the reader to prove this.

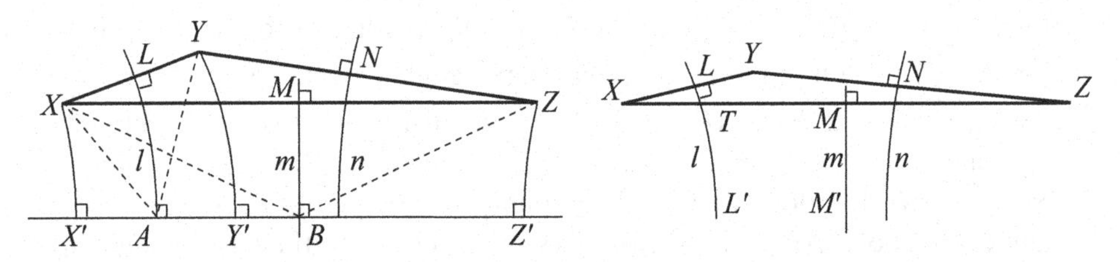

Suppose that l and m have a common perpendicular $\overleftrightarrow{AB}$. Let $L = l \cap XY$, $M = m \cap XZ$, and $N = n \cap YZ$, the midpoints of the segments XY, XZ, and YZ. Construct segments XX', YY', and ZZ' perpendicular to $\overleftrightarrow{AB}$ through X, Y, and Z, respectively. Join X to B and Z to B. Then by Side–Angle–Side, $\Delta BMZ \cong \Delta BMX$. It follows from $\angle XBM \cong \angle ZBM$ that $\angle XBX' \cong \angle ZBZ'$. By Angle–Angle–Side we find $\Delta XBX' \cong \Delta ZBZ'$ and so $XX' \cong ZZ'$. By a similar construction joining X to A and Y to A we find $XX' \cong YY'$. Thus $Y'YZZ'$ is a Saccheri quadrilateral and the line n perpendicular to YZ through the midpoint of YZ meets the side $\overleftrightarrow{Y'Z'} = \overleftrightarrow{AB}$ in a right angle. Thus l, m, and n lie in $\mathbf{P}^{\perp}_{\overleftrightarrow{AB}}$, the pencil of lines perpendicular to $\overleftrightarrow{AB}$.

Finally, if l and m are parallel, then n cannot meet either l or m nor will n share a common perpendicular with l or m. It remains to show that l, m, and n are parallel in the same direction. By Pasch's Axiom, a line entering a triangle along a side must exit through another side. The midpoint on the longest side separates the other lines from each other and so all the lines pass through the longest side. Let $l = \overleftrightarrow{LL'}$ meet XZ at T. Since $\angle LTZ$ is exterior to the right triangle ΔTLX, this angle is obtuse by the Exterior Angle Theorem. Therefore the supplement $\angle L'TZ$ is acute. Since l is parallel to m and m is perpendicular to XZ, $\angle L'TZ = \Pi(TM)$ and so l and m are parallel away from XZ in the direction of L' and M'. The analogous argument shows n parallel to m in the same direction. Thus l, m, and n are elements of $\mathbf{P}_l$. ∎

Definition 4.5 (Gauss 1831). *Given a pencil of lines* **P**, *two points X and Y are said* **to correspond with respect to P** *if the line in the pencil that passes through* *the midpoint of segment XY is perpendicular to XY. When the pencil of lines is the set of lines parallel to a given line, the set of all points corresponding to a given point X is called a* **horocycle** *through X (*Parazykel *in Gauss).*

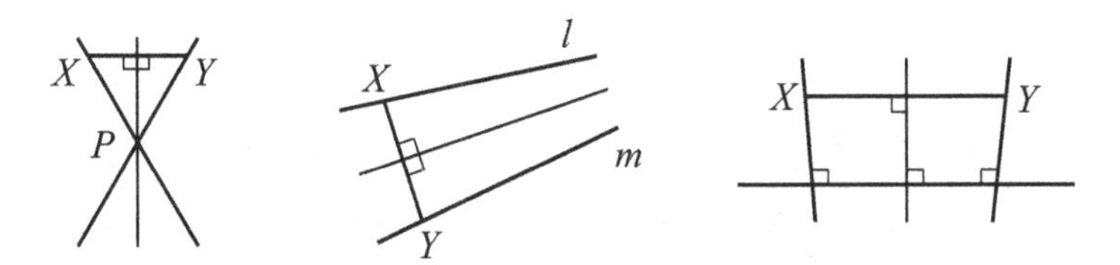

Points that correspond are related by a reflection of the plane across a line in the pencil. For a pencil of lines through a given point, a direct consequence of the proof of Euclid's Proposition I.6 (the converse of *Pons Asinorum*) is that the set of points corresponding to a given point forms a circle. From this point of view, a horocycle may be thought of as a "circle with center at infinity." The set of points corresponding to a given point with respect to $\mathbf{P}_m^{\perp}$ is the curve of points equidistant to the line m. The properties of the Saccheri quadrilateral may be used to explore the properties of this curve.

We leave the proof of the following proposition to the reader.

Proposition 4.6. *A horocycle is a line if and only if* **HRA** *holds.*

Thus, a new feature of the plane with **HAA** is the set of horocycles in the plane. An immediate consequence of Theorem 4.4 is the independence of the choice of the point X in the definition of a horocycle through X.

Corollary 4.7. *If Y and Z lie on the horocycle through X determined by $\mathbf{P}_l$, then X and Y lie on the horocycle through Z determined by $\mathbf{P}_l$.*

The points on a horocycle can be characterized by the angle of parallelism: If X and Y are on the horocycle, the perpendicular bisector m of the segment XY lies in the pencil of parallel lines that determine the horocycle. The corresponding angles at X and Y have measure $\Pi(XM) = \Pi(YM)$, where M is the midpoint of XY.

If m meets the horocycle at a point Q, then the tangent to the horocycle at Q is perpendicular to m as can be seen by letting the chord XY shrink to Q. Thus, as in the case of radii to a circle, the lines in the pencil $\mathbf{P}_l$ meet the horocycle at right angles.

We say that two horocycles are **concentric** if they are determined by the same pencil of parallel lines. Segments of horocycles have lengths that are, as usual, limits of polygonal approximations.

Lemma 4.8.

(1) *To equal chords there correspond equal arcs of the horocycle. To the longer chord there corresponds a longer arc.*

(2) *Segments of lines in the pencil bounded by two concentric horocycles are congruent.*
(3) *The arcs of concentric horocycles bounded by two lines in the pencil decrease in the direction of parallelism.*

PROOF:

(1) Suppose X, Y and X', Y' lie on a horocycle determined by $\mathbf{P}_l$ and that $XY \cong X'Y'$. Denote the arc along a horocycle between points X, and Y by $\overset{\frown}{XY}$. The goal is to show that the horocycle segment $\overset{\frown}{XY}$ is congruent to $\overset{\frown}{X'Y'}$. Let M be the midpoint of XY and m the line in $\mathbf{P}_l$ through M which is perpendicular to XY and suppose m meets the horocycle at Q. By construction $\triangle XMQ \cong \triangle YMQ$ and $XQ \cong YQ$. The angle $\angle XQM$ is congruent to $\Pi(XQ/2)$ because X and Q correspond.

Let M' be the midpoint of $X'Y'$ and m' the line in $\mathbf{P}_l$ through M' perpendicular to $X'Y'$. On m' let Q' be away from the direction of parallelism with $M'Q' \cong MQ$. The triangles $\triangle XMQ$ and $\triangle X'M'Q'$ are congruent so $XQ \cong X'Q'$ and $\angle X'Q'M' \cong \Pi(XQ/2) \cong \Pi(X'Q'/2)$ and so the line through the midpoint of $X'Q'$ perpendicular to $X'Q'$ lies in $\mathbf{P}_l$. Thus Q' corresponds to X' and Q' lies on the horocycle. The arc $\overset{\frown}{XY}$ is approximated by the line segments XQ and QY and $\overset{\frown}{X'Y'}$ by the congruent $X'Q'$ and $Q'Y'$. To complete the proof, take the midpoints of the approximating segments and make the comparisons. Iterated halving gives the arcs in the limit.

(2) Suppose $\overleftrightarrow{XA}$ and $\overleftrightarrow{YB}$ are in $\mathbf{P}_l$ and X and Y lie on one horocycle and A and B on a concentric horocycle determined by $\mathbf{P}_l$. We want to show that $XA \cong YB$. Let m be the line through the midpoint M of XY and perpendicular to XY. Suppose m meets AB at a point N. Join X to N and Y to N to form triangles $\triangle XMN$ and $\triangle YMN$. These triangles are congruent by Side–Angle–Side and so $XN \cong YN$. It follows that:

$$\angle NXA \cong \Pi(XM) - \angle MXN \cong \Pi(YM) - \angle NYB \cong \angle NYB.$$

It is also the case that $\angle NAX \cong \pi - \Pi(AB/2) \cong \angle NBY$. By the congruence criterion Angle–Angle–Side, triangles $\triangle NAX$ and $\triangle NBY$ are congruent and so $XA \cong YB$. It also follows that m is the perpendicular bisector of AB.

(3) This statement follows from the approximations in part 1 and the fact that the angle of parallelism decreases with increasing length (Property 2 of $\Pi(x)$). ■

With these ideas we arrive at the following result.

Theorem 4.9. *The ratio of the lengths of concentric arcs of two horocycles intercepted between two lines in the pencil is expressible in terms of an exponential function of the distance between these arcs.*

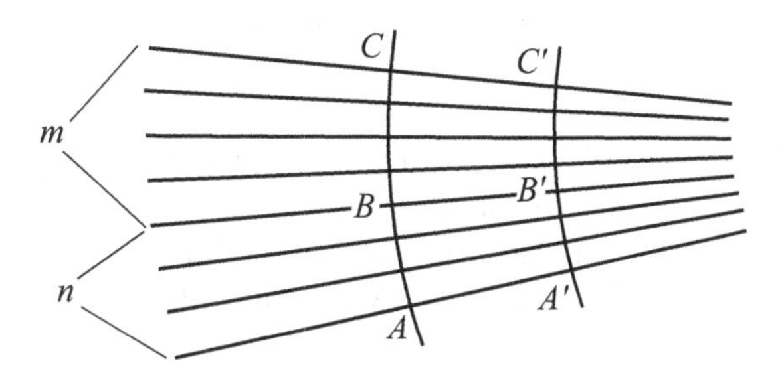

PROOF: We first show that the ratio depends only on the distance. Denote the length of an arc along the horocycle between points A, B by $l(\overset{\frown}{AB})$. First suppose $l(\overset{\frown}{BC}) = (m/n)l(\overset{\frown}{AB})$. Then the congruent segments along the parallels AA', BB', and CC' induce equal subdivisions of $\overset{\frown}{A'B'}$ and $\overset{\frown}{B'C'}$. We then have $l(\overset{\frown}{B'C'}) = (m/n)l(\overset{\frown}{A'B'})$, and so:

$$\frac{l(\overset{\frown}{AB})}{l(\overset{\frown}{A'B'})} = \frac{(m/n)l(\overset{\frown}{BC})}{(m/n)l(\overset{\frown}{B'C'})} = \frac{l(\overset{\frown}{BC})}{l(\overset{\frown}{B'C'})};$$

therefore, the ratio $l(\overset{\frown}{AB})/l(\overset{\frown}{A'B'})$ depends only on the length of AA'. If $l(\overset{\frown}{AB})$ and $l(\overset{\frown}{BC})$ are related by an irrational number, a limiting process obtains the desired ratio.

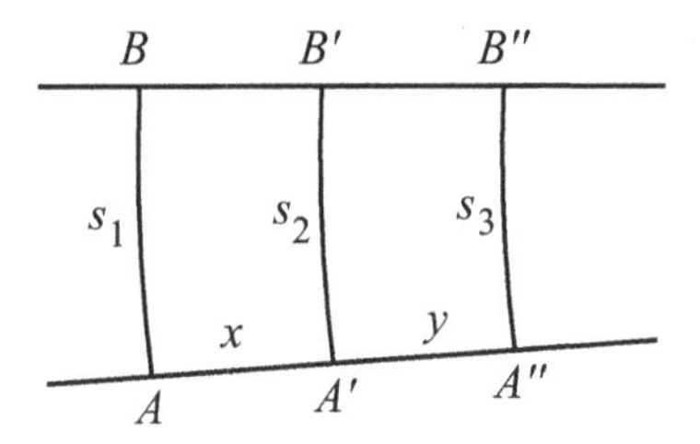

Suppose $s_1 = l(\overset{\frown}{AB})$, $s_2 = l(\overset{\frown}{A'B'})$, and $s_3 = l(\overset{\frown}{A''B''})$. We can write $s_1/s_2 = f(x)$ and $s_2/s_3 = f(y)$, where x and y denote the lengths of the line segments AA' and $A'A''$, respectively. It follows that $s_1/s_3 = f(x+y)$ and we obtain the equation:

$$f(x+y) = f(x)\cdot f(y).$$

Assuming the differentiability of $f(x)$ for the moment we get

$$f'(x) = \lim_{\Delta x\to 0}\frac{f(x+\Delta x)-f(x)}{\Delta x} = f(x)\cdot \lim_{\Delta x\to 0}\frac{f(\Delta x)-1}{\Delta x}.$$

By the definition of $f(x)$, as Δx goes to 0, $f(\Delta x)$ goes to 1, and we write, for some value of k,

$$\lim_{\Delta x\to 0}\frac{f(\Delta x)-1}{\Delta x} = \frac{1}{k}.$$

Then $f'(x) = (1/k)f(x)$ and so $f(x) = e^{x/k}$ or $l(\widehat{AB}) = l(\widehat{A'B'})e^{l(AA')/k}$. ∎

The value $1/k$ is chosen to anticipate the more general development in Chapter 12 where it shows up again. Theorem 4.9 offers us a first step into the analytic properties of the plane under **HAA**. The factor of the exponential function will lead us to the *hyperbolic trigonometric functions*, the elementary properties of which we review next.

In the study of the integral giving the area under a semicircle, $\int \sqrt{a^2 - x^2}\,dx$, the familiar trigonometric functions $\sin(x)$, $\cos(x)$, and $\tan(x)$ play a role as parametrizing functions. Analogously, the functions $\sinh(x)$, $\cosh(x)$, and $\tanh(x)$ play the same roles in the study of the integral giving the area under a hyperbola, $\int \sqrt{x^2 - a^2}\,dx$. In 1768, Lambert produced a comprehensive study of the hyperbolic trigonometric functions containing the work of Euler and the Bernoullis (Lambert 1768). Similar developments were made by V. RICCATI (1701–75) with Saladini for studying the solutions to cubics.

To define the hyperbolic trigonometric functions, formally introduce a complex variable into the Taylor series for e^x to get:

$$\begin{aligned} e^{iy} &= 1 + iy - \frac{y^2}{2} - \frac{iy^3}{3} + \frac{y^4}{4} - \cdots \\ &= \left(1 - \frac{y^2}{2} + \frac{y^4}{4} - \cdots\right) + i\left(y - \frac{y^3}{3} + \frac{y^5}{5} - \cdots\right) \\ &= \cos y + i \sin y. \end{aligned}$$

It follows from these formulas that:

$$\cos y = \frac{e^{iy} + e^{-iy}}{2}, \quad \sin y = \frac{e^{iy} - e^{-iy}}{2i} \quad \text{and} \quad \tan y = \frac{\sin y}{\cos y} = \frac{1}{i}\frac{e^{iy} - e^{-iy}}{e^{iy} + e^{-iy}}.$$

By analogy with these formulas, we define the real-valued functions:

$$\cosh x = \frac{e^x + e^{-x}}{2}, \quad \sinh x = \frac{e^x - e^{-x}}{2}, \quad \text{and} \quad \tanh x = \frac{\sinh x}{\cosh x} = \frac{e^x - e^{-x}}{e^x + e^{-x}}.$$

From the power series, notice that $\cos(iy) = \cosh y$, and $\sin(iy) = i \sinh(y)$. The fundamental relation for the hyperbolic trigonometric functions is given by:

$$\cosh^2 x - \sinh^2 x = 1.$$

The convergence theorems that support our formal manipulations are classical and can be found in any good complex analysis book, for example, Ahlfors (1979).

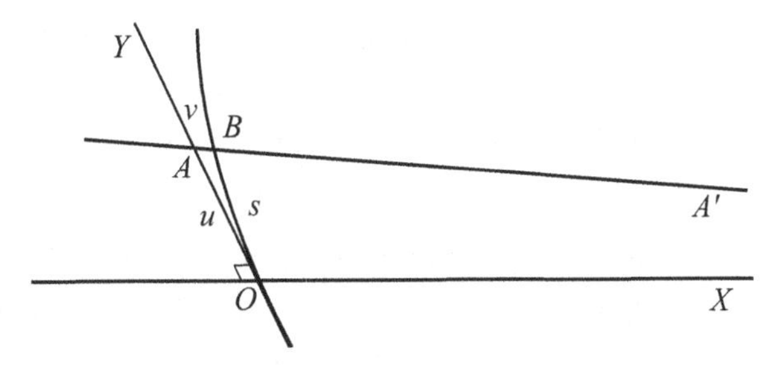

Under **HAA**, a horocycle differs from a straight line. We next seek to measure how much it differs. Fix some line segment RR_1 as a unit length with which we can assign a positive real number length to all other line segments. Let $\overleftrightarrow{OX} = l$ denote a line and consider the horocycle through O determined by the pencil of parallels $\mathbf{P}_l$ in the direction $\overrightarrow{OX}$. Let $\overleftrightarrow{OY}$ denote the line perpendicular to l through O. If we measure a length u along OY to a point A, then there is a line $\overleftrightarrow{AA'}$ in $\mathbf{P}_l$ with A' toward X and a point B on the line $\overleftrightarrow{AA'}$ and on the horocycle through O. Denote the length of AB by v. We apply Theorem 4.9 to relate u and v.

Lemma 4.10. *In the notation of the previous paragraph,* $e^{v/k} = \cosh(u/k)$.

PROOF: In order to apply Theorem 4.9, we need a family of strategically chosen concentric horocycles. To construct them we first obtain a reference line: Among the properties of the angle of parallelism, it was shown that any acute angle is $\Pi(x)$ for some x. Let OC' denote a segment of length x_0 for which $\Pi(x_0) = \pi/4$. Then there is a line $\overleftrightarrow{TT'}$ that is parallel to the perpendicular lines $\overleftrightarrow{OX}$ and $\overleftrightarrow{OY}$. Also $\overleftrightarrow{TT'}$ is in $\mathbf{P}_l$. The horocycle through O determined by $\mathbf{P}_l$ meets $\overleftrightarrow{TT'}$ at a point C. Denote the length of $\overset{\frown}{OC}$ by σ.

Consider the diagram in which we use reflections across lines to construct various comparable angles and figures.

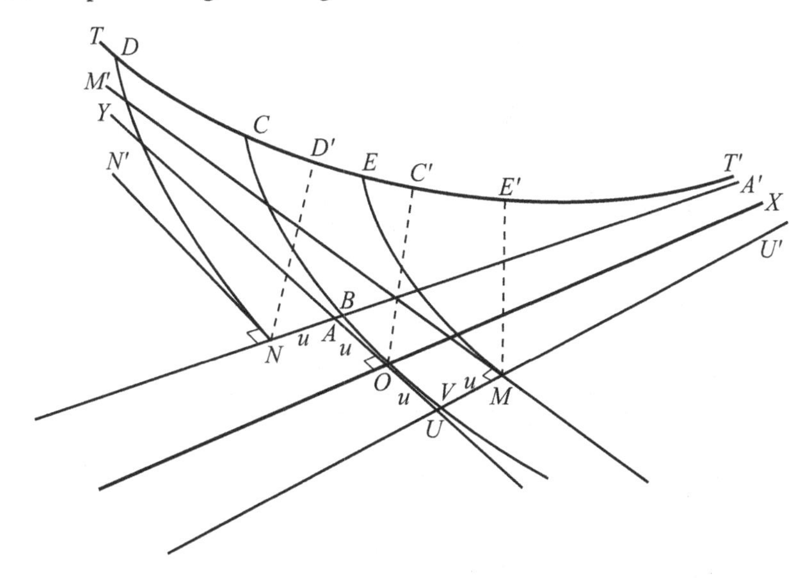

Let $\overleftrightarrow{NN'}$ be perpendicular to $\overleftrightarrow{AA'}$ and parallel to $\overrightarrow{OY}$, that is, choose N with $\angle NAY = \Pi(AN)$. Since $\angle OAA' = \Pi(OA) = \Pi(u)$ and $\angle NAY = \angle OAA'$ we have $AN \cong OA$. Similarly let U be on $\overrightarrow{YO}$ with $OU \cong OA$, $U \neq A$, and $\overleftrightarrow{UU'}$ the parallel to $\overrightarrow{OX}$ through U in the direction of X. Let UM have length u and construct $\overleftrightarrow{MM'}$ perpendicular to $\overleftrightarrow{UU'}$ with $\overleftrightarrow{MM'}$ parallel to $\overrightarrow{OY}$. The segments AN and UM determine concentric horocycles $\overset{\frown}{ND}$, $\overset{\frown}{OC}$, and $\overset{\frown}{ME}$. The perpendiculars to $\overleftrightarrow{TT'}$ from N, O, and M, determine points D', C', and E', respectively. By the construction we see that $\overleftrightarrow{TT'}$ is parallel to $\overleftrightarrow{AA'}$ and $\overleftrightarrow{NN'}$ and so $\angle D'NA' \cong \angle D'NN' \cong \Pi(x_0) = \pi/4$. Similarly, $ND' \cong OC' \cong ME'$ are all of length x_0. It follows from Lemma 4.8 that $\overset{\frown}{ND} \cong \overset{\frown}{OC} \cong \overset{\frown}{ME}$. Their common length is σ.

We can apply the formula in Theorem 4.9, that is, the ratio of the lengths of concentric arcs of two horocycles is given by $l(\overset{\frown}{ZW}) = l(\overset{\frown}{Z'W'})e^{-t/k}$, where t is the distance between the arcs. If B denotes the point along $\overleftrightarrow{AA'}$ where $\overset{\frown}{OC}$ crosses, then let V denote the point corresponding to B along $\overleftrightarrow{UU'}$. Then $l(BN) = u + v$. Let $s = l(\overset{\frown}{OB})$, the length between O and B along the horocycle; we then have:

$$l(\overset{\frown}{ND}) = l(\overset{\frown}{OC}) = \sigma = (\sigma - s)e^{(u+v)/k} \quad \text{or} \quad \frac{\sigma - s}{\sigma} = e^{-(u+v)/k}.$$

Horocycle arcs $\overset{\frown}{VC}$ and $\overset{\frown}{ME}$ are concentric and $l(VM) = u - v$. Since $l(\overset{\frown}{VC}) = \sigma + s$ and $l(\overset{\frown}{ME}) = \sigma$, then we also have $\sigma + s = \sigma e^{(u-v)/k}$, or $(\sigma + s)/\sigma = e^{(u-v)/k}$. Thus,

$$2 = \frac{\sigma + s}{\sigma} + \frac{\sigma - s}{\sigma} = e^{(u-v)/k} + e^{-(u+v)/k}.$$

Multiplying through by $e^{v/k}$ gives $e^{v/k} = \dfrac{e^{u/k} + e^{-u/k}}{2} = \cosh(u/k)$. This proves the lemma. ∎

Corollary 4.11. $s = \sigma \tanh(u/k)$.

PROOF: By the proof of Lemma 4.10 we see that:

$$2\frac{s}{\sigma} = \frac{\sigma + s}{\sigma} - \frac{\sigma - s}{\sigma} = e^{(u-v)/k} - e^{-(u+v)/k} = (e^{u/k} - e^{-u/k})e^{-v/k}.$$

Therefore,

$$\frac{s}{\sigma}\cosh(u/k) = \frac{s}{\sigma}e^{v/k} = \frac{e^{u/k} - e^{-u/k}}{2} = \sinh(u/k),$$

and so $s = \sigma \tanh(u/k)$. ∎

These analytical relations will play a key role in building non-Euclidean trigonometry. To develop further tools we introduce three dimensions.

Digression: Neutral space

In Chapter 1 we saw that the trigonometric formulas for triangles on the sphere could be proved by using the fact that the sphere is embedded in three dimensional space. We adopt this approach more generally by considering three dimensional space, where we find points, lines, and planes for which Hilbert's Axioms of Incidence hold, together with the Postulates I–IV of Euclid (Book XI of *The Elements*). We call such a space *neutral space*.

Without using Postulate V, define *parallel lines* in space to be two lines ℓ_1, ℓ_2 that are coplanar, and, in a chosen direction on that plane, satisfy Definition 4.1.

In order to develop the geometry of neutral space we need to explore certain relations between pairs of lines, lines and planes, and pairs of planes. According to the assumptions of an incidence geometry (Definition 2.1), if we are given three noncollinear points, P, Q, and R, or a line l and a point P not on l or two intersecting lines l and m, then each group uniquely determines a plane. We denote this relation by $T = \text{Plane}(P,Q,R)$ or $T = \text{Plane}(P,l)$ or $T = \text{Plane}(l,m)$.

Our goal in this digression is to prove Theorem 4.25, an analogue of Playfair's Axiom but in a higher dimension. For Playfair's Axiom we are given a point P not on a line m, from which we obtain a unique line m' through P with m' parallel to m. One dimension higher, the role of the point is played by a line l, which is parallel to a plane T from which we obtain a unique plane T' containing l and parallel to T. To prove this assertion we generalize a special case of Euclid's proof of Playfair's Axiom, Proposition I.31: First construct the unique perpendicular PA through P to the given line l (Proposition I.12), then construct the unique perpendicular m to PA through P. The line m is parallel to l under the assumption of Postulate V.

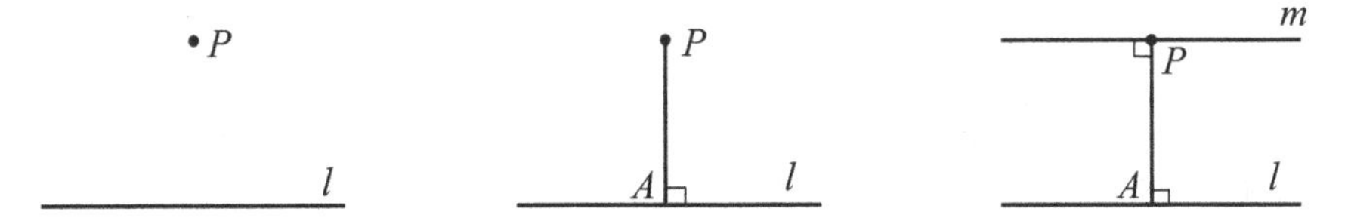

What is remarkable about neutral space is that the same construction, appropriately construed, leads to the analogous result for a line l parallel to a given plane T.

In order to make these constructions in space, we begin by exploring parallelism of lines. Our presentation is inspired by the relevant sections of the book by Borsuk and Smielew (1960).

Lemma 4.12. *Suppose l_1 and l_2 are two lines parallel in space and T_1 and T_2 are two planes with l_1 on T_1, l_2 on T_2, and $T_1 \neq T_2$. If m is the line of intersection of T_1 and T_2, and $l_1 \neq m \neq l_2$, then m is parallel to l_1 and to l_2.*

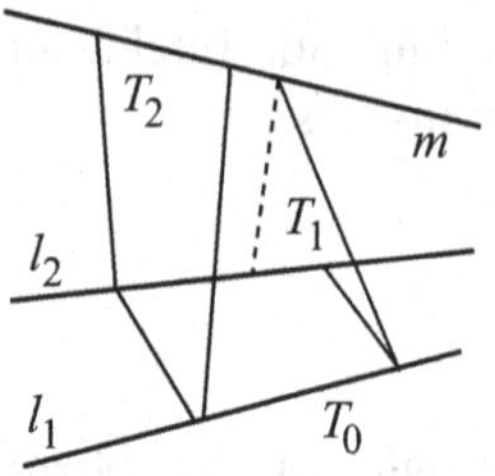

PROOF: We first prove that m does not intersect l_1 or l_2. Let T_0 denote the plane on which l_1 and l_2 lie. Suppose m meets l_1 at a point $P = m \cap l_1$. Since $m \subset T_2$, we have that $T_0 = \text{Plane}(l_1, l_2) = \text{Plane}(P, l_2) = T_2$. It follows that $l_1 = T_0 \cap T_1 = T_2 \cap T_1 = m$, a contradiction. Similarly m does not meet l_2.

Let A lie on $l_1 = \overleftrightarrow{AA'}$, and B lie on $l_2 = \overleftrightarrow{BB'}$, so that A' and B' are in the direction of parallelism. Let Q and Q' lie on m with QA perpendicular to l_1, and QB perpendicular to l_2.

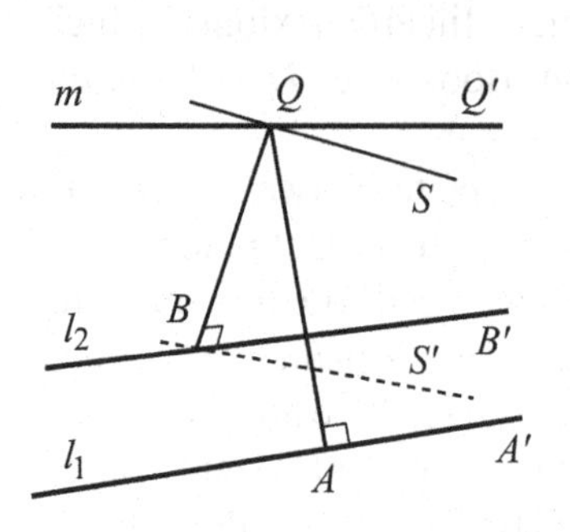

Suppose $\overleftrightarrow{QS}$ is a line through Q on T_1 with $\angle AQS < \angle AQQ'$ and $\overrightarrow{QS}$ lies in $A'AQQ'$. Consider the plane T determined by B, S, and Q. The line $T_0 \cap T = \overrightarrow{BS'}$ lies inside $A'ABB'$, the direction of parallelism of l_1 and l_2. Thus the ray $\overrightarrow{BS'}$ meets l_1 at a point X. But $X = l_1 \cap (T \cap T_0) = T \cap (T_0 \cap T_1) = (T \cap T_1) \cap T_0 = \overrightarrow{QS} \cap T_0$, and so $\overrightarrow{QS}$ meets l_1. Thus m is parallel to l_1. Similarly, m is parallel to l_2. ■

Corollary 4.13. *If l_1, l_2, and l_3 are lines in space, l_1 is parallel to l_2 and l_2 is parallel to l_3, then l_1 is parallel to l_3.*

The proof of the corollary is left as an exercise to the reader. We next develop the relation of perpendicularity between a line and a plane.

Definition 4.14 (Euclid Book XI Definition 3). *A* **line** *l* **is perpendicular to a plane** *T if l meets T at a point P and every line on T passing through P is perpendicular to l. If a line l is perpendicular to a plane T, then we also say that the plane T is perpendicular to the line l.*

Lemma 4.15. *If a line l meets a plane T at a point P, then l is perpendicular to T if and only if there are two lines m_1 and m_2 lying on T, passing through P, and perpendicular to l.*

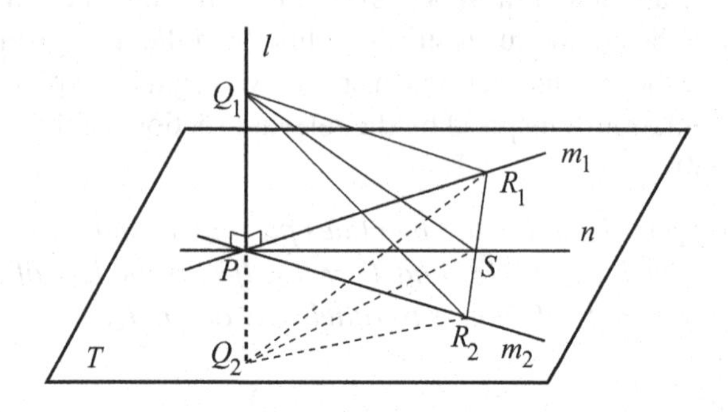

PROOF: Suppose m_1 and m_2 are on T and both perpendicular to l at P. If n is a line lying on T passing through P, choose points R_1 on m_1 and R_2 on m_2 so that n enters the triangle ΔR_1PR_2 at P and exits through a point S on R_1R_2. Choose points Q_1 and Q_2 on l with P between Q_1 and Q_2 and $PQ_1 \cong Q_2P$. Because m_1 and m_2 are both perpendicular to l, the right triangles formed satisfy:

$$\Delta Q_1PR_1 \cong \Delta Q_2PR_1, \text{ and } \Delta Q_1PR_2 \cong \Delta Q_2PR_2.$$

This implies $R_1Q_1 \cong R_1Q_2$ and $R_2Q_1 \cong R_2Q_2$. By the congruence criterion Side–Side–Side, $\Delta R_1Q_1R_2 \cong \Delta R_1Q_2R_2$. This implies that $\angle Q_1R_1R_2 \cong \angle Q_2R_1R_2$ and $\angle Q_1R_2R_1 \cong \angle Q_2R_2R_1$. By Side–Angle–Side, $\Delta R_1SQ_1 \cong \Delta R_1SQ_2$ and so $SQ_1 \cong SQ_2$. Thus, ΔQ_1SQ_2 is an isosceles triangle with the segment PS a bisector of the base Q_1Q_2 (Euclid I.9). Hence $n = \overleftrightarrow{PS}$ is perpendicular to l. The other direction is obvious. ■

The proof of Lemma 4.15 allows us to construct planes perpendicular to a given line and lines perpendicular to a given plane.

Lemma 4.16. *Given a line l and a point P, then there is a unique plane T containing P and perpendicular to l.*

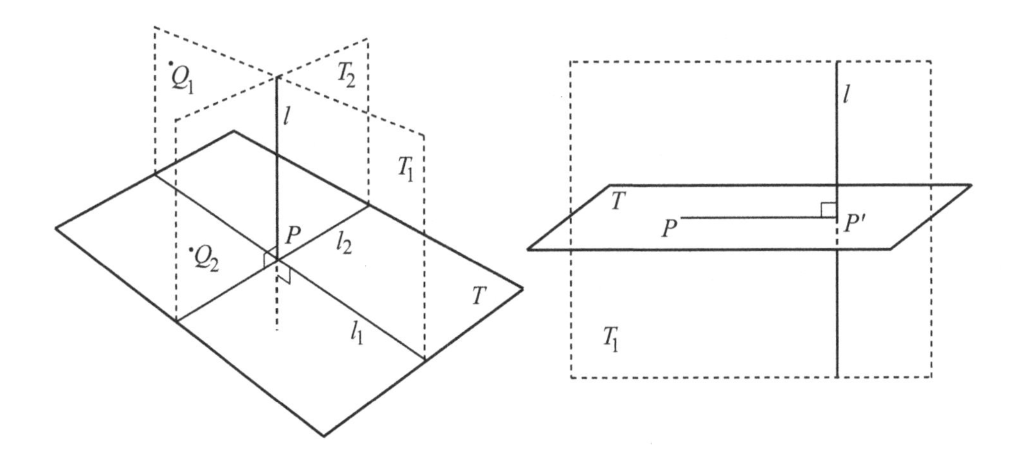

PROOF: We first prove the case when point P lies on line l. Because we have assumed the structure of an incidence geometry, there is some point Q_1 not on the given line l. Let $T_1 = \text{Plane}(l, Q_1)$. On T_1 there is a line l_1 perpendicular to l through P. The assumption of an incidence geometry further implies the existence of another point Q_2 not lying in the plane T_1. Let $T_2 = \text{Plane}(l, Q_2)$. On T_2 let l_2 be the line perpendicular to l through P. This gives the plane $T = \text{Plane}(l_1, l_2)$, which is perpendicular to l.

If T' is another plane containing P and perpendicular to l, then choose a point Q on T' that is not in T. Then the plane $T'' = \text{Plane}(l, Q)$ meets the plane T in a line n and the plane T' in a line n'. However, because Q lies on n', $n \neq n'$. This gives two lines on the plane T'' passing through P both perpendicular to l at P, which is impossible. Thus, T is the unique plane containing P perpendicular to l.

If P is not on l, let $T_1 = \text{Plane}(P,l)$. In T_1, let $\overleftrightarrow{PP'}$ denote the line through P that is perpendicular to l at P'. By the first construction, there is a unique plane T through P' and perpendicular to l. Since $T_1 \cap T$ is perpendicular to l and P' is on T, then $\overleftrightarrow{PP'} = T_1 \cap T$ lies on T, the desired plane. ■

The dual problem of constructing a line perpendicular to a given plane passing through a given point is solved with the help of the following lemma.

Lemma 4.17. *Suppose l is a line perpendicular to a plane T and $P = l \cap T$ and suppose $B \neq P$ is a point on l. If m is any line on T and $\overleftrightarrow{PA}$ is the line on T perpendicular to m with A on m, then the line segment AB and hence* Plane(A,P,B) *are perpendicular to m.*

PROOF: Let $Q \neq A$ be any other point on m and let Q' be on m with A between Q and Q' and $AQ \cong AQ'$. By Side–Angle–Side the triangles $\triangle PAQ$ and $\triangle PAQ'$ are congruent and so $PQ \cong PQ'$. Again by Side–Angle–Side, the right triangles $\triangle PBQ$ and $\triangle PBQ'$ are congruent from which it follows that $BQ \cong BQ'$. By Side–Side–Side triangles $\triangle BAQ$ and $\triangle BAQ'$ are congruent and $\angle BAQ \cong \angle BAQ'$; that is, AB meets m at right angles. Since AB and PA are perpendicular to m, m is perpendicular to Plane(A,P,B). ■

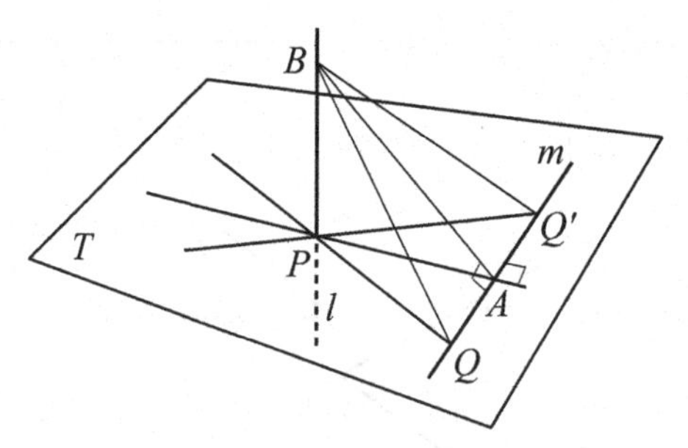

We reverse the roles of the line and plane in Lemma 4.17 to prove:

Corollary 4.18. *Given a plane T and a point P there is a unique line l perpendicular to T and passing through P.*

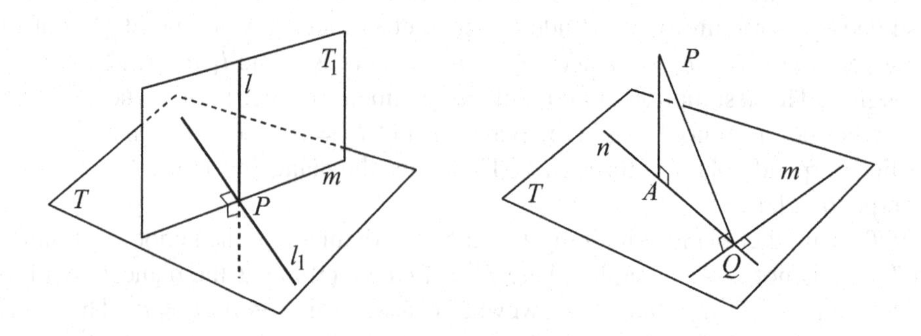

PROOF: If P is in T, then let l_1 be any line on T passing through P. By Lemma 4.16 there is a plane T_1 containing P and perpendicular to l_1. On T_1 let l be the line through P that is perpendicular to the line m given by the intersection of T and T_1. This line is also perpendicular to l_1 and so l is perpendicular to T.

If l' is another line perpendicular to T passing though P, then let T'' denote the plane determined by l and l'. This plane meets T in a line n, which has the property that l and l' are both perpendicular to n at the point P in the plane T''. Since this is impossible, l is unique.

If P is not in T, then suppose m is any line on T. On the plane determined by P and m, form the line segment PQ with Q on m and PQ perpendicular to m. In T construct the line n perpendicular to m through the point Q. In the plane determined by n and P let AP be the line segment perpendicular to n with A on n. By construction, Plane(P,A,Q) is perpendicular to the line m. Apply Lemma 4.17 with m playing the role of given line perpendicular to Plane(n,P) to see that the line $l = \overleftrightarrow{AP}$ is perpendicular to the plane T. The uniqueness of l follows from the uniqueness of the line perpendicular to T through A is in T. ∎

From the relation of perpendicularity of a line and a plane we develop the idea of planes being perpendicular. (See Euclid's Definition 4 in Book XI.)

Definition 4.19. *A plane T is* **perpendicular to a plane** *T' if there is a line l on T that is perpendicular to T'.*

Lemma 4.20. *A plane T is perpendicular to a plane T' if and only if T' is perpendicular to T. If T and T' are perpendicular planes and l is a line on T, then l is perpendicular to T' if and only if l is perpendicular to the line $m = T \cap T'$.*

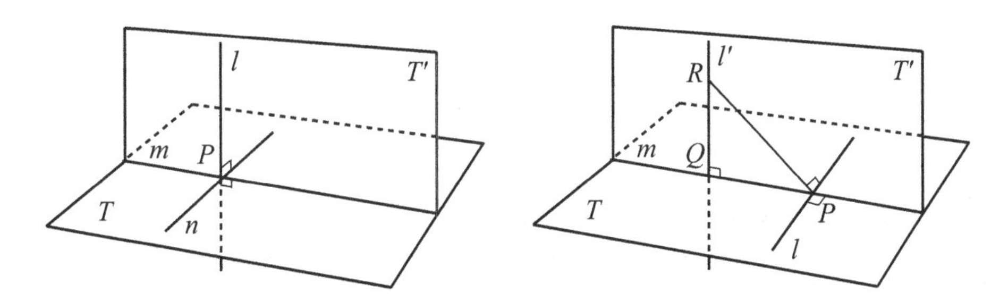

PROOF: Suppose T is perpendicular to T' and n is a line lying in T with n perpendicular to T'. Suppose n meets T' at a point P. Let m denote the intersection $T \cap T'$ and let l be the line on T' passing through P perpendicular to m. Since l is a line through P, it is perpendicular to n. Since the plane T is determined by the lines n and m, l is perpendicular to T and so T' is perpendicular to T.

If l is perpendicular to T' with l in T, then it is perpendicular to any line it meets in T', in particular, to $m = T \cap T'$. Conversely, assume l lies in T and l is perpendicular to m. Since T and T' are perpendicular, there is a line l' in T' with l' perpendicular to T. If l' passes through P, then l' is perpendicular to l and so

l is perpendicular to T' by Lemma 4.15. If l' meets m at Q, then let R be on l' with $R \neq Q$. By Lemma 4.17, RP is perpendicular to l and hence l is perpendicular to T'. ∎

Corollary 4.21. *If l_1 and l_2 are distinct lines both perpendicular to a plane T, then l_1 and l_2 are coplanar and nonintersecting.*

PROOF: Let $l_1 \cap T = P$ and $l_2 \cap T = Q$. Consider the plane $T_1 = \text{Plane}(l_1, Q)$. Then $T_1 \cap T = \overleftrightarrow{PQ} = m$. Let l' be perpendicular to m through Q in T_1. Then both l' and l_2 pass through Q and are perpendicular to m. By Lemma 4.20, both l' and l_2 are perpendicular to T, which implies $l' = l_2$ since only one line can be perpendicular to both m and another line in T through Q. Thus l_2 lies on T_1. ∎

The first construction of the analogue of the Playfair proof requires that, for a line not on a plane T, one can find a plane that contains l and is perpendicular to T.

Proposition 4.22. *Given a line l and a plane T with l not lying on T, there is a unique plane T' containing l and perpendicular to T. Given two distinct planes T_1 and T_2, there is a plane T'' perpendicular to both T_1 and T_2.*

PROOF: Let A be any point on l that does not lie in T. By Corollary 4.18 there is a unique line m passing through A and perpendicular to T. The plane $T' = \text{Plane}(l, m)$ is perpendicular to T. By the uniqueness of the line m and Lemma 4.20, the plane T' is the only plane perpendicular to T and containing l.

From a point A in T_1 that is not in T_2, construct the line l through A perpendicular to T_2. If $l \cap T_2 = B$, then construct the line m through B perpendicular to T_1. If $l = m$, then any plane containing l is perpendicular to both T_1 and T_2. If $l \neq m$, then the plane determined by l and m is perpendicular to T_1 and T_2. ∎

Finally we use perpendicular planes to define what we mean by a line being parallel to a plane and two planes being parallel.

Definition 4.23. *Given a line l and a plane T, if T' is the unique plane containing l and perpendicular to a given plane T, then the line $m = T \cap T'$ is called the* **perpendicular projection** *of l onto T. A line l is* **parallel to a plane** *T if l is parallel to its perpendicular projection onto T. Two planes are* **parallel** *if the intersections with some common perpendicular plane are parallel lines.*

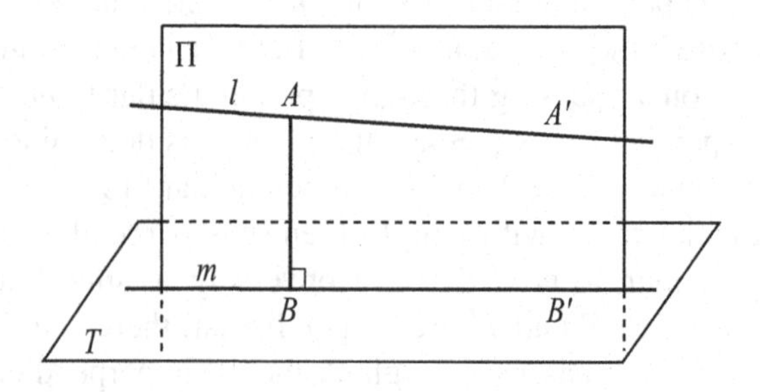

Lemma 4.24. *Parallel planes do not intersect.*

PROOF: If T and T' are parallel planes, then there are lines l on T and m on T' such that l and m are parallel and lie in a plane Π that is perpendicular to both T and T'. Suppose T and T' intersect. If X is a point on $T \cap T'$, the line of intersection, then let XF be the perpendicular to l with F on l and XG be the perpendicular to m with G on m. Since Π is perpendicular to both T and T', it follows that XF, being perpendicular to $l = T \cap \Pi$, is perpendicular to Π. Similarly XG is perpendicular to Π. But then FG is perpendicular to XF and FG is perpendicular to XG, and so the triangle $\triangle XFG$ has two right angles, an impossibility with Postulates I–IV. Thus T and T' do not intersect if parallel. ∎

Finally, we prove the analogue of Playfair's Axiom for lines parallel to a plane.

Theorem 4.25. *Given a line l parallel to a plane T, there is exactly one plane that contains l and is parallel to T.*

PROOF: We leave it to the reader to prove this result under the assumption of **HRA**.

In the case of **HAA**, let A be a point on $l = \overleftrightarrow{AA'}$ with A' in the direction of parallelism. The unique plane Π containing l and perpendicular to T (Proposition 4.22) meets T in the line $l_T^\perp = T \cap \Pi$. Let AB be perpendicular to $l_T^\perp = \overleftrightarrow{BB'}$ with B' in the direction of parallelism.

By Corollary 4.18 there is a unique line m through A and perpendicular to the plane Π. Let $T' = \text{Plane}(l, m)$. Then T' and Π are perpendicular planes. By definition, T and T' are parallel because the perpendicular projection of l is $l_T^\perp$, a parallel line.

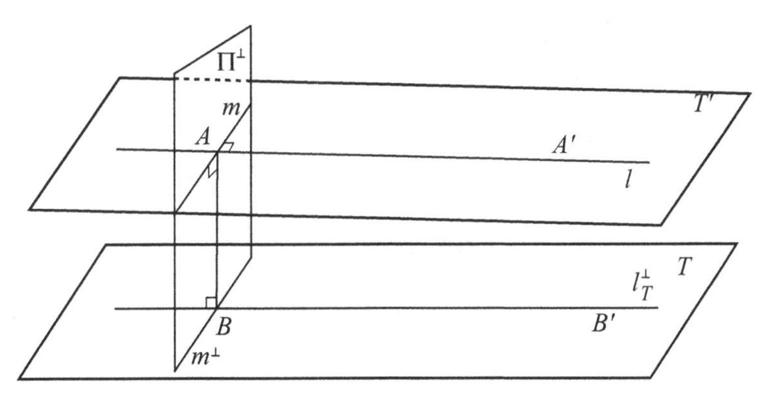

Let $\Pi^\perp = \text{Plane}(m, B)$. Since AB is perpendicular to T, the line $m^\perp = \Pi^\perp \cap T$ is perpendicular to $l_T^\perp$ in T and through B. Also $l_T^\perp$ is perpendicular to AB and perpendicular to $m^\perp$, so $l_T^\perp$ is perpendicular to $\Pi^\perp$.

Suppose $U \neq T'$ is another plane containing l. Let Z be the point on U with BZ perpendicular to U. The line $t = \Pi^\perp \cap U$ divides U into two parts, one of which lies on the side of $\Pi^\perp$ that is the direction of parallelism to l. We claim that Z lies on this side of t. Consider the plane containing Z and $l_T^\perp$, $V = \text{Plane}(Z, l_T^\perp)$. The line $n = V \cap U$ is parallel to l and to $l_T^{\perp}$ by Lemma 4.12. Let $n = \overleftrightarrow{ZZ'}$ with Z' in the direction of parallelism. Let n meet t at a point W. Then BW is line in $\Pi^{\perp}$ through B and hence BW is perpendicular to $l_T^{\perp}$. The angle $\angle BWZ' = \Pi(BW)$ and so is acute. Since $\triangle BWZ$ is a right triangle, Z must lie on the side of U toward Z'.

We next prove that U meets T.

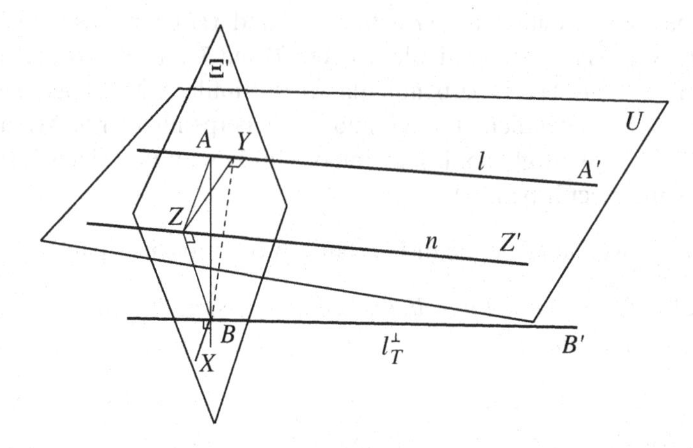

Let ZY denote the line segment on U with Y on l and ZY perpendicular to l. Consider the plane $\Xi = \text{Plane}(A,B,Z)$; Ξ is perpendicular to U because it contains BZ. Let $\overrightarrow{BX}$ be the intersection of Π' and T. If $\angle BAZ < \Pi(AB) = \angle BAY$, then $\overrightarrow{AZ}$ meets $\overrightarrow{BX}$ and U meets T. By construction we have three right triangles, $\triangle AYZ$, $\triangle BZA$, and $\triangle BZY$. Euclid's Proposition I.19, the greater side subtends the greater angle, implies $AY > AZ$ and $BY > BZ$.

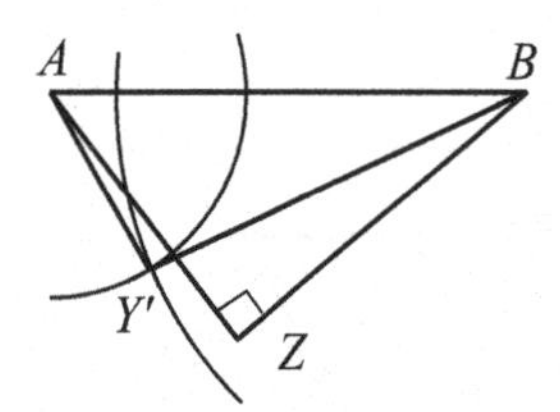

In Ξ consider the triangle $\triangle BZA$. The circles, C_1 centered at B with radius congruent to BY and C_2 centered at A with radius congruent to AY, meet at a point Y' making $\triangle BY'A \cong \triangle BYA$. However, at Z the segments BZ and AZ are shortest to each leg, a consequence of Proposition I.19. Hence Y' lies outside $\triangle BZA$ and $\angle BAY' > \angle BAZ$, so U meets T. ■

This theorem holds whether **HRA** or **HAA** is valid in space. We continue with the development of hyperbolic geometry.

Hyperbolic space

In the digression we derived the properties of neutral space. If we add that **HAA** holds, we call the space *hyperbolic space* or *non-Euclidean space*. In such a space, we fix a line l and the pencil $\mathbf{P}_l$ of all lines parallel to l in a given direction. Two points X and Y in space are said to **correspond** with respect to $\mathbf{P}_l$ if they correspond in the plane determined by the two parallels in $\mathbf{P}_l$ through them. Recall two lines are parallel in space if they are coplanar and parallel in the plane they share. Thus it makes sense to construct the perpendicular bisector of XY in the plane of the parallels in $\mathbf{P}_l$ and ask if it is also in $\mathbf{P}_l$.

Definition 4.26. *The set of points corresponding to a given point in space with respect to a given pencil of parallel lines* $\mathbf{P}_l$ *is called a* **horosphere**.

A horosphere may be thought of as a sphere of infinite radius or as the surface of revolution generated by a horocycle rotated around a line in the pencil $\mathbf{P}_l$. It enjoys a kind of homogeneity regarding the choice of given point that is the spatial version of Corollary 4.7.

Proposition 4.27. *If* $\mathbf{P}_l$ *is a pencil of lines in space parallel to a given line* l *in a fixed direction and* A *corresponds to* B *and to* C *with respect to* $\mathbf{P}_l$*, then* B *corresponds to* C *with respect to* $\mathbf{P}_l$.

PROOF: In Lobachevskiĭ's *Theory of Parallels* this is Proposition 34; it is Proposition 10 in Bolyai's *The Science Absolute of Space* (see Bonola 1955 or Gray 2004). Since A corresponds to B and A corresponds to C, let N be the midpoint of AB and M the midpoint of AC, and the lines $\overleftrightarrow{NN'}$ perpendicular to AB in Plane$(\overleftrightarrow{AA'},\overleftrightarrow{BB'})$ and $\overleftrightarrow{MM'}$ perpendicular to AC in Plane$(\overleftrightarrow{AA'},\overleftrightarrow{CC'})$ are in $\mathbf{P}_l$.

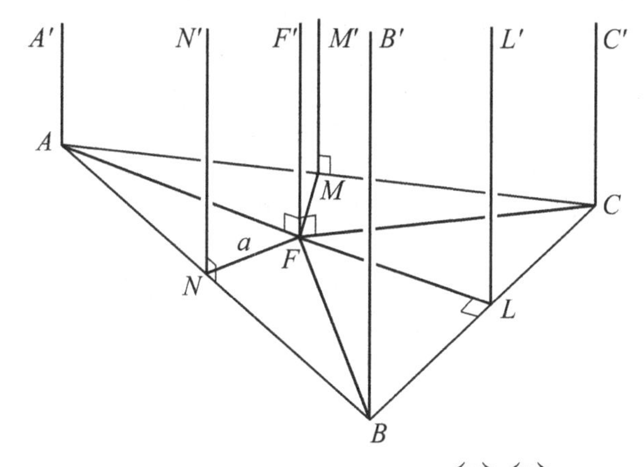

Consider the planes Plane(A,B,C) and Plane$(\overleftrightarrow{AA'},\overleftrightarrow{BB'})$; the dihedral angle between these planes is $\Pi(a)$ for some length a. Let NF be of length a, perpendicular to AB, and lying in Plane(A,B,C). Take the line $\overleftrightarrow{FF'}$ perpendicular to the plane of ABC; it is in the pencil of lines $\mathbf{P}_l$ since it is parallel to $\overleftrightarrow{NN'}$ by the choice of the length of NF. It follows from Corollary 4.13 that $\overleftrightarrow{FF'}$ is parallel to $\overleftrightarrow{MM'}$ and so they are coplanar. The line AC is perpendicular to $\overleftrightarrow{MM'}$, which lies in a plane perpendicular to the plane of ABC containing AC. It follows from Lemma 4.17 that AC is perpendicular to any line through M in the plane of $MM'F'F$, including MF. Construct AF, BF, and CF. By repeated application of Side–Angle–Side we find that $AF \cong BF \cong CF$. Construct the perpendicular bisector from F to BC ($\triangle BFC$ is isosceles) to the point L. Consider the line $\overleftrightarrow{LL'}$ in Plane$(\overleftrightarrow{FF'},L)$ parallel to $\overleftrightarrow{FF'}$.

Since $\overleftrightarrow{LL'}$ is parallel to $\overleftrightarrow{FF'}$, which is parallel to $\overleftrightarrow{BB'}$, $\overleftrightarrow{LL'}$ is parallel to $\overleftrightarrow{BB'}$ and hence they are coplanar. Similarly $\overleftrightarrow{LL'}$ is parallel and coplanar to $\overleftrightarrow{CC'}$ and so $\overleftrightarrow{LL'} = \text{Plane}(\overleftrightarrow{BB'}, \overleftrightarrow{CC'}) \cap \text{Plane}(\overleftrightarrow{FF'}, L)$. Since $BL \cong LC$, $\angle L'LB$ and $\angle L'LC$ are right angles and so B and C correspond. ∎

Proposition 4.27 and the properties of parallelism allow us to take any line in the pencil of parallels as the axis of the horosphere and any point as the initial point to define it, since all other points on it correspond to the initial point.

If we intersect a plane containing one of the parallels in the defining pencil with the horosphere, the intersection is a horocycle. Furthermore, any two distinct points on the horosphere determine a unique horocycle given by the intersection of the horosphere with the plane determined by the parallels through each point. Thus, the horosphere enjoys an incidence geometry of points and horocycles.

Theorem 4.28. *On a horosphere determined by a pencil of parallels* $\mathbf{P}_l$ *in space, the incidence geometry of points and horocycles satisfies Postulate V.*

PROOF: Given a point P on the horosphere there is a unique line m in $\mathbf{P}_l$ through P. A horocycle h lying on this horosphere and not passing through P corresponds to a plane T in space, containing two lines in $\mathbf{P}_l$. By construction m is parallel to T. Playfair's Axiom for lines and planes (Theorem 4.25) holds in neutral space and so there is a plane T' containing m and parallel to T. The intersection of T' with the horosphere is another horocycle h' that does not meet h. Since any other plane containing m meets T, any other horocycle through P meets h. This is equivalent to Postulate V. ∎

We leave it to the reader to establish Postulates I–IV on the horosphere and so prove that the geometry of horocycles and points on the horosphere is Euclidean. This result was established before the works of Lobachevskiĭ and Bolyai. It is due to F. L. WACHTER (1792–1817), who was a student of Gauss in Göttingen (Bonola, 1955, p. 62).

The prospect of comparing the geometry of **HAA** with Euclidean geometry in the same space led Lobachevsky in §36 and Bolyai in §29 of the papers translated in (Bonola 1955) to a formula for the angle of parallelism.

Theorem 4.29 (Lobachevskiĭ–Bolyai Theorem). *For a choice of unit length, when the length of the line segment AP is given by the real number x:*

$$\tan(\Pi(x)/2) = e^{-x/k},$$

where k is the constant determined in Theorem 4.9.

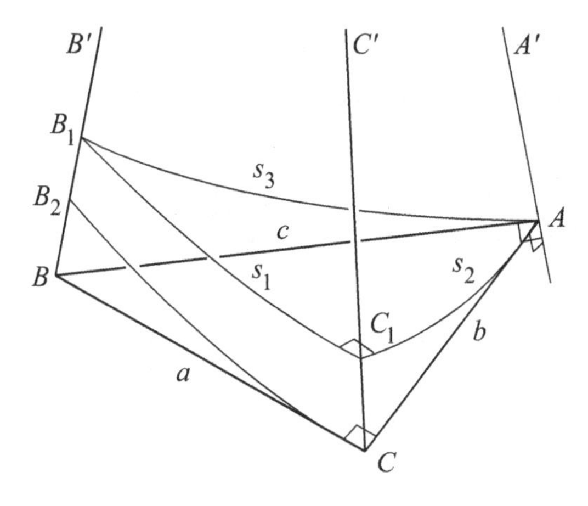

PROOF: Let $\triangle ABC$ be a right triangle in the hyperbolic plane with the right angle at C. Let $\overleftrightarrow{AA'}$ be the unique line in space perpendicular to the plane of $\triangle ABC$ (Corollary 4.18). In the pencil of lines parallel to $\overleftrightarrow{AA'}$ in the direction of A', take the lines $\overleftrightarrow{CC'}$ and $\overleftrightarrow{BB'}$. These lines meet the horosphere through A associated to this pencil at points C_1 and B_1, giving the triangle $\triangle AB_1C_1$ on the horosphere. By Theorem 4.28, ordinary trigonometry applies to the triangle $\triangle AB_1C_1$ of horocycle segments.

In fact, $\triangle AB_1C_1$ is a right triangle on the horosphere. First, notice that Plane(AA', CC') contains AA' and so it is perpendicular to Plane(A, B, C). Since AC and BC are perpendicular, and $\overleftrightarrow{AC} = \text{Plane}(AA', CC') \cap \text{Plane}(A, B, C)$, the line $\overleftrightarrow{BC}$ is perpendicular to Plane(AA', CC') by Lemma 4.17. Thus, Plane(BB', CC') is perpendicular to Plane(AA', CC'). Since the line tangent to a horocycle is perpendicular to the line in the pencil at the point of tangency, the lines tangent to the horocycle segments $\overset{\frown}{AC_1}$ and $\overset{\frown}{C_1B_1}$ are perpendicular to $\overleftrightarrow{CC_1}$. Since $\overleftrightarrow{CC'} = \text{Plane}(AA', CC') \cap \text{Plane}(BB', CC')$, the tangent lines to the horocycle at C_1 are perpendicular, and so $\angle AC_1B_1$ is a right angle. Let $s_1 = l(\overset{\frown}{B_1C_1})$, $s_2 = l(\overset{\frown}{AC_1})$, and $s_3 = l(\overset{\frown}{AB_1})$. Since the geometry on a horosphere is Euclidean, then Euclid's Proposition I.47 implies that $s_3^2 = s_1^2 + s_2^2$; that is, $l(\overset{\frown}{AB_1})^2 = l(\overset{\frown}{B_1C_1})^2 + l(\overset{\frown}{AC_1})^2$.

We next relate s_1, s_2, and s_3 to $a = l(BC)$, $b = l(AC)$, and $c = l(AB)$. By Corollary 4.11 we have $s_2 = \sigma \tanh(b/k)$ and $s_3 = \sigma \tanh(c/k)$, where σ is the length defined in the proof of Lemma 4.10. To express s_1 consider the piece of the concentric horocycle $\overset{\frown}{CB_2}$ in Plane(BB', CC'): By Theorem 4.9,

$$s_1 = l(\overset{\frown}{CB_2})e^{-l(CC_1)/k} = (\sigma \tanh(a/k))e^{-l(CC_1)/k}.$$

For the figure ACC_1 the length of CC_1 corresponds to the length v in Lemma 4.10, and so:

$$e^{l(CC_1)/k} = \cosh(b/k), \text{ and } s_1 = \sigma \frac{\tanh(a/k)}{\cosh(b/k)}.$$

Thus, $s_3^2 = s_1^2 + s_2^2$ implies $\tanh^2(c/k) = \left(\dfrac{\tanh(a/k)}{\cosh(b/k)}\right)^2 + \tanh^2(b/k)$.

Dividing the fundamental relation, $\cosh^2(x) - \sinh^2(x) = 1$, by $\cosh^2(x)$ and rearranging we obtain $\tanh^2(x) = \dfrac{\cosh^2(x) - 1}{\cosh^2(x)}$. Substituting we find:

$$\begin{aligned}\tanh^2(c/k) &= \frac{\cosh^2(a/k) - 1}{\cosh^2(a/k)\cosh^2(b/k)} + \frac{\cosh^2(b/k) - 1}{\cosh^2(b/k)} \\ &= 1 - \frac{1}{\cosh^2(a/k)\cosh^2(b/k)}.\end{aligned}$$

We also know that $\tanh^2(c/k) = 1 - \dfrac{1}{\cosh^2(c/k)}$. Rearranging terms, we have proved:

Theorem 4.30 (Hyperbolic Pythagorean Theorem).

$$\cosh(c/k) = \cosh(a/k)\cosh(b/k).$$

Because the horosphere passes through the point A, and $\overleftrightarrow{AA'}$ is perpendicular to the triangle $\triangle ABC$ at A, $\angle CAB = \angle C_1AB_1$. Thus, $s_2 = s_3\cos(\angle CAB)$, and we have $\sigma\tanh(b/k) = \sigma\tanh(c/k)\cos(\angle CAB)$ or:

$$\cos(\angle CAB) = \frac{\tanh(b/k)}{\tanh(c/k)}.$$

Similarly, had we begun with $\overleftrightarrow{BB'}$ perpendicular to the plane of $\triangle ABC$, we would have gotten:

$$\cos(\angle ABC) = \frac{\tanh(a/k)}{\tanh(c/k)}.$$

Consider the ordinary trigonometric relation $s_1 = s_3\sin(\angle CAB)$. It follows from $s_1 = \sigma\dfrac{\tanh(a/k)}{\cosh(b/k)}$ and $s_3 = \sigma\tanh(c/k)$ that:

$$\sigma\frac{\tanh(a/k)}{\cosh(b/k)} = \sigma\tanh(c/k)\sin(\angle CAB) \text{ or } \sin(\angle CAB) = \frac{\tanh(a/k)}{\cosh(b/k)\tanh(c/k)}.$$

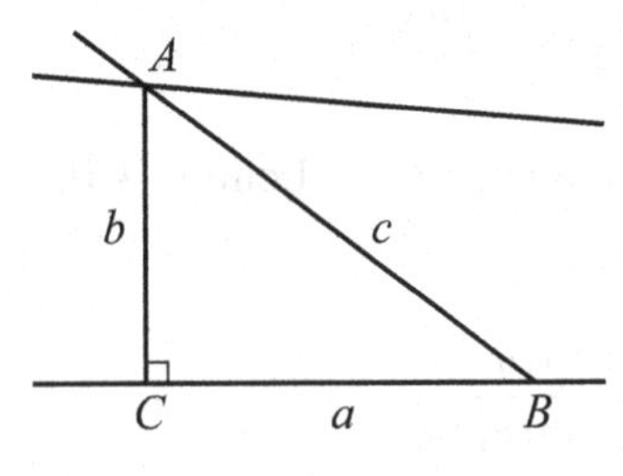

Returning to our right triangle in the hyperbolic plane, as a goes to infinity, so also does c and $\angle CAB$ tends to $\Pi(b)$. Thus $\cos(\Pi(b)) = \lim_{a\to\infty}\cos(\angle CAB)$. Since

$$\lim_{x\to\infty}\tanh x = \lim_{x\to\infty}\frac{e^x - e^{-x}}{e^x + e^{-x}} = 1,$$

we have:

$$\cos\Pi(b) = \lim_{a\to\infty}\frac{\tanh(b/k)}{\tanh(c/k)} = \tanh(b/k),$$

and $\sin \Pi(b) = \lim_{a\to\infty} \dfrac{\tanh(a/k)}{\cosh(b/k)\tanh(c/k)} = \dfrac{1}{\cosh(b/k)}$. Finally, to prove the Lobachevskiĭ–Bolyai Theorem, we apply the half-angle formula $\tan(\theta/2) = \dfrac{1-\cos\theta}{\sin\theta}$:

$$\begin{aligned}\tan(\Pi(b)/2) &= \frac{1-\cos\Pi(b)}{\sin\Pi(b)} = \frac{1-\tanh(b/k)}{(1/\cosh(b/k))} = \cosh(b/k) - \sinh(b/k)\\ &= \frac{e^{b/k}+e^{-b/k}}{2} - \frac{e^{b/k}-e^{-b/k}}{2} = e^{-b/k}.\end{aligned}$$

This proves the theorem. ∎

A nice addition formula follows directly from the Lobachevskiĭ-Bolyai Formula:

$$\tan\left(\frac{\Pi(x+y)}{2}\right) = \tan\left(\frac{\Pi(x)}{2}\right)\tan\left(\frac{\Pi(y)}{2}\right).$$

Also the condition **HRA** corresponds to the case $k = \infty$.

We apply the trigonometry of Lobachevskiĭ and Bolyai and compute the circumference of a circle of radius R in the non-Euclidean plane. The resulting formula can be compared to a formula in Chapter 12 where the mysterious constant k will be determined.

We begin with the non-Euclidean analogue of the spherical sine theorem (Theorem 1.4). This result appears in the 1825 work *Theorie der Parallellinien* of F. A. Taurinus (1794–1874), in the work of Lobachevskiĭ and in a particularly interesting form as Proposition 25 of J. Bolyai's paper (Banola 1955).

Theorem 4.31 (Hyperbolic Law of Sines). *In a triangle $\triangle ABC$ with sides of length $c = l(AB)$, $b = l(AC)$, and $a = l(BC)$ we have:*

$$\frac{\sinh(a/k)}{\sin(\angle A)} = \frac{\sinh(b/k)}{\sin(\angle B)} = \frac{\sinh(c/k)}{\sin(\angle C)}.$$

Proof: Like the proof of the spherical sine theorem (Theorem 1.3), it suffices to prove the case of a right triangle. Suppose that the right angle is at C. In the discussion after the proof of the Hyperbolic Pythagorean Theorem, we proved that:

$$\begin{aligned}\sin(\angle A) &= \frac{\tanh(a/k)}{\cosh(b/k)\tanh(c/k)}\\ &= \frac{\sinh(a/k)}{\cosh(a/k)\cosh(b/k)\tanh(c/k)} = \frac{\sinh(a/k)}{\sinh(c/k)}.\end{aligned}$$

Thus, $\dfrac{\sinh(a/k)}{\sin(\angle A)} = \dfrac{\sinh(c/k)}{\sin(\angle C)}$. Because there is nothing special about choosing the vertex to be A, the other equation holds by relabeling. The general case follows as it does in the spherical case. ∎

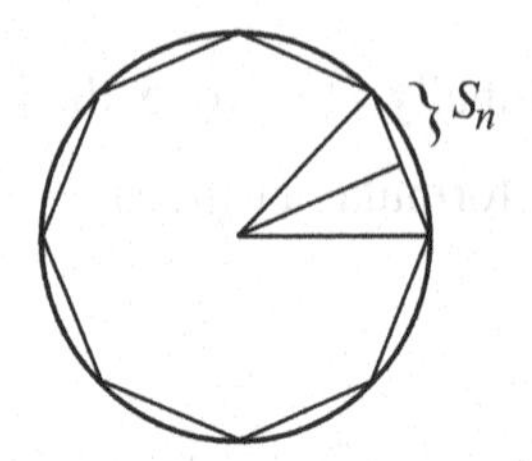

The relations just proved are just what we need to compute the perimeter of a regular n-gon inscribed in a circle of radius R. Let S_n denote half of the length of a side of the regular inscribed n-gon. A right triangle is formed by the center of the circle, the midpoint of a side, and an adjacent vertex. The interior angle at the center is π/n for this triangle, and Theorem 4.31 implies:

$$\frac{\sinh(S_n/k)}{\sin(\pi/n)} = \frac{\sinh(R/k)}{\sin(\pi/2)}.$$

It follows that $\sinh(S_n/k) = \sin(\pi/n)\sinh(R/k)$.

To get the circumference of a circle we follow Archimedes's lead and compute the limit, $\lim_{n\to\infty} 2nS_n$, of the perimeter of the regular inscribed n-gon. The Maclaurin series for $\sinh(x)$ is $x + \frac{x^3}{3} + \frac{x^5}{5} + \cdots$, from which it follows that:

$$\lim_{x\to 0} \frac{\sinh x}{x} = 1.$$

If $\lim_{n\to\infty} 2nS_n$ exists, then it follows from the elementary theory of limits that:

$$\lim_{n\to\infty} 2nS_n = \lim_{n\to\infty} 2nk(S_n/k) = \lim_{n\to\infty} 2nk\sinh(S_n/k).$$

We compute:

$$\lim_{n\to\infty} 2nS_n = \lim_{n\to\infty} 2nk\sinh(S_n/k) = \lim_{n\to\infty} 2nk\sin(\pi/n)\sinh(R/k)$$
$$= 2\pi k\sinh(R/k) \lim_{n\to\infty} \frac{\sin(\pi/n)}{(\pi/n)} = 2\pi k\sinh(R/k).$$

Theorem 4.32 (Gauss 1831). *The circumference of a circle of radius R in the hyperbolic plane is given by $2\pi k\sinh(R/k)$.*

If we denote the circumference of a circle of radius R by circum(R), then we notice that the formula leads to a series expression:

$$\text{circum}(R) = 2\pi k\sinh(R/k) = 2\pi k\left(\frac{R}{k} + \frac{R^3}{6k^3} + \frac{R^5}{120k^5} + \cdots\right)$$
$$= 2\pi R + \frac{\pi R^3}{3k^2} + \cdots.$$

When k goes to infinity, we obtain the familiar formula for the circumference of a circle. The term $\pi R^3/3k^2$ will be identified in Chapter 12.

We pause here to summarize our progress. From a workable definition of parallel lines in the plane and in space we have developed some elementary notions to derive formulas through elementary analysis. These formulas contain an unknown quantity k, and they are analogous to formulas that we found in spherical trigonometry. The formulas also coincide with Euclidean formulas when k goes to ∞. When

$k \neq \infty$, non-Euclidean geometry leads to new phenomena like horocycles and the angle of parallelism. Because the formulas do not lead to contradictions and the analytic picture of this geometry resembles the profile of spherical geometry, we are led to abandon the idea that Postulate V follows from Postulates I through IV. Through the power of the axiomatic method allied with analysis, we have reached a turning point in the 2000-year-long history of Postulate V. What remains to finish the story is a concrete realization of non-Euclidean geometry, a mathematical object consisting of "points," "lines," and "planes" that are related in the manner developed by Gauss, Lobachevskiĭ, and Bolyai. For this goal we turn for support and foundations to deeper analysis and its relation to geometry.

Exercises

4.1 Given three distinct points X, Y, and Z, let l, m, and n be the lines that are the perpendicular bisectors of YZ, XZ, and XY, respectively. If l meets m in a point P, prove that P is the center of a circle passing through X, Y and Z.

4.2 Complete the proof of part (b) in Theorem 4.2 where you consider a ray $\overrightarrow{AC}$ that does not meet the line $\overleftrightarrow{PQ}$.

4.3[†] Assuming **HAA**, show that for every acute angle there is a line parallel to both rays forming the angle.

4.4 Prove Proposition 4.6.

4.5 Prove Corollary 4.13 on the transitivity of parallelism of lines in space.

4.6 Prove Euclid's Proposition XI.5: If a straight line is perpendicular to three straight lines, which are concurrent, at their point of intersection, then the three straight lines lie in the same plane.

4.7 Prove Theorem 4.25 under the assumption of **HRA** for neutral space.

4.8 Lemma 4.24 shows that two planes with a common perpendicular plane whose lines of intersection are parallel do not intersect. What happens if some line l is perpendicular to two distinct planes (Euclid Proposition XI.14)?

4.9 Prove that **HRA** holds if and only if the set of points corresponding to a given point P with respect to a pencil of lines perpendicular to a given line l $\mathbf{P}_l^{\perp}$ with P not on l is a line.

4.10[†] Suppose l, l', and n are coplanar lines with l perpendicular to n, which is also perpendicular to l', and n passes through $l \cap l' = P$. Show that $l = l'$.

4.11 Establish Postulates I–IV on the horosphere with points as points and horocycles as lines.

4.12 The Generalized Pythagorean Theorem takes the form:

$$\cosh(c/k) = \cosh(a/k)\cosh(b/k).$$

Show that as $k \to \infty$ this gives the Pythagorean Theorem. This implies that if a, b, and c are small in comparison to k, then the Pythagorean Theorem is very closely approximated.

4.13 Under **HRA** show that any convex quadrilateral tiles the plane, that is, by repeating the figure the plane can be covered without overlapping. Under **HAA** show that any convex

Solutions to Exercises marked with a dagger appear in the appendix, pp. 328–329.

polygon with an even number of sides with angle sum equal to $2\pi/n$ for some positive integer n tiles the plane. What happens when there is an odd number of sides?

4.15 In the astronomical diagram shown below, let ρ denote the parallax of Sirius, which has been measured to be $1'', 24$. From the Lobachevskiĭ–Bolyai Theorem calculate an upper bound for a/k.

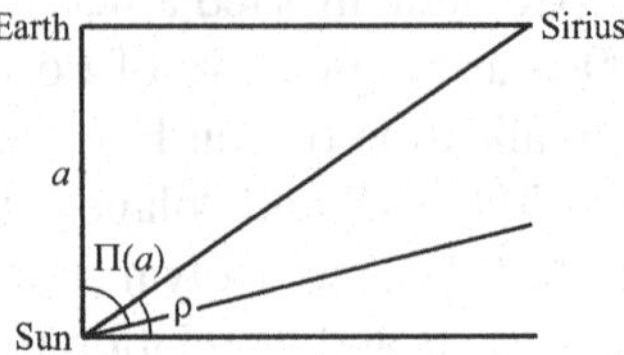

4.16 Prove Lobachevskiĭ's Formula (in Bonola (1955), (5.) in §37) for a triangle $\triangle ABC$ with the length of side denoted by the opposite vertex:

$$\cos(\angle BAC)\cos\Pi(b)\cos\Pi(c)+\frac{\sin\Pi(b)\sin\Pi(c)}{\sin\Pi(a)}=1.$$

4.17[†] Prove Theorem 4.31 for a general triangle, $\triangle ABC$. From Theorem 4.31 deduce the following construction for $\Pi(x)$: Let AE and BD be perpendicular to AB. Let AE have length x. Let DE be perpendicular to AE. If DE meets AE at E, then let $AO\cong DE$ with O on BD. Then $\angle OAE\cong\Pi(x)$.

4.18 If a is a side and γ an interior angle of a regular n-gon prove that $\cos\frac{\pi}{n}=\sin\frac{\gamma}{2}\cosh\frac{a}{2}$. Show that a judicious choice of a gives a right angle as the interior angle.

4.19 Determine the derivative of the angle of parallelism with respect to length, that is, $\frac{d}{dx}(\Pi(x))$.

4.20[†] Under **HAA**, suppose that a figure made up of three mutually parallel lines encloses a finite area A. We prove a theorem of Gauss in this exercise: There is a constant λ such that the area of an arbitrary triangle is $\lambda(\pi-\alpha-\beta-\gamma)$, where α, β, and γ are the measures of the interior angles of the triangle. Following Gauss in a letter to F. Bolyai in 1832, introduce the function $f(\phi)$, which is the area subtended by the supplement to angle ϕ and the parallel to both rays (Figure A). Then $f(\phi)+f(\pi-\phi)=A$ as in Figure B. Figure C shows how $f(\phi)+f(\psi)+f(\pi-\phi-\psi)=A$. Now deduce that $f(\phi)+f(\psi)=f(\phi+\psi)$, which can only be satisfied by a linear function $f(\phi)=\lambda\phi+c$. Since $f(0)=0$, $c=0$. Figure D shows how to introduce the triangle.

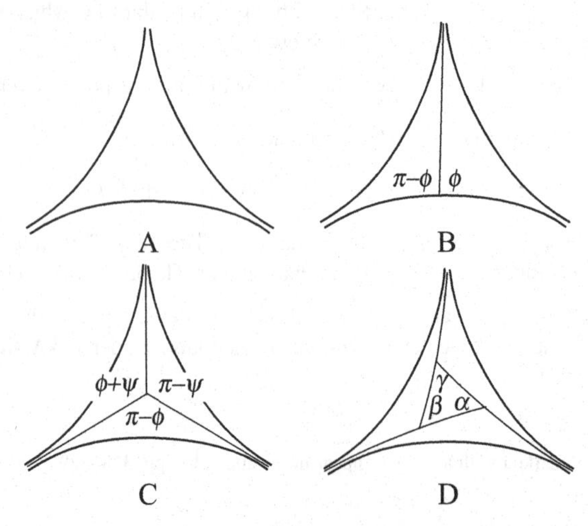

4.21 Another approach to Gauss's Theorem of the previous exercise is to consider *angle defect*. Show that angle defect is proportional to area for triangles by showing that angle defect is additive; that is, if a triangle is decomposed into subtriangles, the angle defect of the large triangle is the sum of the angle defects of the parts. Show that this implies that there is an upper bound to the area of any triangle. In particular, a figure made up of three mutually parallel lines encloses a finite area.

Appendix
The Elements: Selections from Book XI

Definitions

(1) A straight line is perpendicular to a plane when it makes right angles with all the straight lines that meet it and are in the plane.
(2) A plane is at right angles to a plane when the straight lines drawn in one of the planes at right angles to the common section of the planes are at right angles to the remaining plane.
(3) The inclination of a straight line to a plane is obtained by drawing a perpendicular from the line to plane and the line joining the foot of the perpendicular with the intersection of the plane with the given line—the angle contained by the segment in the plane and the given straight line.
(4) The inclination of a plane to a plane is the acute angle contained by two straight lines both perpendicular to the line of intersection of the planes and in each plane.
(5) A plane is similarly inclined to a plane as another is to another when the angles of the inclinations are equal.
(6) Parallel planes are those which do not meet.

Propositions

Proposition XI.1. A part of a straight line cannot be in a plane and inclined to part of the plane.

Proposition XI.2. If two straight lines intersect, they are in the same plane, and every triangle lies in a unique plane.

Proposition XI.3. If two planes meet, their intersection is a straight line.

Proposition XI.4. If a straight line is at right angles to two straight lines at their point of intersection, then the line will be at right angles to the plane of the two straight lines.

Proposition XI.5. If a straight line is at right angles to three straight lines that meet at a single point, then the three lines are coplanar.

Proposition XI.6. If two straight lines are perpendicular to the same plane, then they are parallel.

Proposition XI.7. If two straight lines are parallel and arbitrary points are taken on each, then the line segment joining the points lies in the same plane as the parallel straight lines.

Proposition XI.8. If two straight lines are parallel and one of them is perpendicular to a plane, then the other straight line will also be perpendicular to that plane.

Proposition XI.9. Two lines each parallel to a third line are parallel to each other.

Proposition XI.10. If two straight lines that meet are parallel to two straight lines that meet in another plane, then they will subtend equal angles.

Proposition XI.11. From a point not on a plane to draw a straight line perpendicular to the given plane.

Proposition XI.12. To construct a straight line at right angles to a given plane through a given point on it.

Proposition XI.13. From the same point two straight lines cannot be constructed both perpendicular to the same plane on the same side.

Proposition XI.14. Planes perpendicular to the same straight line are parallel.

Proposition XI.15. If two straight lines that meet are parallel to two straight lines that meet in another plane, then the planes through the pairs of lines are parallel.

Proposition XI.16. If two parallel planes are cut by another plane, their lines of intersection are parallel.

Proposition XI.17. If two straight lines are cut by parallel lines, they are cut in the same ratio.

Proposition XI.18. If a straight line is perpendicular to any plane, then all the planes through it will be perpendicular to the plane.

Proposition XI.19. If two planes meet at right angles to another plane, then their intersection will also be at right angles to that plane.

Proposition XI.20–XI.39. These propositions concern solid angles and the construction of solid figures, especially parallelepipeds.

PART B

Development

Differential geometry

5

Curves in the plane

I have been occupied with a new discovery ... in order to make my clock even more exact. ... What, however, I never had expected I would discover, I have now hit upon, the undoubtedly true shape of curves ... I determined it by geometric reasoning.

C. HUYGENS (1659)

Beginning with the Greek times up to the present day, almost all mathematicians have worked at the assembling of the colorful flock of the diverse, particular algebraic and transcendental curves ...

GINO LORIA (1902)

Differential geometry begins with the subject of curves. Tangents to curves can be found in the works of Euclid (for the circle, see Book III, Definition 2), Archimedes (*Quadrature of the Parabola*), and Appolonius (*The Conics*). The methods of the calculus, as applied to parametrized curves, led to considerable development by the pioneers of analysis—Leibniz, Newton, the Bernoullis, and Euler (Struik 1933).

We begin with the basic properties of curves in $\mathbb{R}^2$ and $\mathbb{R}^3$, Euclidean spaces that are equipped with the dot product and associated norm, distance, and so on. Curves in the plane can arise in various ways: Algebraically, as the solution set of a polynomial in two variables,

$$\mathcal{C} = \{(x,y) \mid f(x,y) = 0\};$$

as the graph of a function

$$g\colon (a,b) \to \mathbb{R} \text{ given by } \mathcal{C} = \{(x, g(x)) \mid x \in (a,b)\};$$

or as a **parametrized curve**,

$$\alpha\colon (a,b) \to \mathbb{R}^2 \text{ for some } -\infty \leq a < b \leq \infty,$$

where $\alpha(t) = (x(t), y(t))$. The coordinate functions $x(t)$ and $y(t)$ are taken to be differentiable of sufficient order in t.

In space, curves can arise as the intersection of two surfaces, as a graph of two functions $y = f(x)$, $z = g(x)$, or as parametrized curves:

$$\alpha(t) = (x(t), y(t), z(t)), \quad t \in (a,b).$$

For such curves we refer to the image of $\alpha(t)$ as its **trace**, the geometric object of interest. We could speak of *curves of class k*, that is, the coordinate mappings all have continuous derivatives up to order k. In each definition and theorem to follow, to obtain the most generality, we could require the minimal order of differentiability. However, to focus more on the geometry than the analysis, we ignore this subtlety and assume curves to be **smooth**, that is, differentiable of all orders. The careful reader can discover the most general formulation by providing the least order of differentiability needed for each proof.

The plane and space are endowed with an inner product or **dot product** of two vectors, given generally on $\mathbb{R}^n$ for $\mathbf{v} = (v_1, \ldots, v_n)$ and $\mathbf{w} = (w_1, \ldots, w_n)$ by the formula:

$$\mathbf{v} \cdot \mathbf{w} = v_1 w_1 + \cdots + v_n w_n.$$

The **norm** on $\mathbb{R}^n$ is given by $\|\mathbf{v}\| = \sqrt{\mathbf{v} \cdot \mathbf{v}}$, and the **distance** between two vectors is defined by $d(\mathbf{v}, \mathbf{w}) = \|\mathbf{v} - \mathbf{w}\|$. From now on, when we write $\mathbb{R}^n$, we mean for $n = 2$ or $n = 3$. The definitions and general remarks also apply to curves in higher dimensional $\mathbb{R}^n$; however, these cases are not the focus of the chapter.

Definition 5.1. *For a parametrized, smooth curve $\alpha\colon (a,b) \to \mathbb{R}^n$, the* **tangent vector** *to the curve at $\alpha(t)$ is given by $\alpha'(t) = (x_1'(t), x_2'(t), \ldots, x_n'(t))$. A curve is called* **regular** *if $\alpha'(t) \neq \mathbf{0}$ for all t in (a,b). The norm $\|\alpha'(t)\|$ is called the* **speed** *of α at $\alpha(t)$. The* **unit tangent vector** *to α, a regular curve, is given by $T(t) = \alpha'(t)/\|\alpha'(t)\|$.*

Notice that the speed of a regular curve is always greater than zero. For two parametrized curves, α, $\beta\colon (a,b) \to \mathbb{R}^n$, the derivative of the dot product $\alpha(t) \cdot \beta(t)$ satisfies the Leibniz Rule with respect to the dot product:

$$\frac{d}{dt}[\alpha(t) \cdot \beta(t)] = \alpha'(t) \cdot \beta(t) + \alpha(t) \cdot \beta'(t).$$

EXAMPLES:

(1) Let $\alpha\colon \mathbb{R} \to \mathbb{R}^3$ be given by $\alpha(t) = \mathbf{p} + t\mathbf{q}$, for $\mathbf{p}$, $\mathbf{q}$ in $\mathbb{R}^3$, $\mathbf{q} \neq \mathbf{0}$. This curve gives a parametrization of the straight line through $\mathbf{p}$ and $\mathbf{q}$ of speed $\|\mathbf{q}\|$. The unit tangent vector is given by $T(t) = \mathbf{q}/\|\mathbf{q}\|$ for all t.

(2) For $r > 0$, let $\alpha\colon \mathbb{R} \to \mathbb{R}^2$ be given by $\alpha(t) = (r\cos(t), r\sin(t))$. The trace of this curve is the circle of radius r centered at the origin and $\alpha(t)$ has speed r. Let $\beta\colon \mathbb{R} \to \mathbb{R}^2$ be given by $\beta(t) = (r\cos mt, r\sin mt)$, where m is a positive real number. Both curves have the same trace but $\beta(t)$ travels the trace m times faster:

$$\|\beta'(t)\| = \sqrt{m^2r^2\sin^2 mt + m^2r^2\cos^2 mt} = mr = m\|\alpha'(t)\|.$$

The unit tangent vector to $\alpha(t)$ is given by $T(t) = (-\sin(t), \cos(t))$.

(3) Let $\alpha\colon \mathbb{R} \to \mathbb{R}^3$ be given by $\alpha(t) = (a\cos t, a\sin t, bt), a, b \neq 0$. This curve is called the **right helix** on the cylinder of radius a of **pitch** $2\pi b$.

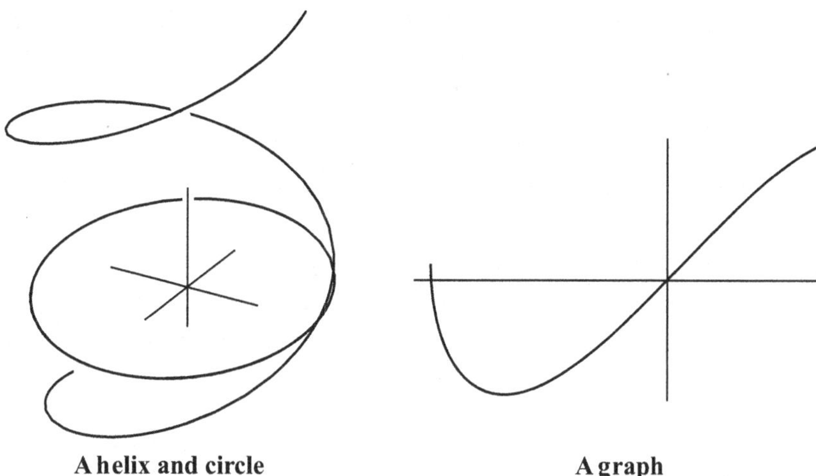

A helix and circle **A graph**

(4) Let $\alpha\colon \mathbb{R} \to \mathbb{R}^2$ be given by $\alpha(t) = (t^3, t^2)$. The trace of this curve satisfies the algebraic equation $x^2 = y^3$ and it is called the **semicubical parabola**. It is *not* regular because $\alpha'(0) = (3t^2, 2t)\big|_{t=0} = (0,0)$. The trace has a cusp at the origin.

(5) If $g\colon (a,b) \to \mathbb{R}$ is a smooth function, then the graph of g is the regular curve given by $\alpha(t) = (t, g(t))$ of speed $\sqrt{1+(g'(t))^2}$.

Definition 5.2. *Suppose $\alpha\colon (a,b) \to \mathbb{R}^n$ is a regular, parametrized, smooth curve and $a < a_0 < b$. The* **arc length** *function $s\colon (a,b) \to \mathbb{R}$ is defined by:*

$$s(t) = \int_{a_0}^{t} \|\alpha'(r)\|\, dr = \int_{a_0}^{t} \sqrt{\alpha'(r)\cdot\alpha'(r)}\, dr.$$

From elementary calculus, $s(t_0)$ equals the distance along the curve $\alpha(t)$ from $\alpha(a_0)$ to $\alpha(t_0)$, oriented by the direction a goes to b and the choice of a_0.

EXAMPLE: The parametrization $\alpha(t) = (a\cos t, b\sin t)$, for $b \neq a$, determines an ellipse in the plane. The tangent vector and speed are given by $\alpha'(t) = (-a\sin t, b\cos t)$ and:

$$\|\alpha'(t)\| = \sqrt{a^2\sin^2 t + b^2\cos^2 t} = \sqrt{a^2 + (b^2 - a^2)\cos^2 t}.$$

The resulting arc length function,

$$s(t) = \int_0^t \sqrt{a^2 + (b^2 - a^2)\cos^2 r}\, dr,$$

is generally not expressible in terms of elementary functions; it is an example of an elliptic integral. The example shows that the arc length function can be difficult to express explicitly. However, it plays a key role in simplifying certain calculations.

The trace of a parametrized curve has many parametrizations. In particular, once a parametrization $\alpha\colon (a,b) \to \mathbb{R}^n$ is chosen, it may be altered as follows.

Definition 5.3. *Let $\alpha\colon (a,b) \to \mathbb{R}^n$ be a regular, smooth, parametrized curve and $g\colon (c,d) \to (a,b)$ a real valued function. If g is smooth and has a smooth inverse, then we say that $g(s)$ is a* **reparametrization** *of $\alpha(t)$ with the reparametrized curve being $\beta(s) = (\alpha \circ g)(s)$.*

An analytic property of arc length leads to a convenient choice of parametrization.

Proposition 5.4. *The arc length function $s(t)$ is independent of reparametrization.*

PROOF: Suppose $g\colon (c,d) \to (a,b)$ is a reparametrization and $g(c_0) = a_0$. Let $t = g(u)$ and $\beta(u) = \alpha(g(u))$. Then, since $|dg/dv|\,dv = \pm dr$,

$$s(u) = \int_{c_0}^{u} \|\beta'(v)\|\,dv = \int_{c_0}^{u} \|\alpha'(r)\| \cdot \left|\frac{dg}{dv}\right| dv = \pm \int_{a_0}^{t} \|\alpha'(r)\|\,dr = s(t).$$

In the case that g is decreasing (for example, $(0,1) \xrightarrow{g} (0,1)$ given by $u \mapsto 1-u$), the sign $\pm$ is negative. However, in that case we are integrating in the opposite direction of the integral determined by α and so the negatives cancel. ∎

When $\alpha(t)$ is a regular, smooth, parametrized curve, we use the arc length to obtain a particular parametrization of the curve. Let (c,d) denote the image of $s\colon (a,b) \to \mathbb{R}$. Then,

$$\frac{ds}{dt} = \frac{d}{dt}\int_{a_0}^{t} \|\alpha'(r)\|\,dr = \|\alpha'(t)\| \neq 0.$$

By the familiar theorems of elementary calculus, $s(t)$ is one-to-one, onto, and smooth. Let $g\colon (c,d) \to (a,b)$ denote the inverse function of $s(t)$, that is, $t = g(s(t))$. Since $s(g(s)) = s$, we have $\dfrac{dg}{ds} = \dfrac{1}{ds/dt}$, and g is smooth. Let $\beta(s) = \alpha(g(s))$. It follows that:

$$\int_0^s \|\beta'(w)\|\,dw = \int_0^s \|\alpha'(r)\| \cdot \left|\frac{dg}{ds}\right| ds = \int_{a_0}^{g(s)} \|\alpha'(r)\|\,dr = s(g(s)) = s.$$

Thus $\beta(s)$ has its arc length as parameter. We call $\beta(s)$ the **arc length parametrization** of $\alpha(t)$.

Observe that a curve parametrized by arc length has constant speed:

$$\|\beta'(s)\| = \|\alpha'(t)\| \cdot \left|\frac{dg}{ds}\right| = \frac{\|\alpha'(t)\|}{|ds/dt|} = \frac{\|\alpha'(t)\|}{\|\alpha'(t)\|} = 1,$$

that is, the tangent vector to $\beta(s)$ has length 1 everywhere. A curve parametrized by arc length is called a **unit speed curve**. We reserve the variable s for the arc length parameter when it is convenient and t for an arbitrary parameter.

(1) If $\alpha(t) = \mathbf{p} + t\mathbf{q}$, then $\beta(s) = \mathbf{p} + s\frac{\mathbf{q}}{\|\mathbf{q}\|}$.

(2) If $\alpha(t) = (r\cos t, r\sin t)$, then $\|\alpha'(t)\| = r$, and $s(t) = r(t - a_0)$. The reparametrization is $g(s) = (s/r) + a_0$, and if we take $a_0 = 0$, then $\beta'(s) = (r\cos(s/r), r\sin(s/r))$.

Early work on plane curves (Huygens, Leibniz, Newton, and Euler)

According to elementary calculus, the tangent to a curve in the plane is found by choosing two points near a given point P on the curve, taking the unique line passing through them, and then finding the line that is the limit of the constructed line as the two points come together at P. If we have a curve that does not contain portions of a line, then choosing three points near a point P on the curve determines a unique circle in the plane, which is characterized by its center and radius. If we allow the three points to come together at P, the limiting circle approximates the curve near P.

A circle may be understood to "bend uniformly" and this construction obtains the same circle as an approximating curve. To measure how much the curve is "bending" at each point we assign a value that is modeled on lines and circles. For a straight line, we want the measure of bending to satisfy $\kappa_{\pm} \equiv 0$; for a circle, let $\kappa_{\pm} = 1/r$, where r is the radius of the circle. Thus, small circles bend more than large circles. Generalizing this value to each point of an arbitrary curve was begun implicitly in work of APPOLONIUS (third century B.C.E.) and in work of JOHANNES KEPLER (1571–1630). It was taken up by ISAAC NEWTON (1642–1727) in his *Methods of Series and Fluxions* (1671), and by GOTTFRIED LEIBNIZ (1646–1716) in the paper *Meditation nova de natura anguli contactus et osculi* (1686) that was developed further by JAKOB BERNOULLI (1654–1705) and JOHANN BERNOULLI (1667–1748). The most significant early treatment of curves and surfaces (see Chapter 8) is due to LEONHARD EULER (1707–83) (Struik1933).

Let $\alpha\colon (a,b) \to \mathbb{R}^2$ be a regular plane curve, parametrized by arc length. For s in (a,b), take s_1, s_2, and s_3 near s so that $\alpha(s_1)$, $\alpha(s_2)$, and $\alpha(s_3)$ are noncollinear, which is possible as long as α is not part of a line near $\alpha(s)$.

Let $C_\alpha(s_1,s_2,s_3)$ denote the center of the circle determined by the three points. We use the dot product to define a function approximating the square of the radius of this circle,

$$\rho(s) = (\alpha(s) - C_\alpha(s_1,s_2,s_3)) \cdot (\alpha(s) - C_\alpha(s_1,s_2,s_3)).$$

If $\alpha(s)$ is smooth, then so is $\rho(s)$. Since $\rho(s_1) = \rho(s_2) = \rho(s_3)$, by Rolle's Theorem, there are points $t_1 \in (s_1,s_2)$ and $t_2 \in (s_2,s_3)$ with $\rho'(t_1) = \rho'(t_2) = 0$. Applying Rolle's Theorem again to $\rho'(t)$ we get a value $u \in (t_1,t_2)$ with $\rho''(t) = 0$. Computing the derivatives, we find:

$$\rho'(s) = 2\alpha'(s) \cdot (\alpha(s) - C_\alpha(s_1, s_2, s_3)), \text{and}$$
$$\rho''(s) = 2[\alpha''(s) \cdot (\alpha(s) - C_\alpha(s_1, s_2, s_3)) + \alpha'(s) \cdot \alpha'(s)].$$

from which the condition $\rho''(u) = 0$ implies that $\alpha''(u) \cdot (\alpha(u) - C_\alpha(s_1, s_2, s_3)) = -\alpha'(u) \cdot \alpha'(u) = -1$.

Take the limit as s_1, s_2, s_3 go to s; $C_\alpha(s_1, s_2, s_3)$ converges to a point $C_\alpha(s)$, and t_1, t_2 go to s, so $\rho'(s) = 0$ and $\alpha'(s) \cdot (\alpha(s) - C_\alpha(s)) = 0$. Finally, we obtain the condition $\alpha''(s) \cdot (\alpha(s) - C_\alpha(s)) = -1$.

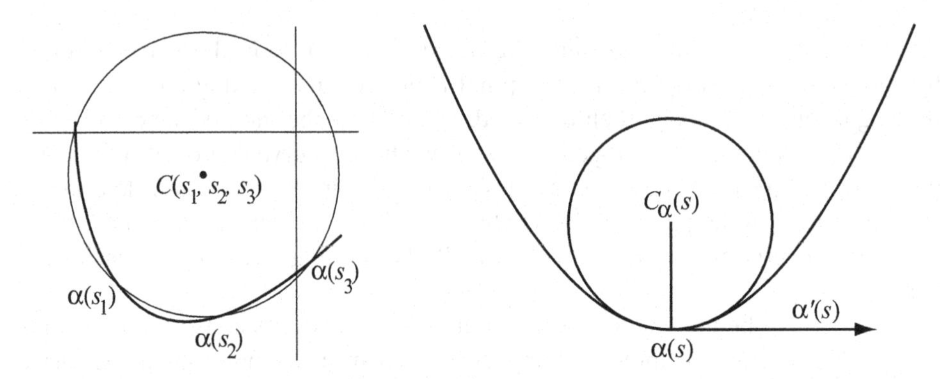

The circle centered at $C_\alpha(s)$ with radius $\|\alpha(s) - C_\alpha(s)\|$ shares the point $\alpha(s)$ with the curve α and, furthermore, the tangent to the circle at $\alpha(s)$ is a multiple of $\alpha'(s)$. The circle is called the **osculating circle** (Leibniz 1686; *osculare*, to kiss) and is tangent to $\alpha(s)$ in the sense that it is a limit of the circles determined by choices of three nearby points on the curve.

Definition 5.5. *The point $C_\alpha(s)$ is called the* **center of curvature** *of $\alpha(s)$, and the curve given by $C_\alpha(s)$ is called the* **curve of centers of curvature**. *The (unsigned)* **plane curvature** *of $\alpha(s)$ at s is given by the reciprocal of the radius of the osculating circle,*

$$\kappa_\pm(\alpha; s) = \frac{1}{\|\alpha(s) - C_\alpha(s)\|}.$$

This geometric construction relates directly to the coordinate expression for $\alpha(s)$.

Theorem 5.6 (Euler 1736). *For a smooth curve $\alpha(s)$, parametrized by arc length,*

$$\kappa_\pm(\alpha; s) = \|\alpha''(s)\|.$$

PROOF: Differentiating $\alpha'(s) \cdot \alpha'(s) = 1$ gives us that $\alpha'(s) \cdot \alpha''(s) = 0$, that is, $\alpha''(s)$ is perpendicular to $\alpha'(s)$. In our construction of $C_\alpha(s)$, the derivative of the function

$\rho(s)$ vanishes in the limit giving $\alpha'(s)\cdot(\alpha(s)-C_\alpha(s))=0$ and so $\alpha(s)-C_\alpha(s)$ is also perpendicular to $\alpha'(s)$. It follows that

$$\alpha(s)-C_\alpha(s)=\mu\alpha''(s), \text{ for some } \mu\in\mathbb{R}.$$

However, $-1=\alpha''(s)\cdot(\alpha(s)-C_\alpha(s))=\alpha''(s)\cdot\mu\alpha''(s)=\mu\|\alpha''(s)\|^2$, and so

$$\frac{1}{\kappa_\pm(\alpha;s)}=\|\alpha(s)-C_\alpha(s)\|=|\mu|\cdot\|\alpha''(s)\|=\frac{1}{\|\alpha''(s)\|^2}\cdot\|\alpha''(s)\|=\frac{1}{\|\alpha''(s)\|}. \quad\blacksquare$$

We observe immediately that the definition and theorem generalize our conventions for the circle. Given the arc length parametrization of the circle of radius r, $\alpha(s)=(r\cos(s/r), r\sin(s/r))$, it follows that

$$\alpha'(s)=(-\sin(s/r),\cos(s/r)) \quad\text{and}\quad \alpha''(s)=(-(1/r)\cos(s/r),-(1/r)\sin(s/r)),$$

so $\|\alpha''(s)\|=1/r$ as desired. For a straight line, $\alpha(s)=\mathbf{p}+s(\mathbf{q}/\|\mathbf{q}\|)$, $\alpha''(s)=0$, so $\kappa_\pm(\alpha;s)=0$.

As the example of an ellipse indicates, it is not always easy to reparametrize a curve by arc length. However, we are able to compute the plane curvature of any regular curve $\alpha(t)$.

Proposition 5.7. *The plane curvature of a regular curve $\alpha(t)=(x(t),y(t))$ is given by*

$$\kappa_\pm(\alpha;t)=\left|\frac{x'(t)y''(t)-y'(t)x''(t)}{((x'(t))^2+(y'(t))^2)^{3/2}}\right|.$$

PROOF: Let $s(t)$ denote the arc length and, risking confusion, denote its inverse by $t(s)$. The reparametrization of $\alpha(t)$ by arc length is given by:

$$\beta(s)=\alpha(t(s))=(x(t(s)),y(t(s))).$$

If $s_0=s(t_0)$, computing $\kappa_\pm(\alpha;t_0)$ is the same as computing $\kappa_\pm(\beta;s_0)$, which denotes the plane curvature of $\beta(s)$ at s_0. To wit,

$$\beta'(s)=\alpha'(t(s))\frac{dt}{ds} \quad\text{and}\quad \beta''(s)=\alpha''(t(s))\left(\frac{dt}{ds}\right)^2+\alpha'(t(s))\frac{d^2t}{ds^2}.$$

Since $\dfrac{ds}{dt}=\sqrt{x'(t)^2+y'(t)^2}$, we can write:

$$\frac{dt}{ds}=\frac{1}{\sqrt{x'(t)^2+y'(t)^2}} \quad\text{and}\quad \frac{d^2t}{ds^2}=-\left[\frac{x'(t)x''(t)+y'(t)y''(t)}{(x'(t)^2+y'(t)^2)^2}\right].$$

By Euler's Formula (Theorem 5.6), we find:

$$\kappa_\pm(\alpha;t_0)=\kappa_\pm(\beta;s_0)=\|\beta''(s_0)\|=\left\|\alpha''(t_0)\left(\frac{dt}{ds}\right)^2+\alpha'(t_0)\frac{d^2t}{ds^2}\right\|$$

$$=\left\|\frac{1}{x'(t)^2+y'(t)^2}(x''(t),y''(t))-\frac{x'(t)x''(t)+y'(t)y''(t)}{(x'(t)^2+y'(t)^2)^2}(x'(t),y'(t))\right\|$$

$$= \left\| \frac{(((x')^2+(y')^2)x''-(x')^2x''-x'y'y'',((x')^2+(y')^2)y''-x'x''y'-(y')^2y'')}{((x'(t))^2+(y'(t))^2)^2} \right\|$$

$$= \left| \frac{x'(t)y''(t)-y'(t)x''(t)}{(x'(t)^2+y'(t)^2)^2} \right| \|(-y'(t),x'(t))\|$$

$$= \left| \frac{x'(t)y''(t)-y'(t)x''(t)}{(x'(t)^2+y'(t)^2)^{3/2}} \right|. \qquad \blacksquare$$

For example, consider the ellipse, $\alpha(t) = (a\cos t, b\sin t)$. By Proposition 5.7 we have:

$$\kappa_{\pm}(\alpha;t) = \left| \frac{ab}{(a^2\sin^2 t + b^2\cos^2 t)^{3/2}} \right| = \left| \frac{ab}{\sqrt{(a^2+(b^2-a^2)\cos^2 t)^3}} \right|.$$

When $b > a$, notice that the curvature achieves a maximum at $t = \pm\pi/2$ and a minimum at $t = 0$ or π.

The tractrix

A curve that will be important in later discussions was discovered in the seventeenth century. The Paris physician CLAUDIUS PERRAULT (1613–88) set the problem of describing the curve followed by a weight (a reluctant dog?) being dragged on the end of a fixed straight length, the other end of which moves along a fixed straight line. Leibniz and Huygens understood that such a curve had the property that at every point the tangent line met the fixed line a constant distance away. Huygens named such curves *Traktorien* (in a 1693 letter to Leibniz). The name that has stuck is the **tractrix** (Loria1902).

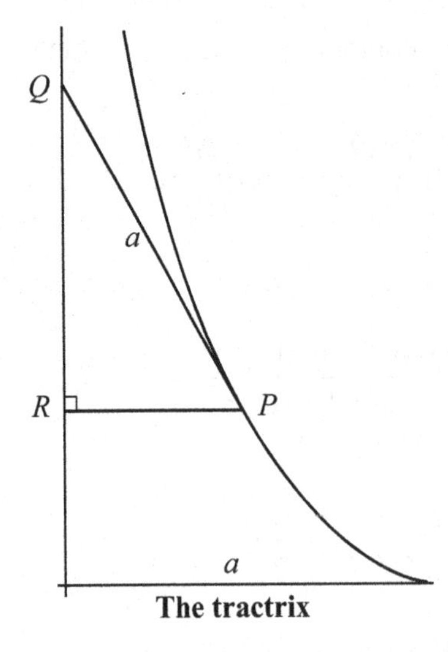

The tractrix

Let us suppose the y-axis to be the fixed line, the fixed length to be a, and the curve to begin at $(a,0)$ on the x-axis. By the analytic characterization, the tractrix $\hat{\Theta}(t) = (x(t), y(t))$ satisfies the equation:

$$\frac{dy}{dx} = \frac{y'}{x'} = \frac{QR}{RP} = \frac{-\sqrt{a^2-x^2}}{x}.$$

Squaring both sides of the equation and manipulating the terms we find:

$$(x')^2 + (y')^2 = \left(\frac{a}{x}\right)^2 (x')^2.$$

Taking the derivative of both sides of the first equation yields $\dfrac{x'y''-y'x''}{(x')^2} = \dfrac{a^2x'}{x^2\sqrt{a^2-x^2}}$, that is,

$$-\frac{x''y'-y''x'}{((x')^2+(y')^2)^{3/2}}=\frac{a^2(x')^3}{x^2\sqrt{a^2-x^2}}\frac{1}{(ax'/x)^3}.$$

This equation gives the plane curvature as the function of $x(t)$ and $x'(t)$:

$$\kappa_\pm(\hat{\Theta};t)=\left|\frac{a^2(x')^3}{x^2\sqrt{a^2-x^2}}\frac{x^3}{a^3(x')^3}\right|=\left|\frac{x(t)}{a\sqrt{a^2-(x(t))^2}}\right|.$$

To obtain a parametrization of the tractrix as a graph of a function we can integrate $\frac{dy}{dx}$ with respect to x:

$$y(x)=\int-\frac{\sqrt{a^2-x^2}}{x}\,dx.$$

This integral suggests a trigonometric substitution that leads to another parametrization. Let $x=a\cos\theta$ and we get:

$$\int-\frac{\sqrt{a^2-x^2}}{x}\,dx=\int\frac{a\sin\theta}{-a\cos\theta}(-a\sin\theta)d\theta=a\int\frac{\sin^2\theta}{\cos\theta}d\theta$$
$$=a\int\frac{1-\cos^2\theta}{\cos\theta}d\theta=a\int\sec\theta\,d\theta-a\int\cos\theta\,d\theta.$$

And so $\hat{\Theta}(\theta)=(a\cos\theta,a\ln|\sec\theta+\tan\theta|-a\sin\theta)$. Substituting $x=a\cos\theta$ into our formula for plane curvature gives $\kappa_\pm(\hat{\Theta};\theta)=\left|\frac{\cot\theta}{a}\right|$.

We can obtain an arc length parametrization of the tractrix by setting $(x')^2+(y')^2=1$. Then $\left(\frac{ax'}{x}\right)^2=1$, which implies that $x'=\pm(1/a)x$. Since $x(s)$ begins at a and moves toward zero, we take the minus sign giving $x(s)=ae^{-s/a}$, and:

$$\frac{dy}{ds}=-\left[\frac{\sqrt{a^2-x^2}}{x}\right]\frac{dx}{ds}=\sqrt{1-e^{-2s/a}}.$$

A routine integration yields $y(s)$ and the arc length parametrization of the tractrix:

$$\hat{\Theta}(s)=\left(e^{-s/a},a\ln(e^{s/a}+\sqrt{e^{2s/a}-1})-a\frac{\sqrt{e^{2s/a}-1}}{e^{s/a}}\right),\quad\text{for}\quad s\geq 0.$$

In this parametrization the plane curvature can be computed directly:

$$\kappa_\pm(\hat{\Theta};s)=\|\hat{\Theta}''(s)\|=\left\|\frac{d}{ds}(x'(s),y'(s))\right\|=\left\|\frac{d}{ds}(-e^{-s/a},\sqrt{1-e^{-2s/a}})\right\|$$
$$=\left\|\left(\frac{e^{-s/a}}{a},\frac{e^{-2s/a}}{a\sqrt{1-e^{-2s/a}}}\right)\right\|=\frac{e^{-s/a}}{a\sqrt{1-e^{-2s/a}}}.$$

The tractrix plays a key role generating a special surface in Chapter 12.

Oriented curvature

Our definition of plane curvature is the reciprocal of a length. If we examine a curve like the graph of the sine function we see that $\kappa_{\pm}(\alpha;t)$ can have equal values at two points but present very different pictures.

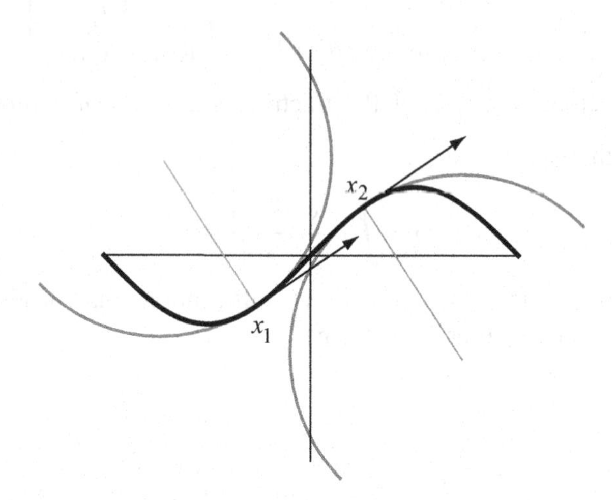

To distinguish these cases we add a sign that encodes whether "the osculating circle lies on the left of the tangent direction" or "on the right." A desirable feature would be to distinguish a curve $\alpha(t)$ from its **reverse**, $\hat{\alpha}(t) = \alpha(a_0 - t)$. We also want to distinguish convexity on either side of a point where the curvature vanishes, for example, around a point of inflection. To define our sign we apply the *right hand rule*.

Definition 5.8. *An ordered pair of nonzero vectors* $[\mathbf{u}, \mathbf{v}]$ *in* $\mathbb{R}^2$ *is in* **standard orientation** *if the matrix* $(\mathbf{u}^t\ \mathbf{v}^t)$ *has positive determinant.*

For example, if $\mathbf{u} = (1,1)$ and $\mathbf{v} = (-1,3)$, then $A = (\mathbf{u}^t\ \mathbf{v}^t) = \begin{pmatrix} 1 & -1 \\ 1 & 3 \end{pmatrix}$ has positive determinant. Thus $[(1,1),(-1,3)]$ is in standard orientation.

Notice that if $[\mathbf{u}, \mathbf{v}]$ is in standard orientation, then $[\mathbf{v}, \mathbf{u}]$ is not; switching columns of a matrix changes the sign of the determinant. Also, if $[\mathbf{u}, \mathbf{v}]$ is in standard orientation, the set $\{\mathbf{u}, \mathbf{v}\}$ is linearly independent.

Given any unit vector $\mathbf{u}_0 = (u_1, u_2) \in \mathbb{R}^2$ there is a unique unit vector $\mathbf{v}_0 = (-u_2, u_1)$ such that $[\mathbf{u}_0, \mathbf{v}_0]$ is in standard orientation, $\mathbf{u}_0$ perpendicular to $\mathbf{v}_0$, that is, $\mathbf{u}_0 \cdot \mathbf{v}_0 = 0$. We can write $\mathbf{u}_0 = (\cos\theta, \sin\theta)$ and so $\mathbf{v}_0 = (-\sin\theta, \cos\theta) = \left(\cos(\theta + \frac{\pi}{2}), \sin(\theta + \frac{\pi}{2})\right)$; that is, $\mathbf{v}_0$ is obtained from $\mathbf{u}_0$ by rotation through $\pi/2$ (counterclockwise, which is the right-hand rule). If we represented points in the plane as complex numbers, $\mathbf{u}_0 = u_1 + iu_2$, then $\mathbf{v}_0 = i\mathbf{u}_0$. See Zwikker (1950) for a thorough development of the charms of this framework. We write $\mathbf{v}_0 = J\mathbf{u}_0$, following the convention in the study of (almost) complex structures (Hopf 1948).

If $\alpha(t)$ is a regular curve, then denote the (unit length) tangent vector $\alpha'(t)/\|\alpha'(t)\|$ by $T(t)$. Let $N(t) = JT(t)$ denote the unique unit vector perpendicular to $T(t)$ with standard orientation $[T(t), N(t)]$. Alternatively we can write $T(t) = (\cos\theta(t), \sin\theta(t))$ for a choice of the *turning angle* $\theta(t)$ that the direction $T(t)$ makes with the x-axis. Note that:

$$T'(t) = \theta'(t)(-\sin\theta(t), \cos\theta(t)) = \theta'(t)J(\cos\theta(t), \sin\theta(t)) = \theta'(t)N(t).$$

Definition 5.9. *The* **oriented curvature** $\kappa_\alpha(t_0)$ *of a curve* $\alpha(t)$ *at* t_0 *is given by the identity*

$$T'(t_0) = (\kappa_\alpha(t_0) \cdot \|\alpha'(t_0)\|)N(t_0).$$

If the turning angle of the curve $\theta(t)$ *is known, then* $\kappa_\alpha(t) = \dfrac{\theta'(t)}{\|\alpha'(t)\|}$.

In the case that a curve is parametrized by arc length, $T'(s) = \alpha''(s)$ and

$$\kappa_\pm(\alpha; s) = \|T'(s)\| = \|\alpha''(s)\| = |\kappa_\alpha(s)| \cdot \|\alpha'(s)\| = |\kappa_\alpha(s)|.$$

Thus, the plane curvature and the oriented curvature differ by at most a sign for unit speed curves. More generally, we have the formula:

Theorem 5.10. *If* $\alpha(t) = (x(t), y(t))$ *is a regular curve in* $\mathbb{R}^2$, *then*

$$\kappa_\alpha(t) = \frac{x'y'' - x''y'}{((x')^2 + (y')^2)^{3/2}} = \frac{\det\begin{bmatrix} x' & x'' \\ y' & y'' \end{bmatrix}}{((x')^2 + (y')^2)^{3/2}}.$$

PROOF: We can write $T(t) = \alpha'(t)/\|\alpha'(t)\| = (\cos\theta(t), \sin\theta(t))$ and so:

$$T'(t) = \theta'(t)(-\sin\theta(t), \cos\theta(t)) = \theta'(t)N(t).$$

Thus, $\kappa_\alpha(t) = \theta'(t)/\|\alpha'(t)\|$.

The tangent of the turning angle satisfies $\tan\theta(t) = y'(t)/x'(t)$. Taking derivatives, we find:

$$\theta'(t)\sec^2\theta(t) = \frac{x'y'' - x''y'}{(x')^2} = \frac{x'y'' - x''y'}{\cos^2\theta(t)((x')^2 + (y')^2)}.$$

Since $(x', y') = \sqrt{(x')^2 + (y')^2}(\cos\theta(t), \sin\theta(t))$, it follows that $\dfrac{\theta'(t)}{\|\alpha'(t)\|} = \dfrac{x'y'' - x''y'}{((x')^2 + (y')^2)^{3/2}} = \kappa_\alpha(t)$. ∎

Proposition 5.7 implies that $|\kappa_\alpha(t)| = \kappa_\pm(\alpha; t)$ for any regular curve. We can interpret the sign by considering the curve given by the graph of a function, $\alpha(t) = (t, f(t))$. Here, $\kappa_\alpha(t) = \dfrac{f''(t)}{(1 + (f'(t))^2)^{3/2}}$. The sign of the curvature is determined by the second derivative $f''(t)$, which is positive if $f(t)$ is concave up, negative if $f(t)$

is concave down. Since any curve in $\mathbb{R}^2$ is locally the graph of a function, we see that the oriented curvature at a point is positive if the curve turns to the left of the tangent vector $T(t)$, negative if to the right.

In a treatise on the tractrix, Euler (1736) suggested that a sometimes useful expression for a curve was in terms of its natural parameters—its arc length s and its oriented curvature $\kappa_\alpha(s)$ at each point. For the tractrix, Euler showed that the relation between s and $\kappa_{\hat{\Theta}}(s)$ is algebraic, while the relation between rectangular coordinates is transcendental. In general, these parameters determine the rectangular coordinates of the curve $\alpha(s) = (x(s), y(s))$, and so the oriented curvature of a curve fixes $\alpha(s)$ up to congruence of the plane.

Theorem 5.11 (Fundamental Theorem for Plane Curves). *Given any continuous function $\kappa\colon (a,b) \to \mathbb{R}$, there is a curve $\alpha\colon (a,b) \to \mathbb{R}^2$, which is parametrized by arc length, such that $\kappa(s) = \kappa_\alpha(s)$ is the oriented curvature of $\alpha(s)$ for all $s \in (a,b)$. Furthermore, any other unit speed curve $\bar{\alpha}\colon (a,b) \to \mathbb{R}^2$ satisfying $\kappa_{\bar{\alpha}}(s) = \kappa(s)$ differs from $\alpha(s)$ by a rotation followed by a translation.*

PROOF: Under the assumption that $\alpha(s)$ is parametrized by arc length, $\|\alpha'(s)\| = 1$ and the formula for oriented curvature give $\kappa_\alpha(s) = \theta'(s)$, where $\theta(s)$, is the turning angle. We obtain $\theta(s)$ by integration:

$$\theta(s) = \int \kappa(s)\,ds,$$

and $\theta(s)$ is determined up to addition of a constant. The curve $\alpha(s) = (x(s), y(s))$ has a derivative $\alpha'(s) = (x'(s), y'(s)) = (\cos\theta(s), \sin\theta(s))$ and so:

$$x(s) = \int \cos\theta(s)\,ds, \qquad y(s) = \int \sin\theta(s)\,ds.$$

Both $x(s)$ and $y(s)$ are each determined up to the addition of a constant.

When $\bar{\alpha}(s) = (\bar{x}(s), \bar{y}(s))$ is another solution for which $\kappa_{\bar{\alpha}}(s) = \kappa(s)$, we have $\bar{\alpha}'(s) = (\cos\bar{\theta}(s), \sin\bar{\theta}(s))$ and $\kappa(s) = \bar{\theta}'(s) = \theta'(s)$. Therefore, $\bar{\theta}(s) = \theta(s) + \beta$, from which it follows that:

$$\begin{aligned}\bar{x}(s) &= \int \cos\bar{\theta}(s)\,ds = \int \cos(\theta(s)+\beta)\,ds = \int (\cos\beta\cos\theta(s) - \sin\beta\sin\theta(s))\,ds \\ &= (\cos\beta)x(s) - (\sin\beta)y(s).\end{aligned}$$

Similarly, $\bar{y}(s) = (\sin\beta)x(s) + (\cos\beta)y(s)$, each determined up to addition of a constant. We can express this relation in matrix form:

$$\begin{pmatrix}\bar{x}(s)\\ \bar{y}(s)\end{pmatrix} = \begin{pmatrix}\cos\beta & -\sin\beta\\ \sin\beta & \cos\beta\end{pmatrix}\begin{pmatrix}x(s)\\ y(s)\end{pmatrix} + \begin{pmatrix}a_0\\ b_0\end{pmatrix}.$$

Thus $\bar{\alpha}(s)$ differs from $\alpha(s)$ by a rotation, through angle β, followed by a translation, by the vector $\begin{pmatrix}a_0\\ b_0\end{pmatrix}$. ■

EXAMPLE: Suppose that a regular unit speed curve in $\mathbb{R}^2$ has constant positive curvature, $r^2 > 0$. By the unit speed assumption we have $T(s) \cdot T'(s) = 0$ and $T'(s) = r^2 N(s)$. Writing $T(s) = (x(s), y(s))$, we obtain the differential equations:

$$xx' + yy' = 0, \text{ and } (x', y') = r^2(-y, x).$$

This implies that $x'' = -r^4 x$, which has $x(s) = a\cos(r^2 s + t_0) + b\sin(r^2 s + t_0)$ as solution. Since we only need to find one solution (and move it about by rotations and translations), we can take $x(s) = (1/r^2)\cos(r^2 s)$. Then $y(s) = (1/r^2)\sin(r^2 s)$ by the condition $x'^2 + y'^2 = 1$. This is a parametrization of the circle of radius $1/r^2$ and so the condition of constant positive plane curvature forces the curve to be part of the circle.

The appearance of the congruences of the plane, rotations followed by translation in the oriented case, as a consequence of constants of integration, makes the idea that analytic methods support a consistent picture of geometry attractive. This idea played a key role in the thinking of Bolyai, Lobachevskiĭ, and Gauss.

Involutes and evolutes

Developments in algebra and physics during the seventeenth century brought new energy to the study of curves and how they are constructed (Bos 2001). Some of these developments were motivated by the study of gravity and by the desire to build more accurate clocks (Yoder). Practical designs were proposed that were based on the motion of a pendulum, requiring the careful study of motion due to gravity first carried out by Galileo, Descartes, and Mersenne. These studies led to the work of CHRISTIAAN HUYGENS (1629–95) and his 1673 treatise *Horologium oscillatorium sive de motu pendulorum ad horologia aptato demonstrationes geometricae*. Some of the ideas introduced in Huygens's classic work, such as the involute and evolute of a curve, are part of our current geometric language. For a thorough historical study of Huygens's work on curves, see Yoder (1988).

It was Galileo who proposed the design of a clock using a pendulum as regulator (Yoder 1988) but his plans were never realized in his lifetime. Huygens pursued an analysis of accelerated motion that led him to his design and the construction of a pendulum clock.

The first practical issue to overcome is the determination of the curve along which the pendulum would travel. On a rigid pendulum, the time to reach the bottom of the swing depends on the initial angle made with the vertical. A better choice would be to constrain the ball of the pendulum to travel along a curve for which the time to reach the bottom of the swing is independent of the initial point of departure. Such a curve is called a **tautochrone** (or it is said to be *isochronal*), and a clock based on this motion would deliver a reliable period, making it the choice of an accurate and seaworthy clock.

Without loss of motivating details, we introduce the following classical curve, named and studied by Galileo.

Definition 5.12. *A* **cycloid** *is the curve swept out by a fixed point on a circular disk as it rolls without slipping along a straight line. If the point is on the circumference, the curve is called a right cycloid; if the point is inside the disk, the curve is a curtate cycloid, and if the point is outside the disk, a prolate cycloid.*

For Galileo, the family of analogous curves called *epitrochoids*, for which the circular disk rolls without slipping on a circle, provided models for astronomical motion. For us the important case has a circular wheel rolling "under" the line, and the point chosen on the circle. Referring to the figure, take the radius of the circle to be r and $-\pi \leq \theta \leq \pi$. Let P be the point on the circle that began at the point $O = (0,0)$ after rolling through θ radians. Consider the radius LC' perpendicular to the x-axis and the associated diameter LK. Let U denote the point $(0,2r)$. Since the circle rolls without slipping, UL has length $r\theta$. The arc $\overset{\frown}{PK}$ also has length $r\theta$, and so $\angle PC'K$ has measure θ. By projecting onto the diameter LK from the point P to the point N, we obtain the equations for $P = (x(\theta), y(\theta))$,

$$x(\theta) = r(\theta + \sin\theta), \quad y(\theta) = r(1 - \cos\theta).$$

We denote this particular parametrization of the cycloid by $\zeta(\theta) = (x(\theta), y(\theta))$. Taking the derivatives we find:

$$\zeta'(\theta) = r(1 + \cos\theta, \sin\theta), \text{ and } \|\zeta'(\theta)\| = r\sqrt{2 + 2\cos\theta}.$$

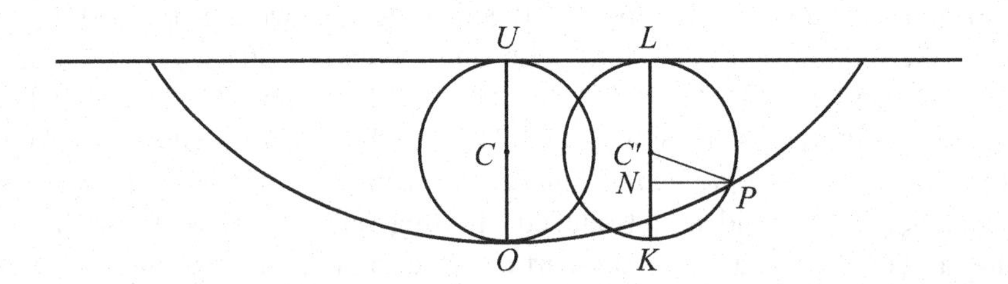

Note that $\zeta(\theta)$ is not a unit speed curve. Our interest in the cycloid is based on the following remarkable property.

Theorem 5.13. *The cycloid is a tautochrone.*

PROOF SKETCH: We will not justify the physics in what follows as it would take us too far afield; for more details see Dugas (1988) or Gindikin (1988). In this sketch we will not try to reproduce Huygens's proof; see Yoder (1988) for a discussion of his proof.

Suppose we restrict a particle to move without friction on the cycloid. Choose a point $P_0 = (x(\theta_0), y(\theta_0)) = (x_0, y_0)$ and the particle moves from P_0 to $O = (0,0)$ under the influence of gravity. Since it starts from rest at P_0, the law of conservation of energy requires kinetic energy to be equal to potential energy, which implies:

$$\frac{mv^2}{2} = mg(y_0 - y),$$

where g is the gravitational constant. This relation implies the velocity $v = \sqrt{2g(y_0 - y)}$. For the curve $(x(\theta(t)), y(\theta(t)))$ we have:

$$v = \sqrt{(x')^2 + (y')^2}\frac{d\theta}{dt} = \sqrt{2rx'}\frac{d\theta}{dt} = \sqrt{2g(y_0 - y)}.$$

Rearranging we get the differential relation $dt = \dfrac{\sqrt{rx'}}{\sqrt{g(y_0 - y)}}\,d\theta$.

The time spent along the path is given by

$$T_0 = \int_{\theta_0}^{0} dt = \int_{\theta_0}^{0}\sqrt{\frac{rx'}{g(y_0 - y)}}\,d\theta = \int_{\theta_0}^{0}\sqrt{\frac{r}{g}}\sqrt{\frac{1+\cos\theta}{\cos\theta - \cos\theta_0}}\,d\theta.$$

The trigonometric identities $\cos\theta = 2\cos^2(\theta/2) - 1 = 1 - 2\sin^2(\theta/2)$ obtain:

$$\int_{\theta_0}^{0}\sqrt{\frac{1+\cos\theta}{\cos\theta - \cos\theta_0}}\,d\theta = \int_{\theta_0}^{0}\sqrt{\frac{2\cos^2(\theta/2)}{2\sin^2(\theta_0/2) - 2\sin^2(\theta/2)}}\,d\theta$$

$$= \int_{\theta_0}^{0}\frac{\cos(\theta/2)}{\sqrt{\sin^2(\theta_0/2) - \sin^2(\theta/2)}}\,d\theta.$$

Make the substitution of $w = \dfrac{\sin(\theta/2)}{\sin(\theta_0/2)}$ giving $dw = \dfrac{\cos(\theta/2)}{\sin(\theta_0/2)}\dfrac{d\theta}{2}$ and we transform the integral to:

$$\int_{\theta_0}^{0}\sqrt{\frac{1+\cos\theta}{\cos\theta - \cos\theta_0}}\,d\theta = \int_{1}^{0}\frac{2\,dw}{\sqrt{1-w^2}}.$$

Since this value is independent of the choice of θ_0, the theorem follows. ∎

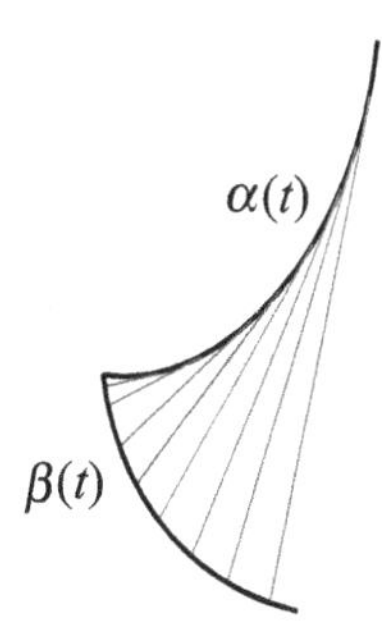

Having determined the curve that we would like the *ideal* pendulum to follow, we face the problem of forcing a pendulum to follow the prescribed curve. Fix a curve $\alpha\colon (a,b) \to \mathbb{R}^2$ and chose a point $\alpha(t_0)$ on it. Consider the associated curve $\beta\colon (a,b) \to \mathbb{R}^2$ given by unwinding a taut string starting at $\alpha(t_0)$ of length given by the arc length along $\alpha(t)$ stretched in the direction tangent to $\alpha(t)$ at each point.

From the description, the curve $\beta(t)$ satisfies the equation:

$$\beta(t) = \alpha(t) \pm \left(\int_{t_0}^{t}\|\alpha'(u)\|\,du\right)\frac{\alpha'(t)}{\|\alpha'(t)\|}.$$

Furthermore, the condition that the string must be taut implies that $\beta'(t)$ is perpendicular to $\alpha'(t)$ for all values of t. A routine calculation shows that at $\pm$ we need a minus sign to achieve $\beta'(t)\cdot\alpha'(t) = 0$.

Definition 5.14. *Fix a curve* $\alpha\colon (a,b) \to \mathbb{R}^2$. *A curve* $\beta\colon (a,b) \to \mathbb{R}^2$ *is called an* **involute** *of* $\alpha(t)$ *if* $\beta(t)$ *lies along the tangent line to* $\alpha(t)$ *and* $\beta'(t)\cdot\alpha'(t) = 0$. *When* $\beta(t)$ *is an involute of* $\alpha(t)$, *we call* $\alpha(t)$ *an* **evolute** *of* $\beta(t)$.

A pendulum with a flexible string constrained to unwind along a fixed curve $\alpha(t)$ follows a path given by an involute of $\alpha(t)$. The practical problem of building the *ideal* pendulum clock requires that we construct the evolute of a cycloid. A beautiful property of the cycloid leads to a remarkable solution. This property was known to Descartes (Loria, 1902, p. 464) and perhaps earlier to Fermat. A nice alternative proof may be found in (Lawlor 1996).

Lemma 5.15. *The line tangent to a cycloid at a point P passes through the lowest point on the generating circle that determines P. Similarly, the normal passes through the highest point of the circle.*

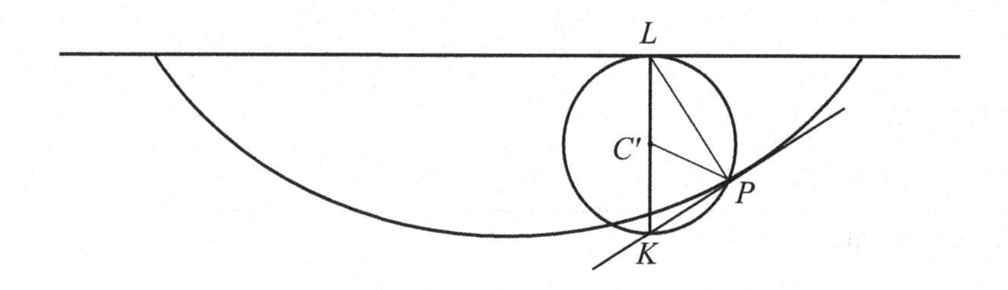

PROOF: Let $P = (x(\theta), y(\theta)) = (r\theta + r\sin\theta, r - r\cos\theta)$ and $K = (r\theta, 0)$. The line PK has slope given by:

$$m_{PK} = \frac{r - r\cos\theta}{r\sin\theta} = \frac{1-\cos\theta}{\sin\theta} = \frac{1-\cos^2\theta}{(\sin\theta)(1+\cos\theta)} = \frac{\sin\theta}{1+\cos\theta} = \frac{y'}{x'} = \frac{dy}{dx}(\theta).$$

Thus, PK is the line with slope dy/dx at θ and through P, hence the tangent line. The line PL is perpendicular to PK since LK is a diameter of a circle. ∎

We can apply this fact to obtain an evolute of the cycloid by using two circles to generate a pair of cycloids. The consequences of this construction are explored in (Brooks, Push 2002) for Huygens's clock.

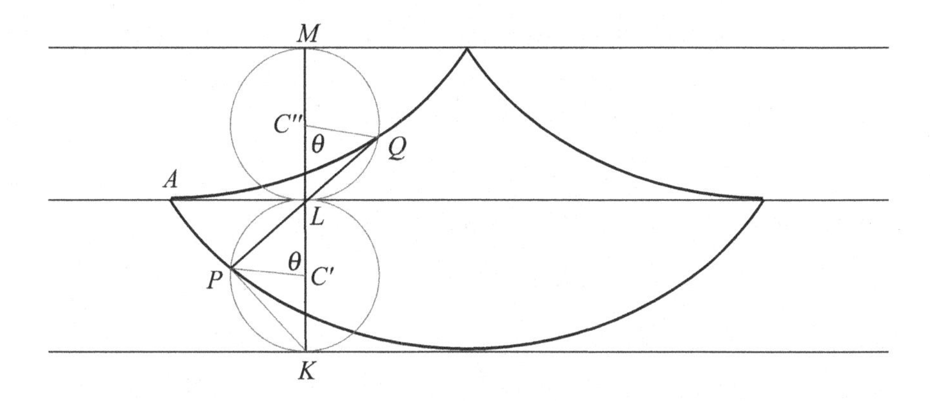

Theorem 5.16. *The cycloid has a congruent cycloid as its evolute.*

PROOF: The two circles that generate the cycloids both begin at $A = (0,0)$, one rolling above the x-axis and one below. After rolling both circles—the one below clockwise and the one above counterclockwise the associated points P and Q on each circle may be joined by a line PQ and, because the central angles turned are the same, PQ passes through the point L of intersection of the two circles. By Lemma 5.15, QL is tangent to the upper cycloid and PL is normal to the lower one. We only need to show how PQ depends upon θ to see that PQ has length given by the arc length along the cycloid.

The arc length along the upper cycloid from A to Q is given by:

$$\text{arc length}(AQ) = \int_0^\theta ds = \int_0^\theta r\sqrt{2+2\cos\theta}\,d\theta = \int_0^\theta 2r\cos(\theta/2)\,d\theta = 4r\sin(\theta/2).$$

The angle of rotation for each circle is $\theta = \angle PC'L$ and so, by Euclid's Proposition III.20—in a circle the angle at the center is double of the angle at the circumference – we have $\theta/2 = \angle PKL$. Since ΔKPL is a right triangle, $\sin\angle PKL = PL/KL = PL/2r$. It follows that arc length$(AQ) = 4r\sin(\theta/2) = 4r(PL/2r) = 2PL$. Since $\angle QC''L \cong \angle PC'L$, $PL \cong QL$, $PQ =$ arc length(AQ). The congruent cycloid above the x-axis is the evolute of the cycloid below. ∎

Huygens applied his research and had pendulum clocks built according to the theorem, using a pair of plates curved to follow a cycloid between which the pendulum would swing. Thus, the bob followed the path of a tautochrone. His proof of Theorem 5.16 used synthetic and infinitesimal ideas (Yoder 1988): he did *not* have the calculus at his disposal. The example of a physical problem leading to developments in differential geometry is repeated throughout the history of the subject.

The notions of involutes and evolutes apply more generally.

Proposition 5.17. *$\beta(s)$ is an involute of a unit speed curve $\alpha\colon (a,b) \to \mathbb{R}^2$ if and only if, for some constant c, $\beta(s) = \alpha(s) + (c-s)\alpha'(s)$.*

PROOF: By the definition, if $\alpha(s)$ is unit speed, the involute takes the form:

$$\beta(s) = \alpha(s) - \left(\int_{s_0}^{s} \|\alpha'(u)\| \, du \right) \alpha'(s) = \alpha(s) + (c-s)\alpha'(s).$$

Suppose $\beta(s) = \alpha(s) + (c-s)\alpha'(s)$. Then $\beta(s)$ lies along the tangent line to $\alpha(s)$ at each point. Computing the tangent to $\beta(s)$ we obtain:

$$\beta'(s) = \alpha'(s) - \alpha'(s) + (c-s)\alpha''(s) = (c-s)\alpha''(s).$$

Since $\alpha(s)$ is unit speed, $\alpha''(s)$ is perpendicular to $\alpha'(s)$, and so $\beta'(s)$ is perpendicular to $\alpha'(s)$. Since $c - s = -\int_{s_0}^{s} du = \int_{s_0}^{s} \|\alpha'(u)\| \, du$, we have $\beta(s)$ is an involute of $\alpha(s)$. ■

For a unit speed curve $\alpha(s)$ with $\kappa_\alpha(s) \neq 0$ for all s, there is a natural evolute associated to $\alpha(s)$. Recall that to any curve $\beta(t)$ we can associate another curve – the locus of centers of curvature, $C_\beta(t)$.

Proposition 5.18. *The locus of centers of curvature $C_\alpha(s)$ associated to a unit speed curve $\alpha(s)$ is an evolute of $\alpha(s)$.*

PROOF: Recall some of the properties of $C_\alpha(s)$:

(1) $\kappa_\alpha(s) = \dfrac{1}{\|\alpha(s) - C_\alpha(s)\|}$,

(2) $(\alpha(s) - C_\alpha(s)) \cdot \alpha''(s) = -1$.

Without an explicit parametrization, we cannot reparametrize $C_\alpha(s)$ to be a unit speed curve. However, an involute is characterized by the properties that a point of the curve lies on the tangent of the evolute and, at associated points, the tangents are orthogonal. To prove the proposition then, it suffices to show that $C'_\alpha(s)$ is perpendicular to $\alpha'(s)$ and that $\alpha(s) = C_\alpha(s) - \mu(s)C'_\alpha(s)$.

Since $\alpha(s)$ is a unit speed curve, we associate to each point of the curve the orthonormal pair $[T(s), N(s)]$, where $T(s) = \alpha'(s)$ and $[T(s), N(s)]$ are in standard orientation. By Euler's Formula (Theorem 5.7), we know that $T'(s) = \kappa_\alpha(s)N(s)$. We take this relation one derivative further to prove the following relation.

Lemma 5.19. $N'(s) = -\kappa_\alpha(s)T(s)$.

PROOF: Since $N(s)$ has unit length, $N(s)$ is perpendicular to $N'(s)$. Thus $N'(s) = \nu(s)T(s)$. Since $N(s) \cdot T(s) = 0$, we can differentiate this relation to obtain:

$$\begin{aligned} 0 &= \frac{d}{ds}(N(s) \cdot T(s)) = N'(s) \cdot T(s) + N(s) \cdot T'(s) \\ &= \nu(s)(T(s) \cdot T(s)) + \kappa_\alpha(s)(N(s) \cdot N(s)) = \nu(s) + \kappa_\alpha(s). \end{aligned}$$

The lemma follows. ■

Since $(\alpha(s) - C_\alpha(s))$ is perpendicular to $\alpha'(s)$ we have that $\alpha(s) - C_\alpha(s) = \mu(s)N(s)$. Since $(\alpha(s) - C_\alpha(s)) \cdot \alpha''(s) = -1$, it follows that:

$$-1 = (\alpha(s) - C_\alpha(s)) \cdot \alpha''(s) = \mu(s)N(s) \cdot \kappa_\alpha(s)N(s),$$

and so $\mu(s) = -1/\kappa_\alpha(s)$. This calculation also implies that:

$$C_\alpha(s) = \alpha(s) + \frac{1}{\kappa_\alpha(s)}N(s).$$

Taking the derivative we get:

$$C'_\alpha(s) = \alpha'(s) - \frac{\kappa'_\alpha(s)}{(\kappa_\alpha(s))^2}N(s) - \frac{1}{\kappa_\alpha(s)}\kappa_\alpha(s)\alpha'(s) = -\frac{\kappa'_\alpha(s)}{(\kappa_\alpha(s))^2}N(s).$$

Thus, $C'_\alpha(s)$ is perpendicular to $\alpha'(s)$ and:

$$\alpha(s) = C_\alpha(s) - \frac{(\kappa_\alpha(s))^2}{\kappa'_\alpha(s)}C'_\alpha(s).$$

To complete the proof we need to show that the distance between $C_\alpha(s)$ and $\alpha(s)$ is arc length along $C_\alpha(s)$ from some point. To see this we compute:

$$\int_{s_0}^{s} \|C'_\alpha(u)\|\,du = \int_{s_0}^{s} \frac{\kappa'_\alpha(u)}{(\kappa_\alpha(u))^2}\,du = \frac{1}{\kappa_\alpha(s_0)} - \frac{1}{\kappa_\alpha(s)} = c - \|\alpha(s) - C_\alpha(s)\|.$$

So $\|\alpha(s) - C_\alpha(s)\| = c-$ arc length along $C_\alpha(s)$, and $C_\alpha(s)$ is an evolute of $\alpha(s)$. ■

EXAMPLE: Let's work out the evolute of the tractrix:

$$\hat{\Theta}(s) = \left(ae^{-s/a}, a\ln\left(e^{s/a} + \sqrt{e^{2s/a} - 1}\right) - a\frac{\sqrt{e^{2s/a} - 1}}{e^{s/a}}\right).$$

Since this parametrization is unit speed, we have:

$$T(s) = \hat{\Theta}'(s) = (-e^{-s/a}, \sqrt{1 - e^{-2s/a}}), \qquad N(s) = (-\sqrt{1 - e^{-2s/a}}, -e^{-s/a}).$$

We can compute the oriented curvature directly from $T'(s) = \kappa_{\hat{\Theta}}(s)N(s)$:

$$T'(s) = \left(\frac{e^{-s/a}}{a}, \frac{e^{-2s/a}}{a\sqrt{1 - e^{-2s/a}}}\right) = -\left[\frac{e^{-s/a}}{a\sqrt{1 - e^{-2s/a}}}\right]N(s).$$

The locus of centers of curvature is given by:

$$\begin{aligned}
C_{\hat{\Theta}}(s) &= \hat{\Theta}(s) + \frac{1}{\kappa_{\hat{\Theta}}(s)}N(s) = \hat{\Theta}(s) - \frac{a\sqrt{1 - e^{-2s/a}}}{e^{-s/a}}(-\sqrt{1 - e^{-2s/a}}, -e^{-s/a}) \\
&= \hat{\Theta}(s) + \left(\frac{a(1 - e^{-2s/a})}{e^{-s/a}}, a\sqrt{1 - e^{-2s/a}}\right) \\
&= a\left(e^{s/a}, \ln\left(e^{s/a} + \sqrt{e^{2s/a} - 1}\right)\right).
\end{aligned}$$

To identify the curve $C_{\hat{\Theta}}(s)$, we introduce the inverse hyperbolic trigonometric function arccosh(x).

Lemma 5.20. $\operatorname{arccosh}(x) = \ln(x + \sqrt{x^2 - 1})$.

PROOF: Let $t = \operatorname{arccosh}(x)$; that is, $x = \cosh(t) = \dfrac{e^t + e^{-t}}{2}$. It follows that $2x = e^t + e^{-t}$, which implies $2xe^t = e^{2t} + 1$. Rearranging we can write $e^{2t} - 2xe^t + 1 = 0$. Solving for e^t we get:

$$e^t = \frac{2x \pm \sqrt{4x^2 - 4}}{2} = x \pm \sqrt{x^2 - 1}.$$

Since $x = e^{s/a}$ in our case, $x \geq 1$ and we get $\operatorname{arccosh}(x) = t = \ln(x + \sqrt{x^2 - 1})$. ∎

It follows that $C_{\hat{\Theta}}(s) = a(e^{s/a}, \operatorname{arccosh}(e^{s/a}))$, which can be written equivalently as $C_{\hat{\Theta}}(y) = a(\cosh(y), y)$, a parametrization of a *catenary* curve, the classical curve that describes the shape of a freely hanging chain, first identified and named by Huygens in a letter to Leibniz (Loria 1902).

Suppose $\alpha(t) = (x(t), y(t))$ is a regular curve, not necessarily unit speed. The locus of centers of curvature is given by:

$$\begin{aligned}
C_\alpha(t) &= \alpha(t) + \frac{1}{\kappa_\alpha(t)} J\left(\frac{\alpha'(t)}{\|\alpha'(t)\|}\right) = \alpha(t) + \frac{1}{\kappa_\alpha(t)} \frac{(-y'(t), x'(t))}{\sqrt{(x'(t))^2 + (y'(t))^2}} \\
&= (x(t), y(t)) + \frac{(x'(t))^2 + (y'(t))^2}{x'y'' - x''y'}(-y'(t), x'(t)).
\end{aligned}$$

Applied to the ellipse $\eta(t) = (a\cos t, b\sin t)$, we have $\eta'(t) = (-a\sin t, b\cos t)$ and $\eta''(t) = (-a\cos t, -b\sin t)$, and so $x'y'' - x''y' = ab$:

$$\begin{aligned}
C_\eta(t) &= (a\cos t, b\sin t) - \frac{a^2\sin^2 t + b^2\cos^2 t}{ab}(b\cos t, a\sin t) \\
&= \left(a\cos t - \frac{a^2 + (b^2 - a^2)\cos^2 t}{a}\cos t, b\sin t - \frac{(a^2 - b^2)\sin^2 t + b^2}{b}\sin t\right) \\
&= \left(\frac{a^2 - b^2}{a}\cos^3 t, \frac{b^2 - a^2}{b}\sin^3 t\right).
\end{aligned}$$

These coordinates satisfy an equation of the sort $(x/r)^{2/3} + (y/s)^{2/3} = 1$, which is the equation for an *astroid*, named by J. J. LITTROW (1781–1840) in 1838 (Loria, 1902, p. 224).

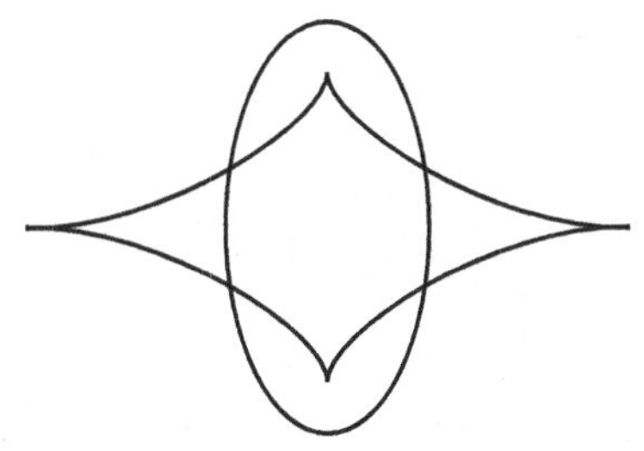

Exercises

5.1 If $\alpha(t)\colon \mathbb{R} \to \mathbb{R}^2$ is such that $\alpha''(t) = \mathbf{0}$, what can be said of $\alpha(t)$?

5.2 Show that $f(t) = \tan(\pi t/2), f\colon (-1,1) \to (-\infty, \infty)$, is a reparametrization. Is $g\colon (0,\infty) \to (0,1)$ given by $g(t) = t^2/(t^2+1)$ a reparametrization?

5.3 Show that the arc length function associated to a regular curve is one-to-one, onto its image, and differentiable.

5.4 Consider the curve in $\mathbb{R}^2$ given by the graph of the sine function $t \mapsto (t, \sin t)$. Determine the directed curvature at each point of this curve.

5.5† Suppose that a plane curve is given in polar coordinates $\rho = \rho(\theta), a \leq \theta \leq b$. Show that the arc length along the curve is given by $s(t) = \int_a^t \sqrt{\rho^2 + (\rho')^2}\, d\theta$, where $\rho' = \dfrac{d\rho}{d\theta}$. Also show that the directed curvature is given by the formula:

$$\kappa(\theta) = \frac{2(\rho')^2 - \rho\rho'' + \rho^2}{[(\rho')^2 + \rho^2]^{3/2}}.$$

5.6 Take any circle of radius R with center along the x-axis. Show that the circle meets the tractrix along this axis at right angles.

5.7 Compute the arc length for the cycloid $\zeta(t) = (t + \sin t, 1 - \cos t)$. One branch of the cycloid is determined by $t \in [-\pi, \pi]$. What is the length of this branch?

5.8† When a cable hangs freely only supporting its own weight, it follows a *catenary* curve. To derive an equation for the curve, notice that the weight is not uniformly distributed horizontally but it is uniformly distributed along its length. Thus, since $dy/dx = \tan\theta$ where θ is the angle the tangent to the curve makes with the horizontal, when we resolve the forces involved, we get:

$$\frac{dy}{dx} = cs(x),$$

where c is a constant and $s(x)$ is the arc length at x where the bottom of the cable passes through $(0,0)$. From this formulation, derive the equation for the catenary curve. (The curve was investigated by Huygens, Leibniz, Newton, Jacob, and Johann Bernoulli around 1690.)

Solutions to Exercises marked with a dagger appear in the appendix, pp. 329–331.

5.9 Determine the involutes of the circle of radius 1 and the parabola $y = x^2$ for $-1 \leq x \leq 1$. Is the involute of a parabola asymptotic to a horizontal line?

5.10 The other versions of the cycloid are the *curtate cycloid*, which is generated by a point in the interior of the disk as it rolls along a line, and the *prolate cycloid*, which is generated by a point exterior to the disk but moving with the disk as it rolls along a line. Determine the equations of these species of cycloid, together with their oriented curvature.

5.11 More versions of the cycloid are obtained by making the generating circle roll without slipping along another circle. Determine the equations of these species of cycloid, the *epicycloid*, if the circle rolls outside the circle, and *hypocycloid*, if inside. Compute their oriented curvatures.

5.12† Parametrize the parabola $y^2 = 2ax$ and determine all of the geometrical apparatus associated to it: its unit tangent and normal, its oriented curvature, and its involute from $(0,0)$ and its evolute.

6

Curves in space

The sections of curved lines which do not lie completely in the same plane and are customarily called lines of double curvature by geometers are found to have been defined long ago. But the analysis through which they were investigated depends on figures of such great complexity that they require not only the utmost attention but also the greatest caution, lest the representation of differential quantities, and likewise differentials of the second grade, confuse the imagination and lead it astray into errors.

L. EULER (1782)

A regular curve in space is a function $\alpha(t) = (x(t), y(t), z(t))$ for $a < t < b$ and $x'(t)$, $y'(t)$, and $z'(t)$ not simultaneously all zero. To generalize the successes for plane curves like the fundamental theorem (Theorem 5.11) we look for notions like the oriented curvature $\kappa_\alpha(t)$. This measure of bending of a curve is derivable from the derivative of the unit tangent vector to the curve. The pair $[T(t), N(t)]$ were easily found in the plane. In three dimensions there are many choices of a vector perpendicular to the tangent, and a curve can rise out of the plane spanned by the tangent and a choice of a normal. Furthermore, to span all directions from a point, three vectors are required. These complications are tamed by a special feature of $\mathbb{R}^3$.

Recall that the **cross product** of two vectors in $\mathbb{R}^3$, $\mathbf{u} = (u_1, u_2, u_3)$ and $\mathbf{v} = (v_1, v_2, v_3)$, is given by

$$\mathbf{u} \times \mathbf{v} = (u_2 v_3 - u_3 v_2, u_3 v_1 - u_1 v_3, u_1 v_2 - u_2 v_1).$$

The principal properties of the cross product are:

(1) For $\mathbf{u} \neq a\mathbf{v}$, $\mathbf{u} \times \mathbf{v}$ is perpendicular to $\mathbf{u}$ and to $\mathbf{v}$, that is, $\mathbf{u} \cdot (\mathbf{u} \times \mathbf{v}) = 0 = \mathbf{v} \cdot (\mathbf{u} \times \mathbf{v})$.
(2) $\mathbf{u} \times \mathbf{v} = -(\mathbf{v} \times \mathbf{u})$.
(3) $\mathbf{u} \times (a\mathbf{v} + \mathbf{w}) = a(\mathbf{u} \times \mathbf{v}) + (\mathbf{u} \times \mathbf{w})$.
(4) $\|\mathbf{u} \times \mathbf{v}\|^2 = \|\mathbf{u}\|^2 \|\mathbf{v}\|^2 - (\mathbf{u} \cdot \mathbf{v})^2$.
(5) If $\alpha, \beta \colon (a, b) \to \mathbb{R}^3$ are differentiable curves, then:

$$\frac{d}{dt}(\alpha \times \beta)(t) = \alpha'(t) \times \beta(t) + \alpha(t) \times \beta'(t);$$

that is, the Leibniz Rule holds for the derivative of the cross product of differentiable functions.

As in the plane, regular curves may be parametrized by arc length and taken to be unit speed. This condition is not simple to obtain, but it is convenient to assume. After some development, we derive the relevant formulas for curves that are not unit speed.

Definition 6.1. *Suppose $\alpha(s)$ is a unit speed curve in $\mathbb{R}^3$. Let $T(s) = \alpha'(s)$ denote the unit tangent vector. Define the* **curvature** *of $\alpha(s)$ to be $\kappa_\alpha(s) = \|\alpha''(s)\| = \|T'(s)\|$. When $\kappa_\alpha(s) \neq 0$, let $N(s) = T'(s)/\kappa_\alpha(s)$ denote the* **unit normal vector** *to $\alpha(s)$. Finally, define the* **unit binormal vector** *to $\alpha(s)$ by $B(s) = T(s) \times N(s)$.*

Since $T(s)$ has constant length, $T(s)$ is perpendicular to $T'(s)$, and so $N(s)$ is perpendicular to $T(s)$. This definition fixes the choice of unit normal; it also generalizes the definition of curvature for planar curves. Notice that $\kappa_\alpha(s) \geq 0$ for regular curves in $\mathbb{R}^3$.

From the properties of the cross product, the ordered triple $[T(s), N(s), B(s)]$ forms a **frame**; that is, an orthonormal basis for all directions in $\mathbb{R}^3$ through $\alpha(s)$.

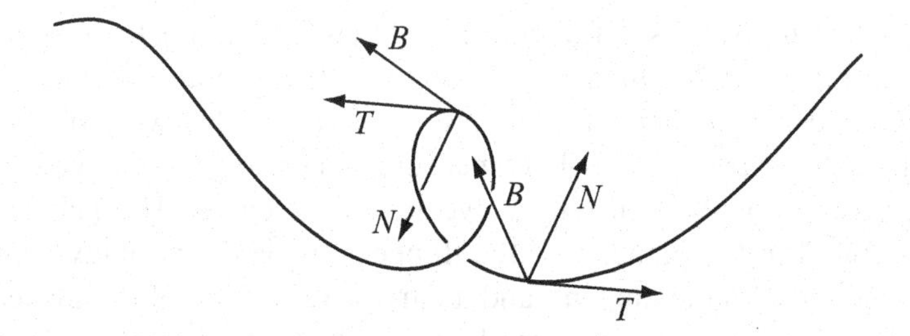

To see this, notice that by choice, $\|N(s)\| = 1$, and further:

$$\|B(s)\|^2 = \|T(s)\|^2 \|N(s)\|^2 - (T(s) \cdot N(s))^2 = 1,$$

so $\|B(s)\| = 1$. The matrix with columns $T(s)^t$, $N(s)^t$, and $B(s)^t$ has positive determinant. When this condition holds we say that the oriented basis $[T(s), N(s), B(s)]$ is **right-hand oriented**. The right-hand orientation is a property of the cross product; in fact, $[\mathbf{u}, \mathbf{v}, \mathbf{u} \times \mathbf{v}]$ always satisfies the right-hand rule as a basis of $\mathbb{R}^3$. We summarize these properties by saying that the triple $[T(s), N(s), B(s)]$ forms a right-hand oriented frame.

Since three noncollinear points on a curve in $\mathbb{R}^3$ determine a plane and a circle in that plane, the limiting process used to define the plane curvature determines an *osculating circle* lying in an *osculating plane* associated to the space curve. The osculating plane at $\alpha(s)$ is spanned by $T(s)$ and $N(s)$. Since the cross product is orthogonal to the tangent and normal vectors, changes in the binormal measure how much a curve fails to remain in its osculating plane.

Proposition 6.2. *Suppose $\alpha(s)$ is a unit speed curve. Then $B(s)$ is a constant vector if and only if $\alpha(s)$ is planar.*

PROOF: If $\alpha(s)$ lies in a plane, then the vectors $T(s)$ and $N(s)$ lie in that plane for all s. Since the unit normal to the plane is determined up to sign, $B(s)$ is a constant multiple of that vector.

Conversely, suppose $B(s) = \mathbf{q}$, with $\|\mathbf{q}\| = 1$. It follows immediately that $B'(s) = \mathbf{0}$. Consider the function

$$f(s) = (\alpha(s) - \alpha(0)) \cdot B(s).$$

The derivative satisfies:

$$\frac{df}{ds} = \alpha'(s) \cdot B(s) + (\alpha(s) - \alpha(0)) \cdot B'(s) = T(s) \cdot B(s) = 0.$$

Therefore, $f(s)$ is a constant function and, since $f(0) = 0$, we have $f(s) = 0$. Thus, $\alpha(s)$ lies entirely in the plane perpendicular to $\mathbf{q}$ passing through $\alpha(0)$. ∎

Since $[T(s), N(s), B(s)]$ consists of unit vectors that are pairwise perpendicular, we obtain various relations on taking derivatives. For example, $B(s) \cdot B'(s) = 0$, and from the relation $B(s) \cdot T(s) = 0$ we deduce:

$$\begin{aligned} 0 &= \frac{d}{dt}(B(s) \cdot T(s)) = B'(s) \cdot T(s) + B(s) \cdot T'(s) \\ &= B'(s) \cdot T(s) + B(s) \cdot \kappa_\alpha(s) N(s) \\ &= B'(s) \cdot T(s), \end{aligned}$$

and so $B'(s)$ is perpendicular to $T(s)$. It follows that $B'(s)$ is a multiple of $N(s)$. By analogy with the definition of curvature we introduce:

Definition 6.3. *$B'(s) = -\tau_\alpha(s) N(s)$ and the function $\tau_\alpha(s)$ is called the* **torsion** *of the curve at $\alpha(s)$.*

EXAMPLES:

(1) *The circle*: Let $\alpha(s) = (r\cos(s/r), r\sin(s/r), 0)$ denote the unit speed parametrization of the circle centered at the origin of radius r and lying in the x–y plane. The tangent vector is given by $T(s) = (-\sin(s/r), \cos(s/r), 0)$ and:

$$T'(s) = (-(1/r)\cos(s/r), -(1/r)\sin(s/r), 0),$$

so $\kappa_\alpha(s) = (1/r)$. It follows that $N(s) = (-\cos(s/r), -\sin(s/r), 0)$. To compute $B(s)$, we make the cross product:

$$B(s) = T(s) \times N(s) = (0,\ 0,\ 1)$$

Thus $B'(s) = \mathbf{0}$, $\tau_\alpha(s) = 0$, and $\alpha(s)$ is planar, as we already knew.

(2) *The helix*: Consider the unit speed right helix:

$$\alpha(s) = \left(a\cos\left(\frac{s}{\sqrt{a^2+b^2}}\right), a\sin\left(\frac{s}{\sqrt{a^2+b^2}}\right), b\frac{s}{\sqrt{a^2+b^2}}\right).$$

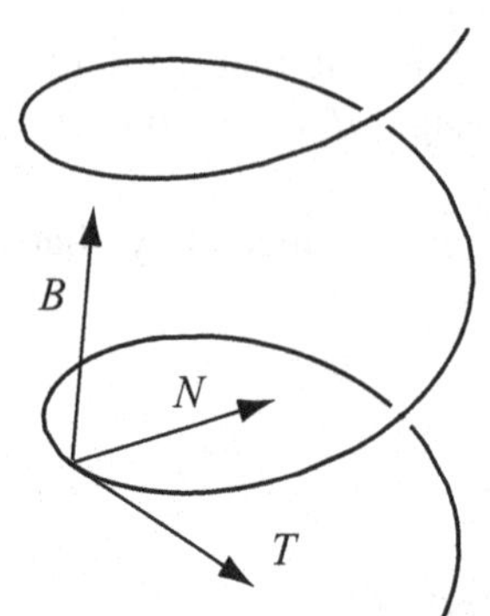

Let ω denote $\dfrac{1}{\sqrt{a^2+b^2}}$. Then:

$$T(s) = (-a\omega\sin\omega s, a\omega\cos\omega s, \omega b),$$

and $T'(s) = (-a\omega^2\cos\omega s, -a\omega^2\sin\omega s, 0)$, so $\kappa_\alpha(s) = a\omega^2$. It follows that $N(s) = (-\cos\omega s, -\sin\omega s, 0)$ and the binormal is given by:

$$B(s) = T(s)\times N(s) = (b\omega\sin\omega s, -b\omega\cos\omega s, a\omega).$$

Thus $B'(s) = (b\omega^2\cos\omega s, b\omega^2\sin\omega s, 0) = -b\omega^2 N(s)$, and $\tau_\alpha(s) = b\omega^2$. The constant nonzero torsion of the helix describes how it lifts out of the osculating plane at a constant rate.

Definition 6.4. *To a regular unit speed curve $\alpha(s)$, the collection associated to each point on the curve $\{\kappa_\alpha(s), \tau_\alpha(s), T(s), N(s), B(s)\}$ is called the* **Frenet-Serret apparatus**. *The orthonormal basis $[T(s), N(s), B(s)]$ is called a* **moving frame** *or moving trihedron along the curve.*

It was Euler (1782) who gave the first systematic investigation of curves in space, where he introduced the mapping of a curve $\alpha(t)$ to an associated curve on the sphere given by $T(t)$. The moving trihedron and its curvature and torsion were introduced in the Toulouse thesis of F. Frenet (1816–68) in 1847. Independently, J. A. Serret (1819–92) published similar results in the *Journal de Liouville* (1851) before Frenet's results had received wide recognition. Frenet published his results in the same journal in 1852. There is evidence (Lumiste 1997) that the moving trihedron and its relations to curvature and torsion were known to K. E. Senff (1810–1917) in 1831.

When such an apparatus comes from a curve we have the following classical theorem.

Theorem 6.5 (Frenet-Serret Theorem). *If $\alpha(s)$ is a unit speed curve with nonzero curvature $\kappa_\alpha(s)$, then:*

$$T'(s) = \kappa_\alpha(s)N(s),\ N'(s) = -\kappa_\alpha(s)T(s) + \tau_\alpha(s)B(s),\ \text{and}\ B'(s) = -\tau_\alpha(s)N(s).$$

Proof: By definition of curvature and torsion we have already that $T'(s) = \kappa_\alpha(s)N(s)$ and $B'(s) = -\tau_\alpha(s)N(s)$. Recall that if $[\mathbf{w}_1, \mathbf{w}_2, \mathbf{w}_3]$ is any orthonormal frame in $\mathbb{R}^3$, then an arbitrary vector $\mathbf{v}\in\mathbb{R}^3$ can be written:

$$\mathbf{v} = (\mathbf{v}\cdot\mathbf{w}_1)\mathbf{w}_1 + (\mathbf{v}_1\cdot\mathbf{w}_2)\mathbf{w}_2 + (\mathbf{v}\cdot\mathbf{w}_3)\mathbf{w}_3.$$

Using the moving frame we apply this fact to $N'(s)$ to get:

$$N'(s) = (T(s)\cdot N'(s))T(s) + (N(s)\cdot N'(s))N(s) + (B(s)\cdot N'(s))B(s).$$

Differentiating $N(s)\cdot T(s) = 0$ gives us that $N'(s)\cdot T(s) = -N(s)\cdot(\kappa_\alpha(s)N(s)) = -\kappa_\alpha(s)$. We already know that $N(s)\cdot N'(s) = 0$.

Finally, $N(s)\cdot B(s)=0$ implies that $N'(s)\cdot B(s)+N(s)\cdot B'(s)=0$, and so $N'(s)\cdot B(s)=-(-\tau_\alpha(s)N(s))\cdot N(s)=\tau_\alpha(s)$. Thus, we have $N'(s)=-\kappa_\alpha(s)T(s)+\tau_\alpha(s)B(s)$. ■

We express this information more succinctly in the matrix equation:

$$\begin{pmatrix} 0 & \kappa_\alpha(s) & 0 \\ -\kappa_\alpha(s) & 0 & \tau_\alpha(s) \\ 0 & -\tau_\alpha(s) & 0 \end{pmatrix}\begin{pmatrix} T(s) \\ N(s) \\ B(s)\end{pmatrix}=\begin{pmatrix} T'(s) \\ N'(s) \\ B'(s)\end{pmatrix}.$$

As usual, it would be helpful to be able to compute the curvature and torsion when the curve is regular but not unit speed. We derive the relevant formulas next.

Proposition 6.6. *For a regular curve $\alpha\colon (a,b)\to\mathbb{R}^3$, not necessarily unit speed, the curvature and torsion are given by:*

$$\kappa_\alpha(t)=\frac{\|\alpha'(t)\times\alpha''(t)\|}{\|\alpha'(t)\|^3} \quad \text{and} \quad \tau_\alpha(t)=\frac{(\alpha'(t)\times\alpha''(t))\cdot\alpha'''(t)}{\|\alpha'(t)\times\alpha''(t)\|^2}.$$

PROOF: As in the planar case, we use the arc length parametrization $\beta(s)=\alpha(t(s))$, where $s(t)$ is the arc length function and $t(s)$ is its inverse. Here $\dfrac{ds}{dt}=\|\alpha'(t)\|$ and so $\dfrac{dt}{ds}=\dfrac{1}{\|\alpha'(t)\|}$. Furthermore,

$$\frac{d^2t}{ds^2}=\frac{d}{ds}\frac{1}{(\alpha'(t)\cdot\alpha'(t))^{1/2}}=-\frac{\alpha'(t)\cdot\alpha''(t)}{(\alpha'(t)\cdot\alpha'(t))^{3/2}}\frac{dt}{ds}=-\frac{\alpha'(t)\cdot\alpha''(t)}{(\alpha'(t)\cdot\alpha'(t))^2}.$$

Suppose $s_0=s(t_0)$ and we want to compute the curvature and torsion at $\alpha(t_0)$. Then the chain rule gives:

$$\begin{aligned}\kappa_\alpha(t_0)^2=\|\beta''(s_0)\|^2&=\left\|\alpha''(t_0)\left(\frac{dt}{ds}\right)^2+\alpha'(t_0)\frac{d^2t}{ds^2}\right\|^2\\
&=\left\|\frac{\alpha''(t_0)}{\|\alpha'(t_0)\|^2}-\left(\frac{\alpha'(t_0)\cdot\alpha''(t_0)}{\|\alpha'(t_0)\|^4}\right)\alpha'(t_0)\right\|^2\\
&=\frac{\|\alpha''(t_0)\|^2}{\|\alpha'(t_0)\|^4}-2\frac{(\alpha'(t_0)\cdot\alpha''(t_0))^2}{\|\alpha'(t_0)\|^6}+\frac{(\alpha'(t_0)\cdot\alpha''(t_0))^2}{\|\alpha'(t_0)\|^6}\\
&=\frac{\|\alpha'(t_0)\|^2\|\alpha''(t_0)\|^2-(\alpha'(t_0)\cdot\alpha''(t_0))^2}{\|\alpha'(t_0)\|^6}=\frac{\|\alpha'(t_0)\times\alpha''(t_0)\|^2}{\|\alpha'(t_0)\|^6},\end{aligned}$$

and the expression for $\kappa_\alpha(t)$ is established.

By the Frenet-Serret Theorem we can compute $\tau_\alpha(t_0)$ by:

$$\tau_\alpha(t_0)=B(s(t_0))\cdot N'(s(t_0))=(T(s_0)\times N(s_0))\cdot N'(s_0).$$

For the arc length parametrized curve $\beta(s)$ we have:

$$B(s_0) = \frac{\beta'(s_0) \times \beta''(s_0)}{\kappa_\alpha(s_0)} \text{ and } N'(s_0) = \frac{\beta'''(s_0)}{\kappa_\alpha(s_0)} - \frac{\beta''(s_0)}{\kappa_\alpha(s_0)^2}\frac{d\kappa_\alpha}{ds}.$$

From these expressions we find:

$$\tau_\alpha(t(s_0)) = \frac{(\beta'(s_0) \times \beta''(s_0)) \cdot \beta'''(s_0)}{\kappa_\alpha(s_0)^2}.$$

Since $\beta'(s_0) = T(s_0) = \dfrac{\alpha'(t_0)}{\|\alpha'(t_0)\|}$ and $\beta''(s_0) = \dfrac{\alpha''(t_0)}{\|\alpha'(t_0)\|^2} + \alpha'(t_0)\dfrac{d^2t}{ds^2}$, we obtain:

$$\beta'(s_0) \times \beta''(s_0) = \frac{\alpha'(t_0) \times \alpha''(t_0)}{\|\alpha'(t_0)\|^3}.$$

Similarly, we have $\beta'''(s_0) = \dfrac{\alpha'''(t_0)}{\|\alpha'(t_0)\|^3} + C_2\alpha''(t_0) + C_1\alpha'(t_0)$ for some constants C_1 and C_2 and so:

$$\tau_\alpha(t_0) = \frac{(\beta'(s_0) \times \beta''(s_0)) \cdot \beta'''(s_0)}{\kappa_\alpha(s_0)^3} = \frac{(\alpha'(t_0) \times \alpha''(t_0)) \cdot \alpha'''(t_0)}{\|\alpha'(t_0)\|^6} \cdot \frac{\|\alpha'(t_0)\|^6}{\|\alpha'(t_0) \times \alpha''(t_0)\|^2},$$

and the proposition is proved. ■

For example, consider the curve:

$$\alpha(t) = (a\cos t, ac\cos t - bd\sin t, ad\cos t + bc\sin t),$$

where $c^2 + d^2 = 1$. Then,

$$\begin{aligned}
\alpha'(t) &= (-a\sin t, -ac\sin t - bd\cos t, -ad\sin t + bc\cos t),\\
\alpha''(t) &= (-a\cos t, -ac\cos t + bd\sin t, -ad\cos t - bc\sin t),\\
\alpha'''(t) &= (a\sin t, ac\sin t + bd\cos t, ad\sin t - bc\cos t)\\
&= -\alpha'(t).
\end{aligned}$$

It follows immediately that $(\alpha'(t) \times \alpha''(t)) \cdot \alpha'''(t) = 0$ and so $\tau_\alpha(t) = 0$ for all t. Therefore, $\alpha(t)$ is planar. After a tedious computation, we also find:

$$\kappa_\alpha(t) = \frac{\|\alpha'(t) \times \alpha''(t)\|}{\|\alpha'(t)\|^3} = \frac{\sqrt{2}ab}{(2a^2\sin^2 t + b^2\cos^2 t)^{3/2}} \neq 0.$$

This curve has the same curvature as an ellipse in the standard plane. We relate this curve to an ellipse in the plane after we prove an analogue in $\mathbb{R}^3$ of the Fundamental Theorem for Curves in $\mathbb{R}^2$ (Theorem 5.11).

The Frenet-Serret apparatus associated to a regular curve $\alpha(t)$, is the collection

$$\{\kappa_\alpha(t), \tau_\alpha(t), T(t), N(t), B(t)\}.$$

For regular curves, not necessarily unit speed, the Frenet-Serret Theorem takes the form:

Theorem 6.7. *If $\alpha(t)$ is a regular curve with nonzero curvature $\kappa_\alpha(t)$, then:*

$$T'(t) = \|\alpha'(t)\|\kappa_\alpha(t)N(t)$$
$$N'(t) = -\|\alpha'(t)\|\kappa_\alpha(t)T(t) + \|\alpha'(t)\|\tau_\alpha(t)B(t)$$
$$\text{and } B'(t) = -\|\alpha'(t)\|\tau_\alpha(t)N(t).$$

PROOF: If $\alpha(t)$ is a regular curve, not necessarily unit speed, then we can reparametrize using arc length, $s(t) = \int_{t_0}^{t} \|\alpha'(u)\|\,du$. The inverse of $s(t)$ is denoted by $t(s)$, and we let $\beta(s) = \alpha(t(s))$. Since the unit tangents to $\alpha(t)$ and $\beta(s(t))$ are the same, $T_\alpha(t) = T_\beta(s(t))$ and so:

$$\frac{d}{dt}T_\alpha(t) = \frac{d}{dt}T_\beta(s(t)) = \frac{dT_\beta}{ds}\frac{ds}{dt} = \kappa_\beta(s(t))N_\beta(s(t))\frac{ds}{dt} = \kappa_\alpha(t)N_\alpha(t)\|\alpha'(t)\|.$$

The formulas for $N'(t)$ and $B'(t)$ follow analogously. ∎

The analogous matrix equation in which we have written the vectors in the Frenet-Serret apparatus as row vectors is given by:

$$\begin{pmatrix} T'(t) \\ N'(t) \\ B'(t) \end{pmatrix} = \|\alpha'(t)\| \begin{pmatrix} 0 & \kappa_\alpha(t) & 0 \\ -\kappa_\alpha(t) & 0 & \tau_\alpha(t) \\ 0 & -\tau_\alpha(t) & 0 \end{pmatrix} \begin{pmatrix} T(t) \\ N(t) \\ B(t) \end{pmatrix}.$$

The skew symmetry of this matrix, that is, the condition $A = -A^t$, where A^t denotes the *transpose* of the matrix A, plays an important role in the next result from which we generalize the Fundamental Theorem for Plane Curves to curves in space. We first introduce an existence result from the theory of differential equations. For a proof we refer the reader to Boyce, Diprima (2008) or Conrad (2002).

Lemma 6.8. *If $\mathfrak{g}(t)$ is a continuous $(n \times n)$-matrix-valued function on an interval (a,b), then there exist solutions, $F\colon (a,b) \to \mathbb{R}^n$, to the differential equation $F'(t) = \mathfrak{g}(t)F(t)$.*

In the case of interest, we would like to study differential equations like the Frenet-Serret Equations. These are really three equations of the type covered by Lemma 6.8, however, they are related because $[T,N,B]$ forms an oriented orthonormal basis for $\mathbb{R}^3$. In this case, there is a general result.

Lemma 6.9. *If $\mathfrak{g}(t)$ is a continuous (3×3)-matrix-valued function on an interval (a,b) and $\mathfrak{g}(t)$ is skew-symmetric, then the matrix differential equation:*

$$\frac{d}{dt}\begin{pmatrix} \mathbf{V}_1(t) \\ \mathbf{V}_2(t) \\ \mathbf{V}_3(t) \end{pmatrix} = \mathfrak{g}(t) \begin{pmatrix} \mathbf{V}_1(t) \\ \mathbf{V}_2(t) \\ \mathbf{V}_3(t) \end{pmatrix}$$

has a solution. Furthermore, if the initial conditions $[\mathbf{V}_1(0), \mathbf{V}_2(0), \mathbf{V}_3(0)]$ form a frame for $\mathbb{R}^3$, then the solution $[\mathbf{V}_1(t), \mathbf{V}_2(t), \mathbf{V}_3(t)]$ gives a frame for all t.

PROOF: Lemma 6.8 gives the existence of the solutions. It suffices to study the orthogonality relations. Consider the matrix:

$$A = \begin{pmatrix} \mathbf{V}_1(t) \\ \mathbf{V}_2(t) \\ \mathbf{V}_3(t) \end{pmatrix}.$$

The differential equation becomes $\dfrac{dA}{dt} = \mathfrak{g}(t)A$. The dot products between the various vectors are gathered in the matrix A^tA.

$$\begin{aligned} \frac{d}{dt}(A^tA) &= \left(\frac{dA}{dt}\right)^t A + A^t\left(\frac{dA}{dt}\right) \\ &= (\mathfrak{g}(t)A)^tA + A^t(\mathfrak{g}(t)A) = (A^t\mathfrak{g}(t)^t)A + A^t(\mathfrak{g}(t)A) \\ &= -(A^t\mathfrak{g}(t))A + A^t(\mathfrak{g}(t)A) = \mathbb{O}, \text{ the zero matrix.} \end{aligned}$$

Thus $A(t)^tA(t)$ is a matrix of constants. The initial conditions determined by the frame $[\mathbf{V}_1(0), \mathbf{V}_2(0), \mathbf{V}_3(0)]$ satisfy $A(0)^tA(0) = \mathrm{id}$, and so $A(t)^tA(t) = \mathrm{id}$ for all t. This matrix equation is equivalent to the conditions $\mathbf{V}_i \cdot \mathbf{V}_j = \delta_{ij}$, where δ_{ij} denotes the *Kronecker delta*, given by the entries in the identity matrix

$$\delta_{ij} = \begin{cases} 1, & \text{if } i = j, \\ 0, & \text{if } i \neq j. \end{cases}$$

Hence $[\mathbf{V}_1(t), \mathbf{V}_2(t), \mathbf{V}_3(t)]$ is a frame for all t. ■

From these results about differential equations, the analogue of Theorem 5.11 for space curves follows.

Theorem 6.10 (Fundamental Theorem for Curves in $\mathbb{R}^3$). *Let $\bar{\kappa}, \bar{\tau} \colon (a,b) \to \mathbb{R}^3$ be continuous functions with $\bar{\kappa}(s) > 0$ for all s. Then there is a curve $\alpha \colon (a,b) \to \mathbb{R}^3$, parametrized by arc length, whose curvature and torsion satisfy $\kappa_\alpha(s) = \bar{\kappa}(s)$ and $\tau_\alpha(s) = \bar{\tau}(s)$, respectively. Furthermore, any two such curves differ by a proper rigid motion, that is, if $\beta(s)$ is any other such curve, then there is a linear mapping $M \colon \mathbb{R}^3 \to \mathbb{R}^3$, which preserves arc length, and a vector $\mathbf{v} \in \mathbb{R}^3$ such that $\beta(s) = M \cdot \alpha(s) + \mathbf{v}$.*

PROOF: The Frenet-Serret Equations take the form of a matrix differential equation in Lemma 6.9 where the skew-symmetric matrix $\mathfrak{g}(s)$ is given by:

$$\mathfrak{g}(s) = \begin{pmatrix} 0 & \bar{\kappa}(s) & 0 \\ -\bar{\kappa}(s) & 0 & \bar{\tau}(s) \\ 0 & -\bar{\tau}(s) & 0 \end{pmatrix}.$$

It suffices to choose the initial conditions. We may take $[T(0), N(0), B(0)]$ to be a choice of an oriented, orthonormal at $s = 0$.

To obtain the curve $\alpha(s)$ we integrate:

$$\alpha(s) = \int_0^s T(t)\,dt.$$

We leave it to the reader to verify that $\alpha(s)$ has the correct curvature and torsion. Adding a vector $\mathbf{v}$ to $\alpha(s)$ does not change $[T(0), N(0), B(0)]$ and so we obtain another solution by translation. We can also choose a different initial frame. For two such choices, $[T_0(0), N_0(0), B_0(0)]$ and $[T_1(0), N_1(0), B_1(0)]$, there is a matrix M taking T_0 to T_1, N_0 to N_1, and B_0 to B_1. Then the matrix M is a rigid motion of $\mathbb{R}^3$ because it takes a frame to a frame. (The reader can prove this.) Finally,

$$\beta(s) = \int_0^s T_1(t)\,dt = \int_0^s MT_0(t)\,dt = M \cdot \alpha(s) + \mathbf{v}.$$

Thus two solutions to the differential equations given by the Frenet-Serret Equations differ by a rigid motion and a translation. ∎

The assumption that $\kappa(s) \neq 0$ is necessary in order to obtain uniqueness in the sense of the theorem. If $\kappa(s) = 0$ along a subinterval (a_1, b_1) of (a, b), then a line segment in $\mathbb{R}^3$ realizes this part of the curve. To see how uniqueness fails, suppose M is the midpoint of the line segment between $\alpha(a_1)$ and $\alpha(b_1)$ of a solution curve with a given $\kappa(s)$ and $\tau(s)$. Suppose $\epsilon > 0$ is some small value such that $\alpha(s)$ lies on one side of the plane through M perpendicular to the line segment for $a_1 - \epsilon < s < a_1$ and on the other side of this plane for $b_1 < s < b_1 + \epsilon$. Rotate the solution around the axis given by the line segment in the half-space containing $\alpha(b_1)$ through some nonzero angle while leaving the other half-space fixed. This operation preserves the curvature and torsion but there is no way to superimpose this new curve on the one given by $\alpha(s)$, and the fundamental theorem fails to hold.

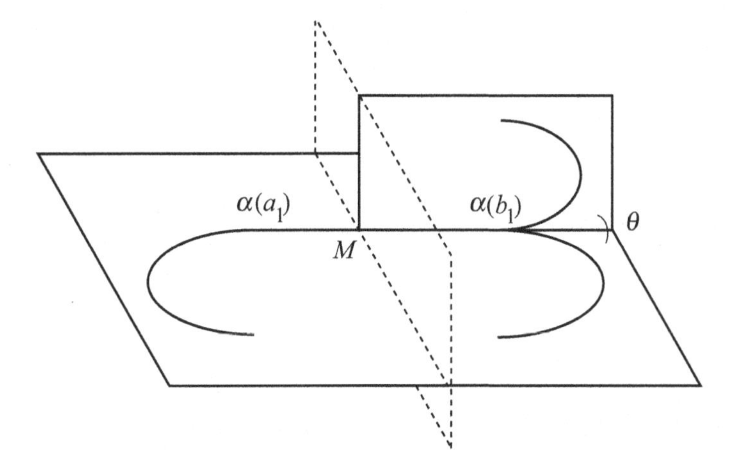

As with plane curves, we get all the relevant geometric data about the curve from its curvature (when nonzero) and torsion. Here are some examples of this idea.

EXAMPLES:

(1) As we have already seen for a regular curve $\alpha(t)$ in $\mathbb{R}^3$, $\alpha(t)$ is a plane curve if and only if $\tau_\alpha(t)=0$ for all t.
(2) If $\alpha(s)$ is unit speed and regular, $\kappa_\alpha(s)$ is constant, and $\tau_\alpha(s)=0$, then $\alpha(s)$ is part of a circle of radius $1/\kappa_\alpha$ lying in some plane in $\mathbb{R}^3$.

The power of the Frenet-Serret apparatus lies in the manner in which it organizes the principal analytic properties of a curve. Applications of this viewpoint are found in the exercises. As a final example, we introduce a class of curves that generalizes the helix.

Definition 6.11. *A curve $\alpha(s)$ is a* **general helix** *if there is some vector* $\mathbf{v}$ *such that the tangent $T(s)$ and* $\mathbf{v}$ *maintain a constant angle with respect to each other. The vector* $\mathbf{v}$ *is called the* **axis** *of the helix. We can take* $\mathbf{v}$ *to be unit length.*

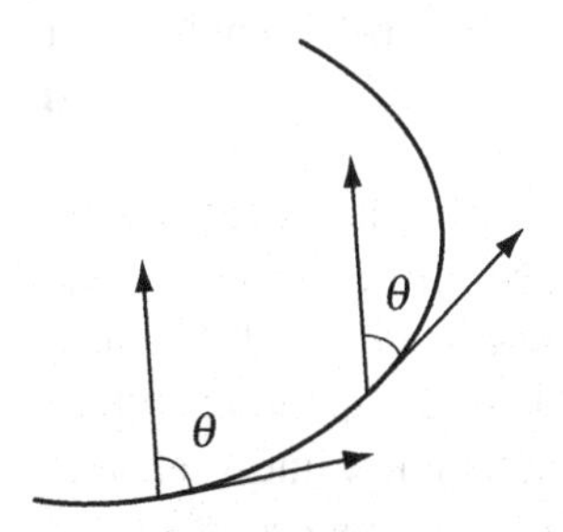

For example, $\mathbf{v}=(0,0,1)$ for the helix $\alpha(s)=(\cos s,\sin s,s)$. Observe that if $\alpha(s)$ is a general helix, then $T(s)\cdot\mathbf{v}=\cos\theta$, where θ is the angle between $T(s)$ and $\mathbf{v}$. From the definition we also see that all plane curves are general helices with the binormal $B(s)$ playing the role of the constant vector $\mathbf{v}$. We can characterize general helices using the Frenet-Serret apparatus.

Proposition 6.12. *A unit speed curve $\alpha(s)$ is a general helix if and only if there is a constant c such that $\tau_\alpha(s)=c\kappa_\alpha(s)$ for all s.*

PROOF: If $\mathbf{v}$ is the axis of a general helix $\alpha(s)$, then $T(s)\cdot\mathbf{v}=\cos\theta$, a constant, and this implies $T'(s)\cdot\mathbf{v}=0$, or $\kappa_\alpha(s)N(s)\cdot\mathbf{v}=0$, that is, $\mathbf{v}$ is perpendicular to $N(s)$. Since $[T(s),N(s),B(s)]$ is an orthonormal frame, $\|\mathbf{v}\|=1$, and $\mathbf{v}$ is in the plane of $T(s)$ and $B(s)$, we get:

$$\begin{aligned}\mathbf{v}&=(T(s)\cdot\mathbf{v})T(s)+(N(s)\cdot\mathbf{v})N(s)+(B(s)\cdot\mathbf{v})B(s)\\&=(\cos\theta)T(s)+(B(s)\cdot\mathbf{v})B(s)=(\cos\theta)T(s)+(\sin\theta)B(s).\end{aligned}$$

This implies further that:

$$\mathbf{0}=\frac{d}{ds}(\mathbf{v})=(\cos\theta)T'(s)+(\sin\theta)B'(s)=((\cos\theta)\kappa_\alpha(s)-(\sin\theta)\tau_\alpha(s))N(s).$$

Since $N(s)\neq 0$, it follows that $\tau_\alpha(s)=(\cot\theta)\kappa_\alpha(s)$.

Conversely, suppose $\kappa_\alpha(s)=c\tau_\alpha(s)$ for some constant $c=\cot\theta$ for some angle θ. Let $V(s)=(\cos\theta)T(s)+(\sin\theta)B(s)$. Notice that $\|V(s)\|=1$. Differentiating we have:

$$\frac{d}{ds}V(s)=(\cos\theta)T'(s)+(\sin\theta)B'(s)=((\cos\theta)\kappa_\alpha(s)-(\sin\theta)\tau_\alpha(s))N(s),$$

and $(\cos\theta)\kappa_\alpha(s) - (\sin\theta)\tau_\alpha(s) = 0$ by the assumption. Thus $V(s)$ is a constant vector $\mathbf{v}$. Furthermore $T(s) \cdot \mathbf{v} = \cos\theta$ and we have proved that $\alpha(s)$ is a general helix. ∎

Much more can be done in developing special properties of curves from the Frenet-Serret apparatus. The successful classification of the smooth curves in $\mathbb{R}^3$ via curvature ($\kappa \neq 0$) and torsion establishes a paradigm that one could hope to generalize to other geometric objects, for example, two-dimensional objects.

Appendix
On Euclidean rigid motions

A missing ingredient in our discussion of curves in $\mathbb{R}^n$ is a notion of equivalence. Given two curves $\alpha(t)$ and $\beta(t)$, when should we consider them geometrically equivalent? Reparametrization is one reduction; certainly a curve and any oriented reparametrization (one for which the derivative is positive) ought to be considered equivalent. Because we can choose the arc length parametrization, we focus on the trace of the curve with a direction along it. Consider the directed trace of a curve as a rigid object in $\mathbb{R}^n$ and move it about in space by a motion of $\mathbb{R}^n$ preserving rigid bodies. For example, translation of $\mathbb{R}^3$ by the vector $(1,0,0)$ corresponds to "moving one unit to the right," and we can expect such a motion to preserve a rigid body. In this appendix we determine those mappings of $\mathbb{R}^n$ to itself that are the rigid motions. Those properties of curves that remain unchanged by such motions are the properties rightfully called *geometric*.

To clarify the notion of a rigid motion, we may make the following definition, which translates rigidity into the idea of distance preservation.

Definition 6.13. *A mapping* $f\colon \mathbb{R}^n \to \mathbb{R}^n$ *is an* **isometry** *if, for all* $\mathbf{v},\mathbf{w} \in \mathbb{R}^n$*:*

$$\|f(\mathbf{v}) - f(\mathbf{w})\| = \|\mathbf{v} - \mathbf{w}\|.$$

The most familiar examples are *translations*, $t_{\mathbf{w}}(\mathbf{v}) = \mathbf{v} + \mathbf{w}$. In $\mathbb{R}^2$ the *rotations* are given by:

$$\rho_\theta(\mathbf{v}) = \begin{pmatrix} \cos\theta & -\sin\theta \\ \sin\theta & \cos\theta \end{pmatrix} \begin{pmatrix} v_1 \\ v_2 \end{pmatrix}.$$

We show that these mappings in the standard basis are isometries. Since ρ_θ is linear, $\rho_\theta(\mathbf{v}) - \rho_\theta(\mathbf{w}) = \rho_\theta(\mathbf{v} - \mathbf{w})$ and so it suffices to show that $\|\rho_\theta(\mathbf{v})\| = \|\mathbf{v}\|$.

$$\begin{aligned} \|\rho_\theta(\mathbf{v})\|^2 &= \|((\cos\theta)v_1 - (\sin\theta)v_2, (\sin\theta)v_1 + (\cos\theta)v_2)\|^2 \\ &= (\cos\theta)v_1 - (\sin\theta)v_2)^2 + ((\sin\theta)v_1 + (\cos\theta)v_2)^2 \\ &= (\cos\theta)^2 v_1 - 2(\cos\theta\sin\theta)v_1v_2 + (\sin\theta)^2 v_2 \\ &\quad + (\sin\theta)^2 v_1^2 + 2(\cos\theta\sin\theta)v_1v_2 + (\cos\theta)^2 v_2 \\ &= v_1^2 + v_2^2 = \|\mathbf{v}\|^2. \end{aligned}$$

Given an isometry $f\colon \mathbb{R}^n \to \mathbb{R}^n$, we can alter f by a translation to obtain an isometry that fixes the origin, namely, $F(\mathbf{v}) = f(\mathbf{v}) - f(\mathbf{0})$. These isometries enjoy an additional property.

Proposition 6.14. *An isometry* $F\colon \mathbb{R}^n \to \mathbb{R}^n$ *that fixes the origin is a linear mapping.*

PROOF: We want to show that $F(a\mathbf{v}+b\mathbf{w}) = aF(\mathbf{v})+bF(\mathbf{w})$ for all a, b in $\mathbb{R}$ and $\mathbf{v}$, $\mathbf{w}$ in $\mathbb{R}^n$. First notice that $\|F(\mathbf{v})\| = \|F(\mathbf{v}) - F(\mathbf{0})\| = \|\mathbf{v}-\mathbf{0}\| = \|\mathbf{v}\|$, and so F preserves the length of vectors.

We leave it to the reader to prove the following **polarization identity**:

$$\mathbf{v}\cdot\mathbf{w} = \frac{1}{2}(\|\mathbf{v}\|^2 + \|\mathbf{w}\|^2 + \|\mathbf{v}-\mathbf{w}\|^2).$$

Applying this identity we find

$$\begin{aligned} F(\mathbf{v})\cdot F(\mathbf{w}) &= \frac{1}{2}(\|F(\mathbf{v})\|^2 + \|F(\mathbf{w})\|^2 - \|F(\mathbf{v}) - F(\mathbf{w})\|^2) \\ &= \frac{1}{2}(\|\mathbf{v}\|^2 + \|\mathbf{w}\|^2 - \|\mathbf{v}-\mathbf{w}\|^2) \\ &= \mathbf{v}\cdot\mathbf{w}. \end{aligned}$$

Thus $F(\mathbf{v})\cdot F(\mathbf{w}) = \mathbf{v}\cdot\mathbf{w}$ for all $\mathbf{v}$, $\mathbf{w}$ in $\mathbb{R}^n$, that is, F preserves the inner product on $\mathbb{R}^n$. In particular, we find that an orthonormal basis $[\mathbf{v}_1,\ldots,\mathbf{v}_n]$ is carried by F to another orthonormal basis $[F(\mathbf{v}_1),\ldots,F(\mathbf{v}_n)]$.

Suppose $\mathbf{v}$ and $\mathbf{w}$ are in $\mathbb{R}^n$. Since $[\mathbf{v}_1,\ldots,\mathbf{v}_n]$ is an orthonormal basis, we can write: $\mathbf{v} = a_1\mathbf{v}_1+\cdots+a_n\mathbf{v}_n$ and $\mathbf{w} = b_1\mathbf{v}_1+\cdots b_n\mathbf{v}_n$.

Because $[F(\mathbf{v}_1),\ldots,F(\mathbf{v}_n]$ is also an orthonormal basis, we can write:

$$F(\mathbf{v}) = \sum_{i=1}^{n}(F(\mathbf{v}_i)\cdot F(\mathbf{v}))F(\mathbf{v}_i) = \sum_{i=1}^{n}(\mathbf{v}_i\cdot\mathbf{v})F(\mathbf{v}_i) = \sum_{i=1}^{n} a_iF(\mathbf{v}_i).$$

It follows that:

$$\begin{aligned} F(a\mathbf{v}+b\mathbf{w}) &= F((aa_1+bb_1)\mathbf{v}_1+\cdots+(aa_n+bb_n)\mathbf{v}_n) \\ &= (aa_1+bb_1)F(\mathbf{v}_1)+\cdots+(aa_n+bb_n)F(\mathbf{v}_n) \\ &= a(a_1F(\mathbf{v}_1)+\cdots+a_nF(\mathbf{v}_n))+b(b_1F(\mathbf{v}_1)+\cdots+b_nF(\mathbf{v}_n)) \\ &= aF(\mathbf{v})+bF(\mathbf{w}). \end{aligned}$$

Thus we have proved the proposition. ∎

An isometry is always a one-to-one mapping: If $F(\mathbf{v}) = F(\mathbf{w})$, then $0 = \|F(\mathbf{v}) - F(\mathbf{w})\| = \|\mathbf{v}-\mathbf{w}\|$ and so $\mathbf{v} = \mathbf{w}$. An isometry fixing the origin, a linear isometry, has a linear isometry as its inverse. We denote by $O(n)$ the **orthogonal group** of all linear isometries of $\mathbb{R}^n$. Choosing a basis for $\mathbb{R}^n$, we can write every element of $O(n)$ as a matrix.

Proposition 6.15. *$O(n)$ consists of all $n\times n$ matrices A satisfying the equation $A^{-1} = A^t$.*

PROOF: Choose the canonical basis for $\mathbb{R}^n$, $[\mathbf{e}_1,\ldots,\mathbf{e}_n]$, where:

$$\mathbf{e}_i = (0,\ldots,0,1,0,\ldots,0),\ 1 \text{ in the } i\text{th place.}$$

In this orthonormal basis we can compute the entries in a matrix A representing a linear transformation in $O(n)$ by:

$$a_{ij} = A\mathbf{e}_i \cdot \mathbf{e}_j.$$

Since the inverse of an isometry is also an isometry, we obtain the relation:

$$A\mathbf{e}_i \cdot \mathbf{e}_j = A^{-1}A\mathbf{e}_i \cdot A^{-1}\mathbf{e}_j = \mathbf{e}_i \cdot A^{-1}\mathbf{e}_j.$$

Identifying the dot product with matrix multiplication we also have:

$$A\mathbf{e}_i \cdot \mathbf{e}_j = (A\mathbf{e}_i)^t\mathbf{e} = \mathbf{e}_i^t A^t \mathbf{e}_j = \mathbf{e}_i \cdot A^t\mathbf{e}_j.$$

Thus the matrix representing A^{-1} is the transpose of A. ∎

The determinant satisfies $\det(A) = \det(A^t)$ and so we find:

$$1 = \det \mathrm{id} = \det(AA^{-1}) = \det(AA^t) = \det(A)\det(A^t) = (\det(A))^2.$$

Thus, the elements of $O(n)$ have determinant ± 1.

Suppose that $n = 3$ and consider the effect of applying $A \in O(3)$ to a unit speed curve $\alpha(s)$. Let $\beta(s) = A \circ \alpha(s)$. Because A is a linear mapping we have $\beta'(s) = A \circ \alpha'(s)$ and similarly for all higher derivatives of $\beta(s)$. This implies that:

$$\kappa_\beta(s) = \|\beta''(s)\| = \|A \circ \alpha''(s)\| = |\det(A)|\|\alpha''(s)\| = \|\alpha''(s)\|,$$

that is, $\alpha(s)$ and $\beta(s)$ have the same curvature. Proposition 6.4 implies that $\alpha(s)$ and $\beta(s)$ have the same torsion. Thus, the curvature and torsion of a space curve are unchanged under rigid motions. The fundamental theorem for curves in $\mathbb{R}^3$ shows how curvature and torsion determine a space curve up to isometry.

Exercises

6.1 Establish the properties of the cross product on $\mathbb{R}^3$ listed at the beginning of the chapter. Also prove the *Lagrange Formula*:

$$(\mathbf{t} \times \mathbf{u}) \cdot (\mathbf{v} \times \mathbf{w}) = (\mathbf{t} \cdot \mathbf{v})(\mathbf{u} \cdot \mathbf{w}) - (\mathbf{t} \cdot \mathbf{w})(\mathbf{u} \cdot \mathbf{v}).$$

6.2 Show that the following statements characterize the straight lines as curves in $\mathbb{R}^3$:

(1) All tangent lines to the curve pass through a fixed point.
(2) All tangent lines are parallel to a given line.
(3) $\kappa = 0$.
(4) The curve $\alpha(t)$ satisfies $\alpha'(t)$ and $\alpha''(t)$ being linearly dependent.

Solutions to Exercises marked with a dagger appear in the appendix, pp. 331–332.

6.3 Consider the curve $\beta(s) = ((4/5)\cos s, 1 - \sin s, (-3/5)\cos s)$. Determine its Frenet-Serret apparatus. What is the nature of the curve?

6.4 Prove the polarization identity $\mathbf{v} \cdot \mathbf{w} = \frac{1}{2}(\|\mathbf{v}\|^2 + \|\mathbf{w}\|^2 - \|\mathbf{v} - \mathbf{w}\|^2)$. Can you find other related expressions that give the dot product in terms of the norm?

6.5 Let $\alpha(s)$ be a unit speed curve with domain $(-\epsilon, \epsilon)$, and with $\kappa_\alpha > 0$ and $\tau_\alpha > 0$. Let:

$$\beta(s) = \int_0^s B(u)\,du.$$

(1) Prove that $\beta(s)$ is a unit speed curve.
(2) Show that the Frenet-Serret apparatus for $\beta(s)$ is $\{\kappa_\beta, \tau_\beta, \bar{T}, \bar{N}, \bar{B}\}$, where $\kappa_\beta = \tau_\alpha$, $\tau_\beta = \kappa_\alpha$, $\bar{N} = -N$, and $\bar{B} = T$, and $\{\kappa_\alpha, \tau_\alpha, T, N, B\}$ is the Frenet-Serret apparatus for $\alpha(s)$.

6.6† If $\alpha(s)$ is a unit speed curve with nonzero curvature, find a vector $\mathbf{w}(s)$ such that:

$$T' = \mathbf{w} \times T, \quad N' = \mathbf{w} \times N, \quad B' = \mathbf{w} \times B.$$

This vector $\mathbf{w}(s)$ is called the **Darboux vector**. Determine its length.

6.7 For a sufficiently differentiable unit speed curve $\alpha(s)$, show that the Taylor series of the curve at a point $\alpha(s_0)$ can be given by:

$$\alpha(s_0 + \Delta s) = \alpha(s_0) + T(s)\Delta s\left(1 - \frac{\kappa_\alpha^2}{6}(\Delta s)^2\right) + N(s)(\Delta s)^2\left(\frac{\kappa_\alpha}{2} + \frac{\kappa_\alpha'}{6}\Delta s\right)$$
$$+ B(s)(\Delta s)^3\frac{\kappa_\alpha\tau_\alpha}{6} + R_4,$$

where the remainder R_4 satisfies $\lim_{\Delta s \to 0} \frac{R_4}{(\Delta s)^3} = 0$. Ignoring the remainder R_4 by projecting into the various planes determined by $\{T, N\}, \{T, B\}, \{N, B\}$, determine the algebraic curves on which the curvature and torsion of a curve must lie.

6.8† In 1774, Charles Tinseau (1749–1822), a student of Monge, proved that an orthogonal projection of a space curve onto a plane has a point of inflection, that is, a point where $\kappa_\pm(\beta, t) = 0$ if the plane is perpendicular to the osculating plane of the curve. Prove this.

6.9 The tangent vector $T(s)$ to a unit speed curve $\alpha(s)$ determines a curve on the unit sphere in $\mathbb{R}^3$. Suppose that its arc length is given by the function $\tilde{s}(s)$. Make sense of the equation $\kappa_\alpha(s) = \frac{d\tilde{s}}{ds}$, where s denotes the arc length function of the curve $\alpha(s)$. This equation has a generalization that we will see in Chapter 8. What does one get by doing the analogous procedure for $N(s)$ and $B(s)$?

6.10† **Intrinsic equations** of a curve are given by equations in quantities unchanged by an isometry, such as $\kappa(s)$ and $\tau(s)$. By the fundamental theorem for curves in $\mathbb{R}^3$, these functions determine the curve up to isometry. Furthermore, relations between the functions κ and τ may be used to characterize curves. Prove that a nonplanar curve lies entirely on a sphere if and only if the following equation holds between the curvature and torsion:

$$\frac{\tau}{\kappa} = \left(\frac{\kappa'}{\tau\kappa^2}\right)'.$$

6.11 Given a unit speed curve $\alpha(s)$, determine the equation satisfied by the osculating plane to the curve as a function of s. Show that a curve is planar if and only if the osculating plane contains some nonzero point for all s.

6.12 For a regular curve $\alpha(t) = (x(t), y(t), z(t))$, not necessarily unit speed, prove the following formula for the torsion of α:

$$\tau_\alpha = \frac{1}{\kappa_\alpha^2 \|\alpha'(t)\|^6} \det \begin{pmatrix} x' & y' & z' \\ x'' & y'' & z'' \\ x''' & y''' & z''' \end{pmatrix}.$$

6.13 Take the theory of the previous chapter of involutes and evolutes and generalize it to curves in $\mathbb{R}^3$. In particular, show that an evolute $\beta(s)$ of a curve $\alpha(s)$ with Frenet-Serret apparatus $\{\kappa_\alpha, \tau_\alpha, T, N, B\}$ satisfies the equation:

$$\beta(s) = \alpha(s) + \lambda N + \mu B,$$

where $\lambda = 1/\kappa_\alpha$ and $\mu = (1/\kappa_\alpha)\cot(\int \tau_\alpha \, dt + C)$ for some constant C.

6.14† Denote the line through $\alpha(t)$ in the direction of the normal to α by:

$$l_{\alpha(t)}^{\perp} = \{\alpha(t) + uN_\alpha(t) \mid u \in \mathbb{R}\}.$$

A pair of distinct curves $\alpha(t)$ and $\beta(t)$ are **Bertrand mates** if $l_{\alpha(t)}^{\perp} = l_{\beta(t)}^{\perp}$ for all t. It follows immediately that $\beta(t) = \alpha(t) + r(t)N_\alpha(t)$ for all t. Prove that $r(t)$ is a constant r given by the constant distance between $\alpha(t)$ and $\beta(t)$ for each t. Show that the angle between $T_\alpha(t)$ and $T_\beta(t)$ is constant. Show that, when $\kappa_\alpha \tau_\alpha \neq 0$, $r\kappa_\alpha(t) + r\cot\theta\,\tau_\alpha(t) = 1$ for all t when $\alpha(t)$ has a Bertrand mate. Finally, show that this condition is sufficient for a curve $\alpha(t)$ to have a Bertrand mate.

6.15 Suppose $F\colon \mathbb{R}^3 \to \mathbb{R}^3$ is a linear mapping that takes an initial orthonormal frame $[\mathbf{u}_0, \mathbf{v}_0, \mathbf{w}_0]$ to another orthonormal frame $[\mathbf{u}_1, \mathbf{v}_1, \mathbf{w}_1]$. Prove that F is a rigid motion, an isometry.

6.16 Suppose that A is an element of $O(3)$ with determinant -1, and $\alpha(s)$ is a unit speed curve in $\mathbb{R}^3$. Compute the curvature and torsion of $A \circ \alpha$ in terms of the curvature and torsion of $\alpha(s)$.

6.17 Suppose $L\colon \mathbb{R}^n \to \mathbb{R}^n$ is a linear transformation. Define the **adjoint** L^* of L to be the transformation of $\mathbb{R}^n$ determined by $L(\mathbf{v}) \cdot \mathbf{w} = \mathbf{v} \cdot L^*(\mathbf{w})$. Show that for a particular orthonormal basis, L^* is represented by the matrix that is the transpose of the matrix representing L. Define the **Rayleigh quotient** of a nonzero vector $\mathbf{v}$ as:

$$R(\mathbf{v}) = \frac{L(\mathbf{v}) \cdot \mathbf{v}}{\mathbf{v} \cdot \mathbf{v}}.$$

Show that R attains a maximum and minimum on $\mathbb{R}^n - \{\mathbf{0}\}$. Show that R is differentiable with gradient given by:

$$\nabla R(\mathbf{v}) = \frac{L(\mathbf{v}) - L^*(\mathbf{v}) - 2R(\mathbf{v})\mathbf{v}}{\mathbf{v} \cdot \mathbf{v}}.$$

Finally, use these properties to show that any linear isometry $F\colon \mathbb{R}^n \to \mathbb{R}^n$ has a subspace $K \subset \mathbb{R}^n$ such that (a) $\dim K = 1$ or 2, and (b) $L(\mathbf{v}) \in K$ for all $\mathbf{v} \in K$, that is, F has an invariant subspace K of dimension 1 or 2.

6.19 Following the ideas of the previous exercise, suppose K is a linear subspace of $\mathbb{R}^n$. Let $K^{\perp}$ be the subspace of $\mathbb{R}^n$ satisfying $K^{\perp} = \{\mathbf{v} \in \mathbb{R}^n \mid \mathbf{v} \cdot \mathbf{w} \text{ for all } \mathbf{v} \in K\}$. Show that if K is an invariant subspace of $\mathbb{R}^n$ with respect to a linear isometry F, then $K^{\perp}$ is also an invariant subspace. Show also that $F|_K$ is an isometry. If K has dimension 2 and $F|_K$ is a rotation, then K is called a *rotation plane*. Prove that for $n \geq 3$ a linear isometry has a rotation plane. It is also possible that $F|_K$ is a reflection across a line contained in K or across the origin. Prove that any linear isometry of $\mathbb{R}^n$ can be expressed as a composition of n or fewer reflections in hyperplanes. (See the article by Lee Rudolph, The structure of orthogonal transformations, *American Mathematical Monthly* **98**(1991), 349–52.)

7

Surfaces

And because of the nature of surfaces any coordinate function ought to be of two variables.

EULER, OPERA POSTHUMA (VOL. 1, P. 494)

Although geometers have given much attention to general investigations of curved surfaces and their results cover a significant portion of the domain of higher geometry, this subject is still so far from being exhausted, that it can well be said that, up to this time, but a small portion of an exceedingly fruitful field has been cultivated.

GAUSS, Abstract to *Disquisitiones* (1827)

Definition 5 of Book I of Euclid's *The Elements* tells us that "a surface is that which has length and breadth only." Analytic geometry turns this definition into functions on a surface that behave like length and breadth, namely, a pair of independent coordinates that determine uniquely each point on the surface. For example, spherical coordinates, longitude and colatitude (chapter 1), apply to the surface of a sphere, and they permit new arguments via the calculus, revealing the geometry of a sphere.

Our goal in developing the classical topics of differential geometry is to discover the surfaces on which non-Euclidean geometry holds. On the way to this goal the major themes of differential geometry emerge, which include the notions of intrinsic properties, curvature, geodesics, and abstract surfaces. In this chapter we develop the analogues of notions for curves such as parameters and their transformations, tangent directions, and lengths.

We begin with the theory of surfaces in $\mathbb{R}^3$, founded by Euler in his 1760 paper *Recherches sur la courbure des surfaces*. He introduced the general notion of coordinates on a surface (in the special case of developable surfaces), and the definition of a surface evolved from the work of Euler and Gauss. Euler's ideas were developed extensively by the French school of geometry led by GASPARD MONGE (1746–1818). We take a more modern tone in our presentation as a short cut to the objects of study. The historical development of the definitions chosen here is a fascinating story, told well by Scholz (1999).

The main example of a surface is a sphere that we can describe in several ways:

(1) As the set of points (x,y,z) in $\mathbb{R}^3$, satisfying the algebraic equations $x^2+y^2+z^2=1$.

(2) As the union of the graphs of the functions $z = \pm f(x,y) = \pm\sqrt{1-x^2-y^2}$, defined on the unit disk in $\mathbb{R}^2$.
(3) As the image of the spherical coordinate mapping, for $0 \leq \theta \leq 2\pi$, $0 \leq \phi \leq \pi$,

$$(\theta,\phi) \mapsto (\cos\theta\sin\phi, \sin\theta\sin\phi, \cos\phi).$$

The modern definition of a surface includes all of these viewpoints—algebraic, analytic, and via coordinates. Recall that a subset $U \subset \mathbb{R}^n$ is **open** if, for every $\mathbf{u} \in U$, there is an $\epsilon > 0$, dependent on $\mathbf{u}$, such that the *open ball*:

$$B_\epsilon(\mathbf{u}) = \{\mathbf{v} \in \mathbb{R}^n \mid \|\mathbf{u}-\mathbf{v}\| < \epsilon\}$$

is contained in U, that is, $B_\epsilon(\mathbf{u}) \subset U$.

Definition 7.1. *A subset S of $\mathbb{R}^3$ is a* **(regular) surface** *if, for each point p in S, there is an open set V of $\mathbb{R}^3$ and a function $x\colon (U \subset \mathbb{R}^2) \to V \cap S$, where U is an open set in $\mathbb{R}^2$, satisfying the following conditions:*

(1) x is differentiable, that is, writing $x(u,v) = (f_1(u,v), f_2(u,v), f_3(u,v))$, the functions f_i have partial derivatives of all orders.
(2) x is a homeomorphism, that is, x is continuous, one-to-one, and onto $V \cap S$, and x has an inverse $x^{-1}\colon V \cap S \to U$ that is also continuous. This means that x^{-1} is the restriction to $V \cap S$ of a continuous function $W \to \mathbb{R}^2$, for some open set $W \subset \mathbb{R}^3$ that contains $V \cap S$.
(3) The Jacobian matrix

$$J(x)(u,v) = \begin{pmatrix} \dfrac{\partial f_1}{\partial u}(u,v) & \dfrac{\partial f_1}{\partial v}(u,v) \\ \dfrac{\partial f_2}{\partial u}(u,v) & \dfrac{\partial f_2}{\partial v}(u,v) \\ \dfrac{\partial f_3}{\partial u}(u,v) & \dfrac{\partial f_3}{\partial v}(u,v) \end{pmatrix}$$

has rank 2. That is, for each value of $(u,v) \in U$, $J(x)(u,v)\colon \mathbb{R}^2 \to \mathbb{R}^3$, as a linear mapping, is one-to-one, or equivalently, the null space of the matrix $J(x)(u,v)$ is $\{\mathbf{0}\}$. This condition on x is called **regularity**.

The function $x\colon U \to S$ is called a **coordinate chart**.

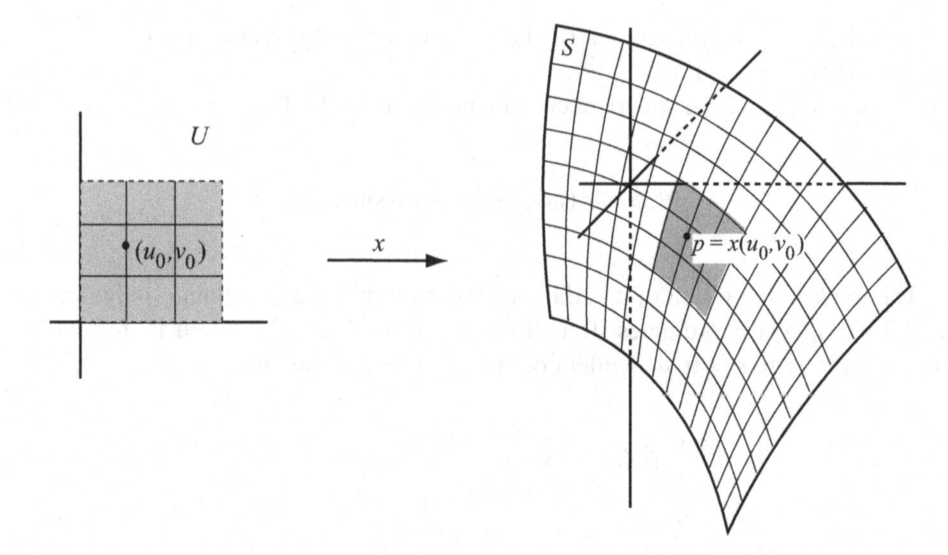

The functions $x\colon (U \subset \mathbb{R}^2) \to S$ assign coordinates to each point on a surface in an open set containing a given point. Hence, a surface is *locally two dimensional*, that is, the open set $V \cap S$ is identifiable with $U \subset \mathbb{R}^2$. Colloquially a coordinate chart $x\colon (U \subset \mathbb{R}^2) \to S$ is the correspondence on a surface that a cartographic map would make with the portion of the surface of the Earth it describes.

EXAMPLES:

(1) For $\mathbf{u}$ and $\mathbf{q} \neq \mathbf{0}$ fixed vectors in $\mathbb{R}^3$, let $\Pi_{\mathbf{u},\mathbf{q}}$ denote the *plane in* $\mathbb{R}^3$ *normal to the vector* $\mathbf{q} = (q_1, q_2, q_3)$ *and containing the point* $\mathbf{u} = (u_1, u_2, u_3)$. Suppose that $q_3 \neq 0$. A coordinate chart for $\Pi_{\mathbf{u},\mathbf{q}}$ is given by $x\colon \mathbb{R}^2 \to \mathbb{R}^3$, defined by:

$$x(r,s) = \left(r, s, \frac{(u_1 - r)q_1 + (u_2 - s)q_2}{q_3} + u_3 \right).$$

The mapping x satisfies the conditions $x(u_1, u_2) = \mathbf{u}$ and $(x(r,s) - \mathbf{u}) \cdot \mathbf{q} = 0$. Thus, x maps $\mathbb{R}^2$ to $\Pi_{\mathbf{u},\mathbf{q}}$. Since x is an affine mapping, it is differentiable. The inverse of x is given by $x^{-1}(r,s,t) = (r,s)$, which is continuous. It follows that x is one-to-one and maps $\mathbb{R}^2$ onto $\Pi_{\mathbf{u},\mathbf{q}}$. Finally, the Jacobian of x is given by:

$$\begin{pmatrix} \dfrac{\partial f_1}{\partial u} & \dfrac{\partial f_1}{\partial v} \\ \dfrac{\partial f_2}{\partial u} & \dfrac{\partial f_2}{\partial v} \\ \dfrac{\partial f_3}{\partial u} & \dfrac{\partial f_3}{\partial v} \end{pmatrix} = \begin{pmatrix} 1 & 0 \\ 0 & 1 \\ -\dfrac{q_1}{q_3} & -\dfrac{q_2}{q_3} \end{pmatrix},$$

which has rank 2. Thus the planes in $\mathbb{R}^3$ are regular surfaces. (For planes that are perpendicular to a vector $\mathbf{q} \neq \mathbf{0}$ with $q_3 = 0$, a change of names of the axes allows us to apply the same argument.)

(2) Let $S^2 = \{(x,y,z) \in \mathbb{R}^3 \mid x^2+y^2+z^2 = 1\}$ denote the unit sphere in $\mathbb{R}^3$, and let $\mathbb{D} = \{(u,v) \in \mathbb{R}^2 \mid u^2+v^2 < 1\}$ denote the open unit disk in $\mathbb{R}^2$. As mentioned earlier, we can define the coordinate chart $x\colon \mathbb{D} \to S^2$ by:

$$x(u,v) = (u, v, \sqrt{1-u^2-v^2}).$$

The image of this chart only covers the upper hemisphere of S^2. The mapping is differentiable and has the inverse $x^{-1}(u_1,u_2,u_3) = (u_1,u_2)$. The Jacobian is of rank 2 and so the patch is regular. In order to cover all of S^2 with similar hemispheric charts, we need five more patches like this one corresponding to pairs of hemispheres: "upper-lower," "back-front," and "east-west." We discuss other coordinate charts on S^2 in the next chapter.

(3) Both of the previous examples are special cases of the following construction: Suppose $f\colon (U \subset \mathbb{R}^2) \to \mathbb{R}$ is a smooth function, where U is an open set in $\mathbb{R}^2$. Consider the **graph** of f, the set $\Gamma(f) = \{(u,v,f(u,v)) \mid (u,v) \in U\} \subset \mathbb{R}^3$, as a surface with a single coordinate chart $x\colon (U \subset \mathbb{R}^2) \to \Gamma(f)$ given by $x(u,v) = (u,v,f(u,v))$. Notice that derivatives of all orders exist for the function x, which is one-to-one and onto $\Gamma(f)$. Finally, the Jacobian:

$$J(x) = \begin{pmatrix} 1 & 0 \\ 0 & 1 \\ \dfrac{\partial f}{\partial u} & \dfrac{\partial f}{\partial v} \end{pmatrix} \text{ has rank 2.}$$

Thus, smooth, real-valued functions of two variables provide us with many examples of surfaces.

Constructions on curves in the plane and in space also provide many examples of surfaces.

(4) Suppose that $g\colon (a,b) \to \mathbb{R}$ is a smooth function satisfying $g(t) > 0$ for all $t \in (a,b)$. We obtain a surface in $\mathbb{R}^3$ by rotating the graph of g around the x-axis. Define a coordinate chart for this surface, $x\colon (a,b) \times (0,2\pi) \to \mathbb{R}^3$, by:

$$x(u,v) = (u, g(u)\cos v, g(u)\sin v).$$

Since g is smooth, x is differentiable. The Jacobian:

$$J(x)(u,v) = \begin{pmatrix} 1 & 0 \\ (\cos v)\partial g/\partial u & -g(u)\sin v \\ (\sin v)\partial g/\partial u & g(u)\cos v \end{pmatrix}$$

contains square submatrices with determinants $g(u)\partial g/\partial u$, $-g(u)\sin v$, and $g(u)\cos v$. For each (u,v) among these three expressions, at least one is nonzero, so $J(x)(u,v)$ has rank 2. Since $g(u) > 0$ for all $u \in (a,b)$, the function $v \mapsto (g(u)\cos v, g(u)\sin v)$ is one-to-one as a function of v, and so x is one-to-one. To construct the inverse of x, suppose (a,b,c) is in the image of the chart x and solve for v:

$$v = \begin{cases} \arccos(b/\sqrt{b^2+c^2}), & c \geq 0, \\ 2\pi - \arccos(b/\sqrt{b^2+c^2}), & c < 0. \end{cases}$$

The reader can check that this expression is continuous at the only problematic point, that is, $c = 0$. In this case, $\sin v = 0$ and hence $\cos v = -1$, and $v = \pi$, giving a continuous function.

More generally, if $\alpha\colon (a,b) \to \mathbb{R}^2$ is a regular, parametrized, differentiable curve that does not intersect the x-axis and does not intersect itself, then we obtain a surface by rotating the trace of $\alpha(t)$ around the x-axis. A natural coordinate chart for the surface $x\colon (a,b) \times (0,2\pi) \to \mathbb{R}^3$ is given by $x(t,s) = (u(t), \cos(s)v(t), \sin(s)v(t))$ where $\alpha(t) = (u(t), v(t))$. In fact, any line in the plane may be chosen as the axis of rotation generating a surface from a curve that does not intersect itself or the chosen line.

(5) Let $\alpha\colon (a,b) \to \mathbb{R}^3$ be a regular curve ($\alpha'(t) \neq \mathbf{0}$ for all t) and consider the union of the tangent lines to the curve. This collection of points in $\mathbb{R}^3$ may be coordinatized by the chart $x\colon (a,b) \times (c,d) \to \mathbb{R}^3$ given by $x(t,u) = \alpha(t) + u\alpha'(t)$, where $0 \notin (c,d)$. When (c,d) is chosen so that x is one-to-one, the surface determined by x is called a *tangent developable surface* whose properties we will develop later. Euler investigated such surfaces in (Euler 1771).

Examples 3, 4, and 5 show that the class of regular surfaces in $\mathbb{R}^3$ is large. To generate another class of surfaces, we recall two important theorems from the calculus of several variables (see Spivak (1971) for details).

Inverse Function Theorem. *Let U be an open set in $\mathbb{R}^n$, and $F\colon (U \subset \mathbb{R}^n) \to \mathbb{R}^n$ a differentiable mapping. Suppose, at $\mathbf{u} \in U$, the Jacobian matrix $J(F)(\mathbf{u}) = \left(\dfrac{\partial F_i}{\partial x_j}(\mathbf{u})\right)$ is a linear isomorphism $\mathbb{R}^n \to \mathbb{R}^n$. Then there is an open set $V \subset U$, containing $\mathbf{u}$, and an open set $W \subset \mathbb{R}^n$ containing $F(\mathbf{u})$ such that $F\colon V \to W$ has a differentiable inverse $F^{-1}\colon W \to V$.*

Implicit Function Theorem. *Suppose $f\colon \mathbb{R}^n \times \mathbb{R}^m \to \mathbb{R}^m$ is differentiable in an open set around $(\mathbf{u},\mathbf{v})$ and $f(\mathbf{u},\mathbf{v}) = \mathbf{0}$. Let M be the $m \times m$ matrix given by:*

$$M = \left(\frac{\partial f_i(\mathbf{u},\mathbf{v})}{\partial x_{n+j}}\right), \quad 1 \leq i, j \leq m.$$

If $\det M \neq 0$, there is an open set $U \subset \mathbb{R}^n$ containing $\mathbf{u}$ and an open set V containing $\mathbf{v}$ so that, for each $\mathbf{r} \in U$, there is a unique $\mathbf{s} \in V$ such that $f(\mathbf{r},\mathbf{s}) = \mathbf{0}$. If we define $g\colon U \to V$ by $g(\mathbf{r}) = \mathbf{s}$, then g is differentiable and $f(\mathbf{r}, g(\mathbf{r})) = \mathbf{0}$.

These theorems are equivalent and they provide us with another method for generating surfaces.

Definition 7.2. *Given a differentiable function $f\colon (U \subset \mathbb{R}^n) \to \mathbb{R}^m$ with U an open set in $\mathbb{R}^n$ and $n \geq m$, and denote the Jacobian of f by $J(f) = (\partial f_i/\partial x_j)$. A point $\mathbf{u} \in \mathbb{R}^n$ is a* **critical point** *of f if the linear mapping $J(f)(\mathbf{u})\colon \mathbb{R}^n \to \mathbb{R}^m$ is not onto. The point $f(\mathbf{u}) \in \mathbb{R}^m$ is called a* **critical value** *of f. If $\mathbf{v} \in \mathbb{R}^m$ is not a critical value of f, then $\mathbf{v}$ is called a* **regular value** *of f.*

Theorem 7.3. *If $f\colon (U \subset \mathbb{R}^3) \to \mathbb{R}$ is a differentiable function and $r \in f(U)$ is a regular value, then $f^{-1}(\{r\}) = \{\mathbf{u} \in U \mid f(\mathbf{u}) = r\}$ is a regular surface in $\mathbb{R}^3$.*

PROOF: Let $\mathbf{p} = (x_0, y_0, z_0) \in f^{-1}(\{r\})$. Since r is a regular value of f, we can assume that $\dfrac{\partial f}{\partial z}(\mathbf{p}) \neq 0$. Define $F\colon (U \subset \mathbb{R}^3) \to \mathbb{R}^3$ by $F(x,y,z) = (x,y,f(x,y,z))$. The Jacobian of F has the form:

$$J(F) = \begin{pmatrix} 1 & 0 & 0 \\ 0 & 1 & 0 \\ \dfrac{\partial f}{\partial x} & \dfrac{\partial f}{\partial y} & \dfrac{\partial f}{\partial z} \end{pmatrix}$$

and so $\det J(F)(\mathbf{p}) \neq 0$. Apply the Inverse Function Theorem to obtain open sets around $\mathbf{p}$ and $F(\mathbf{p})$, say V and W, respectively, such that $F\colon V \to W$ is invertible and the inverse $F^{-1}\colon W \to V$ is differentiable. To construct our coordinate charts take $U' = W \cap \{z = r\}$ and let $x\colon U' \to V \cap f^{-1}(\{r\})$ be given by $x(u,v) = F^{-1}(u,v,r)$. This coordinate chart determines a regular surface. ∎

Corollary 7.4. *For a, b, and c nonzero, the following functions have 0 as a regular value and so determine a regular surface in $\mathbb{R}^3$, namely, $f^{-1}(\{0\})$:*

Ellipsoids: $\quad f(x,y,z) = \dfrac{x^2}{a^2} + \dfrac{y^2}{b^2} + \dfrac{z^2}{c^2} - 1;$

Hyperboloids of 1 or 2 sheets: $\quad f(x,y,z) = \dfrac{x^2}{a^2} \pm \dfrac{y^2}{b^2} - \dfrac{z^2}{c^2} - 1;$

For $0 < r < a$, the torus: $\quad f(x,y,z) = z^2 + (\sqrt{x^2+y^2} - a)^2 - r^2.$

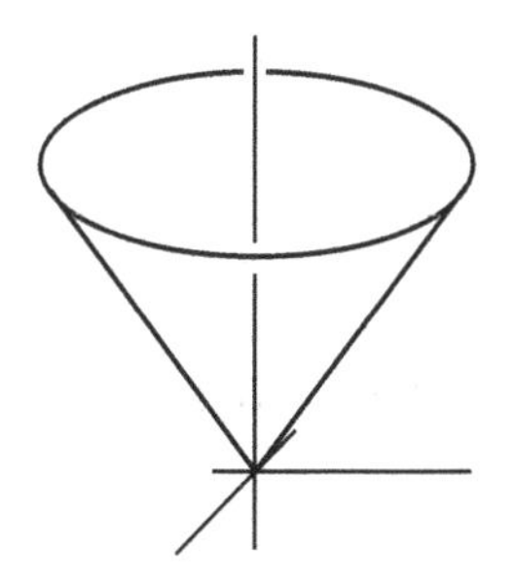

An instructive nonexample of a surface is the *cone*:

$$\{(u,v,w) \mid u^2 + v^2 = w^2 \text{ and } w \geq 0\}.$$

There is no way to introduce coordinates to a neighborhood of $(0,0,0)$ in a smooth manner, and so the cone is not a regular surface. If we weaken the requirements on a surface and substitute continuity for differentiability, then we get the definition of a *topological surface*, of which the cone is an example.

We get another nonexample by considering how the pages of a book come together. Away from the spine the pages are locally two dimensional, but if three pages come together in a line segment, no point on the spine has a neighborhood that is locally two dimensional.

Having defined surfaces, we develop the calculus on a regular surface. In particular, we define what it means for a real-valued function, $f\colon S \to \mathbb{R}$, or a function between surfaces, $\phi\colon S_1 \to S_2$, to be differentiable. We can test the differentiability of a function $f\colon S \to \mathbb{R}$ using coordinate charts $x\colon (U \subset \mathbb{R}^2) \to S$, where we would like the composite $U \xrightarrow{x} S \xrightarrow{f} \mathbb{R}$ to be differentiable. However, this presents a problem – there may be infinitely many different coordinate patches containing a given point to check for differentiability at the point. This problem is rendered unimportant by the following proposition.

Proposition 7.5 (Change of Coordinates). *Suppose S is a regular surface and $x\colon (U \subset \mathbb{R}^2) \to \mathbb{R}^3$ and $y\colon (V \subset \mathbb{R}^2) \to \mathbb{R}^3$ are coordinate charts with $W = x(U) \cap y(V) \subset S$ a nonempty set. Then the* **change of coordinates***:*

$$h = y^{-1} \circ x\colon x^{-1}(W) \to y^{-1}(W)$$

is differentiable, one-to-one, and onto, and its inverse $h^{-1} = x^{-1} \circ y$ is differentiable.

PROOF: We know that h is one-to-one, onto, and continuous since it is a composite of homeomorphisms. Since h^{-1} and h are symmetric in x and y, we only need to show that h is differentiable. Let $(r_0, s_0) \in x^{-1}(W)$ and write $p = x(r_0, s_0)$. Consider the following trick: y is a coordinate chart that we write as:

$$y(u,v) = (f_1(u,v), f_2(u,v), f_3(u,v)),$$

and so

$$J(y) = \begin{pmatrix} \partial f_1/\partial u & \partial f_1/\partial v \\ \partial f_2/\partial u & \partial f_2/\partial v \\ \partial f_3/\partial u & \partial f_3/\partial v \end{pmatrix} \text{ has rank 2.}$$

We assume that $\dfrac{\partial f_1}{\partial u}\dfrac{\partial f_2}{\partial v} - \dfrac{\partial f_1}{\partial v}\dfrac{\partial f_2}{\partial u} \neq 0$ at $(u_0, v_0) = h(r_0, s_0) = y^{-1}(p)$. Extend y to the function:

$$F\colon V \times \mathbb{R} \to \mathbb{R}^3, \quad F(u,v,t) = y(u,v) + (0,0,t).$$

The function F is differentiable and $F(u,v,0) = y(u,v)$. Consider the Jacobian of F:

$$J(F) = \begin{pmatrix} \partial f_1/\partial u & \partial f_1/\partial v & 0 \\ \partial f_2/\partial u & \partial f_2/\partial v & 0 \\ \partial f_3/\partial u & \partial f_3/\partial v & 1 \end{pmatrix}.$$

At $(u_0, v_0, 0)$ this matrix has a nonzero determinant and so, by the Inverse Function Theorem, there is an open set containing:

$$F(u_0, v_0, 0) = y(u_0, v_0) = y(h(r_0, s_0)) = x(r_0, s_0),$$

on which F^{-1} exists and is differentiable. We can write:

$$(h(r_0, s_0), 0) = (u_0, v_0, 0) = F^{-1} \circ x(r_0, s_0),$$

that is, h is obtained from the composite of F^{-1} and x, ignoring the third coordinate which is zero. Because these mappings are differentiable, so is h. ∎

Definition 7.6. *If p is a point in S and $f\colon S \to \mathbb{R}$, then f is said to be* **differentiable at** *p if, for any coordinate chart $x\colon (U \subset \mathbb{R}^2) \to S$ with $p \in x(U)$, the mapping $f \circ x\colon (U \subset \mathbb{R}^2) \to \mathbb{R}$ is differentiable at $x^{-1}(p)$. We say f is* **differentiable** *if it is differentiable at every point of S.*

Change of coordinates implies that if $f\colon S \to \mathbb{R}$ is differentiable at a point in one coordinate chart, then it is differentiable at that point for any other chart containing the point. The proposition also allows us to define what it means for a function between surfaces $\phi\colon S_1 \to S_2$ to be differentiable.

Definition 7.7. *A function $\phi\colon S_1 \to S_2$ between surfaces S_1 and S_2 is* **differentiable** *at p in S_1, if there are coordinate charts $x\colon (U \subset \mathbb{R}^2) \to S_1$ and $y\colon (V \subset \mathbb{R}^2) \to S_2$ with $p \in x(U)$, $\phi(p) \in y(V)$, such that the composite function $y^{-1} \circ \phi \circ x\colon U \to V$ is differentiable at $x^{-1}(p)$. We say $\phi\colon S_1 \to S_2$ is* **differentiable** *if it is differentiable at each point in S_1. A function $\phi\colon S_1 \to S_2$ is a* **diffeomorphism** *if it is differentiable, one-to-one, and onto, and has a differentiable inverse function. Two surfaces S_1 and S_2 are said to be* **diffeomorphic** *if there is a diffeomorphism $\phi\colon S_1 \to S_1$.*

EXAMPLE: The ellipsoid $\left\{(x,y,z) \in \mathbb{R}^3 \mid \frac{x^2}{a^2} + \frac{y^2}{b^2} + \frac{z^2}{c^2} = 1\right\}$ and the sphere S^2 are diffeomorphic via the diffeomorphism $\phi\colon S^2 \to$ the ellipsoid, given by:

$$\phi(x,y,z) = (ax, by, cz).$$

The notion of diffeomorphism determines an equivalence relation on the set of all regular surfaces in $\mathbb{R}^3$. Two diffeomorphic surfaces are equivalent in the sense that they share equivalent sets of differentiable functions and any construction involving the calculus on one carries over via the diffeomorphism to the other. This equivalence relation is the basis for the differential topology of surfaces.

The tangent plane

The derivative of a function is the best linear approximation of the function at a point in the domain. A geometric realization of linear approximation is the tangent plane—a linear subspace of $\mathbb{R}^3$ that is the best approximation to a surface at a point. To define the tangent plane we restrict our attention to a single coordinate patch, $x\colon (U \subset \mathbb{R}^2) \to S$. Suppose $p \in x(U)$ and $p = x(u_0, v_0)$. Since U is an open set containing (u_0, v_0), there are values $\epsilon > 0$ and $\eta > 0$ with $(u_0 - \epsilon, u_0 + \epsilon) \times (v_0 - \eta, v_0 + \eta) \subset U$. These conditions determine two curves,

$$u \mapsto x(u, v_0) \text{ and } v \mapsto x(u_0, v),$$

called the **coordinate curves** on S through p. Define:

$$x_u = \frac{\partial}{\partial u}(x(u, v_0))\Big|_{(u_0,v_0)} = \frac{\partial x}{\partial u}(u_0, v_0),$$

$$\text{and } x_v = \frac{\partial}{\partial v}(x(u_0, v))\Big|_{(u_0,v_0)} = \frac{\partial x}{\partial v}(u_0, v_0);$$

x_u and x_v denote the **tangent vectors** in $\mathbb{R}^3$ to the coordinate curves at p.

Proposition 7.8. *A mapping $x\colon (U \subset \mathbb{R}^2) \to S$ satisfies the regularity condition if and only if the cross product $x_u \times x_v$ is nonzero.*

PROOF: The coordinate expressions for x_u and x_v are the columns of the Jacobian $J(x)$. Since $x_u \times x_v = \mathbf{0}$, if and only if x_u is a scalar multiple of x_v, if and only if $J(x)(u_0, v_0)$ has a rank less than two, we have proved the proposition. ■

It follows that for each coordinate patch on a regular surface in $\mathbb{R}^3$ we can define a **normal direction**, that is, $x_u \times x_v$. We define the **unit normal vector** to the surface S at a point p as

$$N(p) = \frac{x_u \times x_v}{\|x_u \times x_v\|}.$$

In order for this notion to be well-defined, we need to show that it is independent of the choice of patch containing p. As it turns out, this may be too much to ask for.

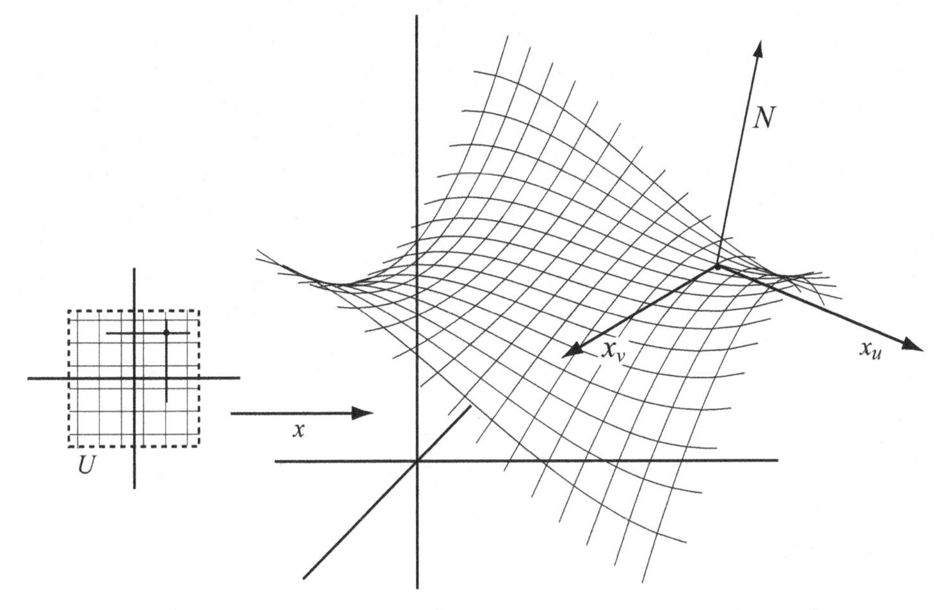

Proposition 7.9. *If p is a point in a regular surface S, and $x\colon (U \subset \mathbb{R}^2) \to S$ and $y\colon (V \subset \mathbb{R}^2) \to S$ are two coordinate charts containing p, then at p we have:*

$$\frac{x_u \times x_v}{\|x_u \times x_v\|} = \pm \frac{y_{\bar{u}} \times y_{\bar{v}}}{\|y_{\bar{u}} \times y_{\bar{v}}\|}.$$

PROOF: Let $W = x(U) \cap y(V)$. The subsets $x^{-1}(W) \subset U$ and $y^{-1}(W) \subset V$ are related by the diffeomorphism:

$$y^{-1} \circ x\colon x^{-1}(W) \to y^{-1}(W), \quad y^{-1} \circ x(u,v) = (\bar{u}(u,v), \bar{v}(u,v)).$$

The coordinate curves in the chart x are $x(u, v_0)$ and $x(u_0, v)$. They are expressed in the chart y by using $x(u,v) = y(\bar{u}(u,v), \bar{v}(u,v))$. The tangent vectors become:

$$x_u = y_{\bar{u}} \frac{\partial \bar{u}}{\partial u} + y_{\bar{v}} \frac{\partial \bar{v}}{\partial u} \quad \text{and} \quad x_v = y_{\bar{u}} \frac{\partial \bar{u}}{\partial v} + y_{\bar{v}} \frac{\partial \bar{v}}{\partial v}.$$

We can compute $x_u \times x_v$ directly to get:

$$\begin{aligned} x_u \times x_v &= \left(y_{\bar{u}} \frac{\partial \bar{u}}{\partial u} + y_{\bar{v}} \frac{\partial \bar{v}}{\partial u} \right) \times \left(y_{\bar{u}} \frac{\partial \bar{u}}{\partial v} + y_{\bar{v}} \frac{\partial \bar{v}}{\partial v} \right) \\ &= \left(\frac{\partial \bar{u}}{\partial u} \frac{\partial \bar{v}}{\partial v} - \frac{\partial \bar{v}}{\partial u} \frac{\partial \bar{u}}{\partial v} \right) y_{\bar{u}} \times y_{\bar{v}} = \det J(y^{-1} \circ x) y_{\bar{u}} \times y_{\bar{v}}. \end{aligned}$$

It follows that:

$$\frac{x_u \times x_v}{\|x_u \times x_v\|} = \frac{\det J(y^{-1} \circ x)}{|\det J(y^{-1} \circ x)|} \frac{y_{\bar{u}} \times y_{\bar{v}}}{\|y_{\bar{u}} \times y_{\bar{v}}\|} = \pm \frac{y_{\bar{u}} \times y_{\bar{v}}}{\|y_{\bar{u}} \times y_{\bar{v}}\|}. \qquad \blacksquare$$

On a given coordinate chart, the normal is simply $\dfrac{x_u \times x_v}{\|x_u \times x_v\|}$. By interchanging u and v we get a chart with normal pointing in the opposite direction. We can "translate" the normal over the surface by moving patch to patch, where we choose the variables u and v so that the unit normal vectors agree on the overlaps. However, for surfaces such as the Möbius band, this cannot be done. To physically construct the Möbius band, take a long rectangular band of paper, make a half-twist, and glue the ends of the band together. The mathematical version of this construction gives a surface when we ignore the boundary (see the exercises). When we move the normal at a point around the surface, we arrive back at the first chart from which the translation began, and the normal has changed sign. Thus the sign in Proposition 7.9 is unavoidable. We distinguish the class of surfaces for which a coherent normal can be chosen.

Definition 7.10. *We say that a surface S is* **orientable** *if there is a collection of charts $\{x_\alpha : (U_\alpha \subset \mathbb{R}^2) \to S \mid \alpha \in A\}$ that cover S, that is, $S = \bigcup_{\alpha \in A} x_\alpha(U_\alpha)$, and if $p \in S$ lies in the overlap of two charts, $p \in x_\beta(U_\beta) \cap x_\gamma(U_\gamma)$, then the Jacobian $J(x_\gamma^{-1} \circ x_\beta)$ at p has positive determinant.*

Since the sign of the determinant of the Jacobian determines whether the unit normals at a point in the overlap of coordinate charts agree, the definition of an orientable surface lets us translate a unit normal at any given point over the rest of the surface.

Even without assuming orientability of a regular surface, $\pm N(p)$ does determine a plane.

Definition 7.11. *Let p be a point in a regular surface S, and $x : (U \subset \mathbb{R}^2) \to S$ any coordinate chart containing p. The vector subspace of $\mathbb{R}^3$ of vectors normal to $N(p) = \dfrac{x_u \times x_v}{\|x_u \times x_v\|}$ is called the* **tangent plane** *to S at p and is denoted by $T_p(S)$.*

Notice that $T_p(S)$ is a plane through the origin and so does not necessarily contain the point p. Translating this plane by adding the vector determined by p we get the plane tangent to S at p. It is visually satisfying to think of $T_p(S)$ in this fashion. However, it is more important to think of $T_p(S)$ as a set of directions, admitting the operations of addition and multiplication by a scalar, that is, a vector space.

An immediate consequence of the definition is that $\{x_u, x_v\}$ is a basis for $T_p(S)$. In fact, $T_p(S)$ contains the direction of the tangent vector to any regular curve on the surface S through p.

Proposition 7.12. *A vector $\mathbf{v} \in \mathbb{R}^3$ is an element of $T_p(S)$ if and only if there is a differentiable curve $\alpha : (-\epsilon, \epsilon) \to S$ with $\alpha(0) = p$ and $\mathbf{v} = \alpha'(0)$.*

PROOF: Suppose that $\alpha: (-\epsilon,\epsilon) \to S$ is a curve lying on S with $\alpha(0) = p$. Choose $\epsilon > 0$ small enough so that the image of α lies entirely in the image of a coordinate chart $x: (U \subset \mathbb{R}^2) \to S$ containing p. We write the composite $x^{-1} \circ \alpha: (-\epsilon,\epsilon) \to U$ as $(x^{-1} \circ \alpha)(t) = (u(t), v(t))$, and so $\alpha(t) = x(u(t), v(t))$. By the Chain Rule,

$$\alpha'(t) = \frac{\partial x}{\partial u}\frac{du}{dt} + \frac{\partial x}{\partial v}\frac{dv}{dt} = x_u \frac{du}{dt} + x_v \frac{dv}{dt},$$

that is, $\alpha'(t)$ is a linear combination of x_u and x_v. Thus $\alpha'(0)$ lies in $T_p(S)$.

If $\mathbf{v} \in T_p(S)$, then we can write $\mathbf{v} = ax_u + bx_v$. Suppose $p = x(u_0, v_0)$. Consider the curve $\beta(t) = x(u_0 + at, v_0 + bt)$ for $-\epsilon < t < \epsilon$ small enough so that $(u_0 + at, v_0 + bt) \in U$. Then $\beta'(0) = \mathbf{v}$. Thus every vector in $T_p(S)$ is realized as a tangent vector to some curve on the surface. ∎

A differentiable mapping between surfaces induces a mapping of tangent planes at corresponding points. To define the mapping, suppose $F: S_1 \to S_2$ is a differentiable function and $p \in S_1$. A tangent vector $\mathbf{v} \in T_p(S_1)$ corresponds to $\alpha'(0)$ for some curve $\alpha: (-\epsilon,\epsilon) \to S_1$. Composition with F determines another curve $F \circ \alpha: (-\epsilon,\epsilon) \to S_2$. Define $dF_p(\mathbf{v}) = dF_p(\alpha'(0)) = \left.\frac{d}{dt}(F \circ \alpha)\right|_{t=0}$.

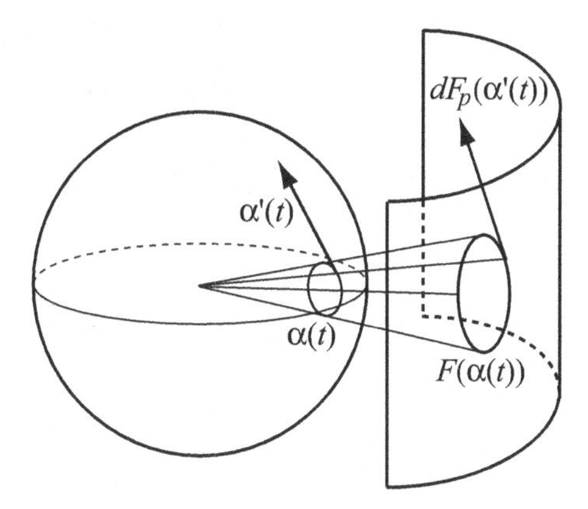

Proposition 7.13. *The mapping $dF_p: T_p(S_1) \to T_p(S_2)$ is linear.*

PROOF: Suppose p and $F(p)$ lie in the images of coordinate charts x on S_1 and y on S_2. Consider the composite:

$$(U \subset \mathbb{R}^2) \xrightarrow{x} S_1 \xrightarrow{F} S_2 \xrightarrow{y^{-1}} (V \subset \mathbb{R}^2).$$

In these local coordinates F can be rewritten as $F \circ x(u,v) = y(F_1(u,v), F_2(u,v))$. If a curve $\alpha: (-\epsilon,\epsilon) \to S_1$ is given by $\alpha(t) = x(u(t), v(t))$, then we can write:

$$F \circ \alpha(t) = y(F_1(u(t), v(t)), F_2(u(t), v(t))).$$

Suppose $\mathbf{v} = \alpha'(0) \in T_p(S_1)$; with respect to the basis $\{x_u, x_v\}$ we can write $\mathbf{v} = x_u \frac{du}{dt} + x_v \frac{dv}{dt}$. With all this in place we compute:

$$dF_p(\mathbf{v}) = \frac{d}{dt}(F \circ \alpha)\bigg|_{t=0} = \left(\frac{\partial F_1}{\partial u}\frac{du}{dt} + \frac{\partial F_1}{\partial v}\frac{dv}{dt}, \frac{\partial F_2}{\partial u}\frac{du}{dt} + \frac{\partial F_2}{\partial v}\frac{dv}{dt}\right)\bigg|_{t=0}$$

$$= \begin{pmatrix} \frac{\partial F_1}{\partial u} & \frac{\partial F_1}{\partial v} \\ \frac{\partial F_2}{\partial u} & \frac{\partial F_2}{\partial v} \end{pmatrix} \begin{pmatrix} \frac{du}{dt} \\ \frac{dv}{dt} \end{pmatrix}\Bigg|_{t=0}$$

Thus, in the basis $\{x_u, x_v\}$, dF_p is given by multiplication by the matrix determined by the Jacobian $J(y^{-1} \circ F \circ x)(p)$. ∎

Notice that the previous argument applies to the identity mapping $\mathrm{id}\colon S \to S$ and two different coordinate charts containing a given point p. The change of basis between the bases $\{x_u, x_v\}$ and $\{y_{\bar{u}}, y_{\bar{v}}\}$ is given by the Jacobian of the composite $y^{-1} \circ x$. This fact is useful when we check certain constructions for invariance under change of coordinates.

The first fundamental form

In the study of linear algebra, a vector space possesses a geometry when there is an inner product defined on it. On $\mathbb{R}^3$ the usual dot product provides such an inner product.

Definition 7.14. *For a point p in $S \subset \mathbb{R}^3$ a regular surface, the* **first fundamental form**

$$\mathrm{I}_p\colon T_p(S) \times T_p(S) \to \mathbb{R}$$

is the inner product on $T_p(S)$ induced by the dot product on $\mathbb{R}^3$. We can write $\mathrm{I}_p(\mathbf{v}, \mathbf{w}) = \mathbf{v} \cdot \mathbf{w}$. *We also denote* $\mathrm{I}_p(\mathbf{v}, \mathbf{w})$ *by* $\langle \mathbf{v}, \mathbf{w} \rangle_p$.

Notice that the definition is dependent on the fact that S is a subset of $\mathbb{R}^3$. This fact appears undesirable at first as the inner product seems to be an artifact of the way in which S lies in $\mathbb{R}^3$ and not a geometric feature of the surface. Later we will show that important geometric properties of the surfaces can be developed from the fundamental form, properties that are independent of the embedding.

In order to calculate with the first fundamental form, it is useful to have a local expression, that is, a formula in terms of a coordinate chart. For $p \in x(U) \subset S$, the tangent plane $T_p(S)$ is spanned by x_u and x_v. Tangent vectors $\mathbf{v} = \alpha'(0)$ and $\mathbf{w} = \beta'(0)$ in $T_p(S)$ are associated to curves $\alpha\colon (-\epsilon, \epsilon) \to S$ and $\beta\colon (-\eta, \eta) \to S$ with $\alpha(0) = \beta(0) = p$. If we write $\alpha(t) = x(u(t), v(t))$ and $\beta(t) = x(\bar{u}(t), \bar{v}(t))$, then we can compute $\mathrm{I}_p(\mathbf{v},\mathbf{w})$ by:

$$\mathrm{I}_p(\mathbf{v},\mathbf{w}) = \mathrm{I}_p(\alpha'(0),\beta'(0)) = \mathrm{I}_p\left(x_u\frac{du}{dt}+x_v\frac{dv}{dt}, x_u\frac{d\bar{u}}{dt}+x_v\frac{d\bar{v}}{dt}\right)$$
$$= \mathrm{I}_p(x_u,x_u)\frac{du}{dt}\frac{d\bar{u}}{dt} + \mathrm{I}_p(x_u,x_v)\left(\frac{du}{dt}\frac{d\bar{v}}{dt}+\frac{d\bar{u}}{dt}\frac{dv}{dt}\right) + \mathrm{I}_p(x_v,x_v)\frac{dv}{dt}\frac{d\bar{v}}{dt}.$$

We introduce the functions, $E, F, G\colon (U \subset \mathbb{R}^2) \to \mathbb{R}$, dependent on the choice of coordinate chart $x\colon (U \subset \mathbb{R}^2) \to S$, given by:

$$E(u,v) = \mathrm{I}_p(x_u,x_u), \quad F(u,v) = \mathrm{I}_p(x_u,x_v), \text{ and } G(u,v) = \mathrm{I}_p(x_v,x_v).$$

These functions are differentiable and called the **component functions of the metric** on S. If we write tangent vectors as column vectors with respect to the basis $\{x_u, x_v\}$, then $\mathbf{v} = \begin{pmatrix} du/dt \\ dv/dt \end{pmatrix}$ and $\mathbf{w} = \begin{pmatrix} d\bar{u}/dt \\ d\bar{v}/dt \end{pmatrix}$, and we can express the first fundamental form by the matrix equation:

$$\mathrm{I}_p(\mathbf{v},\mathbf{w}) = \mathbf{v}^t \begin{pmatrix} E & F \\ F & G \end{pmatrix} \mathbf{w},$$

where $\mathbf{v}^t$ is the transpose of $\mathbf{v}$, that is, $\mathbf{v}$ as a row vector.

Notice that the determinant of the matrix giving the first fundamental form is given by:

$$\det\begin{pmatrix} E & F \\ F & G \end{pmatrix} = \det\begin{pmatrix} x_u \cdot x_u & x_u \cdot x_v \\ x_u \cdot x_v & x_v \cdot x_v \end{pmatrix} = \|x_u\|^2\|x_v\|^2 - (x_u \cdot x_v)^2 = \|x_u \times x_v\|^2$$

by Lagrange's formula. For a regular surface $\|x_u \times x_v\|^2 > 0$, and so the matrix of component functions of the metric is always nonsingular.

EXAMPLES:

(1) The uv-plane, $\Pi_{(0,0,0),(0,0,1)} = \{(u,v,0) \in \mathbb{R}^3\}$, has coordinates given by the single chart $x\colon \mathbb{R}^2 \to \mathbb{R}^3$, $x(u,v) = (u,v,0)$. For this chart $x_u = (1,0,0)$ and $x_v = (0,1,0)$. The components of the metric are given by the functions:

$$E(u,v) = 1 = G(u,v), \text{ and } F(u,v) = 0.$$

If we remove the origin and the positive u-axis, there is another chart given by polar coordinates (r,θ). The chart takes the form:

$$y\colon (0,\infty) \times (0,2\pi) \to \mathbb{R}^3, \quad y(r,\theta) = (r\cos\theta, r\sin\theta, 0).$$

Then $y_r = (\cos\theta, \sin\theta, 0)$ and $y_\theta = (-r\sin\theta, r\cos\theta, 0)$. The components of the metric are given by:

$$E(r,\theta) = 1, \; F(r,\theta) = 0, \; G(r,\theta) = r^2.$$

(2) Standard spherical coordinates on S^2 (see chapter 1) are given by

$$x\colon (0,2\pi) \times (0,\pi) \to S^2, \quad x(\theta,\phi) = (\cos\theta\sin\phi, \sin\theta\sin\phi, \cos\phi).$$

For this chart,

$$x_\theta = (-\sin\theta\sin\phi, \cos\theta\sin\phi, 0) \text{ and } x_\phi = (\cos\theta\cos\phi, \sin\theta\cos\phi, -\sin\phi)$$

and so the components of the metric are given by:

$$E(\theta,\phi) = \sin^2\theta,\ F(\theta,\phi) = 0, \text{ and } G(\theta,\phi) = 1.$$

In the language of linear algebra, the inner product on the vector space $T_p(S)$ given by the first fundamental form is **positive definite**, that is, $\mathrm{I}_p(\mathbf{v},\mathbf{v}) \geq 0$ and zero if and only if $\mathbf{v} = \mathbf{0}$. Furthermore, I_p is **nonsingular**, that is, $\mathrm{I}_p(\mathbf{v}_0, \mathbf{w}) = 0$ for all $\mathbf{w} \in T_p(S)$ if and only if $\mathbf{v}_0 = \mathbf{0}$. These properties are inherited from the dot product on $\mathbb{R}^3$, and can also be established from the local expressions and the components of the metric. These properties suggest possible generalizations that we will study later (chapter 13).

Lengths, angles, and areas

The first fundamental form allows us to calculate the lengths of curves on a surface. If $\alpha\colon (a,b) \to S$ is a curve on S, then the length of the curve between $\alpha(t_0)$ and $\alpha(t)$ is given by:

$$s(t) = \int_{t_0}^{t} \|\alpha'(r)\|\,dr = \int_{t_0}^{t} \sqrt{\mathrm{I}_p(\alpha'(r),\alpha'(r))}\,dr.$$

In local variables, write $\alpha(t) = x(u(t), v(t))$ and $s(t)$ can be expressed by the expression:

$$\int_{t_0}^{t} \sqrt{E(u(r),v(r))\left(\frac{du}{dr}\right)^2 + 2F(u(r),v(r))\frac{du}{dr}\frac{dv}{dr} + G(u(r),v(r))\left(\frac{dv}{dr}\right)^2}\,dr.$$

It is customary to abbreviate this expression as:

$$ds^2 = E\,du^2 + 2F\,du\,dv + G\,dv^2,$$

and call ds the **element of arc length** or the **line element** on S. This expression is shorthand for:

$$\frac{ds}{dt} = \sqrt{E\left(\frac{du}{dt}\right)^2 + 2F\frac{du}{dt}\frac{dv}{dt} + G\left(\frac{dv}{dt}\right)^2}.$$

For example, if $E = G = 1$ and $F = 0$, then $ds^2 = du^2 + dv^2$, an infinitesimal version of the Pythagorean Theorem.

The rules for transforming the local expressions for a line element follow from the calculus, and conveniently they are what we would expect algebraically of such expressions.

EXAMPLES:

(1) We have computed the line element for the uv-plane $\{(u,v,0) \in \mathbb{R}^3\}$ in the usual coordinate chart and polar coordinates:

$$ds^2 = du^2 + dv^2 = dr^2 + r^2 d\theta^2.$$

Writing $u = r\cos\theta$ and $v = r\sin\theta$, then $du = -r\sin\theta d\theta + \cos\theta dr$, $dv = r\cos\theta d\theta + \sin\theta$. Squaring and summing we have:

$$\begin{aligned} du^2 + dv^2 &= r^2\cos^2\theta\, d\theta^2 - 2r\sin\theta\cos\theta\, d\theta\, dr + \cos^2\theta\, dr^2 \\ &\qquad + r^2\sin^2\theta\, d\theta^2 + 2r\sin\theta\cos\theta\, d\theta\, dr + \sin^2\theta\, dr^2 \\ &= dr^2 + r^2\, d\theta^2. \end{aligned}$$

This example shows how the algebra works.

(2) More generally, consider a plane in $\mathbb{R}^3$ that is parallel to the plane spanned by the basis $\{\mathbf{w}_1, \mathbf{w}_2\}$ containing the given point $\mathbf{p}_0$. We can parametrize this plane by the coordinate chart $x\colon \mathbb{R}^2 \to \mathbb{R}^3$, $x(u,v) = \mathbf{p}_0 + u\mathbf{w}_1 + v\mathbf{w}_2$. This coordinate chart has $x_u = \mathbf{w}_1$ and $x_v = \mathbf{w}_2$. The associated line element is:

$$ds^2 = \|\mathbf{w}_1\|^2 du^2 + 2(\mathbf{w}_1 \cdot \mathbf{w}_2) du dv + \|\mathbf{w}_2\|^2 dv^2.$$

Taking $\mathbf{w}_1$ and $\mathbf{w}_2$ to be unit vectors with $\mathbf{w}_1$ perpendicular to $\mathbf{w}_2$, we get $ds^2 = du^2 + dv^2$, once again, the *Euclidean* line element.

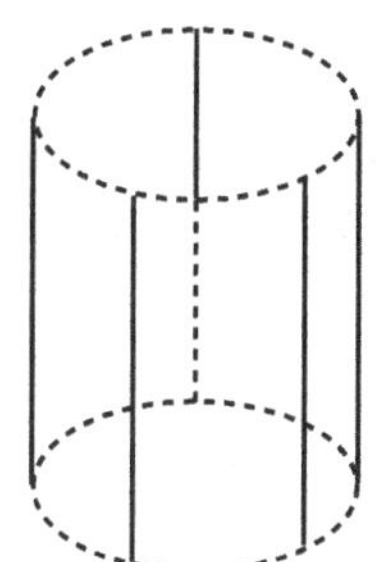

(3) Consider the cylinder over the unit circle with coordinate chart

$x\colon (0, 2\pi) \times \mathbb{R} \to \mathbb{R}^3$ given by $x(\theta, v) = (\cos\theta, \sin\theta, v)$.

Here, $x_\theta = (-\sin\theta, \cos\theta, 0)$ and $x_v = (0,0,1)$ and so

$E(\theta, v) = \sin^2\theta + \cos^2\theta = 1,\ F = 0,$ and $G = 1$.

Once again we have $ds^2 = d\theta^2 + dv^2$. To illustrate how this line element is the same as the Euclidean line element, consider the cylindrical rollers used to print a newspaper – no distortion of the print takes place as the roller transfer print onto a planar page. The cylinder is a piece of the plane rolled up without altering any geometric relations between points.

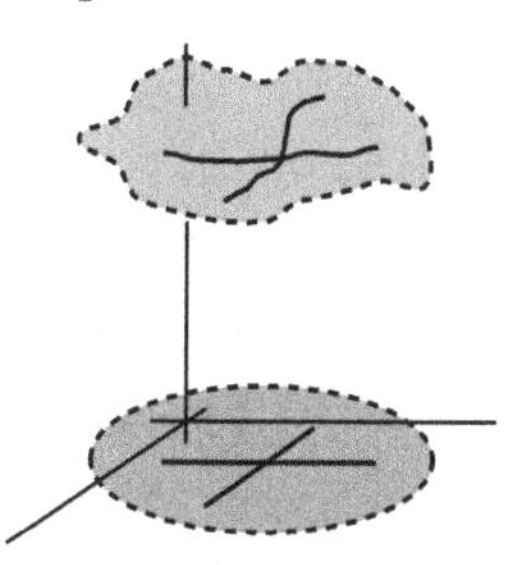

(4) Let $f\colon (U \subset \mathbb{R}^2) \to \mathbb{R}$ be a differentiable function and consider the surface $\Gamma(f) = \{(u, v, f(u,v)) \mid (u,v) \in U\}$, the graph of f, with the coordinate chart defined by $x(u,v) = (u, v, f(u,v))$. Then,

$$x_u = (1, 0, \partial f/\partial u) \text{ and } x_v = (0, 1, \partial f/\partial v).$$

It follows that the line element is given by:

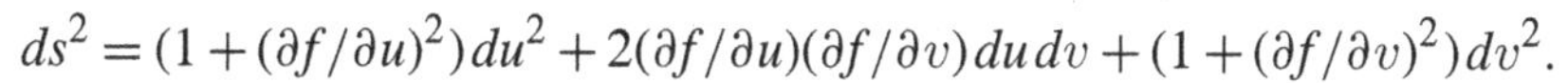

$$ds^2 = (1 + (\partial f/\partial u)^2) du^2 + 2(\partial f/\partial u)(\partial f/\partial v) du dv + (1 + (\partial f/\partial v)^2) dv^2.$$

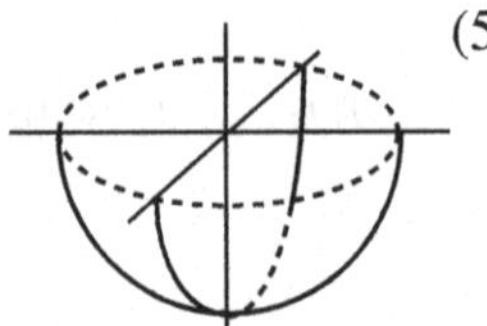

(5) Let S_R^2 denote the sphere of radius R centered at the origin in $\mathbb{R}^3$. Consider the coordinate chart given by $x\colon \{(u,v) \mid u^2+v^2<R\} \to S_R^2$ given by:

$$x(u,v) = (u, v, -\sqrt{R^2-u^2-v^2}).$$

The tangent vectors to the coordinate curves are $x_u = (1, 0, \dfrac{u}{\sqrt{R^2-u^2-v^2}})$ and $x_v = (0, 1, \dfrac{v}{\sqrt{R^2-u^2-v^2}})$. When we compute the dot products and simplify, we get the line element:

$$ds^2 = \frac{R^2-v^2}{R^2-u^2-v^2}\,du^2 + \frac{2uv}{R^2-u^2-v^2}\,du\,dv + \frac{R^2-u^2}{R^2-u^2-v^2}\,dv^2.$$

The expression for the arc length of a curve in a surface, given locally by the line element, is something that should depend only on the curve and not on the choice of coordinates. How does a line element transform under a change of coordinates? Suppose $\alpha\colon (-\epsilon,\epsilon) \to S$ is a curve on S lying in the intersection of two coordinate charts:

$$\alpha(t) = x(u(t), v(t)) = y(\bar{u}(t), \bar{v}(t)).$$

Notice that $(\bar{u}(t), \bar{v}(t)) = y^{-1} \circ x(u(t), v(t))$. Expressing the tangent vector to the curve $\alpha(t)$ in these different coordinates we get:

$$\alpha'(t) = x_u \frac{du}{dt} + x_v \frac{dv}{dt} = y_{\bar{u}} \frac{d\bar{u}}{dt} + y_{\bar{v}} \frac{d\bar{v}}{dt}.$$

Write the component functions for the metric associated to x as E, F, and G, and those associated to y as $\bar{E}$, $\bar{F}$, and $\bar{G}$. The change of coordinates $x(u,v) = y(\bar{u}(u,v), \bar{v}(u,v))$ implies:

$$x_u = y_{\bar{u}} \frac{\partial \bar{u}}{\partial u} + y_{\bar{v}} \frac{\partial \bar{v}}{\partial u}, \quad x_v = y_{\bar{u}} \frac{\partial \bar{u}}{\partial v} + y_{\bar{v}} \frac{\partial \bar{v}}{\partial v}$$

$$E = x_u \cdot x_u = \bar{E}\left(\frac{\partial \bar{u}}{\partial u}\right)^2 + 2\bar{F}\frac{\partial \bar{u}}{\partial u}\frac{\partial \bar{v}}{\partial u} + \bar{G}\left(\frac{\partial \bar{v}}{\partial u}\right)^2$$

$$F = x_u \cdot x_v = \bar{E}\frac{\partial \bar{u}}{\partial u}\frac{\partial \bar{u}}{\partial v} + \bar{F}\left(\frac{\partial \bar{u}}{\partial u}\frac{\partial \bar{v}}{\partial v} + \frac{\partial \bar{u}}{\partial v}\frac{\partial \bar{v}}{\partial u}\right) + \bar{G}\frac{\partial \bar{v}}{\partial u}\frac{\partial \bar{v}}{\partial v}$$

$$G = x_v \cdot x_v = \bar{E}\left(\frac{\partial \bar{u}}{\partial v}\right)^2 + 2\bar{F}\frac{\partial \bar{u}}{\partial v}\frac{\partial \bar{v}}{\partial v} + \bar{G}\left(\frac{\partial \bar{v}}{\partial v}\right)^2.$$

We can write these equations in a convenient matrix form:

$$\begin{pmatrix} E & F \\ F & G \end{pmatrix} = \begin{pmatrix} \partial\bar{u}/\partial u & \partial\bar{v}/\partial u \\ \partial\bar{u}/\partial v & \partial\bar{v}/\partial v \end{pmatrix} \begin{pmatrix} \bar{E} & \bar{F} \\ \bar{F} & \bar{G} \end{pmatrix} \begin{pmatrix} \partial\bar{u}/\partial u & \partial\bar{u}/\partial v \\ \partial\bar{v}/\partial u & \partial\bar{v}/\partial v \end{pmatrix}$$

$$= J(y^{-1} \circ x)^t \begin{pmatrix} \bar{E} & \bar{F} \\ \bar{F} & \bar{G} \end{pmatrix} J(y^{-1} \circ x).$$

It follows that the change of component functions of the metric is made by multiplying the given matrix of the component functions on the right by the Jacobian of the transformation $(\bar{u},\bar{v}) = y^{-1} \circ x(u,v)$ and multiplying again by its transpose on the left.

Line elements may be compared by noticing from these computations that:

$$\begin{pmatrix} du/dt \\ dv/dt \end{pmatrix} = J(y^{-1} \circ x) \begin{pmatrix} d\bar{u}/dt \\ d\bar{v}/dt \end{pmatrix}.$$

This leads to the equation:

$$\begin{aligned} ds^2 &= (du\ dv)\begin{pmatrix} E & F \\ F & G \end{pmatrix}\begin{pmatrix} du \\ dv \end{pmatrix} = (du\ dv) J(y^{-1} \circ x)^t \begin{pmatrix} \bar{E} & \bar{F} \\ \bar{F} & \bar{G} \end{pmatrix} J(y^{-1} \circ x)\begin{pmatrix} du \\ dv \end{pmatrix} \\ &= \left(J(y^{-1} \circ x)\begin{pmatrix} du \\ dv \end{pmatrix}\right)^t \begin{pmatrix} \bar{E} & \bar{F} \\ \bar{F} & \bar{G} \end{pmatrix}\left(J(y^{-1} \circ x)\begin{pmatrix} du \\ dv \end{pmatrix}\right) = (d\bar{u}\ d\bar{v})\begin{pmatrix} \bar{E} & \bar{F} \\ \bar{F} & \bar{G} \end{pmatrix}\begin{pmatrix} d\bar{u} \\ d\bar{v} \end{pmatrix}. \end{aligned}$$

Thus, the line element is unchanged by a change of coordinates. The length of a curve on a surface is a geometric value, independent of the choice of coordinate chart in which it is computed. The transformations between $\{E,F,G\}$ and $\{\bar{E},\bar{F},\bar{G}\}$, and between (du,dv) and $(d\bar{u},d\bar{v})$ led to the invariance of ds. Similar expressions that do not change with the choice of coordinates may be interpreted as truly geometric expressions on the surface. The general formalism for this kind of invariance is developed in the tensor calculus (see chapter 15).

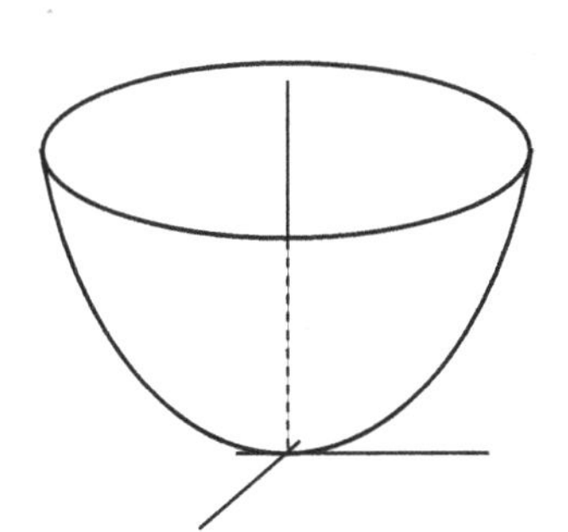

EXAMPLE: In practice, the transformation of the line element is carried out like a change of variable in integration. Consider the surface S given by the graph u^2+v^2. One coordinate chart is given by:

$$x\colon \mathbb{R}^2 \to S, \quad x(u,v) = (u,v,u^2+v^2).$$

This paraboloid also has polar coordinates:

$$y\colon (0,\infty)\times(0,2\pi) \to S, \quad y(r,\theta) = (r\cos\theta, r\sin\theta, r^2).$$

Being a graph, the surface has its line element given by:

$$ds^2 = (1+4u^2)du^2 + (8uv)\,du\,dv + (1+4v^2)dv^2.$$

To change to polar coordinates, we write $u = r\cos\theta$, $v = r\sin\theta$, which implies:

$$du = \cos\theta\,dr - r\sin\theta\,d\theta, \quad dv = \sin\theta\,dr + r\cos\theta\,d\theta.$$

Substitute these expressions for u, v, du, and dv into the given line element to obtain:

$$\begin{aligned} ds^2 &= (1+4r^2\cos^2\theta)(\cos^2\theta\,dr^2 - 2r\sin\theta\cos\theta\,dr\,d\theta + r^2\sin^2\theta\,d\theta^2) \\ &\quad + 8r^2\cos\theta\sin\theta(\cos\theta\sin\theta\,dr^2 + r(\cos^2\theta - \sin^2\theta)dr\,d\theta - r^2\sin\theta\cos\theta\,d\theta^2) \\ &\quad + (1+4r^2\sin^2\theta)(\sin^2\theta\,dr^2 + 2r\sin\theta\cos\theta\,dr\,d\theta + r^2\cos^2\theta\,d\theta^2) \\ &= (1+4r^2)dr^2 + r^2\,d\theta^2. \end{aligned}$$

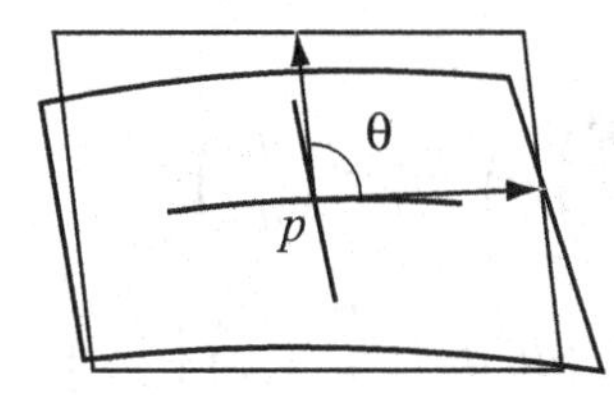

Given two curves $\alpha\colon(-\epsilon,\epsilon)\to S$ and $\beta\colon(-\epsilon,\epsilon)\to S$ on a surface S that intersect at a point $p=\alpha(0)=\beta(0)$, the angle between the curves is given by the angle between the tangent vectors $\alpha'(0)$ and $\beta'(0)$ in $T_p(S)$. Using the first fundamental form on $T_p(S)$, the usual formula for the angle between vectors applies: Suppose θ denotes the angle between $\alpha(t)$ and $\beta(t)$ at p, then:

$$\cos\theta = \frac{\mathrm{I}_p(\alpha'(0),\beta'(0))}{\sqrt{\mathrm{I}_p(\alpha'(0),\alpha'(0))\,\mathrm{I}_p(\beta'(0),\beta'(0))}}.$$

For example, given a coordinate chart on any surface, the angle between the coordinate curves is the angle θ between x_u and x_v at a point p, giving:

$$\cos\theta = \frac{\mathrm{I}_p(x_u,x_v)}{\sqrt{\mathrm{I}_p(x_u,x_u)\,\mathrm{I}_p(x_v,x_v)}} = \frac{F}{\sqrt{EG}}.$$

It follows that x_u and x_v are orthogonal when $F=0$. We call such a coordinate chart an **orthogonal parametrization**.

EXAMPLE: In the earlier computation for the paraboloid $\{(u,v,u^2+v^2)\}$, the rectangular coordinates give:

$$\frac{F}{\sqrt{EG}} = \frac{4uv}{\sqrt{(1+4u^2)(1+4v^2)}}.$$

Coordinate curves for this surface are representable as the intersections of the surface with planes that are vertical and parallel to the u- or v-axis. The resulting curves meet at a point in the surface in an angle given by $\arccos\left(\dfrac{F}{\sqrt{EG}}\right)$. The polar coordinates, $\{(r\cos\theta, r\sin\theta, r^2)\}$, on the other hand, give an orthogonal parametrization.

The first fundamental form provides the definition of the arc length of a curve and the angle between curves. We use the component functions of the metric to define area, another fundamental geometric quantity.

Definition 7.15. *By a* **bounded region** *R in a surface S in $\mathbb{R}^3$, we mean a subset of S that is contained in some ball of finite radius in $\mathbb{R}^3$.*

The following theorem is a standard result of multivariable calculus.

Theorem 7.16. *Suppose S is a regular surface and $x\colon (U \subset \mathbb{R}^2) \to \mathbb{R}^3$ is a coordinate chart. If $R \subset S$ is a bounded region and $R \subset x(U)$, then the area of the region R is given by:*

$$\text{area}(R) = \iint_R dA = \iint_{x^{-1}(R)} \|x_u \times x_v\| \, du\, dv.$$

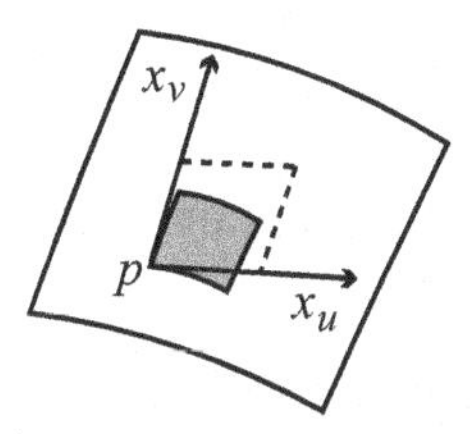

Area may be defined as the limit of a Riemann sum associated to a system of partitions. At a point p in the region R, a small rectangle on the surface is approximated by a small parallelogram in $T_p(S)$. To prove the theorem one shows that the contribution in the small parallelogram by this approximation is given by a factor of $\|x_u \times x_v\|$. For details, see your favorite multivariable calculus book (Spivak 1971).

Lagrange's formula $\|x_u \times x_v\|^2 = \|x_u\|\,\|x_v\| - (x_u \cdot x_v)^2$ leads to a local expression for area, given by the integral:

$$\text{area}(R) = \iint_{x^{-1}(R)} \sqrt{E(u,v)G(u,v) - F(u,v)^2}\, du\, dv.$$

The area of a bounded region of a surface should be independent of the choice of coordinate chart. In order to see that our local expression has this property, we consider how it changes with respect to a change of coordinates.

Proposition 7.17. *Let $x\colon (U \subset \mathbb{R}^2) \to S$ be a coordinate chart and R a bounded region of S with $R \subset x(U)$. Let $h\colon V \to U$ be an orientation-preserving diffeomorphism of open sets of $\mathbb{R}^2$ (that is, $\det J(h) > 0$) and let $y = x \circ h$. Then:*

$$\iint_{y^{-1}(R)} \|y_{\bar u} \times y_{\bar v}\| \, d\bar u\, d\bar v = \iint_{x^{-1}(R)} \|x_u \times x_v\| \, du\, dv.$$

PROOF: The mapping $y\colon (V \subset \mathbb{R}^2) \to S$ is also a coordinate chart, because h is a diffeomorphism. Denote the coordinates in the chart y by $\bar u$ and $\bar v$, the coordinates in the chart x by u and v. The diffeomorphism h takes $(\bar u, \bar v)$ to (u, v) and so we can write $(u,v) = h(\bar u, \bar v) = (u(\bar u, \bar v), v(\bar u, \bar v))$. Then $y(\bar u, \bar v) = x(u(\bar u, \bar v), v(\bar u, \bar v))$ and:

$$y_{\bar u} = x_u \frac{\partial u}{\partial \bar u} + x_v \frac{\partial v}{\partial \bar u} \qquad y_{\bar v} = x_u \frac{\partial u}{\partial \bar v} + x_v \frac{\partial v}{\partial \bar v}.$$

It follows then that $y_{\bar u} \times y_{\bar v} = \det(J(h))\, x_u \times x_v$ by the properties of the cross product. Furthermore, the expression $du\,dv$ transforms to $\det J(h)\, d\bar u\, d\bar v$ by the properties of integrals (or differential forms, that is, because $d\bar{u}\,d\bar{u} = 0 = d\bar{v}\,d\bar{v}$ and $d\bar{u}\,d\bar{v} = -d\bar{v}\,d\bar{u}$). Thus,

$$\iint_{y^{-1}(R)} \|y_{\bar{u}} \times y_{\bar{v}}\| d\bar{u}\,d\bar{v} = \iint_{y^{-1}(R)} \|x_u \times x_v\| \det J(h)\, d\bar{u}\,d\bar{v}$$
$$= \iint_{x^{-1}(R)} \|x_u \times x_v\|\, du\,dv. \qquad \blacksquare$$

For the change of coordinates $h = x^{-1} \circ y$, we have shown that the component functions of the metric transform in such a way that:

$$\det\begin{pmatrix} \bar{E} & \bar{F} \\ \bar{F} & \bar{G} \end{pmatrix} = \det\begin{pmatrix} E & F \\ F & G \end{pmatrix} (\det J(x^{-1} \circ y))^2,$$

and so $\sqrt{\bar{E}\bar{G} - \bar{F}^2} = |\det J(x^{-1} \circ y)|\sqrt{EG - F^2}$.

As an example we compute the area of the lower hemisphere of S^2. Parametrize the lower hemisphere from the unit disk with polar coordinates by:

$$x\colon (0,1) \times (0, 2\pi) \to \mathbb{R}^3, \quad x(r,\theta) = (r\cos\theta, r\sin\theta, -\sqrt{1-r^2}).$$

In this system of coordinates the tangent vectors to the coordinate curves are given by $x_r = (\cos\theta, \sin\theta, (r/\sqrt{1-r^2}))$ and $x_\theta = (-r\sin\theta, r\cos\theta, 0)$. It follows that $E = 1/(1-r^2)$, $F = 0$, and $G = r^2$. The area of the hemisphere is given by

$$\iint_R \sqrt{EG - F^2}\, dr\, d\theta = \int_0^{2\pi} \int_0^1 \frac{r}{\sqrt{1-r^2}}\, dr\, d\theta$$
$$= 2\pi.$$

As another example, suppose R is a bounded region on a cylinder and $x\colon (0, 2\pi) \times \mathbb{R} \to \mathbb{R}^3$ is the coordinate chart $x(\theta, v) = (\cos\theta, \sin\theta, v)$. Then $\text{area}(R) = \text{area}(x^{-1}(R))$ since $E = G = 1$ and $F = 0$. The expression $EG - F^2$ is a measure of how much the surface differs (or *bends*) from its parametrizing plane.

Exercises

7.1 Determine if the following functions $x\colon (-1,1) \times (-1,1) \to \mathbb{R}^3$ qualify as coordinate charts for a regular surface in $\mathbb{R}^3$.

(1) $x(u,v) = (u^2, u - v, v^2)$.
(2) $x(u,v) = (\sin\pi u, \cos\pi u, v)$.
(3) $x(u,v) = (u, u + v, v^{3/2})$.

7.2 Let $x\colon \mathbb{R}^2 \to \mathbb{R}^3$ be given by $x(u,v) = (u - v, u + v, u^2 - v^2)$. Show that the image of x is a regular surface S and describe S.

Solutions to Exercises marked with a dagger appear in the appendix, p. 332.

7.3 Prove that "S_1 is diffeomorphic to S_2" is an equivalence relation on regular surfaces in $\mathbb{R}^3$. Show that geometric details, such as length of curves, angles between curves, and area are not necessarliy preserved by diffeomorphisms.

7.4† Let $\alpha\colon (a,b) \to \mathbb{R}^3$ be a unit-speed curve with nonzero curvature. Let $x\colon (a,b) \times (\mathbb{R} - \{0\} \to \mathbb{R}^3$ denote the coordinate chart for a tangent developable surface (when x is one-to-one),

$$x(u,v) = \alpha(u) + v\alpha'(u).$$

Determine the component functions of the metric for such a surface.

7.5 Consider the coordinate charts $x, \bar{x}\colon (0, 2\pi) \times (-1, 1) \to \mathbb{R}^3$ defined by:

$$x(u,v) = \left(\left(2 - v\sin\frac{u}{2}\right)\sin u, \left(2 - v\sin\frac{u}{2}\right)\cos u, v\cos\frac{u}{2}\right),$$

$$\bar{x}(\bar{u},\bar{v}) = \left(\left(2 - \bar{v}\sin\left(\frac{\pi}{4} + \frac{\bar{u}}{2}\right)\right)\cos\bar{u}, -\left(2 - \bar{v}\sin\left(\frac{\pi}{4} + \frac{\bar{u}}{2}\right)\right)\sin\bar{u}, \bar{v}\cos\left(\frac{\pi}{4} + \frac{\bar{u}}{2}\right)\right).$$

Show that these charts cover the Möbius band as a regular surface in $\mathbb{R}^3$. Determine the change of coordinates between charts and show that the Möbius band is *not* orientable.

7.6† Suppose that a surface S is given by a smooth function $\Phi\colon \mathbb{R}^3 \to \mathbb{R}$ for which 0 is a regular value. At a point $p = (x_0, y_0, z_0) \in S$, give an expression for the component functions of the metric E, F, and G in terms of the function Φ. (*Hint*: What does the implicit function theorem tell you?)

7.7† Show that the tangent plane $T_p(S)$ at a point p of a surface S is given by:

$$T_p(S) = J(x)(p)(\mathbb{R}^2) \subset \mathbb{R}^3,$$

where $J(x)(p)$ is the linear mapping determined by the Jacobian evaluated at p.

7.8 Consider the following coordinate chart for fixed values a, b, $c > 0$ and u, $v \in \mathbb{R}$:

$$x(u,v) = (a\sinh u \sinh v, b\sinh u \sinh v, c\sinh u).$$

Determine the first fundamental form and metric coefficients for the chart x.

7.9 Determine the component functions of the metric for a surface of revolution generated by the graph of a function $f\colon (a,b) \to \mathbb{R}$, that is, the set $\{(u,v) \mid v = f(u)\}$.

7bis

Map projections

You are quite right that the essential condition in every map projection is the infinitesimal similarity; a condition which should be neglected only in very special cases of need.

C. F. GAUSS (TO HANSEN, 1825)

The world is round; maps are flat.

PORTER W. MCDONNELL (1979)

The quote from McDonnell poses an interesting problem for geometers. Coordinate charts for surfaces were motivated by and named after cartographic maps. In this chapter (a side trip on our journey) we consider the classical representations of the Earth, thought of as a sphere. Cartography is almost as old as geometry itself; for some of its history, see Snyder (1993).

The fundamental problem of cartography is to represent a portion (a region) R of the globe, idealized as the unit sphere S^2, on a flat surface, that is, to give a mapping $Y\colon (R \subset S^2) \to \mathbb{R}^2$ that is one-one and differentiable, and has a differentiable inverse on $Y(R)$. Such a mapping is called a **map projection**. The connection to differential geometry is that any coordinate chart for the sphere, $x\colon (U \subset \mathbb{R}^2) \to S^2$, determines a map projection $Y = x^{-1}$ on the region $R = x(U)$, and conversely, a map projection determines a coordinate chart.

In fact, the Earth is not a sphere but an oblate spheroid, a surface of revolution with the generating curve an ellipse. The history of the true shape of the Earth is fascinating and the subject of the discipline of *geodesy*. The geometry of an ellipsoid of revolution is considerably more complicated than that of the sphere, and it remains a topic of current research; see, for example, Meyer (2010). A classic history of the shape of the Earth was written by Todhunter (1873).

The purposes of cartography—navigation, government—determine the properties of interest of a map projection. An *ideal map projection* is one for which all the relevant geometric features of the sphere—lengths, angles, and areas—are preserved in the image. An ideal map projection carries these values on the sphere to identical values in the Euclidean plane for an appropriate choice of unit. Such a mapping between arbitrary surfaces is called an *isometry*. Many geographical features of interest could be found on an ideal map; for example, shortest distances between points on the sphere would be represented by line segments. Navigation would be as simple as a ruler and protractor. Also the relative sizes of land areas would be apparent on an ideal map.

One of the first important results in the study of mathematical cartography is due to Euler in his 1778 paper *De repraesentatione superficiei sphaericae super plano.*

Theorem 7bis.1. *There are no ideal map projections.*

PROOF: Euler's proof is focused on the line elements for the sphere and the plane. If there were an isometry, at corresponding points we would have the equation:

$$dx^2 + dy^2 = \sin^2 v\, du^2 + dv^2,$$

where (x,y) are the coordinates of the plane, and (u,v) the coordinates on the sphere. Since these are coordinate charts, we can write $x = x(u,v)$ and $y = y(u,v)$, which leads to the transformations:

$$dx = p\,du + q\,dv, \text{ and } dy = r\,du + s\,dv.$$

Since dx and dy are differentials of functions, there are integrability conditions, namely,

$$\frac{\partial p}{\partial v} = \frac{\partial q}{\partial u}, \text{ and } \frac{\partial r}{\partial v} = \frac{\partial s}{\partial u}.$$

Furthermore, the equation $dx^2 + dy^2 = \sin^2 v\, du^2 + dv^2$ implies $p^2 + r^2 = \sin^2 v$, $pq + rs = 0$ and $q^2 + s^2 = 1$.

Write the quotient $-r/p = \tan\phi = q/s$, where $\phi = \phi(u,v)$. The relations imply:

$$p = \cos\phi \sin v, \quad q = \sin\phi, \quad r = -\sin\phi \sin v, \quad s = \cos\phi.$$

Thus, $dx = \cos\phi \sin v\, du + \sin\phi\, du$ and $dy = -\sin\phi \sin v\, du + \cos\phi\, dv$. The integrability conditions imply:

$$\frac{\partial p}{\partial v} = -\sin\phi \sin v \frac{\partial \phi}{\partial v} + \cos\phi \cos v = \cos\phi \frac{\partial \phi}{\partial u} = \frac{\partial q}{\partial u},$$
$$\frac{\partial r}{\partial v} = -\cos\phi \sin v \frac{\partial \phi}{\partial v} - \sin\phi \cos v = -\sin\phi \frac{\partial \phi}{\partial u} = \frac{\partial s}{\partial u}.$$

Multiplying the first equation by $\sin\phi$ and the second by $\cos\phi$, and adding the equations gives the relation $-\sin\phi \dfrac{\partial \phi}{\partial v} = 0$. Multiplying the first equation by $-\cos\phi$ and the second by $\sin\phi$ and adding the equations gives $\dfrac{\partial \phi}{\partial u} = -\cos v$. However, the first of these relations implies $\phi(u,v) = \phi(u)$, while the second relation cannot hold if $\phi(u,v) = \phi(u)$. Thus, an ideal map projection does not exist. Euler's ingenious argument follows from basic analytic principles.

Alternatively, there is a synthetic argument. Let $\triangle ABC$ be a triangle on S^2 formed of segments of great circles and lying in the region R. From chapter 1, we know to think of great circles as lines because they minimize lengths. An ideal map projection would take the great circle segments AB, BC, and CA to line segments in the plane, which are the curves in $\mathbb{R}^2$ that minimize length. However, the angle sum of the spherical triangle exceeds π, while the triangle in the plane has angle sum π. Thus, such a mapping cannot be angle preserving and is not an ideal map projection. ∎

When a map projection displays particular properties to suit the purpose of the map, we must settle for the failure of the other properties. Two properties play a major role in map making.

Definition 7bis.2. *A diffeomorphism $f\colon S_1 \to S_2$ of surfaces (in particular, a map projection) is* **conformal** *if it preserves angles; that is, given two curves $\alpha\colon (-\epsilon,\epsilon) \to S_1$ and $\beta\colon (-\eta,\eta) \to S_1$ with $\alpha(0) = p = \beta(0)$, the angle between $\alpha'(0)$ and $\beta'(0)$ in $T_p(S_1)$ is the same as the angle between $df_p(\alpha'(0))$ and $df_p(\beta'(0))$ in $T_{f(p)}(S_2)$. A mapping is* **equiareal** *if the areas of any region and its image are the same.*

Navigation requires conformal map projections, while civil authorities may use equiareal maps. Before discussing the classical examples, we develop general criteria for a diffeomorphism to be conformal and to be equiareal.

Proposition 7bis.3. *Suppose $f\colon S_1 \to S_2$ is a diffeomorphism. Then f is conformal if and only if there is a nonzero function $\rho\colon S_1 \to \mathbb{R}$ such that, for all $p \in S_1$ and $\mathbf{v}$, $\mathbf{w} \in T_p(S_1)$,*

$$\mathrm{I}_p(\mathbf{v},\mathbf{w}) = (\rho(p))^2 \mathrm{I}_{f(p)}(df_p(\mathbf{v}), df_p(\mathbf{w})).$$

PROOF: Suppose $f\colon S_1 \to S_2$ is a diffeomorphism and there is a nonzero function $\rho\colon S_1 \to \mathbb{R}$ satisfying the condition of the theorem. We compare the angle θ_1 between the two curves on S_1 and the angle θ_2 between the corresponding curves on S_2:

$$\begin{aligned}
\cos\theta_1 &= \frac{\mathrm{I}_p(\mathbf{v},\mathbf{w})}{\sqrt{\mathrm{I}_p(\mathbf{v},\mathbf{v})\mathrm{I}_p(\mathbf{w},\mathbf{w})}} \\
&= \frac{(\rho(p))^2 \mathrm{I}_{f(p)}(df_p(\mathbf{v}), df_p(\mathbf{w}))}{\sqrt{(\rho(p))^2 \mathrm{I}_{f(p)}(df_p(\mathbf{v}), df_p(\mathbf{v}))(\rho(p))^2 \mathrm{I}_{f(p)}(df_p(\mathbf{w}), df_p(\mathbf{w}))}} \\
&= \frac{\mathrm{I}_{f(p)}(df_p(\mathbf{v}), df_p(\mathbf{w}))}{\sqrt{\mathrm{I}_{f(p)}(df_p(\mathbf{v}), df_p(\mathbf{v}))\mathrm{I}_{f(p)}(df_p(\mathbf{w}), df_p(\mathbf{w}))}} = \cos\theta_2.
\end{aligned}$$

Thus, if $\mathrm{I}_p(\mathbf{v},\mathbf{w}) = (\rho(p))^2 \mathrm{I}_{f(p)}(df_p(\mathbf{v}), df_p(\mathbf{w}))$, the mapping is conformal.

If we have a conformal mapping $f\colon S_1 \to S_2$, since f is a diffeomorphism, a chart $x\colon (U \subset \mathbb{R}^2) \to S_1$ determines a chart on S_2, namely, $y = f \circ x\colon (U \subset \mathbb{R}^2) \to S_2$. Thus we can apply the same domain of coordinates (u,v) on each surface. Furthermore, the local expression for a curve on S_1, $\alpha(t) = x(u(t), v(t))$ carries over to S_2 via the mapping f as $f \circ \alpha(t) = y(u(t), v(t))$. Thus,

$$df_p\left(x_u \frac{du}{dt} + x_v \frac{dv}{dt}\right) = y_u \frac{du}{dt} + y_v \frac{dv}{dt}.$$

The component functions of the metric at corresponding points $p = x(u,v)$ and $f(p) = y(u,v)$ may be written as $E(u,v)$ and $\bar{E}(u,v)$; $F(u,v)$ and $\bar{F}(u,v)$; $G(u,v)$ and $\bar{G}(u,v)$.

Suppose that f is conformal, and that $\mathbf{v}$ and $\mathbf{w}$ are orthogonal unit vectors in $T_p(S_1)$. Then $\mathrm{I}_p(\mathbf{v},\mathbf{w}) = 0$ implies that $\mathrm{I}_{f(p)}(df_p(\mathbf{v}), df_p(\mathbf{w})) = 0$. Write $\mathbf{V} = df_p(\mathbf{v})$ and $\mathbf{W} = df_p(\mathbf{w})$, and suppose that $\|\mathbf{V}\| = c_1$, and $\|\mathbf{W}\| = c_2$. By the linearity of the inner product we find:

$$\begin{aligned}\frac{1}{\sqrt{2}} &= \frac{\mathrm{I}_p(\mathbf{v},\mathbf{v}+\mathbf{w})}{\sqrt{\mathrm{I}_p(\mathbf{v},\mathbf{v})\mathrm{I}_p(\mathbf{v}+\mathbf{w},\mathbf{v}+\mathbf{w})}} \\ &= \frac{\mathrm{I}_{f(p)}(\mathbf{V},\mathbf{V}+\mathbf{W})}{\sqrt{\mathrm{I}_{f(p)}(\mathbf{V},\mathbf{V})\mathrm{I}_{f(p)}(\mathbf{V}+\mathbf{W},\mathbf{V}+\mathbf{W})}} = \frac{c_1^2}{c_1\sqrt{c_1^2+c_2^2}}.\end{aligned}$$

This equation implies that $c_1 = c_2$. If $\mathbf{u} \in T_p(S_1)$ is any other unit tangent vector at p, then $\mathbf{u} = (\cos\beta)\mathbf{v} + (\sin\beta)\mathbf{w}$ and so $df_p(\mathbf{u}) = (\cos\beta)\mathbf{V} + (\sin\beta)\mathbf{W}$ from which it follows that $\|df_p(\mathbf{u})\| = c_1$. Define the function $\rho\colon S_1 \to \mathbb{R}$ given by $\rho(p) = c_1$. At $p \in S_1$ write $x_u = a\mathbf{v} + b\mathbf{w}$. Then $df_p(x_u) = a\mathbf{V} + b\mathbf{W}$ and:

$$\begin{aligned}\bar{E} &= \mathrm{I}_{f(p)}(df_p(x_u), df_p(x_u)) = a^2\mathbf{V}\cdot\mathbf{V} + 2ab\mathbf{V}\cdot\mathbf{W} + b^2\mathbf{W}\cdot\mathbf{W} \\ &= (\rho(p))^2(a^2+b^2) = (\rho(p))^2 E.\end{aligned}$$

Similarly $\bar{F} = (\rho(p))^2 F$ and $\bar{G} = (\rho(p))^2 G$. Varying the orthonormal basis pointwise in a small neighborhood around p we obtain this relationship between component functions of the metric at all points near p. This proves the proposition. ∎

Let us fix a map projection by choosing a particular coordinate chart on S^2 with which we can compare other projections. The representation of choice dates back to the second century B.C.E. and is standard in geography. A point on S^2 is determined by its **longitude** λ, measured from the Greenwich meridian, and its **latitude** ϕ, measured from the equatorial plane. These coordinates are like spherical coordinates for S^2 but differ in the choice of reference plane for the latitude. In the language of chapter 7 we get a coordinate chart:

$$x\colon (-\pi,\pi)\times\left(-\frac{\pi}{2},\frac{\pi}{2}\right) \to S^2, x(\lambda,\phi) = (\cos\lambda\cos\phi, \sin\lambda\cos\phi, \sin\phi).$$

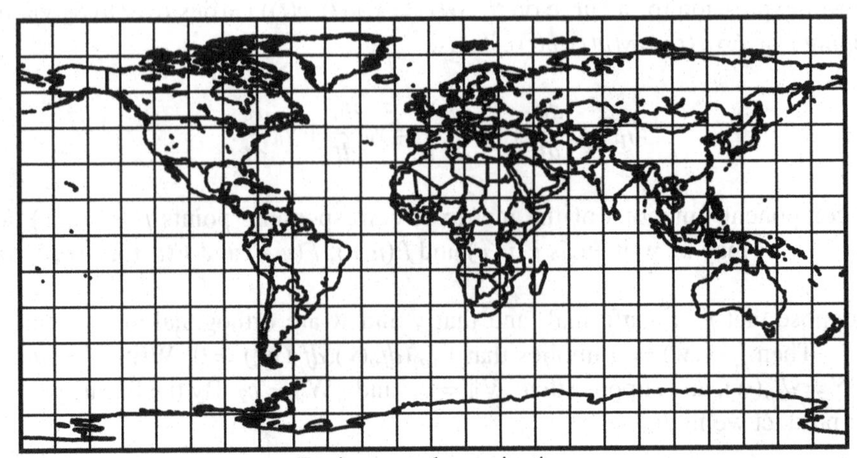
Equirectangular projection

The coordinate curves $x(\lambda_0, \phi)$ and $x(\lambda, \phi_0)$ are called **meridians** and **parallels**, respectively. This chart covers most of S^2, leaving out the poles and the *International Date Line*. By rotating the chart we could obtain other charts to cover S^2 completely. For our purposes this will not be necessary.

We compute the first fundamental form directly from the expression for the chart:

$$x_\lambda = (-\sin\lambda\cos\phi, \cos\lambda\cos\phi, 0), \quad x_\phi = (-\cos\lambda\sin\phi, -\sin\lambda\sin\phi, \cos\phi).$$

Then $E = x_\lambda \cdot x_\lambda = \cos^2\phi$, $F = x_\lambda \cdot x_\phi = 0$, and $G = x_\phi \cdot x_\phi = 1$. The line element is given by:

$$ds^2 = (\cos^2\phi)\,d\lambda^2 + d\phi^2.$$

With this explicit metric, Proposition $7^{bis}.3$ implies:

Corollary 7bis.4. *Suppose $Y\colon (R \subset S^2) \to \mathbb{R}^2$ is a map projection and $U = x^{-1}(R)$. If $d\bar{s}^2 = \bar{E}du^2 + 2\bar{F}\,d\bar{u}\,d\bar{v} + \bar{G}d\bar{v}^2$ is the line element associated to the composite coordinate chart:*

$$Y \circ x\colon \left(U \subset (-\pi, \pi) \times \left(-\frac{\pi}{2}, \frac{\pi}{2}\right)\right) \to \mathbb{R}^2,$$

then Y is a conformal map projection if $\bar{F} = 0$, $\bar{E} = (\rho(\lambda,\phi))^2\cos^2\phi$, and $\bar{G} = (\rho(\lambda,\phi))^2$ for some nonzero function ρ on U.

To derive a criterion for equiareal map projections we use the area formula in Theorem 7.16. The trick to apply the formula is to think of $\mathbb{R}^2 \subset \mathbb{R}^3$ as the subset of points $(u, v, 0)$. Then $\mathbb{R}^2$ is a surface and we can define different coordinate charts for it. For example, if Y is a map projection, then the composite:

$$(-\pi, \pi) \times \left(-\frac{\pi}{2}, \frac{\pi}{2}\right) \xrightarrow{x} S^2 \xrightarrow{Y} \mathbb{R}^2$$

gives a coordinate chart for a subset of $\mathbb{R}^2 \subset \mathbb{R}^3$. Suppose S is a bounded region in S^2 (away from the poles and the International Date Line) and that $S = x(T)$. Then the area of S is given by:

$$\text{area}(S) = \iint_T \sqrt{EG - F^2}\, d\lambda\, d\phi = \iint_T \cos\phi\, d\lambda\, d\phi.$$

Let $\bar{E}$, $\bar{F}$, and $\bar{G}$ denote the component functions for the metric associated to the chart $Y \circ x$ on $\mathbb{R}^2$. The area of $Y(S)$ is given by

$$\text{area}(Y(S)) = \iint_T \sqrt{\bar{E}\bar{G} - \bar{F}^2}\, d\lambda\, d\phi.$$

This expression leads to a criterion for a map projection to be equiareal.

Proposition 7bis.5. *A map projection is equiareal if and only if the induced coordinate chart on* $\mathbb{R}^2 \subset \mathbb{R}^3$ *satisfies* $\bar{E}\bar{G} - \bar{F}^2 = \cos^2\phi$.

More generally, a diffeomorphism of surfaces is equiareal if the induced metric satisfies $EG - F^2 = \bar{E}\bar{G} - \bar{F}^2$. These criteria are fashioned from the basic stuff of the local differential geometry of surfaces—the first fundamental form and the component functions of the metric. We turn to the examples; the world maps shown were produced using `G.Projector`, software developed by Robert Schmunk and freely available on the web from NASA's Godard Institute of Space Studies.

Stereographic projection

This map projection as applied to star charts was known to HIPPARCHUS OF NICAE (190–125 B.C.E.) who is believed to have introduced the ideas of longitude and latitude. The term *stereographic projection* was first used by F. D'AIGUILLON (1566–1617) (see Snyder 1993).

To describe the projection let $T = (0,0,1)$ denote the North Pole and P a point on the sphere with $T \neq P$. The line TP meets the plane of the Equator in a point $Y(P)$. The mapping $P \mapsto Y(P)$ is *stereographic projection.* In coordinates the plane through the Equator is $\mathbb{R}^2 = \{(u,v,0) \mid u,v \in \mathbb{R}\}$. Let $P = (x,y,z)$ denote a point on S^2, so $x^2 + y^2 + z^2 = 1$. The ray $\overrightarrow{TP}$ meets the plane $z = 0$ in a point $Q = Y(P) = (u,v,0)$. Regarded as vectors in $\mathbb{R}^3$, $P - T$ and $Q - T$ are linearly dependent and so $(P - T) \times (Q - T) = \mathbf{0}$, where $\times$ denoted the cross product.

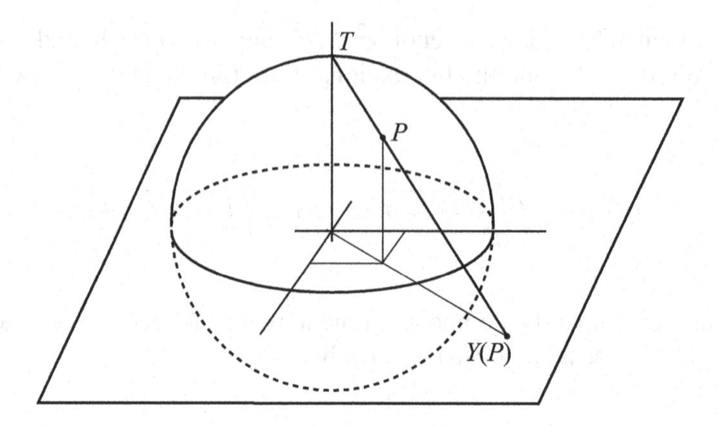

The vanishing of the cross product leads to the relations:

$$(x,y,z-1)\times(u,v,-1)=(-y-v(z-1),x+u(z-1),xv-yu)=(0,0,0).$$

Thus, $u=\dfrac{x}{1-z}$, $v=\dfrac{y}{1-z}$. In geographic coordinates the point P may be written $(x,y,z)=(\cos\lambda\cos\phi,\sin\lambda\cos\phi,\sin\phi)$, and the formula for stereographic projection is given by:

$$Y\circ x(\lambda,\phi)=\left(\frac{\cos\lambda\cos\phi}{1-\sin\phi},\frac{\sin\lambda\cos\phi}{1-\sin\phi},0\right).$$

In 1695, EDMOND HALLEY (1656–1742) of comet fame proved the following feature of stereographic projection.

Proposition 7bis.6. *Stereographic projection is conformal.*

PROOF: We view $\mathbb{R}^2\subset\mathbb{R}^3$ as a surface with a coordinate chart given by $Y\circ x$. The tangent vectors to the coordinate curves are given by:

$$(Y\circ x)_\lambda=\left(\frac{-\sin\lambda\cos\phi}{1-\sin\phi},\frac{\cos\lambda\cos\phi}{1-\sin\phi},0\right),\quad (Y\circ x)_\phi=\left(\frac{\cos\lambda}{1-\sin\phi},\frac{\sin\lambda}{1-\sin\phi},0\right),$$

and the component functions of the metric are:

$$\bar{E}(\lambda,\phi)=\frac{2\cos^2\phi}{(1-\sin\phi)^2},\quad \bar{F}(\lambda,\phi)=0,\ \text{ and }\ \bar{G}(\lambda,\phi)=\frac{2}{(1-\sin\phi)^2}.$$

The function $\rho(\lambda,\phi)=\dfrac{\sqrt{2}}{1-\sin\phi}$ is nonzero for $\phi\neq\pi/2$ (the North Pole), and so by Corollary 7bis.4 stereographic projection is conformal. ∎

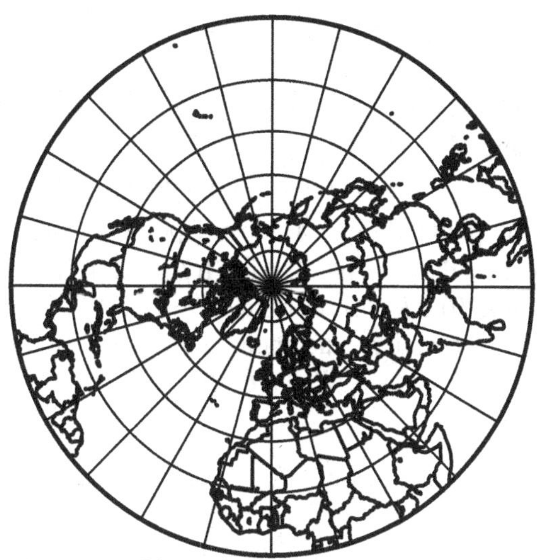

Stereographic Projection from the South Pole

Notice that the previous computations also show that stereographic projection is not equiareal. The metric coefficients associated to $Y \circ x$ satisfy:

$$\bar{E}\bar{F} - \bar{F}^2 = \frac{16\cos^2\phi}{(1-\sin\phi)^4} \neq \cos^2\phi.$$

More directly, one can see that the image of stereographic projection is all of $\mathbb{R}^2$ and such a mapping simply cannot be equiareal.

Stereographic projection comes up often in other parts of mathematics, especially in uses of the complex plane. Euler, Lagrange, and others introduced complex variables into the study of map projections where the new point of view proved useful. Stereographic projection also has a remarkable geometric property that will play a role in chapter 14.

Proposition 7bis.7. *If a circle on S^2 does not pass through the North Pole, its image under stereographic projection is a circle in the plane. If a circle on S^2 does pass through the North Pole, then its image is a straight line in the plane.*

PROOF: Any circle on S^2 is the intersection of a plane in $\mathbb{R}^3$ with the sphere (Exercise 1.2). Consider three cases: (1) a circle passing through the North Pole; (2) a circle that does not pass through the North Pole and is not a great circle; and (3) a great circle that does not pass through the North Pole.

In the first case, the image of the circle under stereographic projection is the intersection of the plane giving the circle with the x-y-plane, that is, a line.

In the second case, the planes tangent to the sphere at points on the circle C meet at a point S that is the vertex of a cone tangent to the sphere and containing

C. The ray $\overrightarrow{TS}$ is not perpendicular to the z-axis and so it meets the plane $z = 0$ in a point M.

Suppose P is a point on C and $P' = Y(P)$ is the image of P under stereographic projection. In the plane $\Pi(T,P',M)$, consider $\angle TP'M$ and $\angle P'PS$. Extend $\overrightarrow{SP}$ to meet the line given by the intersection of the plane tangent to the sphere through the North Pole and the plane $\Pi(T,P',M)$ in the point R. Since the lines $\overleftrightarrow{MP'}$ and $\overleftrightarrow{TR}$ are in parallel planes, they are parallel in $\Pi(T,P',M)$. By the Alternate Angles Theorem (Proposition I.29) $\angle RTP$ is congruent to $\angle MP'P$. Since the segments TR and RP are tangent to the sphere and meet at R, we have $TR \cong RP$, giving an isosceles triangle $\triangle TRP$. Thus, $\angle RTP \cong \angle RPT \cong \angle SPP'$.

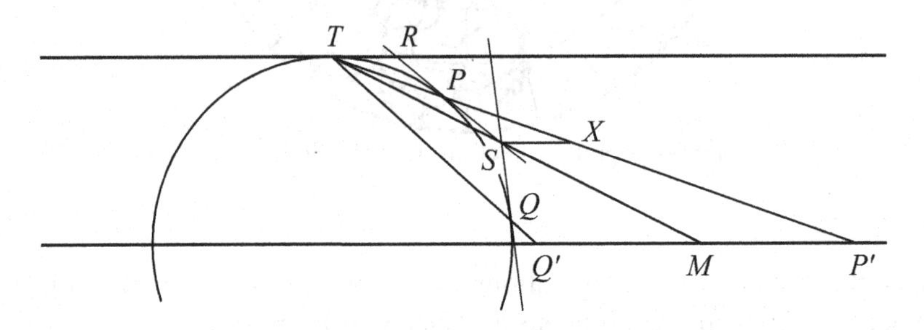

Suppose X is the point on TP' with $\angle SXT \cong \angle MP'T$. Then $\triangle TXS$ is similar to $\triangle TP'M$. By the previous discussion, $\triangle SPX$ has congruent angles at P and X and so $SP \cong SX$. The similarity of $\triangle TXS$ and $\triangle TP'M$ implies the following ratios:

$$\frac{TS}{TM} = \frac{SX}{MP'} = \frac{SP}{MP'}.$$

It follows that $MP' = \dfrac{SP \cdot TM}{TS}$. Since SP has the same length for every point P on the circle C, and TM and TS are constant as well, we see that MP' is constant for all points P on C, and so the image of the circle C is a circle C' in the plane with center M and radius $P'M$.

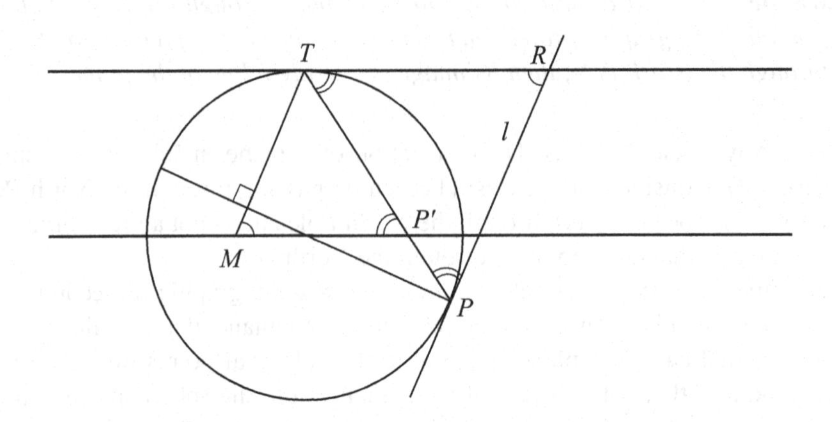

In the case of a great circle not passing through the North Pole, the argument just given fails – there is no tangent cone and hence no cone point S. To remedy this situation consider the unique line through T that is perpendicular to the plane that determines the great circle. This line meets the plane $z = 0$ at a point M. Let P be on the great circle and $P' = Y(P)$ and consider the intersection of the plane $\Pi(T,P',M)$ with the sphere. Since TM is perpendicular to the plane in which the great circle lies, TM is parallel to the tangent plane at P and so parallel to the line l lying in the plane $\Pi(T,P',M)$, which is tangent to the sphere at P. The line l meets the line through T tangent to the sphere at the North Pole and lying in $\Pi(T,P',M)$ at a point R. The pairs of alternate interior angles $\angle PRT$ and $\angle TMP'$ and $\angle PTR$ and $\angle TP'M$ are congruent. Since segments PR and TR are tangent to a circle, they are congruent, and so $\angle PTR \cong \angle TPR$. Because:

$$\angle TMP' + \angle MP'T + \angle P'TM = \angle TPR + \angle PRT + \angle RTP,$$

and $\angle TMP' \cong \angle TRP$, and $\angle MP'T \cong PTR$, then we have that $\angle TP'M \cong \angle TPR \cong \angle P'TM$ and so $P'M \cong TM$. Thus the image of a great circle is a circle of radius congruent to TM. ∎

Central (gnomonic) projection

Stereographic projection takes most great circles to circles in the plane. For navigation purposes it would be most convenient to represent any great circle route as a straight line on a map. This attractive feature is enjoyed uniquely by central projection, also known as gnomic projection.

To construct the projection fix the tangent plane at a point R on the sphere and join a point P on the sphere in the hemisphere adjacent to the tangent plane to the center of the sphere. Extend this segment to the tangent plane to get the image of the point P. Central projection is defined on the open hemisphere with R as center. The analytic expression for this map projection is easily derived when R is taken to be the south pole:

$$Y(\lambda,\phi) = (-\cos\lambda\cot\phi, -\sin\lambda\cot\phi, -1).$$

The main property – that great circle segments map to line segments in the plane – will play an important role in Beltrami's discovery of a model of non-Euclidean geometry (chapter 14).

Proposition 7bis.8. *Under central projection, great circle segments map to line segments.*

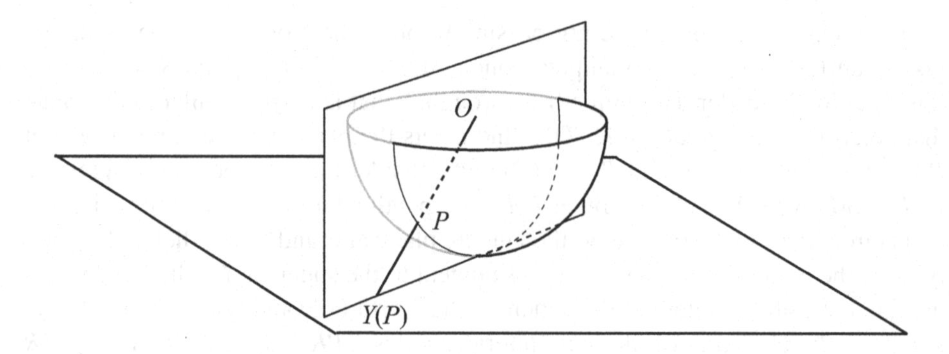

PROOF: A great circle is determined by the intersection of the sphere and a plane through the center of the sphere. The image of the great circle under central projection is the intersection of this plane with the tangent plane, which determines a straight line. ∎

Central projection with the North Pole as center

Cylindrical projections

Stereographic and central projections map portions of the sphere to a plane. We could also project the sphere to another surface that then can be mapped to the plane without distortion. For example, take a rectangle and roll it up into a cylinder that fits around the sphere. Projections onto the cylinder can then be unrolled to give a planar map. Such projections are called **cylindrical**. Another surface that can be unrolled onto part of the plane is a cone—such projections are called *conical* (see the exercises for examples).

The next projection is due to Lambert. It is called **orthographic** because the lines of projection are orthogonal to the tangent plane of the target. Lambert's work appeared in his book (1772) in which chapter VI is dedicated to finding the general form of a conformal map projection. His work is among the first mathematical treatises on the subject of cartography. Lambert gave several new projections that bear his name and are in use today.

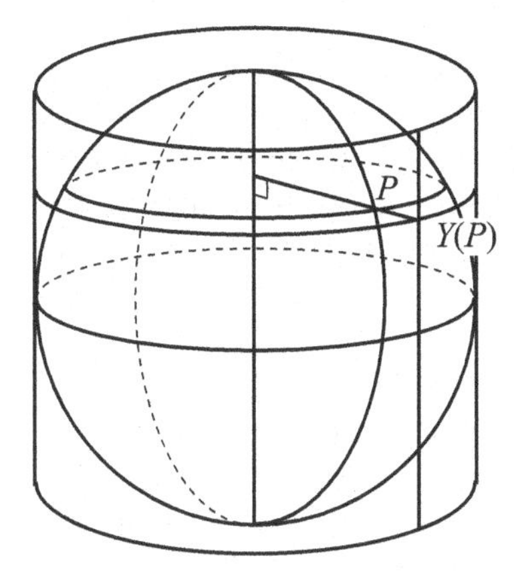

Consider the cylinder of height two and radius one centered at the origin and wrapped around S^2. Join a point P on the sphere to the z-axis by a perpendicular line. Then $Y(P)$ is the point on the cylinder where this line meets the cylinder closest to P.

If we use cylindrical coordinates (λ, t), where λ corresponds to longitude and t the vertical component, then by unrolling the cylinder we obtain the open rectangle $(-\pi,\pi)\times(-1,1)$. Composing with geographic coordinates we have:

$$Y \circ x(\lambda,\phi) = (\lambda, \sin\phi, 0).$$

Proposition 7bis.9. *The map projection* $Y \circ x(\lambda,\phi) = (\lambda, \sin\phi, 0)$ *is equiareal.*

PROOF: Following our criterion for equiareal maps, we compute the component functions of the metric associated to $Y \circ x$ on the surface $(-\pi,\pi)\times(-1,1)\times\{0\}\subset\mathbb{R}^3$:

$$(Y\circ x)_\lambda = (1,0,0), \quad (Y\circ x)_\phi = (0,\cos\phi,0).$$

Thus, $\bar{E}=1$, $\bar{F}=0$, and $\bar{G}=\cos^2\phi$, and $\bar{E}\bar{G}-\bar{F}^2=\cos^2\phi$. By Proposition 7bis.5 the map projection is equiareal. ∎

Notice that this projection has some other nice properties: It takes meridians to vertical lines and parallels to horizontal lines. In fact, it is unique as an equiareal projection among projections of the form $Y\circ x(\lambda,\phi)=(\lambda, y(\phi), 0)$.

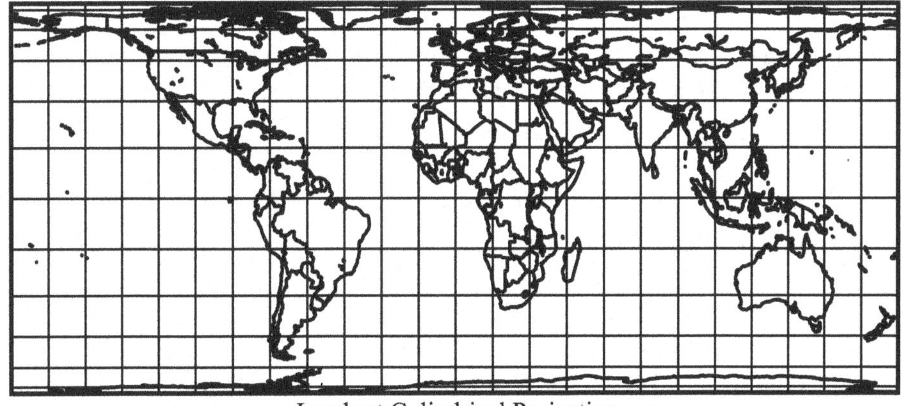

Lambert Cylindrical Projection

The possibility of such a cylindrical equiareal projection may have been known to Archimedes, who had computed the area of a sphere and the area of a cylinder of the same height by use of this projection.

The best known map projection appeared in 1569 and marked the beginning of a new era of cartography. GERARDUS KRÄMER (MERCATOR) (1512–94) was a Flemish cartographer who flourished in the "*Age of Discovery*" when new demands were made of mapmakers because commerce was spreading across the entire globe. To understand the problem solved by the Mercator projection we introduce the following idea.

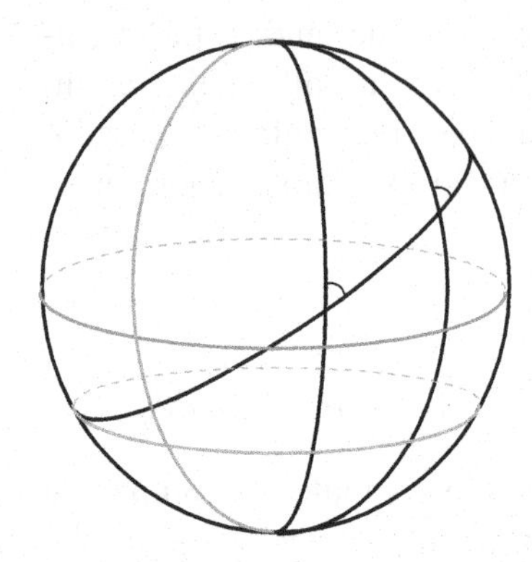

Definition 7bis.10. *A curve on the sphere S^2 is called a* **loxodrome** *(or* **rhumb line***) if it maintains a constant angle with respect to the family of meridians.*

Since the meridians determine the North–South directions, a course set by a navigator on a constant compass heading follows a loxodrome. In general, a loxodrome is not the shortest path between two points on the sphere, but it is the simplest to navigate. Extending most loxodromes determines a spiral path to one of the poles.

The problem solved by Mercator is to find a map projection such that: (1) meridians and parallels project to orthogonal lines; and (2) loxodromes project to line segments.

To construct the Mercator projection we send parallels to themselves, that is, the composite:

$$Y \circ x(\lambda,\phi)\colon (-\pi,\pi)\times\left(-\frac{\pi}{2},\frac{\pi}{2}\right) \to S^2 \to \mathbb{R}^2 \subset \mathbb{R}^3$$

satisfies $Y \circ x(\lambda,\phi) = (\lambda, v(\lambda,\phi), 0)$. The condition $v(\lambda,\phi) = v(\phi)$ guarantees that parallels project to lines. For the second property, we characterize loxodromes.

Proposition 7bis.11. *A unit speed curve $\alpha\colon (-\epsilon,\epsilon)\to S^2$ is a loxodrome if and only if $\alpha(s) = x(\lambda(s),\phi(s))$ where $\lambda'(s) = \dfrac{c_1}{\cos\phi(s)}$, and $\phi'(s) = c_2$ for some constants c_1, c_2.*

PROOF: Suppose $\alpha\colon (-\epsilon,\epsilon)\to S^2$ is a unit speed curve and $\alpha(s) = x(\lambda(s),\phi(s))$ in the geographic coordinate chart. Then we write $\alpha'(s) = \lambda'(s)x_\lambda + \phi'(s)x_\phi$. The condition that the angle made by α with all meridians be constant is equivalent to:

$$\frac{\alpha'(s)\cdot x_\lambda}{\|\alpha'(s)\|\cdot\|x_\lambda\|}=c_1,\text{ a constant, and }\frac{\alpha'(s)\cdot x_\phi}{\|\alpha'(s)\|\cdot\|x_\phi\|}=c_2,\text{ a constant.}$$

Taking dot products we have $\alpha'(s)\cdot x_\lambda=\lambda'(s)x_\lambda\cdot x_\lambda=\lambda'(s)\cos^2\phi$ and $\|x_\lambda\|=\cos\phi$, so we get $\lambda'(s)=\dfrac{c_1}{\cos\phi}$. Also $\alpha'(s)\cdot x_\phi=\phi'(s)x_\phi\cdot x_\phi=\phi'(s)$ and $\|x_\phi\|=1$, so $\phi'(s)=c_2$. Since meridians and parallels are perpendicular, the constants are the cosines of complementary angles so $c_1^2+c_2^2=1$. ∎

Suppose a map projection Y satisfies conditions (1) and (2). Whenever $\alpha(s)=x(\lambda(s),\ \phi(s))$ is a unit speed loxodrome, then $(Y\circ\alpha)(s)=(\lambda(s),v(\phi(s)),0)$ parametrizes a line in $\mathbb{R}^2$. We can write $A\lambda(s)+Bv(\phi(s))+C=0$ for the equation of the line for some A, B, and C. This equation leads to the condition:

$$\frac{v'(\phi(s))\phi'(s)}{\lambda'(s)}=-\frac{B}{A}=a,\text{ a constant.}$$

By the proof of Proposition 7bis. 11:

$$\frac{c_2\,dv/d\phi}{c_1/\cos\phi(s)}=a\text{ or }\frac{dv}{d\phi}=\frac{b}{\cos\phi}.$$

Integrate this differential equation by separating variables. We are free to choose the unknown constant b. However, the following observation leads to an auspicious choice.

Lemma 7bis.12. *A map projection $Y\circ x(\lambda,\phi)=(\lambda,v(\phi),0)$, satisfying $dv/d\phi=b/\cos\phi$, is conformal if and only if $b=\pm1$.*

PROOF: Since $Y\circ x(\lambda,\phi)=(\lambda,v(\phi),0)$, we have:

$$(Y\circ x)_\lambda=(1,0,0)\text{ and }(Y\circ x)_\phi=(0,dv/d\phi,0).$$

Thus the components of the metric induced by $Y\circ x$ are given by $\bar{E}=1$, $\bar{F}=0$, and $\bar{G}=(dv/d\phi)^2=b^2/\cos^2\phi$. If $\cos^2\phi=\rho^2\bar{E}$, we have $\rho^2=\cos^2\phi$. Then $1=\rho^2\bar{G}=b^2$, and so $b=\pm1$. ∎

We finally arrive at the Mercator projection by integrating $dv/d\phi=1/\cos\phi$ with the initial condition $v(0)=0$:

$$Y(\lambda,\phi)=\left(\lambda,\ln\tan\left(\frac{\pi}{4}+\frac{\phi}{2}\right),0\right).$$

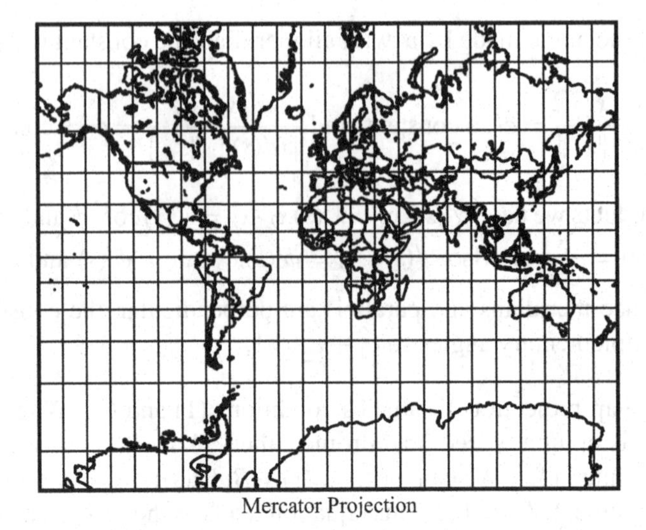
Mercator Projection

Mercator's achievement is considerable as the calculus had not yet been developed in 1569. He was one of the first compilers of logarithm tables and so the infinitesimal corrections needed to discover his projection were computationally at hand.

Sinusoidal projection

A nice feature of the projections discussed so far is that meridians may be easily identified. They are the vertical lines in the Mercator and Lambert equiareal projections, and they are the pencil of lines through the point that is antipodal to the point of projection in stereographic projection. On the Earth the meridians are great circles and the parallels are circles whose radius varies with the latitude. To reduce the distortion we could make each parallel a horizontal line and give it its proper length. These choices change the meridians. The *sinusoidal projection* fixes one meridian as a straight line and bends the other meridians in order to achieve horizontal parallels of the correct length.

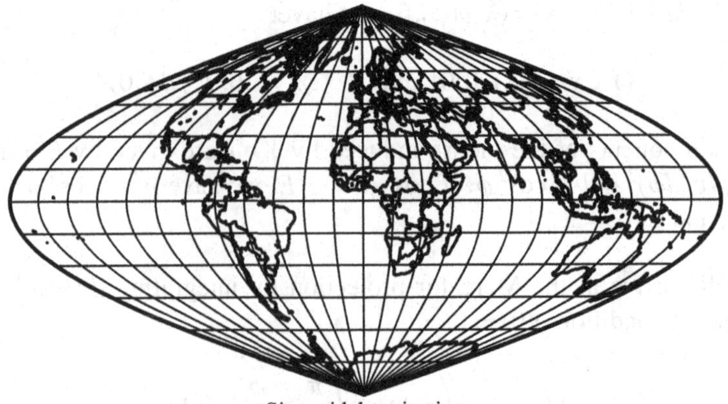
Sinusoidal projection.

To express this analytically, the parallels have constant ϕ-value, so $Y \circ x(\lambda, \phi) = (u(\lambda, \phi), \phi)$. For a fixed value of ϕ_0, the horizontal extent of the given parallel is $2\pi \cos\phi_0$. Since $-\pi < \lambda < \pi$, $u(\lambda, \phi) = \lambda \cos\phi$ achieves the goal. Notice that:

$$(Y \circ x)_\lambda = (\cos\phi, 0, 0), \text{ and } (Y \circ x)_\phi = (-\lambda \sin\phi, 1, 0).$$

It follows that $E = \cos^2\phi$, $F = -\lambda\cos\phi\sin\phi$, and $G = 1 + \lambda^2\sin^2\phi$. Computing $EG - F^2$ we get:

$$EG - F^2 = (\cos^2\phi)(1 + \lambda^2\sin^2\phi)^2 - (-\lambda\cos\phi\sin\phi)^2 = \cos^2\phi,$$

and so we conclude that sinusoidal projection is equiareal.

Azimuthal projection

Map projections are a particular use of change of coordinates. An important choice of coordinates is polar coordinates on the plane. The analogous coordinates on the sphere determine the *azimuthal projection*. Fix a point P on the sphere and a point Q in the plane. Send points on a circle of radius ρ centered at P on S^2 to the points on the circle of radius ρ centered at Q in $\mathbb{R}^2$.

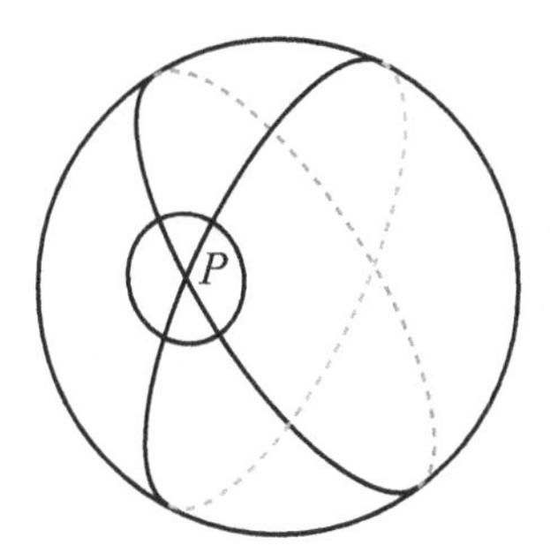

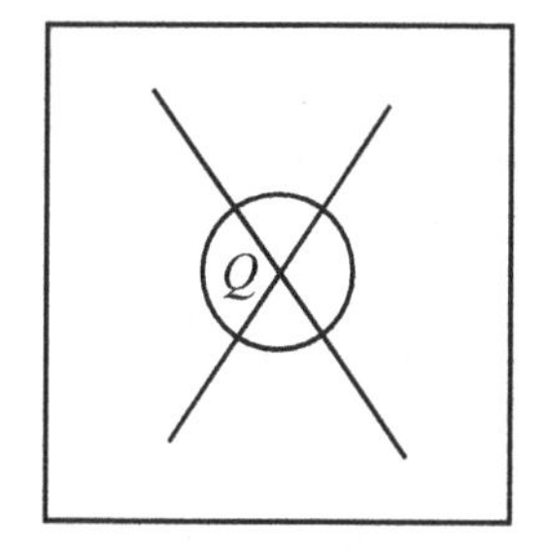

To obtain an analytic expression we choose a convenient point for P, say, the North Pole, and let Q be the origin in $\mathbb{R}^2$. This choice allows us to map the longitude as the angle for polar coordinates in the plane. Distance on the sphere between the North Pole and an arbitrary point (λ, ϕ) is measured along a meridian and is given by $\frac{\pi}{2} - \lambda$ (the *colatitude*). Thus azimuthal projection is simply the change of coordinates $(\lambda, \phi) \mapsto (\lambda, \rho)$ where $\rho = \frac{\pi}{2} - \lambda$. The line element becomes:

$$ds^2 = \cos^2\phi\, d\lambda + d\phi^2 = \sin^2(\pi/2 - \lambda)\, d\lambda^2 + (-d(\pi/2 - \lambda))^2 = \sin^2\rho\, d\lambda^2 + d\rho^2.$$

From this expression we can see that along meridians, that is, along a curve with λ fixed, $ds = d\rho$ and so distances are preserved. If we rotate the sphere to center the projection at an important point, then the distances from the center of the coordinates to other points on the globe are compared easily by following the new set of meridians. Notice that in these rotated coordinates the line element remains $ds^2 = \sin^2 \rho \, d\lambda^2 + d\rho^2$, the **polar line element** on S^2.

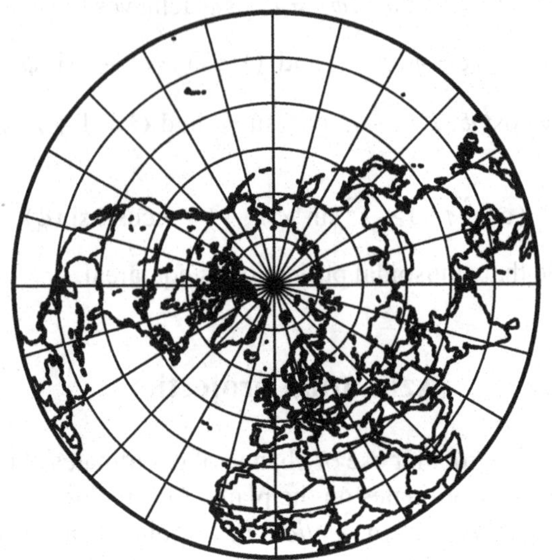

Azimuthal Projection from the North Pole

Exercises

7bis**.1** Complete the set of charts based on longitude-latitude coordinates to cover all of S^2. Do the same with the coordinates determined by the inverse of stereographic projection.

7bis**.2** Suppose $f\colon S_1 \to S_2$ is a diffeomorphism between surfaces that is both conformal and equiareal. Show that f is an isometry, that is, there are coordinate patches around corresponding points on the surfaces for which the line elements satisfy $ds_1 = ds_2$.

7bis**.3** Apply the criterion for conformality to Lambert's cylindrical projection and determine the regions of maximal failure of this property.

7bis**.4** From a satellite, the projection of the Earth onto a plane given by a piece of film in a camera (or a light sensing grid) is orthographic, that is, perpendicular to the plane of the film. This is essentially the map projection given by:

$$(x, y, \sqrt{1 - x^2 - y^2}) \mapsto (x, y, 0).$$

Determine a complete set of charts for S^2 of this sort, and determine if the map projection is equiareal or conformal.

7bis**.5**† One way to find a parametrization of a loxodrome is to exploit properties of map projections. Consider the curve in the plane that makes a constant angle with each line through the origin. Using polar coordinates, show that such a curve is a *logarithmic spiral*, that is, a curve given in polar coordinates by $\rho(\theta) = Ke^{a\theta}$ for K and a constants. Since stereographic projections take the lines through the origin in the plane to meridians, the image of the logarithmic spiral is a loxodrome on the sphere. What parametrization follows from this procedure?

Solutions to Exercises marked with a dagger appear in the appendix, p. 333.

7bis.6 Find the length of a loxodrome on a sphere that makes an angle θ with the meridian at the Equator and ends up by winding around the Pole. Contrast this with the length of the great circle route.

7bis.7 Derive the formula for central projection centered at the South Pole. What is the inverse of central projection? Since the inverse is a coordinate chart, what are the components of the metric associated with this coordinate chart?

7bis.8† The plane and the cylinder were used in map projections because there is an ideal map projection between them. Another *developable* surface (to the plane) is the cone. Give coordinates to a cone as a surface of revolution, and give coordinate charts for it. Show that the coordinate charts you have chosen provide an ideal map projection from the cone to the plane.

7bis.9† In the previous exercise you showed that the cone was developable on the plane. Thus one could fashion a map projection by mapping the sphere to a cone and then to the plane. Choose a parallel ϕ_0 in the Northern Hemisphere and construct the cone tangent to the sphere at this latitude. Project to the cone from the center of the sphere. Give an analytic expression for the resulting map projection.

7bis.10 Consider the following mapping to the cone tangent to the parallel at ϕ_0: Send a point $x(\lambda,\phi)$ to the point on the cone with angle λ for its first spherical coordinate and $f(\phi) = \dfrac{1}{\sqrt{k}}\sqrt{2(C-\sin\phi)}$, where $f(\phi)$ determines the distance of the image point from the vertex of the cone, and $k=\cos\phi_0$ and $C=\dfrac{1+k^2}{2k}$. Using the results of the previous exercise, show that this projection is equiareal.

7bis.11 A measure of the local distortion made by a map projection is given by the **Tissot Indicatrix**, introduced by Nicolas Auguste Tissot (1824–1897) in 1881. The idea of the indicatrix is that a very small circle on the sphere is mapped by the map projection into an ellipse by the linearization of the projection. Show that generally, if $Y\circ x(\lambda,\phi) = (u(\lambda,\phi), v(\lambda,\phi))$, then the ellipse has major and minor axes given by $h=\sqrt{\left(\dfrac{\partial u}{\partial\phi}\right)^2+\left(\dfrac{\partial y}{\partial\phi}\right)^2}$ and $k=\dfrac{1}{\cos\phi}\sqrt{\left(\dfrac{\partial u}{\partial\lambda}\right)^2+\left(\dfrac{\partial y}{\partial\lambda}\right)^2}$. Compute the major and minor axes of the ellipse for stereographic and Mercator projections.

8

Curvature for surfaces

Investigations, in which the directions of various straight lines in space are to be considered, attain a high degree of clarity and simplicity if we employ, as an auxiliary, a sphere of unit radius described about an arbitrary center, and suppose the different points of the sphere to represent the directions of straight lines parallel to the radii ending at these points.

C. F. GAUSS (1827)

If all [accelerated] systems are equivalent, then Euclidean geometry cannot hold in all of them. To throw out geometry and keep [physical] laws is equivalent to describing thoughts without words. We must search for words before we express thoughts. What must we search for at this point? This problem remained insoluble to me until 1912, when I suddenly reaiized that Gauss's theory of surfaces holds the key to unlocking this mystery.

A. EINSTEIN (1922)

In chapter 7, surfaces in $\mathbb{R}^3$ were introduced and basic structures, such as the first fundamental form, were defined. In chapter 6 the notions of curvature and torsion for curves were very successful in organizing the geometry of curves. We next develop their analogues for surfaces in $\mathbb{R}^3$.

How does a surface "curve" at a point? In this chapter we consider two answers, the first introduced by Euler focuses on the intersections of a plane with the surface and the planar curvature of the resulting curves. The second approach is due to Gauss and features the mapping between a surface and a sphere that is determined by a normal direction at each point. Linear algebraic ideas are at the bottom of Gauss's work and we emphasize them in order to present the clearest path to essential results. Although this choice is historically inaccurate, the spirit of the analysis of Euler and Gauss is maintained.

Euler's work on surfaces

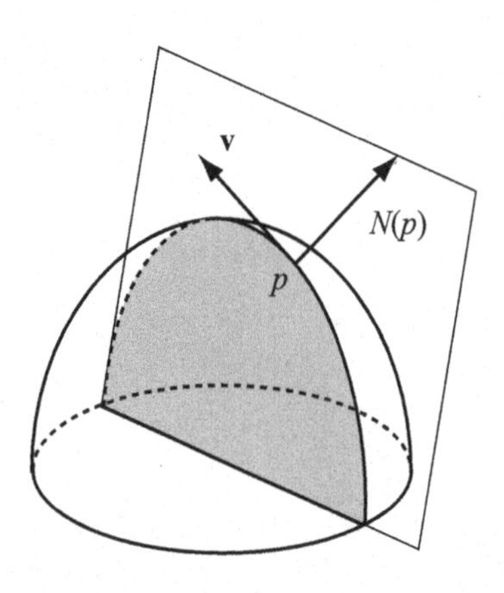

Suppose p is a point in a surface S and $N(p)$ is a choice of unit normal. Let $\mathbf{v}$ be a unit tangent vector in $T_p(S)$, the tangent plane to S at p. The pair $\{\mathbf{v}, N(p)\}$ determines a plane in $\mathbb{R}^3$. Translate this plane so that the origin is at p. The intersection of this plane with S gives a curve on S near p.

Let $c_{\mathbf{v}}(s)$ denote a unit speed parametrization of the curve so that $c'_{\mathbf{v}}(0) = \mathbf{v}$. We define $k_n(\mathbf{v})$, the **normal curvature of S in the v direction** at p, to be the directed curvature $\kappa_{c_{\mathbf{v}}}(s)$ of $c_{\mathbf{v}}(s)$

at $s=0$, in the plane through p parallel to the plane of $\{\mathbf{v}, N(p)\}$. At a fixed point p, $k_n(\mathbf{v})$ is a function of the directions in the tangent plane. In the classic paper *Recherches sur la courbure des surfaces* (1760), Euler proved the following useful theorem.

Theorem 8.1 (Euler 1760). *If the normal curvature $k_n(\mathbf{v})$ is not a constant function of $\mathbf{v}$, then there are precisely two unit tangent vectors X_1 and X_2, such that $k_n(X_1)=k_1$ is maximal and $k_n(X_2)=k_2$ is minimal. Furthermore, X_1 is perpendicular to X_2.*

PROOF: We follow (1825 §8): Choose coordinates for $\mathbb{R}^3$ so that $p=(0,0,0)$ and for which $T_p(S)$ is the x-y plane. By the Implicit Function Theorem (chapter 7), there is an open set U of S containing p that takes the form of the graph of a function, $U=\{(x,y,z) \mid z=f(x,y)\}$. We then have the coordinate chart for S near p given by $x(u,v)=(u,v,f(u,v))$. By rescaling if needed we can take $x_u|_{(0,0)}=(1,0,0)$, $x_v|_{(0,0)}=(0,1,0)$, that is,

$$f(0,0)=0, \quad \left.\frac{\partial f}{\partial x}\right|_{(0,0)}=0, \text{ and } \left.\frac{\partial f}{\partial y}\right|_{(0,0)}=0. \qquad (\star)$$

One further change is needed – we rotate the x-y plane until $\left.\dfrac{\partial^2 f}{\partial x \partial y}\right|_{(0,0)}=0.$ Rotations are given by:

$$\rho_\theta(x,y)=(x\cos\theta-y\sin\theta, x\sin\theta+y\cos\theta),$$

and, being linear, they do not alter the conditions fixed in $(\star)$. Let $f_\theta=f\circ\rho_\theta$. Then $\dfrac{\partial f_\theta}{\partial y}=-\sin\theta\dfrac{\partial f}{\partial x}+\cos\theta\dfrac{\partial f}{\partial y}$, and:

$$\begin{aligned}\frac{\partial^2 f_\theta}{\partial x\partial y}&=-\sin\theta\cos\theta\frac{\partial^2 f}{\partial x^2}+(\cos^2\theta-\sin^2\theta)\frac{\partial^2 f}{\partial x\partial y}+\sin\theta\cos\theta\frac{\partial^2 f}{\partial y^2}\\&=\cos 2\theta\frac{\partial^2 f}{\partial x\partial y}+\left(\frac{\sin 2\theta}{2}\right)\left(\frac{\partial^2 f}{\partial y^2}-\frac{\partial^2 f}{\partial x^2}\right).\end{aligned}$$

To obtain $\dfrac{\partial^2 f}{\partial x\partial y}=0$ when $\dfrac{\partial^2 f}{\partial x^2}\neq\dfrac{\partial^2 f}{\partial y^2}$, solve for θ in the equation:

$$\tan 2\theta=\left.\left(2\frac{\partial^2 f}{\partial x\partial y}\Big/\left(\frac{\partial^2 f}{\partial x^2}-\frac{\partial^2 f}{\partial y^2}\right)\right)\right|_{(0,0)}.$$

If $\dfrac{\partial^2 f}{\partial x^2}=\dfrac{\partial^2 f}{\partial y^2}$, let $\theta=\pi/4$. Rotate the x-y plane through θ as chosen and set f to be f_θ as constructed.

When the surface S intersects the x-z plane, the resulting curve is parametrized (not necessarily unit speed) in the x-z plane by $c_{xz}(t) = (t, f(t,0))$. Since x is in the t-direction at the origin, $x' = 1$ and $z' = \dfrac{\partial f}{\partial x} = 0$. The directed curvature $\kappa_{c_{xy}}(t)$ is given by

$$k_1 = \frac{x'z'' - x''z'}{[(x')^2 + (z')^2]^{3/2}} = \frac{\partial^2 f}{\partial x^2}(0,0).$$

Similarly, the intersection of S with the y-z plane is the curve, parametrized in the y-z plane by $c_{yz}(t) = (t, f(0,t))$, with directed curvature $\kappa_{c_{yz}}(t) = \dfrac{\partial^2 f}{\partial y^2} = k_2$.

If $\mathbf{v}$ is a unit length vector in some direction in the x-y plane, then $\mathbf{v}$ may be written as $(\cos\phi, \sin\phi)$. The curve defined by the intersection of the plane $\{\mathbf{v}, \mathbf{e}_3\}$ with S can be parametrized near p by $c_{\mathbf{v}}(t) = (t, f(t\cos\phi, t\sin\phi))$. The directed curvature is given by:

$$\begin{aligned} k_n(\mathbf{v}) &= \left.\frac{d^2}{dt^2} f(t\cos\phi, t\sin\phi)\right|_{t=0} = \cos^2\phi \frac{\partial^2 f}{\partial x^2}(0,0) + \sin^2\phi \frac{\partial^2 f}{\partial y^2}(0,0) \\ &= (\cos^2\phi)k_1 + (\sin^2\phi)k_2. \end{aligned}$$

As a function of ϕ, for $0 \leq \phi \leq 2\pi$, the value $k_n((\cos\phi, \sin\phi))$ achieves its minimal value when $\phi = \pi/2$ at $k_2 = \kappa_{c_{yz}}(0)$ and its maximal value when $\phi = 0$ at $k_1 = \kappa_{c_{xy}}(0)$. ∎

The values k_1 and k_2 are called the **principal curvatures** of S at p, and the directions X_1 and X_2 the **principal directions**. Together the principal curvatures give a qualitative picture of the surface near a point. From Gauss's proof of Euler's theorem we can write the equation of the surface S near p as a graph $z = f(x,y)$ with:

$$z = k_1 x^2 + k_2 y^2 + R(x,y), \quad \text{where} \lim_{x,y \to 0} \frac{R(x,y)}{x^2 + y^2} = 0.$$

Thus, near p, the surface is approximated, up to order 2, by the quadratic surface $z = k_1 x^2 + k_2 y^2$.

Euler appreciated the scope of Theorem 8.1. He wrote, *"And so the measurement of the curvature of surfaces, however complicated, ... is reduced at each point to the knowledge of two radii of curvature, one the largest and the other the smallest, at that point: these two things entirely determine the nature of the curvature and we can determine the curvatures of all possible sections perpendicular at the given point."* (Euler, 1760, p. 143).

What do the signs of the principal curvatures tell us?

Case 1. $k_1, k_2 > 0$ or $k_1, k_2 < 0$ at p.

Such a point is called **elliptic** and the quadratic surface approximating S at p is a paraboloid $z = \pm((ax)^2 + (by)^2)$. The level sets given by a constant z value are ellipses. From the defining equation for S it follows that if x and y are sufficiently small, $z = f(x,y)$ is nonzero and always positive or always negative. Therefore, near p, S lies on one side of the tangent plane.

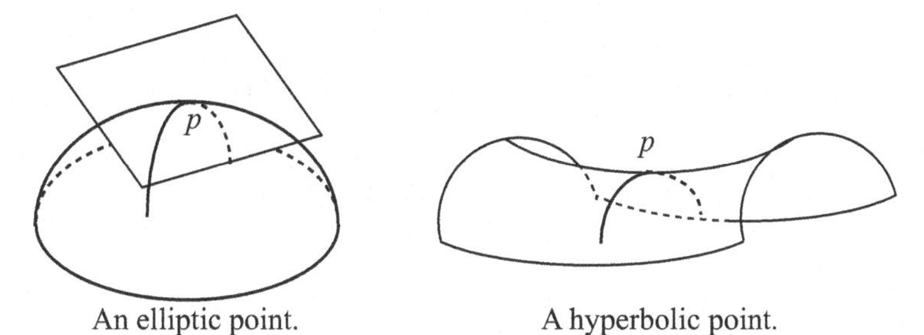

An elliptic point. A hyperbolic point.

Case 2. $k_1 > 0, k_2 < 0$.

Such a point is called **hyperbolic** and the quadratic surface approximating S at p is a hyperbolic paraboloid $z = \pm((ax)^2 - (by)^2)$. The level sets given by a constant z value are hyperbolas. In this case, near p, there are always points of S on either side of the tangent plane.

Case 3. $k_1 > 0, k_2 = 0$ or $k_1 = 0, k_2 < 0$.

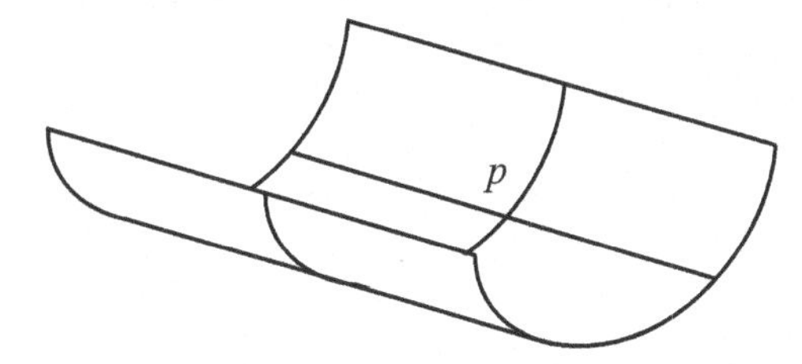

Such a point is called **parabolic** and the quadratic surface approximating S at p is a parabolic cylinder, $z = k_1 x^2$ or $z = k_2 y^2$. Near p, all points lie on the same side of the tangent plane.

Case 4. $k_1 = k_2 = 0$.
Such a point is called **planar**. Examples such as $f(x,y) = x^3$ or $f(x,y) = x^4$ show that the behavior of S with respect to the tangent plane near p may vary.

When $k_1 = k_2 \neq 0$, then every direction is a principal direction. Such a point is called an **umbilic point** and we may take any pair of orthogonal directions at such a point as principal directions. For example, all points on a sphere are umbilic.

We jump forward sixty-eight years to the pioneering work of Gauss on surfaces.

The Gauss map

For planar curves, curvature measures change in the tangent vector, or equivalently the normal vector. This notion was generalized to surfaces by Gauss in his seminal 1828 paper, *Disquisitiones generales circa superficies curvas*. He took his inspiration from astronomy and the idea of the "celestial sphere." Like many other contributions of Gauss, *Disquisitiones* had a profound effect on subsequent work on this subject. The mapping, called the Gauss map, was also introduced earlier by O. Rodrigues (1794–1851) in two papers that appeared in 1815.

Definition 8.2. *Let S be an orientable surface and S^2, the unit sphere in $\mathbb{R}^3$ centered at the origin. Define the* **Gauss map** *to be the mapping $N\colon S \to S^2$ associating to each point p in S the unit normal $N(p)$ at the point.*

Locally we have the formula $N(p) = \dfrac{x_u \times x_v}{\|x_u \times x_v\|}$. Notice that this expression depends on the choice of coordinates. For example, if you switch u and v, the normal changes from $N(p)$ to $-N(p)$. An important feature of the measure of curvature based on the Gauss map is that it will not depend on the choice of coordinates.

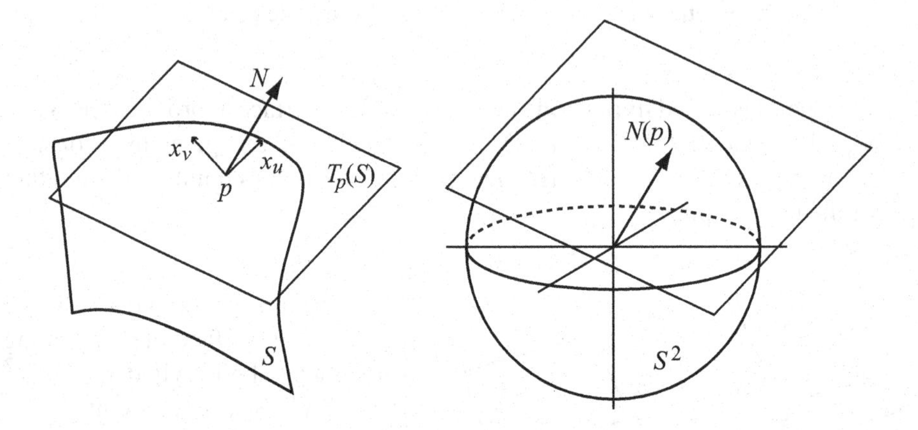

For a regular surface, we leave it to the reader to show that $N\colon S \to S^2$ is differentiable and so it has a differential map $dN_p\colon T_p(S) \to T_{N(p)}(S^2)$. The Gauss map is unusual because the tangent plane to S at p is the plane normal to $N(p)$. It is a particular property of the sphere S^2 that $T_q(S^2)$ is the plane normal to q. Thus $T_p(S) = T_{N(p)}(S^2)$, and dN_p can be taken to be a linear mapping $dN_p\colon T_p(S) \to T_p(S)$.

To compute dN_p we may use curves on S through p. Consider a tangent vector at p, $\mathbf{v} \in T_p(S)$, as the tangent to a curve, that is, there is a curve $\alpha\colon (-\epsilon, \epsilon) \to S$ with $\alpha'(0) = \mathbf{v}$. It follows that $dN_p(\mathbf{v}) = (N \circ \alpha)'(0)$. This expression gives $dN_p(\mathbf{v})$ as the rate of change of the normal vector at p when restricted to $\alpha(t)$.

EXAMPLES:

(1) Let Π be the plane $ax + by + cz + d = 0$ and $p \in \Pi$. The unit normal at p is $N(p) = \dfrac{(a,b,c)}{\sqrt{a^2+b^2+c^2}}$, a constant vector, and so $dN_p \equiv 0$.

(2) Consider the unit sphere S^2 in $\mathbb{R}^3$. For a curve $\alpha(t)$ on the sphere we can write $\alpha(t) = (x(t), y(t), z(t))$, where $(x(t))^2 + (y(t))^2 + (z(t))^2 = 1$. It follows that $2xx' + 2yy' + 2zz' = 0$, that is, $\alpha'(t)$ is perpendicular to $\alpha(t)$. Thus $N(x,y,z) = (x,y,z)$

is a choice for the unit normal vector to S^2 at (x,y,z). It follows immediately that $dN_p = \mathrm{id}$.

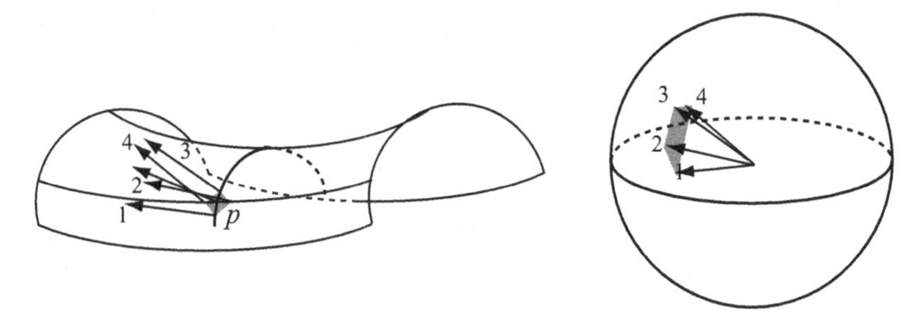

(3) Consider a saddle surface, for example, the graph of $f(x,y) = x^2 - y^2$. Through $(0,0,0)$ the surface includes two parabolas, one up and one down, giving the saddle. Let $x(u,v) = (u, v, u^2 - v^2)$ be the canonical coordinate chart. Then,

$$x_u = (1, 0, 2u), \quad x_v = (0, 1, -2v).$$

The normal to the surface is given by:

$$N(p) = N(x(u_0, v_0)) = \frac{(-2u_0, 2v_0, 1)}{\sqrt{4u_0^2 + 4v_0^2 + 1}}.$$

To obtain an expression for dN_p at $p = (0,0,0)$, let $\alpha\colon (-\epsilon, \epsilon) \to S$ be a unit speed curve with $\alpha(0) = (0,0,0)$; then:

$$\alpha(t) = (u(t), v(t), u^2(t) - v^2(t)), \text{ so } N \circ \alpha(t) = \frac{(-2u(t), 2v(t), 1)}{\sqrt{4u^2(t) + 4v^2(t) + 1}}.$$

Finally, evaluate the derivative at $t = 0$ to get:

$$(N \circ \alpha)'(0) = (-2u'(0), 2v'(0), 0).$$

Since $\alpha'(0) = (u'(0), v'(0), 2u'(0)u(0) - 2v'(0)v(0)) = (u'(0), v'(0), 0)$, we have that

$$dN_{(0,0,0)}(\mathbf{v}) = dN_{(0,0,0)}(u'(0), v'(0), 0) = (-2u'(0), 2v'(0), 0).$$

When we write dN_p in matrix form with respect to the basis x_u, x_v, we get

$$dN_{(0,0,0)} = \begin{pmatrix} -2 & 0 \\ 0 & 2 \end{pmatrix}.$$

In general, the mapping dN_p has a remarkable property with respect to the first fundamental form. To describe this property we introduce an idea from linear algebra. The inner product $\mathrm{I}_p(\mathbf{u}, \mathbf{v})$ is linear in each variable. Consider the following computation applied to the local basis vectors, x_u and x_v. Suppose $p = x(u_0, v_0)$. We can determine $dN_p(x_u)$ and $dN_p(x_v)$ by:

$$dN_p(x_u) = (N \circ x(u, v_0))'(u_0) = \left.\frac{\partial N}{\partial u}\right|_{(u_0, v_0)} = N_u.$$

Similarly, $dN_p(x_v) = N_v$. From the definition of the normal vector it follows that $\mathrm{I}_p(N, x_v) = 0 = \mathrm{I}_p(x_u, N)$. Taking partial derivatives with respect to u and v in turn, we obtain:

$$\mathrm{I}_p(N_v, x_u) + \mathrm{I}_p(N, x_{uv}) = 0 = \mathrm{I}_p(N_u, x_v) + \mathrm{I}_p(N, x_{vu}).$$

By the differentiability of the coordinate patch, $x_{uv} = x_{vu}$. Thus,

$$\mathrm{I}_p(dN_p(x_u), x_v) = \mathrm{I}_p(N_u, x_v) = -\mathrm{I}_p(N, x_{uv}) = \mathrm{I}_p(x_u, N_v) = \mathrm{I}_p(x_u, dN_p(x_v)).$$

More generally, if V is a real vector space with an inner product $\langle\ ,\ \rangle\colon V \times V \to \mathbb{R}$ and $A\colon V \to V$ is a linear mapping, then A is said to be **self-adjoint** if, for all $v, w \in V$,

$$\langle A(v), w\rangle = \langle v, A(w)\rangle.$$

Thus we have shown:

Proposition 8.3. *The differential $dN_p\colon T_p(S) \to T_p(S)$ is self-adjoint with respect to the first fundamental form.*

What is so special about self-adjoint linear mappings? Fix a two-dimensional vector space V with inner product $\langle\ ,\ \rangle$.

Property 1. Suppose $A\colon V \to V$ is a nonzero self-adjoint operator with respect to $\langle\ ,\ \rangle$. Then A has two real, not necessarily distinct, eigenvalues.

PROOF: Suppose $[e, e']$ is an orthonormal basis for V. The transformation A may be represented as the matrix $\begin{pmatrix} a & b \\ c & d \end{pmatrix}$ where $A(e) = ae + be'$ and $A(e') = ce + de'$. Since A is self-adjoint,

$$b = \langle ae + be', e'\rangle = \langle A(e), e'\rangle = \langle e, A(e')\rangle = \langle e, ce + de'\rangle = c.$$

Hence A is represented by the matrix $\begin{pmatrix} a & b \\ b & d \end{pmatrix}$. To compute the eigenvalues of A we solve for the roots of the characteristic polynomial:

$$\det(tI - A) = t^2 - (a + d)t + (ad - b^2).$$

Since the discriminant $(a + d)^2 - 4(ad - b^2) = (a - d)^2 + 4b^2 > 0$, we get two real roots, and so A has two real eigenvalues. ■

Property 2. Suppose $A\colon V \to V$ is a self-adjoint operator. Suppose A has eigenvectors e_1, e_2 of unit length. If the eigenvalues of A are not equal, then e_1 is perpendicular to e_2.

PROOF: The eigenvectors satisfy $Ae_1 = \lambda_1 e_1$ and $Ae_2 = \lambda_2 e_2$ for the eigenvalues λ_1, λ_2. Then $\lambda_1\langle e_1, e_2\rangle = \langle \lambda_1 e_1, e_2\rangle = \langle Ae_1, e_2\rangle = \langle e_1, Ae_2\rangle = \langle e_1, \lambda_2 e_2\rangle = \lambda_2\langle e_1, e_2\rangle$, and if $\lambda_1 \neq \lambda_2$, we must have $\langle e_1, e_2\rangle = 0$, and e_1 is perpendicular to e_2 with respect to $\langle\ ,\ \rangle$. ■

Property 3. Suppose $A\colon V \to V$ is a self-adjoint operator with eigenvalues $\lambda_1 > \lambda_2$ and associated eigenvectors e_1 and e_2. Take e_1 and e_2 to be of unit length. Suppose $v \in V$ is also of unit length. We can write $v = \cos(\theta)e_1 + \sin(\theta)e_2$, and $\langle A(v), v\rangle = \lambda_1 \cos^2(\theta) + \lambda_2 \sin^2(\theta)$. Furthermore, the expression $\langle A(v), v\rangle$ achieves its maximum λ_1 when $\theta = 0$ or π and its minimum λ_2 when $\theta = \pm\pi/2$.

PROOF: Since e_1 is perpendicular to e_2, the set $\{e_1, e_2\}$ is an orthonormal basis for V. It follows that:

$$\begin{aligned}\langle A(v), v\rangle &= \langle A(\cos(\theta)e_1 + \sin(\theta)e_2), \cos(\theta)e_1 + \sin(\theta)e_2\rangle \\ &= \langle \cos(\theta)\lambda_1 e_1 + \sin(\theta)\lambda_2 e_2, \cos(\theta)e_1 + \sin(\theta)e_2\rangle \\ &= \lambda_1 \cos^2(\theta) + \lambda_2 \sin^2(\theta).\end{aligned}$$

The rest of the statement follows as in the proof of Theorem 8.1. ∎

Property 3 resembles the formula of Euler discussed in the proof of Theorem 8.1. We apply these properties of self-adjoint mappings to give a second derivation of Euler's result once we have developed the further properties of the Gauss map. Recall that a **quadratic form** on a real vector space V is a function $B\colon V \to \mathbb{R}$ that satisfies $B(rv) = r^2 B(v)$ for all $v \in V$ and $r \in \mathbb{R}$ and for which the associated mapping:

$$b_B\colon V \times V \to \mathbb{R}, \text{ given by } b_B(v, w) = \frac{1}{2}(B(v+w) - B(v) - B(w))$$

is an inner product on V.

Definition 8.4. *Let p be a point in a surface S. The* **second fundamental form** *on S at p is the quadratic form $\mathrm{II}_p(\mathbf{v})\colon T_p(S) \to \mathbb{R}$. defined by:*

$$\mathrm{II}_p(\mathbf{v}) = -\mathrm{I}_p(dN_p(\mathbf{v}), \mathbf{v}).$$

There is a geometric interpretation of $\mathrm{II}_p(\mathbf{v})$ when $\mathbf{v}$ is a unit tangent vector. As in the proof of Theorem 8.1, let $c_{\mathbf{v}}(s)$ denote the unit speed curve with $c_{\mathbf{v}}(0) = p$ and $c'_{\mathbf{v}}(0) = \mathbf{v}$. The trace of this curve is the intersection of the plane parallel to the plane spanned by $\{\mathbf{v}, N(p)\}$ and passing through p with S. The normal curvature of S at p in the direction $\mathbf{v}$ is denoted by $k_n(\mathbf{v})$ and satisfies $c''_{\mathbf{v}}(0) = k_n(\mathbf{v})N(p)$.

Theorem 8.5. *For $\mathbf{v} \in T_p(S)$ of unit length, $\mathrm{II}_p(\mathbf{v}) = k_n(\mathbf{v})$.*

PROOF: Since $N_{c_{\mathbf{v}}}(s)$ is perpendicular to $c'_{\mathbf{v}}(s)$ for each s, when we take the derivative of $\mathrm{I}_{c_{\mathbf{v}}(s)}(N_{c_{\mathbf{v}}}(s), c_{\mathbf{v}}(s)) = 0$, we get:

$$\mathrm{I}_{c_{\mathbf{v}}(s)}((N \circ c_{\mathbf{v}})'(s), c'_{\mathbf{v}}(s)) + \mathrm{I}_{c_{\mathbf{v}}(s)}(N(c_{\mathbf{v}}(s)), c''_{\mathbf{v}}(s)) = 0.$$

For $s = 0$ we can write $\mathrm{I}_p(dN_p(\mathbf{v}), \mathbf{v}) + \mathrm{I}_p(N(p), c''_{\mathbf{v}}(0)) = 0$. Since $c''_{\mathbf{v}}(0) = k_n(\mathbf{v})N(p)$, it follows that $\mathrm{II}_p(\mathbf{v}) = -\mathrm{I}_p(dN_p(\mathbf{v}), \mathbf{v}) = \mathrm{I}_p(N(p), k_n(\mathbf{v})N(p)) = k_n(\mathbf{v})$. ∎

The second fundamental form derives its properties from the self-adjoint operator dN_p. In particular, Property 3 and Theorem 8.5 give another proof of Theorem 8.1.

There are many other planar curves through p on S with $\mathbf{v}$ as tangent vector. The following generalization of Theorem 8.1 is due to JEAN-BAPTISTE MEUSNIER (1754–93), a student of the leading French geometer of the time, Gaspard Monge. Meusnier published only one paper in mathematics and went on to publish important papers in other scientific fields. He died a hero's death in the siege of Mayence.

Theorem 8.6 (Meusnier 1776). *Suppose* $\mathbf{w} \in \mathbb{R}^3$ *is any unit vector that is perpendicular to* $\mathbf{v} \in T_p(S)$*, and* $\mathbf{w}$ *makes an angle* β *with* $N(p)$ *where* $0 \leq \beta < \pi/2$. *Let* $c_{\mathbf{w}}(s)$ *denote the unit speed curve through* $p = c_{\mathbf{v}}(0)$ *whose trace is the intersection of the plane through p parallel to the plane* Plane$(\mathbf{v},\mathbf{w})$ *and the surface S. Here* $c'_{\mathbf{w}}(0) = \mathbf{v}$. *Then* $c_{\mathbf{w}}(s)$ *has directed curvature* $\kappa_{\mathbf{w}}$ *in* Plane$(\mathbf{v},\mathbf{w})$ *given by:*

$$\kappa_{\mathbf{w}} = \frac{\mathrm{II}_p(\mathbf{v})}{\cos\beta}.$$

PROOF: Since $\mathbf{v} = c'_{\mathbf{w}}(0)$ is perpendicular to $N(p)$, it follows that $\mathrm{I}_p(c'_{\mathbf{w}}(0), N(p)) = 0$. More generally, we can write $\mathrm{I}_{c_{\mathbf{w}}(s)}(c'_{\mathbf{w}}(s), N(c_{\mathbf{w}}(s))) = 0$. Taking the derivative of this expression with respect to s we find:

$$\begin{aligned} 0 &= \frac{d}{ds}\mathrm{I}_{c_{\mathbf{w}}(s)}(c'_{\mathbf{w}}(s), N(c_{\mathbf{w}}(s))) \\ &= \mathrm{I}_{c_{\mathbf{w}}(s)}(c''_{\mathbf{w}}(s), N(c_{\mathbf{w}}(s))) + \mathrm{I}_{c_{\mathbf{w}}(s)}(c'_{\mathbf{w}}(s), \frac{d}{ds}N(c_{\mathbf{w}}(s))). \end{aligned}$$

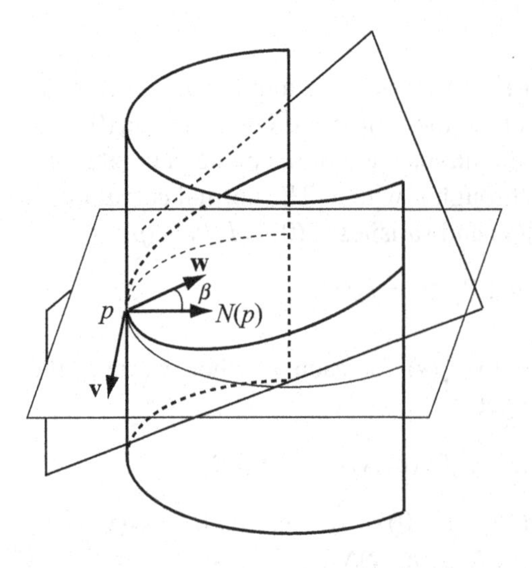

However, by definition,

$$\left.\frac{d}{ds}N(c_{\mathbf{w}}(s))\right|_{s=0} = dN_p(\mathbf{v}).$$

The previous equation implies $\mathrm{II}_p(\mathbf{v}) =$

$$-\mathrm{I}_p(\mathbf{v}, dN_p(\mathbf{v})) = \mathrm{I}_p(c''_{\mathbf{w}}(0), N(p)).$$

Because $c_{\mathbf{w}}(s)$ lies in the plane of $\{\mathbf{v},\mathbf{w}\}$, an orthonormal basis for the plane, it follows that $c''_{\mathbf{w}}(0) = (\kappa_{\mathbf{w}})\mathbf{w}$. Furthermore, $\cos\beta = \mathbf{w} \cdot N(p)$. Putting these facts together we get $\mathrm{II}_p(\mathbf{v}) = \mathrm{I}_p(c''_{\mathbf{w}}(0), N(p)) = \mathrm{I}_p((\kappa_{\mathbf{w}})\mathbf{w}, N(p)) = \kappa_{\mathbf{w}}\cos\beta$. ∎

Let $\{X_1, X_2\}$ be the orthonormal set of principal directions at p in S. The transformation dN_p is represented as a matrix in this basis as:

$$dN_p = \begin{pmatrix} -k_1 & 0 \\ 0 & -k_2 \end{pmatrix}.$$

In §8 of Gauss (1825), "the measure of curvature at a point ..." was introduced:

Definition 8.7. *The* **Gaussian curvature** *of S at p is defined by:*

$$K(p) = k_1 k_2 = \det(dN_p).$$

Observe that $K(p) > 0$ at elliptic points (*convexo-convex* points in Gauss's terminology), $K(p) < 0$ at hyperbolic points (*concavo-convex* points), and $K(p) = 0$ at parabolic and planar points. The nature of the approximating quadratic surface is determined by the sign of the Gaussian curvature. Another important observation is that the Gaussian curvature does not depend on the choice of coordinates around p; any change of coordinates leads to a change of basis for the tangent plane at p, but $K(p)$ remains the same because it is a determinant.

To develop and apply Gaussian curvature, we need a local, that is, calculable, expression for it. Let $p \in S$ be in a coordinate chart $x\colon (U \subset \mathbb{R}^2) \to S$. Then $N(p) = (x_u \times x_v)/\|x_u \times x_v\|$. If we let $\mathbf{w} = \alpha'(0) \in T_p(S)$, for some $\alpha\colon (-\epsilon, \epsilon) \to S$, then:

$$\mathbf{w} = \alpha'(0) = \frac{d}{dt}(x(u(t), v(t)))\bigg|_{t=0} = u'(0)x_u + v'(0)x_v,$$

$$dN_p(\mathbf{w}) = \frac{d}{dt} N \circ x(u(t), v(t))\bigg|_{t=0} = u'(0)N_u + v'(0)N_v.$$

Since N_u and N_v lie in $T_p(S)$, we can express them in the basis $\{x_u, x_v\}$ as:

$$N_u = a_{11}x_u + a_{12}x_v \quad N_v = a_{21}x_u + a_{22}x_v.$$

If we abbreviate $u' = u'(0)$ and $v' = v'(0)$, we can write $dN_p\colon T_p(S) \to T_p(S)$ in matrix form in terms of the basis $\{x_u, x_v\}$ as:

$$dN_p \begin{pmatrix} u' \\ v' \end{pmatrix} = \begin{pmatrix} a_{11} & a_{21} \\ a_{12} & a_{22} \end{pmatrix} \begin{pmatrix} u' \\ v' \end{pmatrix}.$$

We express the a_{ij}s through the second fundamental form. For convenience, write $\mathrm{I}_p(\mathbf{w}, \mathbf{v}) = \langle \mathbf{w}, \mathbf{v} \rangle$. For $\mathbf{w} = u'x_u + v'x_v$,

$$\begin{aligned} \mathrm{II}_p(\mathbf{w}) &= -\langle dN_p(\mathbf{w}), \mathbf{w} \rangle = -\langle u'N_u + v'N_v, u'x_u + v'x_v \rangle \\ &= -\langle N_u, x_u \rangle (u')^2 - (\langle N_u, x_v \rangle + \langle N_v, x_u \rangle) u'v' - \langle N_v, x_v \rangle (v')^2. \end{aligned}$$

Define the following functions on U:

$$\begin{aligned} e &= -\langle N_u, x_u \rangle = \langle N, x_{uu} \rangle, \\ f &= -\langle N_u, x_v \rangle = \langle N, x_{uv} \rangle = \langle N, x_{vu} \rangle = -\langle N_u, x_v \rangle, \\ g &= -\langle N_v, x_v \rangle = \langle N, x_{vv} \rangle. \end{aligned}$$

Thus, $\mathrm{II}_p(\mathbf{w}) = e(u')^2 + 2fu'v' + g(v')^2$. Using the basis $\{x_u, x_v\}$ and matrix multiplication, we can give local expressions for the first and second fundamental forms:

$$\mathrm{I}_p\left(\begin{pmatrix} u_1' \\ v_1' \end{pmatrix}, \begin{pmatrix} u_2' \\ v_2' \end{pmatrix}\right) = (u_1', v_1') \begin{pmatrix} E & F \\ F & G \end{pmatrix} \begin{pmatrix} u_2' \\ v_2' \end{pmatrix}, \quad \mathrm{II}_p\left(\begin{pmatrix} u' \\ v' \end{pmatrix}\right) = (u', v') \begin{pmatrix} e & f \\ f & g \end{pmatrix} \begin{pmatrix} u' \\ v' \end{pmatrix}.$$

Combining these expressions, the second fundamental form becomes:

$$\begin{aligned} -\mathrm{II}_p\left(\begin{pmatrix} u' \\ v' \end{pmatrix}\right) &= \langle dN_p(u'x_u + v'x_v), u'x_u + v'x_v \rangle \\ &= (a_{11}u' + a_{21}v', a_{12}u' + a_{22}v') \begin{pmatrix} E & F \\ F & G \end{pmatrix} \begin{pmatrix} u' \\ v' \end{pmatrix} \\ &= (u', v') \begin{pmatrix} a_{11} & a_{12} \\ a_{21} & a_{22} \end{pmatrix} \begin{pmatrix} E & F \\ F & G \end{pmatrix} \begin{pmatrix} u' \\ v' \end{pmatrix} \end{aligned}$$

Comparing this expression with the new functions e, f, and g we get the matrix equation:

$$-\begin{pmatrix} e & f \\ f & g \end{pmatrix} = \begin{pmatrix} a_{11} & a_{12} \\ a_{21} & a_{22} \end{pmatrix} \begin{pmatrix} E & F \\ F & G \end{pmatrix}$$

from which it follows that:

$$\begin{pmatrix} a_{11} & a_{12} \\ a_{21} & a_{22} \end{pmatrix} = \frac{-1}{EG - F^2} \begin{pmatrix} e & f \\ f & g \end{pmatrix} \begin{pmatrix} G & -F \\ -F & E \end{pmatrix}.$$

Thus we can write the functions a_{ij} explicitly:

Theorem 8.8. *The following equations hold:*

$$a_{11} = \frac{fF - eG}{EG - F^2}, \quad a_{12} = \frac{eF - fE}{EG - F^2}, \quad a_{21} = \frac{gF - fG}{EG - F^2}, \quad a_{22} = \frac{fF - gE}{EG - F^2}.$$

Corollary 8.9. $K(p) = \det dN_p = \det(a_{ij}) = \dfrac{eg - f^2}{EG - F^2}$.

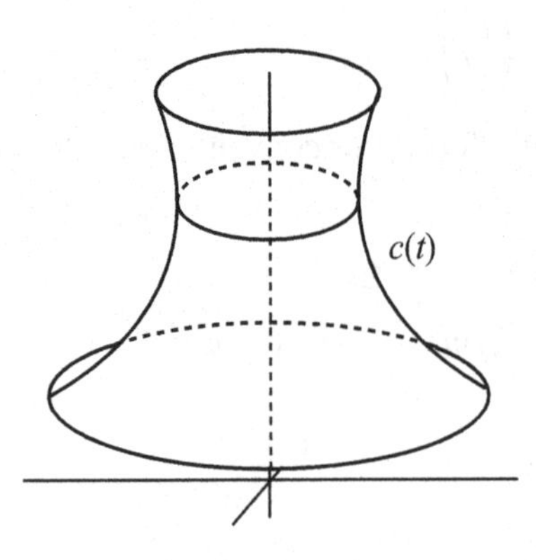

EXAMPLE: Let's calculate the Gaussian curvature for a surface of revolution (a generalization of Example 4 in chapter 7 on page 119). Suppose

$$c(t) = (\lambda(t), \mu(t)) \colon (a, b) \to \mathbb{R}^2$$

is a regular curve in the y-z-plane that does not cross itself or the z-axis. A coordinate chart for the surface of revolution in $\mathbb{R}^3$ obtained by revolving the curve $c(t)$ around the z-axis is given by:

$$x(u, v) = (\lambda(u)\cos v, \lambda(u)\sin v, \mu(u)), \text{ for } a < u < b \text{ and } 0 < v < 2\pi.$$

We begin the calculation with the basis for the tangent vectors:

$$x_u = (\lambda'(u)\cos v, \lambda'(u)\sin v, \mu'(u)), \quad x_v = (-\lambda(u)\sin v, \lambda(u)\cos v, 0)$$
$$x_u \times x_v = (-\mu'(u)\lambda(u)\cos v, -\mu'(u)\lambda(u)\sin v, \lambda(u)\lambda'(u))$$

It follows then that $E = (\lambda')^2 + (\mu')^2$, $F = 0$, and $G = \lambda^2$. The normal determined by the coordinate chart has length $\|x_u \times x_v\| = \sqrt{EG - F^2} = \lambda\sqrt{(\lambda')^2 + (\mu')^2}$. The other auxiliary functions are given by:

$$\begin{aligned} e = \langle N, x_{uu}\rangle &= \left\langle \frac{x_u \times x_v}{\|x_u \times x_v\|}, x_{uu} \right\rangle \\ &= \frac{1}{\lambda\sqrt{(\lambda')^2 + (\mu')^2}} \langle x_u \times x_v, x_{uu}\rangle = \frac{\lambda'\mu'' - \lambda''\mu'}{\sqrt{(\lambda')^2 + (\mu')^2}}. \end{aligned}$$

By similar calculations we obtain $f = 0$ and $g = \dfrac{\lambda\mu'}{\sqrt{(\lambda')^2 + (\mu')^2}}$. In the basis $\{x_u, x_v\}$, $dN_{x(u,v)}$ takes the form:

$$dN_{x(u,v)} = \begin{pmatrix} -\dfrac{\lambda'\mu'' - \lambda''\mu'}{((\lambda')^2 + (\mu')^2)^{3/2}} & 0 \\ 0 & \dfrac{-\mu'}{\lambda\sqrt{(\lambda')^2 + (\mu')^2}} \end{pmatrix}.$$

It is immediate that the principal curvatures are given by the diagonal entries in the matrix. Furthermore, x_u and x_v are the principal directions and so the meridian curves, $x(u, v_0)$ and the parallels, $x(u_0, v)$, on a surface of revolution are the curves of least and greatest normal curvature. The Gaussian curvature is given by the determinant of dN_p:

$$K(p) = \frac{\lambda'\mu'\mu'' - \lambda''(\mu')^2}{\lambda((\lambda')^2 + (\mu')^2)^2}.$$

When $(\lambda')^2 + (\mu')^2 = 1$, it follows that $\lambda'\lambda'' = -\mu'\mu''$ and so we get a simpler formula:

$$K(p) = \left(\frac{\lambda'\mu'\mu''}{\lambda} - \frac{\lambda''(\mu')^2}{\lambda}\right) = -\left(\frac{\lambda''(\mu')^2}{\lambda} + \frac{\lambda''(\lambda')^2}{\lambda}\right) = -\frac{\lambda''}{\lambda}.$$

For example, suppose that $c(t) = (\cos t, \sin t)$ and $-\pi/2 < t < \pi/2$. The surface of revolution is S^2 . The formula gives another proof that $K(p) = 1$ for all $p \in S^2$.

A surface of interest in later chapters is the surface of revolution of the tractrix (see chapter 5). A parametrization for this curve is given by:

$$\hat{\Theta}(t) = (\cos(t), \ln|\sec t + \tan t| - \sin(t)).$$

Applying the formula for Gaussian curvature for surfaces of revolution we find:

$$\begin{aligned}K(p) &= \frac{\lambda'\mu'\mu'' - \lambda''(\mu')^2}{\lambda((\lambda')^2 + (\mu')^2)^2} \\ &= \frac{(-\sin t)(\sec t - \cos t)(\sec t \tan t + \sin t) + (\cos t)(\sec t - \cos t)^2}{(\cos t)((-\sin t)^2 + (\sec t - \cos t)^2)^2} \\ &= \cdots = \frac{-\sin^4 t/\cos^3 t}{(\cos t)(\tan^4 t)} \equiv -1.\end{aligned}$$

Thus, revolving the tractrix leads to a surface of constant Gaussian curvature, like the sphere, but for which the curvature is negative everywhere.

Besides the interpretation of Gaussian curvature in terms of principal curvatures, there is another geometric interpretation, namely, the one introduced by Gauss.

Proposition 8.10. *Let P be a point in a surface S with $K(p) \neq 0$. Then:*

$$K(p) = \lim_{A \to p} \frac{\text{area}\, N(A)}{\text{area}\, A},$$

where A is a connected region around p and the limit is taken through a sequence of such regions that converge to p.

PROOF: There is another formula for that Gaussian curvature that is suited to this application. We can write $N_u = a_{11}x_u + a_{12}x_v$ and $N_v = a_{21}x_u + a_{22}x_v$. Then, by the linearity of the cross product:

$$\begin{aligned}N_u \times N_v &= (a_{11}x_u + a_{12}x_v) \times (a_{21}x_u + a_{22}x_v) \\ &= (a_{11}a_{22} - a_{12}a_{21})(x_u \times x_v) = K(x_u \times x_v).\end{aligned}$$

Thus, $N_u \times N_v = K(x_u \times x_v)$.

In a small enough region to be contained in a coordinate patch, the area of a region A around p is given by the formula:

$$\text{area}(A) = \iint_{x^{-1}(A)} \|x_u \times x_v\|\, du\, dv.$$

Mapping the region to S^2 by the Gauss map we get:

$$\text{area}(N(A)) = \iint_{x^{-1}(A)} \|N_u \times N_v\|\, du\, dv = \iint_{x^{-1}(A)} K\|x_u \times x_v\|\, du\, dv.$$

Here we should have written $|K|\, \|x_u \times x_v\|\, du\, dv$. If p is an elliptic point, this makes no difference. At a hyperbolic point, the Gauss map reverses orientation and so, to obtain the area of $N(A)$, we must multiply by -1 giving $-|K| = K$.

We apply the mean value theorem for double integrals (Lang), which yields:

$$\lim_{A\to p}\frac{\iint_{x^{-1}(A)} K\|x_u\times x_v\|\,du\,dv}{\iint_{x^{-1}(A)}\|x_u\times x_v\|\,du\,dv}=\lim_{A\to p}\frac{1/\text{area}(A)\iint_{x^{-1}(A)} K\|x_u\times x_v\|\,du\,dv}{1/\text{area}(A)\iint_{x^{-1}(A)}\|x_u\times x_v\|\,du\,dv}$$

$$=\lim_{A\to p}\left.\frac{K\|x_u\times x_v\|}{\|x_u\times x_v\|}\right|_{\text{some point in A}}=K(p).$$ ∎

The limit of areas definition of the Gaussian curvature is the first one given in Gauss (1825). In the years 1820–23 and 1828–44, he led the geodetic survey of the principality of Hannover. The analysis of the subsequent data posed basic geometric problems in what Gauss called *higher geodesy*, the solution of which led him to new methods in differential geometry. The astronomical data and the network of geodetic observations provided a practical laboratory for the study of curvature (see Scholz 1980).

Another measure of the manner in which a surface curves is its **mean curvature**. Its study was initiated by SOPHIE GERMAIN (1776–1831) in her formulation of elastic surfaces. Mean curvature is defined as the average of the principal curvatures

$$H=\frac{1}{2}(k_1+k_2).$$

The condition $H\equiv 0$ is equivalent to the property that, for every $p\in S$, there is an open set $p\in U\subset S$ for which U has the least area for its boundary ∂U. Such a surface is called a **minimal surface**. The study of minimal surfaces involves many surprising analytic methods; for example, the coordinate charts satisfy the harmonic condition ($\Delta x=0$). The interested reader should consult Oprea (2000) and Hoffman, Meeks (1990).

Exercises

8.1 Show that, for a regular surface, the Gauss map is differentiable.

8.2 Determine a formula for the Gaussian curvature K for a surface given by the graph of a smooth function $z=f(x,y)$ in terms of the function f.

8.3 Suppose that S is a surface in $\mathbb{R}^3$ and suppose that the principal curvatures k_1 and k_2 satisfy $|k_1|\leq 1$ and $|k_2|\leq 1$ at every point in S. If $\alpha(t)$ is a regular curve in S, does it follow that $|\kappa_\alpha(t)|\leq 1$ for every t in the domain of α?

8.4 Determine the principal curvatures of a cylinder, an ellipsoid, and a saddle.

8.5 Define the **Dupin indicatrix** D_p of a surface S at a point p to be the subset of $T_p(S)$:

$$D_p=\{\mathbf{v}\in T_p(S)\mid \mathrm{II}_p(\mathbf{v})=1\text{ or }-1\}.$$

Solutions to Exercises marked with a dagger appear in the appendix, pp. 333–334.

Prove that if $K(p) > 0$, D_p is an ellipse, if $K(p) < 0$, D_p is a pair of conjugate hyperbolas, and if $K(p) = 0$, then D_p is a pair of parallel lines or is empty.

8.6 A surface is said to be **convex** if it lies on one side of the tangent plane at some point of the surface. Interpret the property of convexity in terms of the principal curvatures and Gaussian curvature.

8.7 Define the **mean curvature** of a surface to be $H(p) = -\frac{1}{2}\text{trace}(dN_p) = \frac{1}{2}(k_1 + k_2)$. Prove that $H^2 \geq K$. When does equality hold?

8.8† Prove the formula for mean curvature:

$$H = \frac{Eg - 2Ff + Ge}{2(EG - F^2)}.$$

8.9 Determine the mean curvature of a cylinder and the plane. Since they differ, we see that mean curvature is *not* intrinsic.

8.10† A surface is minimal if $H \equiv 0$ on the surface. Suppose on a coordinate chart $x(u,v) = (f_1(u,v), f_2(u,v), f_3(u,v))$ and that $ds^2 = \phi\, du^2 + \phi\, dv^2$, for some function ϕ. Show that the image of this coordinate chart is minimal if and only if the functions f_i are harmonic, that is:

$$\frac{\partial^2 f_i}{\partial u^2} + \frac{\partial^2 f_i}{\partial v^2} = 0.$$

8.11 Let M be the surface determined by the coordinate chart:

$$x(u,v) = (a\cosh v \cos u, a\cosh v \sin u, av)$$

for $0 < u < 2\pi$ and $-\infty < v < \infty$. This is the surface of revolution generated by rotating the catenary curve $y = a\cosh(z/a)$ around the z-axis. Prove that M is a minimal surface.

8.12† A curve in a surface S, $\alpha(t)$, is a **line of curvature** if $\alpha'(t)$ is in the direction of one of the principal directions of curvature for each t. Prove *Rodrigues's formula*: $\alpha(t)$ is a line of curvature if and only if:

$$\frac{d}{dt}N(\alpha(t)) = \lambda(t)\alpha'(t)$$

for some differentiable function $\lambda(t)$. Prove further that the principal curvature in this case is $-\lambda(t)$.

9

Metric equivalence of surfaces

It is evident that any finite part whatever of the curved surface will retain the same integral curvature after development upon another surface.

C. F. GAUSS (1827)

The problem of constructing useful flat maps of the Earth led to the mathematical question of the existence of an ideal map projection (chapter 7bis), that is, a mapping $Y\colon (R \subset S^2) \to \mathbb{R}^2$ preserving lengths, angles, and area. The archaic terminology for such a map projection is a *development of the sphere onto the plane.* Euler showed that no such development can exist (Theorem 7bis.1). A natural generalization of this problem is to determine whether there is a development of a given surface onto another.

When a mapping between two surfaces, $\phi\colon S_1 \to S_2$ is a diffeomorphism, the structures defined by the calculus on the surfaces is shared; there is a one-to-one correspondence between the coordinate charts and the differentiable functions on each surface. However, the structures defined by geometry on the two surfaces, for example, arc length and area, may differ. Consider a sphere and an ellipsoid as examples: they are diffeomorphic, but the geometry of the arc length differs at corresponding points. To test if the geometric structures coincide, we focus on the associated first fundamental forms.

Definition 9.1. *A diffeomorphism $\phi\colon S_1 \to S_2$ between surfaces S_1 and S_2 is an* **isometry** *if, for all $p \in S_1$, and $\mathbf{w} \in T_p(S_1)$, we have:*

$$\mathrm{I}_p(\mathbf{w}, \mathbf{w}) = \mathrm{I}_{\phi(p)}(d\phi_p(\mathbf{w}), d\phi_p(\mathbf{w})).$$

Two surfaces are **isometric** *if there is an isometry between them.*

If $\phi\colon S_1 \to S_2$ is an isometry, by the polarization identity,

$$\mathrm{I}_p(\mathbf{v}, \mathbf{w}) = \frac{1}{2}(\mathrm{I}_p(\mathbf{v}+\mathbf{w}, \mathbf{v}+\mathbf{w}) - \mathrm{I}_p(\mathbf{v}, \mathbf{v}) - \mathrm{I}_p(\mathbf{w}, \mathbf{w})),$$

together with the linearity of $d\phi_p$, it follows that, for all $\mathbf{w}_1$, $\mathbf{w}_2 \in T_p(S_1)$,

$$\mathrm{I}_p(\mathbf{w}_1, \mathbf{w}_2) = \mathrm{I}_{\phi(p)}(d\phi_p(\mathbf{w}_1), d\phi_p(\mathbf{w}_2)).$$

The simplest source of isometries is the collection of rigid motions of $\mathbb{R}^3$. Recall from the appendix to chapter 6 that a rigid motion $F\colon \mathbb{R}^3 \to \mathbb{R}^3$ for which $F(\mathbf{0}) = \mathbf{0}$ satisfies $F(\mathbf{v}) \cdot F(\mathbf{w}) = \mathbf{v} \cdot \mathbf{w}$ for all $\mathbf{v}$, $\mathbf{w} \in \mathbb{R}^3$. A rigid motion is determined by multiplication by an invertible matrix, $A\colon \mathbb{R}^3 \to \mathbb{R}^3$ satisfying $A^{-1} = A^t$, plus a translation, that is, $F(\mathbf{v}) = A\mathbf{v} + \mathbf{v}_0$. If $S \subset \mathbb{R}^3$ is a regular surface, then so is $S' = F(S)$ and $F|_S\colon S \to S'$ satisfies $dF_p = dA_p = A$. Furthermore,

$$\mathrm{I}_{F(p)}(dF_p(\mathbf{w}_1), dF_p(\mathbf{w}_2)) = A\mathbf{w}_1 \cdot A\mathbf{w}_2 = A^t A\mathbf{w}_1 \cdot \mathbf{w}_2 = \mathbf{w}_1 \cdot \mathbf{w}_2 = \mathrm{I}_p(\mathbf{w}_1, \mathbf{w}_2).$$

Hence $F|_S$ is an isometry and S and S' are isometric.

If we further restrict our attention to isometries of $\mathbb{R}^3$ that fix the origin, the linear isometries, then restriction of any such mapping to the sphere $S^2 = \{\mathbf{v} \in \mathbb{R}^3 \mid \mathbf{v} \cdot \mathbf{v} = 1\}$ is an isometry $A\colon S^2 \to S^2$. In fact, every isometry of S^2 to itself arises in this manner (see Ryan 1986). This class of mappings includes rotations of the sphere around a line through its center and reflection across a plane containing the center of the sphere.

Not every isometry between surfaces is the restriction of a rigid motion of $\mathbb{R}^3$. For example, the cylinder $(\cos\theta, \sin\theta, z)$ with $0 < \theta < 2\pi$ and $-\infty < z < \infty$ may be "opened up" to an infinite open strip $(\theta, z, 0)$ by the inverse of the coordinate chart $x\colon (0, 2\pi) \times \mathbb{R} \to \mathbb{R}^3$, $x(\theta, z) = (\cos\theta, \sin\theta, z)$ which is an isometry (check the first fundamental forms). This mapping is not the restriction of a linear mapping. The notion of isometry determines an equivalence relation on surfaces.

Isometries make precise what we mean by rigid motions and congruence. Define a **congruence** to be an isometry of a surface with itself $\phi\colon S \to S$. Two *figures*, that is, subsets of S, are **congruent** if there is an isometry with $\phi(\text{figure}_1) = \text{figure}_2$. A figure made up of segments of curves on a surface may be thought of as rods in a configuration and the term *rigid motion* is synonymous with congruence.

It follows from the definition that the inverse of an isometry is also an isometry and so the set of congruences of a surface forms a group. The importance of this observation cannot be overestimated. It is the basis of the approach to geometry via *transformation groups*, initiated by FELIX KLEIN (1849–1925) and SOPHUS LIE (1847–99). (Excellent introductions to this approach are found in Ryan (1986) and Brannan, Esplen, and Gray (1999).)

Properties that are preserved under isometries are the most important to the geometry of surfaces. Such properties are called **intrinsic**. Properties that depend on the particular description of a surface or on the manner in which it lies in $\mathbb{R}^3$ are called **extrinsic**. For example, the fact that the z-axis is asymptotically close to the surface of revolution of the tractrix is an extrinsic property of this surface. We will see later that this surface intrinsically looks the same from almost every point on it.

To investigate intrinsic properties, we will want to argue locally with the apparatus of functions associated to a coordinate chart.

Proposition 9.2. *If $\phi\colon S_1 \to S_2$ is an isometry, and $p \in S_1$, then there are coordinate charts $x\colon (U \subset \mathbb{R}^2) \to S_1$ with $p \in x(U)$ and $\bar{x}\colon (U \subset \mathbb{R}^2) \to S_2$ with $\phi(p) \in \bar{x}(U)$ such that the component functions of the metric associated to x and $\bar{x}$, respectively, satisfy $E = \bar{E}$, $F = \bar{F}$, and $G = \bar{G}$.*

PROOF: Let $x\colon (U \subset \mathbb{R}^2) \to S_1$ be any coordinate chart around p, and define the coordinate chart for S_2 by $\bar{x}\colon (U \subset \mathbb{R}^2) \to S_2$ given by $\bar{x} = \phi \circ x$. This function satisfies the properties of a coordinate chart by virtue of the properties of a diffeomorphism. It follows that $\bar{x}_u = d\phi_p(x_u)$ and $\bar{x}_v = d\phi_p(x_v)$. Since ϕ is an isometry,

$$\bar{E} = \mathrm{I}_{\phi(p)}(\bar{x}_u, \bar{x}_u) = \mathrm{I}_{\phi(p)}(d\phi_p(x_u), d\phi_p(x_u)) = \mathrm{I}_p(x_u, x_u) = E.$$

Similarly $\bar{F} = F$ and $\bar{G} = G$. ∎

The proposition provides an important pointwise property of an isometry: Length, angle, and area are preserved by an isometry. For properties sufficiently local, it is enough to have a local version of isometry to compare surfaces geometrically.

Definition 9.3. *A mapping $\phi\colon (V \subset S_1) \to S_2$ of an open set V containing a point $p \in S_1$ is a* **local isometry at** *p if there are open sets $W \subset V$ containing p and $\bar{W}$ containing $\phi(p)$ with $\phi|_W\colon W \to \bar{W}$ an isometry. Two surfaces are* **locally isometric** *if there is a local isometry for every pair $p \in S_1$ and $q \in S_2$ taking p to q.*

A local isometry may fail to be an isometry. For example, a cylinder is locally isometric to the plane (roll it out). However, because the plane and the cylinder have different topological properties, no single isometry identifies the cylinder with part of the plane.

The next theorem is a partial converse to Proposition 9.2.

Theorem 9.4. *If there are coordinate patches $x\colon (U \subset \mathbb{R}^2) \to S_1$ and $\bar{x}\colon (U \subset \mathbb{R}^2) \to S_2$ such that $E = \bar{E}$, $F = \bar{F}$, and $G = \bar{G}$ on U, then $\phi = \bar{x} \circ x^{-1}\colon (x(U) \subset S_1) \to S_2$ is a local isometry.*

PROOF: Suppose $\alpha(t) = x(u(t), v(t))$ is a curve in S_1 with $\alpha(0) = p$, and $\mathbf{v} = \alpha'(0) = u'(0)x_u + v'(0)x_v \in T_p(S_1)$. Because $\phi \circ \alpha(t) = \bar{x}(u(t), v(t))$, we have $d\phi_p(\alpha'(0)) = u'(0)\bar{x}_u + v'(0)\bar{x}_v$. Since $\mathrm{I}_p(\alpha'(0), \alpha'(0)) = E(u')^2 + 2Fu'v' + G(v')^2 = \bar{E}(u')^2 + 2\bar{F}u'v' + \bar{G}(v')^2 = \mathrm{I}_{\phi(p)}(d\phi_p(\alpha'(0)), d\phi_p(\alpha'(0)))$, we get that $\mathrm{I}_p(\mathbf{v}, \mathbf{v}) = \mathrm{I}_{\phi(p)}(d\phi_p(\mathbf{v}), d\phi_p(\mathbf{v}))$ and ϕ is an isometry. ∎

When are two surfaces isometric? If they share the same geometry at corresponding points, then we expect them to share all other intrinsic properties as well. At any point of a regular surface $S \subset \mathbb{R}^3$ and a choice of coordinate chart there is a (not necessarily orthogonal) basis for the $\mathbb{R}^3$ directions from the point given locally by $[x_u, x_v, N]$. Consider the derivatives with respect to u and v of this trihedron:

$$x_{uu} = \Gamma_{11}^1 x_u + \Gamma_{11}^2 x_v + L_1 N, \qquad N_u = a_{11}x_u + a_{12}x_v,$$

$$x_{uv} = \Gamma_{12}^1 x_u + \Gamma_{12}^2 x_v + L_2 N, \qquad N_v = a_{21}x_u + a_{22}x_v,$$

$$x_{vv} = \Gamma_{22}^1 x_u + \Gamma_{22}^2 x_v + L_3 N.$$

The functions Γ_{ij}^k, defined by these equations, are called the **Christoffel symbols** of S associated to the coordinate chart x. The indices i, j, and k vary over 1 and 2 where 1 corresponds to the variable u and 2 to the variable v. The lower indices ij refer to the second partials of the chart x, and the upper index k to the component in the ordered basis $[x_u, x_v, N]$.

Since $\mathrm{I}_p(x_u, N) = \mathrm{I}_p(x_v, N) = 0$, we find $L_1 = \mathrm{I}_p(L_1 N, N) = \mathrm{I}_p(x_{uu}, N) = e$, one of the component functions associated to the second fundamental form. Similarly, $L_2 = f$, and $L_3 = g$. Writing $\langle\ ,\ \rangle$ for $\mathrm{I}_p(\ ,\ )$, and f_u for $\partial f/\partial u$, f_v for $\partial f/\partial v$ for a function $f(u,v)$, we also have:

$$\begin{aligned} E_u &= 2\langle x_{uu}, x_u\rangle, & E_v &= 2\langle x_{uv}, x_u\rangle, \\ F_u &= \langle x_{uu}, x_v\rangle + \langle x_u, x_{uv}\rangle, & F_v &= \langle x_{uv}, x_v\rangle + \langle x_u, x_{vv}\rangle, \\ G_u &= 2\langle x_{uv}, x_v\rangle, & G_v &= 2\langle x_{vv}, x_v\rangle. \end{aligned}$$

Expanding these inner products in terms of the Christoffel symbols we obtain the following relations, here written as equations and in matrix form:

$$\begin{cases} \Gamma_{11}^1 E + \Gamma_{11}^2 F &= \frac{1}{2}E_u, \\ \Gamma_{11}^1 F + \Gamma_{11}^2 G &= F_u - \frac{1}{2}E_v, \end{cases} \qquad \begin{pmatrix} E & F \\ F & G \end{pmatrix}\begin{pmatrix} \Gamma_{11}^1 \\ \Gamma_{11}^2 \end{pmatrix} = \begin{pmatrix} \frac{1}{2}E_u \\ F_u - \frac{1}{2}E_v \end{pmatrix}$$

$$\begin{cases} \Gamma_{12}^1 E + \Gamma_{12}^2 F &= \frac{1}{2}E_v, \\ \Gamma_{12}^1 F + \Gamma_{12}^2 G &= \frac{1}{2}G_u, \end{cases} \qquad \begin{pmatrix} E & F \\ F & G \end{pmatrix}\begin{pmatrix} \Gamma_{12}^1 \\ \Gamma_{12}^2 \end{pmatrix} = \begin{pmatrix} \frac{1}{2}E_v \\ \frac{1}{2}G_u \end{pmatrix}$$

$$\begin{cases} \Gamma_{22}^1 E + \Gamma_{22}^2 F &= F_v - \frac{1}{2}G_u, \\ \Gamma_{22}^1 F + \Gamma_{22}^2 G &= \frac{1}{2}G_v, \end{cases} \qquad \begin{pmatrix} E & F \\ F & G \end{pmatrix}\begin{pmatrix} \Gamma_{22}^1 \\ \Gamma_{22}^2 \end{pmatrix} = \begin{pmatrix} F_v - \frac{1}{2}G_u \\ \frac{1}{2}G_v \end{pmatrix}.$$

Because the matrix of components of the metric is invertible, the Christoffel symbols Γ_{ij}^k are expressible in terms of the functions E, F, and G and their derivatives. Proposition 9.2 implies that, at corresponding points in isometric surfaces, the Christoffel symbols agree. Furthermore, any expression given in terms of the component functions of the metric, the Christoffel symbols, and their derivatives that is independent of the choice of coordinates is intrinsic.

The most important theorem of Gauss's *Disquisitiones* (1825, §12) is the result that identifies curvature among the intrinsic properties of surfaces, making it one of the principal objects of study for differential geometry.

Theorema Egregium. *Gaussian curvature is intrinsic.*

PROOF: We give two proofs. Each proof shows how the Gaussian curvature K can be expressed entirely in terms of the component functions of the metric, E, F, and G, and their derivatives. Recall the formula for Gaussian curvature from the proof of Proposition 8.10:

$$K(x_u \times x_v) = N_u \times N_v.$$

Take the dot product of both sides with $x_u \times x_v$ and then apply a formula of Lagrange for the dot product of cross products (Exercise 6.1) to get:

$$K(EG-F^2) = (N_u \cdot x_u)(N_v \cdot x_v) - (N_u \cdot x_v)(N_v \cdot x_u).$$

Since $N = \dfrac{x_u \times x_v}{\|x_u \times x_v\|}$, we get:

$$N_u = \frac{(x_{uu} \times x_v) + (x_u \times x_{uv})}{\sqrt{EG-F^2}} + \frac{\partial}{\partial u}\left(\frac{1}{\sqrt{EG-F^2}}\right)(x_u \times x_v),$$
$$N_v = \frac{(x_{uv} \times x_v) + (x_u \times x_{vv})}{\sqrt{EG-F^2}} + \frac{\partial}{\partial v}\left(\frac{1}{\sqrt{EG-F^2}}\right)(x_u \times x_v).$$

The summand with $x_u \times x_v$ contributes zero when we dot with x_u or x_v. Furthermore, any repeated vector, as in the expression $(x_{uu} \times x_v) \cdot x_v$, vanishes, so we can rewrite the identity for K as

$$\begin{aligned} K(EG-F^2)^2 &= (EG-F^2)[(N_u \cdot x_u)(N_v \cdot x_v) - (N_u \cdot x_v)(N_v \cdot x_u)] \\ &= ((x_{uu} \times x_v) \cdot x_u)((x_u \times x_{vv}) \cdot x_v) - ((x_u \times x_{uv}) \cdot x_v)((x_{uv} \times x_v) \cdot x_u). \end{aligned}$$

Recall that $(\mathbf{u} \times \mathbf{v}) \cdot \mathbf{w}$ is equal to the determinant of the 3×3-matrix whose rows are $\mathbf{u}$, $\mathbf{v}$, and $\mathbf{w}$, in that order. It follows that we can write $K(EG-F^2)^2$ in terms of determinants of matrices with rows determined by the coordinate chart:

$$\begin{aligned} K(EG-F^2)^2 &= \det\begin{pmatrix} x_{uu} \\ x_v \\ x_u \end{pmatrix} \det\begin{pmatrix} x_u \\ x_{vv} \\ x_v \end{pmatrix} - \det\begin{pmatrix} x_u \\ x_{uv} \\ x_v \end{pmatrix} \det\begin{pmatrix} x_{uv} \\ x_v \\ x_u \end{pmatrix} \\ &= \det\left[\begin{pmatrix} x_{uu} \\ x_u \\ x_v \end{pmatrix}\begin{pmatrix} x_{vv} \\ x_u \\ x_v \end{pmatrix}^t\right] - \det\left[\begin{pmatrix} x_{uv} \\ x_u \\ x_v \end{pmatrix}\begin{pmatrix} x_{uv} \\ x_u \\ x_v \end{pmatrix}^t\right] \\ &= \det\begin{pmatrix} x_{uu} \cdot x_{vv} & x_{uu} \cdot x_u & x_{uu} \cdot x_v \\ x_u \cdot x_{vv} & E & F \\ x_v \cdot x_{vv} & F & G \end{pmatrix} - \det\begin{pmatrix} x_{uv} \cdot x_{uv} & x_{uv} \cdot x_u & x_{uv} \cdot x_v \\ x_u \cdot x_{uv} & E & F \\ x_v \cdot x_{uv} & F & G \end{pmatrix}. \end{aligned}$$

In this form, notice that the entries in the first rows and first columns, except $x_{uu} \cdot x_{vv}$ and $x_{uv} \cdot x_{uv}$ can be written in terms of Christoffel symbols and the coefficients of the metric. For example, $x_{uu} \cdot x_u = \Gamma_{11}^1 E + \Gamma_{11}^2 F$. Thus, except for the corner entries, everything else in the matrices is expressible in terms of the component functions of the metric and their derivatives.

We play with the properties of the determinant a little more: Notice that expanding along the first row of each matrix, we get terms:

$$(x_{uu} \cdot x_{vv})(EG-F^2) - (x_{uv} \cdot x_{uv})(EG-F^2) = (x_{uu} \cdot x_{vv} - x_{uv} \cdot x_{uv})(EG-F^2).$$

Thus we can compute:

$$\det\begin{pmatrix} x_{uu}\cdot x_{vv} & x_{uu}\cdot x_u & x_{uu}\cdot x_v \\ x_u\cdot x_{vv} & E & F \\ x_v\cdot x_{vv} & F & G \end{pmatrix} - \det\begin{pmatrix} x_{uv}\cdot x_{uv} & x_{uv}\cdot x_u & x_{uv}\cdot x_v \\ x_u\cdot x_{uv} & E & F \\ x_v\cdot x_{uv} & F & G \end{pmatrix}$$

$$= \det\begin{pmatrix} x_{uu}\cdot x_{vv} - x_{uv}\cdot x_{uv} & x_{uu}\cdot x_u & x_{uu}\cdot x_v \\ x_u\cdot x_{vv} & E & F \\ x_v\cdot x_{vv} & F & G \end{pmatrix} - \det\begin{pmatrix} 0 & x_{uv}\cdot x_u & x_{uv}\cdot x_v \\ x_u\cdot x_{uv} & E & F \\ x_v\cdot x_{uv} & F & G \end{pmatrix}.$$

Since the coordinate chart is infinitely differentiable, all mixed partials are equal and so $x_{uvv} = x_{vuv} = x_{vvu}$, which implies:

$$\begin{aligned} x_{uu}\cdot x_{vv} - x_{uv}\cdot x_{uv} &= x_{uu}\cdot x_{vv} + x_u\cdot x_{uvv} - x_u\cdot x_{uvv} - x_{uv}\cdot x_{uv} \\ &= \frac{\partial}{\partial u}(x_u\cdot x_{vv}) - \frac{\partial}{\partial v}(x_u\cdot x_{uv}) \\ &= \frac{\partial}{\partial u}(\Gamma^1_{22}E + \Gamma^2_{22}F)) - \frac{\partial}{\partial v}(\Gamma^1_{12}E + \Gamma^2_{12}F) \\ &= \frac{\partial}{\partial u}(F_v - \frac{1}{2}G_u) - \frac{\partial}{\partial v}(\frac{1}{2}E_v) \\ &= -\frac{1}{2}E_{vv} + F_{uv} - \frac{1}{2}G_{uu}. \end{aligned}$$

Putting this all together, we substitute for the Christoffel symbols the appropriate expressions in the derivatives of E, F, and G, take some further derivatives if needed, and we get the formula for Gaussian curvature:

$$K = \frac{1}{(EG-F^2)^2}\left[\det\begin{pmatrix} -\frac{1}{2}E_{vv} + F_{uv} - \frac{1}{2}G_{uu} & \frac{1}{2}E_u & F_u - \frac{1}{2}E_v \\ F_v - \frac{1}{2}G_u & E & F \\ \frac{1}{2}G_v & F & G \end{pmatrix}\right.$$

$$\left. - \det\begin{pmatrix} 0 & \frac{1}{2}E_v & \frac{1}{2}G_u \\ \frac{1}{2}E_v & E & F \\ \frac{1}{2}G_u & F & G \end{pmatrix}\right].$$

Thus, we have written K entirely in terms of E, F, and G, and their derivatives. ∎

Corollary 9.5. *Isometric surfaces have the same Gaussian curvature at corresponding points.*

The definition of Gaussian curvature relies on many features of the embedding of the surface in $\mathbb{R}^3$ such as the Gauss map. It is remarkable (*egregium*) that this function is independent of these trappings. Euler's Theorem on the nonexistence of an ideal map projection follows directly from *Theorema Egregium* since the sphere and the plane have different curvatures.

There are further relations between the Christoffel symbols that follow from the differentiability of the parametrization. These relations also lead to a second proof of *Theorema Egregium* based on a calculation similar to the one at the heart of the Frenet-Serret Theorem (chapter 6).

The following equations hold by virtue of the differentiability of a coordinate chart:

$$(x_{uu})_v - (x_{uv})_u = 0, \tag{1}$$

$$(x_{uv})_v - (x_{vv})_u = 0, \tag{2}$$

$$N_{uv} - N_{vu} = 0 \tag{3}$$

Equation (1) produces the following flurry of subscripts and superscripts:

$$\begin{aligned}
\frac{\partial}{\partial v} x_{uu} &= (\Gamma_{11}^1)_v x_u + \Gamma_{11}^1 x_{uv} + (\Gamma_{11}^2)_v x_v + \Gamma_{11}^2 x_{vv} + eN_v + e_v N \\
&= (\Gamma_{11}^1)_v x_u + \Gamma_{11}^1 (\Gamma_{12}^1 x_u + \Gamma_{12}^2 x_v + fN) + (\Gamma_{11}^2)_v x_v \\
&\quad + \Gamma_{11}^2 (\Gamma_{22}^1 x_u + \Gamma_{22}^2 x_v + gN) + e(a_{21} x_u + a_{22} x_v) + e_v N \\
&= ((\Gamma_{11}^1)_v + \Gamma_{11}^1 \Gamma_{12}^2 + \Gamma_{11}^2 \Gamma_{22}^1 + e a_{21}) x_u \\
&\quad + ((\Gamma_{11}^2)_v + \Gamma_{11}^1 \Gamma_{12}^2 + \Gamma_{11}^2 \Gamma_{22}^2 + e a_{22}) x_v + (\Gamma_{11}^1 f + \Gamma_{11}^2 g + e_v) N. \\
\frac{\partial}{\partial u} x_{uv} &= (\Gamma_{12}^1)_u x_u + \Gamma_{12}^1 x_{uu} + (\Gamma_{12}^2)_u x_v + \Gamma_{12}^2 x_{uv} + fN_u + f_u N \\
&= (\Gamma_{12}^1)_u x_u + \Gamma_{12}^1 (\Gamma_{11}^1 x_u + \Gamma_{11}^2 x_v + eN) + (\Gamma_{12}^2)_u x_v \\
&\quad + \Gamma_{12}^2 (\Gamma_{12}^1 x_u + \Gamma_{12}^2 x_v + fN) + f(a_{11} x_u + a_{12} x_v) + f_u N \\
&= ((\Gamma_{12}^1)_u + \Gamma_{12}^1 \Gamma_{11}^1 + \Gamma_{12}^2 \Gamma_{12}^1 + f a_{11}) x_u \\
&\quad + ((\Gamma_{12}^2)_u + \Gamma_{12}^1 \Gamma_{11}^2 + \Gamma_{12}^2 \Gamma_{12}^2 + f a_{12}) x_v + (\Gamma_{12}^1 e + \Gamma_{12}^2 f + f_u) N.
\end{aligned}$$

Since x_u, x_v and N are linearly independent, Equation (1) leads to the relations

$$\begin{aligned}
(\Gamma_{11}^1)_v + \Gamma_{11}^1 \Gamma_{12}^2 + \Gamma_{11}^2 \Gamma_{22}^1 + e a_{21} - (\Gamma_{12}^1)_u - \Gamma_{12}^1 \Gamma_{11}^1 - \Gamma_{12}^2 \Gamma_{12}^1 - f a_{11} &= 0, \\
(\Gamma_{11}^2)_v + \Gamma_{11}^1 \Gamma_{12}^2 + \Gamma_{11}^2 \Gamma_{22}^2 + e a_{22} - (\Gamma_{12}^2)_u - \Gamma_{12}^1 \Gamma_{11}^2 - \Gamma_{12}^2 \Gamma_{12}^2 - f a_{12} &= 0, \\
\Gamma_{11}^1 f + \Gamma_{11}^2 g + e_v - \Gamma_{12}^1 e - \Gamma_{12}^2 f - f_u &= 0.
\end{aligned}$$

From Theorem 8.8 we know that $a_{22} = \dfrac{fF - gE}{EG - F^2}$ and $a_{21} = \dfrac{eF - fE}{EG - F^2}$ and so from the second line we obtain the relation:

$$\begin{aligned}
(\Gamma_{11}^2)_v - (\Gamma_{12}^2)_u + \Gamma_{11}^1 \Gamma_{12}^2 + \Gamma_{11}^2 \Gamma_{22}^2 - \Gamma_{12}^1 \Gamma_{11}^2 - \Gamma_{12}^2 \Gamma_{12}^2 &= f a_{12} - e a_{22} \\
&= f \frac{eF - fE}{EG - F^2} - e \frac{fF - gE}{EG - F^2} \\
&= E \frac{eg - f^2}{EG - F^2} = EK.
\end{aligned}$$

Once again we have proved that K is expressible in terms of the component functions of the metric, the Christoffel symbols, and their derivatives.

Does the apparatus of functions $\{E,F,G,e,f,g\}$, defined on the domain $U \subset \mathbb{R}^2$ of a coordinate chart $x\colon (U \subset \mathbb{R}^2) \to S$, and its associated functions, the Christoffel symbols, characterize a surface? For curves, the functions in the Frenet-Serret apparatus $\{\kappa_\alpha, \tau_\alpha\}$ characterized the curve up to placement in $\mathbb{R}^3$ (the fundamental theorem for curves). We ask if the same sort of result holds for surfaces in $\mathbb{R}^3$.

Let the **Gauss equations** refer to the relations among the Christoffel symbols obtained from $(x_{uu})_v - (x_{uv})_u = 0$ and $(x_{uv})_v - (x_{vv})_u = 0$:

$$(\Gamma_{11}^1)_v - (\Gamma_{12}^1)_u + \Gamma_{11}^1\Gamma_{12}^2 + \Gamma_{11}^2\Gamma_{22}^1 - \Gamma_{12}^1\Gamma_{11}^2 - \Gamma_{12}^2\Gamma_{12}^2 = EK, \tag{a}$$

$$(\Gamma_{12}^1)_u - (\Gamma_{11}^1)_v + \Gamma_{12}^2\Gamma_{12}^1 - \Gamma_{11}^2\Gamma_{22}^1 = FK, \tag{b}$$

$$(\Gamma_{22}^1)_u - (\Gamma_{12}^1)_v + \Gamma_{22}^1\Gamma_{11}^1 + \Gamma_{22}^2\Gamma_{12}^1 - \Gamma_{12}^1\Gamma_{12}^1 - \Gamma_{12}^2\Gamma_{22}^1 = GK, \tag{c}$$

$$(\Gamma_{12}^2)_v - (\Gamma_{22}^2)_u + \Gamma_{12}^1\Gamma_{12}^2 - \Gamma_{22}^1\Gamma_{11}^2 = FK. \tag{d}$$

The **Mainardi-Codazzi equations** refer to the relations coming from $N_{uv} - N_{vu} = 0$:

$$e_v - f_u = e\Gamma_{12}^1 + f(\Gamma_{12}^2 - \Gamma_{11}^1) - g\Gamma_{11}^2,$$

$$f_v - g_u = e\Gamma_{22}^1 + f(\Gamma_{22}^2 - \Gamma_{12}^1) - g\Gamma_{12}^2.$$

G. Mainardi (1800–79) and D. Codazzi (1824–75) were colleagues at the University of Pavia during a particularly active period in the development of differential geometry. Mainardi published his paper on the relations among the six functions E,F,G,e,f, and g in 1856. Ten years later, Codazzi published his account of these relations in terms of directional cosines. In this equivalent form, the equations became widely known as part of his second-prize winning memoir submitted to the Paris Academy in 1860. The equations were known in 1853 by K. Peterson (1828–81), a Russian geometer who studied in Dorpat under Senff and Minding (Reich 1973).

The Gauss equations and the Mainardi-Codazzi equations together are called the **compatibility equations** for surfaces. Pierre Ossian Bonnet (1819–92) proved the following version of the fundamental theorems we have proven for curves.

Theorem 9.6 (Fundamental Theorem for Regular Surfaces). *Let E,F,G,e,f, and g be differentiable functions defined on an open set $V \subset \mathbb{R}^2$ with $E > 0$ and $G > 0$. Suppose that these functions satisfy the compatibility equations and $EG - F^2 > 0$. Then for each $q \in V$, there is a open set U with $q \in U \subset V$, and a diffeomorphism $x\colon U \to x(U) \subset \mathbb{R}^3$, such that the regular surface $x(U)$ has E,F,G,e,f, and g as component functions of the first and second fundamental forms. Furthermore, if U is connected and $\bar{x}\colon U \to \bar{x}(U) \subset \mathbb{R}^3$ is another such diffeomorphism, then there is a vector $\mathbf{v} \in \mathbb{R}^3$ and a matrix A in $O(3)$ such that $\bar{x} = A \circ x + \mathbf{v}$.*

The proof, like those of the previous fundamental theorems, is a side trip into the theory of differential equations and would take us too far afield (see Spivak, 1975, Vol. 3, pp. 79–85). It follows from the theorem that the six functions E, F, G, e, f, and g determine the local theory of a surface.

Special coordinates

An important simplifying feature of the study of curves is the arc length parametrization. However, there is no such canonical choice of coordinates for surfaces. This fact is a source of both difficulty and freedom. We use it to our advantage and try to construct coordinate charts whose properties are closer to the geometric features of the surfaces. Such charts can simplify the computation of important invariants such as the Gaussian curvature. Artful choices of coordinates play a crucial role in certain arguments in later chapters. Here are two examples of useful constructions to illustrate these ideas.

A familiar feature of rectangular Cartesian coordinates in elementary geometry is that coordinate lines meet at right angles. On a surface, this condition is equivalent to the second component function of the metric F being identically zero. There are many ways to construct such a coordinate patch for a surface. We consider a particular choice that arises from the surface's geometry.

Definition 9.7. *A curve* $\alpha\colon (a,b) \to S$ *in a regular surface* S *is a* **line of curvature** *if, for all* $t \in (a,b)$, $\alpha'(t)$ *is a nonzero multiple of a principal direction at* $\alpha(t)$.

Euler's Theorem 8.1 showed that when the principal curvatures differ, the principal directions are orthogonal. A patch for which the coordinate curves are lines of curvature is orthogonal, that is, $F = 0$. The principal directions may be identified as the eigenvectors of the differential of the Gauss map, $dN_{\alpha(t)}$, and if x_u and x_v are these eigenvectors, then:

$$dN_p = \begin{pmatrix} a_{11} & a_{21} \\ a_{12} & a_{22} \end{pmatrix} = \begin{pmatrix} a_{11} & 0 \\ 0 & a_{22} \end{pmatrix},$$

since $dN_p(x_u) = a_{11}x_u$ and $dN_p(x_v) = a_{22}x_v$. However, we know that:

$$\begin{pmatrix} a_{11} & a_{12} \\ a_{21} & a_{22} \end{pmatrix} = \frac{-1}{EG - F^2}\begin{pmatrix} e & f \\ f & g \end{pmatrix}\begin{pmatrix} G & -F \\ -F & E \end{pmatrix}.$$

Hence, if the coordinate curves are orthogonal ($F = 0$), then they are lines of curvature if and only if $f = 0$. When $f = 0$, the matrix for dN_p is diagonal in the basis $\{x_u, x_v\}$, and so the coordinate curves are eigenvectors for dN_p. If $F = 0$ and dN_p is diagonal in the basis $\{x_u, x_v\}$, then we get the relation $fE = 0$ from the equation of matrices. Since $E \neq 0$, we obtain $f = 0$.

For example, a surface of revolution has $F = 0 = f$, and hence the coordinate curves, meridians and parallels, are lines of curvature.

Another set of curves on a surface is determined by a condition on the second fundamental form:

Definition 9.8. *A curve $\alpha\colon (a,b) \to S$ in a regular surface S is an* **asymptotic line** *if, for all $t \in (a,b)$, we have $\mathrm{II}_{\alpha(t)}(\alpha'(t)) = 0$.*

Asymptotic lines do not always exist at each point in a surface. To see this, write the differential of the Gauss map in the orthonormal basis of principal directions, X_1, X_2, $dN_p = \begin{pmatrix} -k_1 & 0 \\ 0 & -k_2; \end{pmatrix}$ it follows, for $\mathbf{v} = aX_1 + bX_2 \in T_p(S)$, that:

$$\begin{aligned}\mathrm{II}_p(\mathbf{v}) &= -\mathrm{I}_p(dN_p(aX_1 + bX_2), aX_1 + bX_2) \\ &= -\mathrm{I}_p(-k_1 aX_1 - k_2 bX_2, aX_1 + bX_2) = k_1 a^2 + k_2 b^2.\end{aligned}$$

The condition $\mathrm{II}_p(\mathbf{v}) = 0$ has no nonzero solutions when $k_1 \geq k_2 > 0$. If $k_1 > 0$ and $k_2 < 0$ at each point in a neighborhood of a point in a surface, then there are two asymptotic directions at each point. Thus, if $K(q) < 0$ for all q in an open set containing a given point in a surface, we can try to construct a coordinate chart with asymptotic lines for coordinate curves.

There is a general method for constructing special coordinate patches when the conditions are given in terms of tangent directions at each point. If W is an open subset of a surface S, and $V(q)$ is a smooth assignment, $q \mapsto V(q) \in T_q(S)$, for each $q \in W$, then we call V a **vector field** on W. In a coordinate patch, $x\colon (U \subset \mathbb{R}^2) \to S$, a vector field is expressed as a pair of smooth functions $a(u,v)$, $b(u,v)$ for which $V(x(u,v)) = a(u,v)x_u + b(u,v)x_v$. An **integral curve** for a vector field $V(q)$ is a curve $\alpha(t)$ satisfying the equation $\alpha'(t) = V(\alpha(t))$ for all t.

Given a vector field $V(q)$ on an open set $W \subset S$, we say that a curve $\alpha(t)$ is a **line in the direction of** $V(q)$ if $\alpha'(t) = \lambda(\alpha(t))V(\alpha(t))$ for some nonzero, differentiable function $\lambda\colon W \to \mathbb{R}$. In this terminology, the principal directions determine vector fields on a surface, and lines of curvature are lines in the direction of a principal vector; asymptotic lines are lines in asymptotic directions.

Theorem 9.9. *Suppose p_0 is a point in a regular surface S, and, in an open set $W \subset S$ containing p_0, two linearly independent vector fields $X(q)$ and $Y(q)$ are defined for $q \in W$. Then there is a chart around p_0 with associated coordinates curves given by lines in the direction of X and Y.*

PROOF: Suppose $x\colon (U \subset \mathbb{R}^2) \to S$ is a coordinate chart with $(0,0) \in U$ and $p_0 = x(0,0)$. Suppose that $X(x(u,v)) = a(u,v)x_u + b(u,v)x_v$ and $Y(x(u,v)) = c(u,v)x_u + d(u,v)x_v$. An integral curve through p_0 for X, $\alpha_X(t)$, must satisfy the conditions $\alpha_X(0) = p_0$, $\alpha'_X(t) = X(\alpha_X(t))$. In coordinates, suppose $\alpha_X(t) = x(u(t), v(t))$, then:

$$u'(t) = a(u(t), v(t)), \quad v'(t) = b(u(t), v(t)). \qquad (\heartsuit)$$

This system of differential equations has a solution according to the following general theorem; (see Arnold 1973 for a proof).

Theorem 9.10. *Let $F_1, F_2, \ldots, F_n\colon (R \subset \mathbb{R}^{n+1}) \to \mathbb{R}$ be differentiable functions on $R = (-\epsilon,\epsilon)\times(-a_1,a_1)\times\cdots\times(-a_n,a_n)$ for which $\dfrac{\partial F_i}{\partial x_j}$ is continuous on R for all i, j. For any $(t_0,x_1^0,\ldots,x_n^0) \in R$, there is a unique curve in $\mathbb{R}^n$, $\Phi(t) = (\phi_1(t),\ldots,\phi_n(t))$, defined on $(t_0-\eta,t_0+\eta) \subset (-\epsilon,\epsilon)$, such that, for all i, $\phi_i(t_0) = x_i^0$ and $\phi_i'(t) = F_i(\Phi(t))$. Furthermore, solutions $\Phi(t)$ depend differentiably on the choice of initial condition $(t_0,x_1^0,\ldots,x_n^0)$.*

The existence of an integral curve $\alpha_X(t)$ for X follows from Theorem 9.10 applied to the system of differential equations (♡). Thus there is a curve $\alpha_X\colon(-\epsilon_1,\epsilon_1)\to S$ through p_0 and satisfying $\alpha_X'(t) = X(\alpha_X(t))$. Similarly there is a curve $\alpha_Y\colon(-\epsilon_2,\epsilon_2)\to S$ through p_0 and satisfying $\alpha_Y'(s) = Y(\alpha_Y(s))$.

Varying the initial conditions defines the family of curves:

$$\lambda_Y(t;s) \text{ with } \lambda_Y(t;0) = \alpha_X(t), \text{ and } \frac{\partial \lambda_Y(t;s)}{\partial s} = Y(\lambda_Y(t;s)).$$

By a compactness argument, we can take λ_Y defined on an open rectangle $(-m_1,m_1)\times(-n_1,n_1)$. Thus, for each $t\in(-m_1,m_1)$, the curves $s\mapsto\lambda_Y(t;s)$ is an integral curve for Y. At $(0,0)$, $\dfrac{\partial\lambda_Y(t;s)}{\partial t} = X(p_0)$ and $\dfrac{\partial\lambda_Y(t;s)}{\partial s} = Y(p_0)$. Since $\{X(p_0),Y(p_0)\}$ is linearly independent, the Inverse Function Theorem implies that $\lambda_Y\colon(-m_2,m_2)\times(-n_2,n_2)\to S$ is a coordinate chart for some $0<m_2\le m_1$.

Analogously we could use $\alpha_Y(s)$ as an axis and obtain another coordinate chart $\lambda_X(t;s)$ for $(t,s)\in(-m_3,m_3)\times(-n_3,n_3)$, which satisfies, for $s\in(-n_3,n_3)$, the property that the curve $t\mapsto\lambda_X(t;s)$ is an integral curve for X, and $\lambda_X(0;s) = \alpha_Y(s)$.

We combine these coordinate charts into a coordinate chart $y\colon(-m,m)\times(-n,n)\to S$ for which $y(u,v)$ is the intersection of the curves $s\mapsto\lambda_Y(u;s)$ and $t\mapsto\lambda_X(t;v)$.

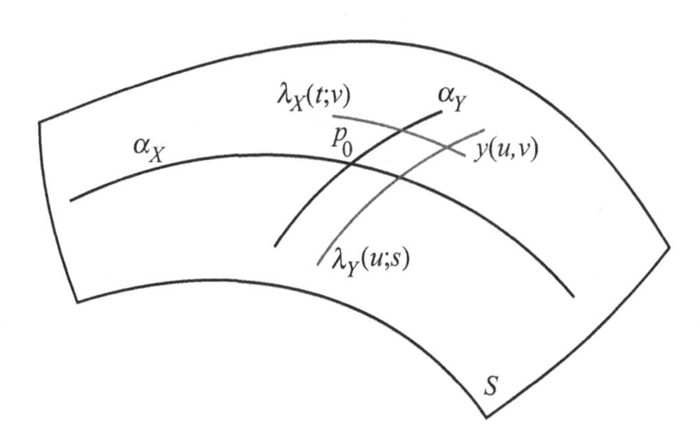

We compute y_u. Since v is fixed, we continue along the curve λ_X:

$$\frac{\partial y}{\partial u} = \lim_{h\to 0} \frac{y(u+h,v)-y(u,v)}{h} = \lim_{h\to 0} \frac{\lambda_X(t_u+h',v)-\lambda_X(t_u;v)}{h}$$
$$= \frac{\partial \lambda_X}{\partial t}\frac{\partial t}{\partial u} = \frac{\partial t}{\partial u}X.$$

Similarly, $y_v = \frac{\partial s}{\partial v}Y$. Since (t,s) and (u,v) are coordinates around p_0, the partial derivatives $\partial t/\partial u$ and $\partial s/\partial v$ are nonzero, and so $y(u,v)$ is a coordinate chart with coordinate curves given by lines in the direction of X and Y at each point. ∎

In general we cannot construct a coordinate chart whose coordinate curves are integral curves for the fields of principal directions – these vector fields $\{X_1, X_2\}$ are orthogonal and of unit length, and so $F = 0$ and $E = G = 1$. Thus, the line element takes the form $ds^2 = du^2 + dv^2$ from which we deduce that the surface has Gaussian curvature $K = 0$, which is not always true.

Corollary 9.11. *Suppose p is a point in a regular surface S. If p is not an umbilic point, then there is a patch around p with coordinate curves that are lines of curvature. If $K < 0$ in a neighborhood around p, then there is a patch around p whose coordinate curves are asymptotic lines.*

PROOF: We leave it to the reader to show that the set of umbilic points in a regular surface is closed. It follows that, around a nonumbilic point p, there is an open set of nonumbilic points. The principal directions determine a pair of linearly independent vector fields on that neighborhood to which Theorem 9.9 applies.

If the curvature in an open set containing p is negative, then there are two linearly independent unit vectors satisfying $\mathrm{II}_q(\mathbf{v}_1) = \mathrm{II}_q(\mathbf{v}_2) = 0$ for q in this open set. These vector fields may be written as smooth linear combinations of the principal directions. We can apply Theorem 9.9 again to obtain a chart with coordinate curves that are asymptotic lines. ∎

If the coordinate curves are asymptotic lines, then $\mathrm{II}_p(x_u) = \mathrm{II}_p(x_v) = 0$. If we express the second fundamental form in terms of the basis $\{x_u, x_v\}$, we get that $e = 0 = g$, which gives the simpler expression for Gaussian curvature $K = \frac{-f^2}{EG-F^2}$.

The simplification of the components of the first and second fundamental forms leads to a simplification of the associated Mainardi–Codazzi equations. If the coordinate curves are lines of curvature, the Mainardi–Codazzi equations take the form:

$$e_v = e\Gamma_{12}^1 - g\Gamma_{11}^2, \qquad -g_u = e\Gamma_{22}^1 - g\Gamma_{12}^2.$$

Rewriting the Christoffel symbols in terms of the components of the metric we get:

$$e_v = \frac{E_v}{2}\left(\frac{e}{E} + \frac{g}{G}\right), \qquad g_u = \frac{G_u}{2}\left(\frac{e}{E} + \frac{g}{G}\right).$$

When the coordinate curves are asymptotic lines, then the Mainardi–Codazzi equations are:

$$f_u = f(\Gamma_{11}^1 - \Gamma_{12}^2), \qquad f_v = f(\Gamma_{22}^2 - \Gamma_{12}^1).$$

In terms of the component functions of the metric, the first equation becomes:

$$\begin{aligned} f_u &= \frac{f}{EG - F^2}\left(\frac{1}{2}GE_u - FF_u + \frac{1}{2}FE_v + \frac{1}{2}FE_v - \frac{1}{2}EG_u\right) \\ &= \frac{f}{EG - F^2}\left(\frac{1}{2}\frac{\partial}{\partial u}(EG - F^2) + FE_v - EG_u\right). \end{aligned}$$

Similarly, the form of f_v is given by:

$$f_v = \frac{f}{EG - F^2}\left(\frac{1}{2}\frac{\partial}{\partial v}(EG - F^2) + FG_u - GE_v\right).$$

We will need the formulas derived here in chapter 12.

Exercises

9.1 Show that a surface S is isometric to a surface of revolution if it has a coordinate chart around each point such that the line element can be expressed:

$$ds^2 = \Lambda(du^2 + dv^2),$$

for Λ some function of u or v alone. Show that this chart determines a conformal mapping of the surface to the plane.

9.2 Determine the Christoffel symbols Γ_{ij}^k for the polar-coordinate parametrization of the plane.

9.3 Suppose that $F = 0$ for some coordinate chart. Prove that, on this chart,

$$K = -\frac{1}{2\sqrt{EG}}\left(\frac{\partial}{\partial v}\left(\frac{\frac{\partial E}{\partial v}}{\sqrt{EG}}\right) + \frac{\partial}{\partial u}\left(\frac{\frac{\partial G}{\partial u}}{\sqrt{EG}}\right)\right).$$

Solutions to Exercises marked with a dagger appear in the appendix, pp. 334–335.

9.4† Looking forward to the generalization of the differential geometry of surfaces to their higher-dimensional analogues (Riemannian manifolds), we present another version of *Theorema Egregium* notation-wise. All indices used vary over the set $\{1,2\}$ corresponding to coordinates $\{u,v\}$, where $u = u_1$ and $v = u_2$. Consider the matrices:

$$(g_{ij}) = \begin{pmatrix} E & F \\ F & G \end{pmatrix}, \qquad (g^{ij}) = (g_{ij})^{-1}.$$

Show that the Christoffel symbols are defined by:

$$\Gamma^i_{jk} = \sum_{l=1,2} \frac{g^{il}}{2}\left(\frac{\partial g_{kl}}{\partial u_j} + \frac{\partial g_{lj}}{\partial u_k} - \frac{\partial g_{jk}}{\partial u_l}\right).$$

Define the **Riemann curvature tensor** to be the function:

$$R^l_{ijk} = \frac{\partial}{\partial u_j}\Gamma^l_{ik} - \frac{\partial}{\partial u_k}\Gamma^l_{ij} + \sum_{m=1,2} \Gamma^m_{ik}\Gamma^l_{mj} - \Gamma^m_{ij}\Gamma^l_{mk},$$

along with the transformed version:

$$R_{ijkl} = \sum_{m=1,2} g_{im}R^m_{jkl}.$$

Show that *Theorema Egregium* becomes $K = \dfrac{R_{1212}}{\det((g_{ij}))}$.

9.5 A surface S in $\mathbb{R}^3$ is **ruled** if through each point p there is a line in $\mathbb{R}^3$ entirely contained in S. Show that the line through p lies along an asymptotic direction. Prove that if a surface is ruled, then $K \leq 0$ at each point. Give an example of a ruled surface that is not a cylinder.

9.6† Show that a surface is minimal if and only if the asymptotic directions are perpendicular.

9.7 Consider the surfaces determined by the coordinates $-a \leq u, u_1 < a$, $0 \leq v, v_1 < 2\pi$:

$$x = u\cos v, \quad y = u\sin v, \quad z = a\operatorname{arccosh}\left(\frac{u}{a}\right), \text{the catenoid;}$$

$$x = u_1\cos v_1, \quad y = u_1\sin v_1, \quad z = av_1, \text{the right helicoid.}$$

Show that the mappings $v = v_1$ and $u = \sqrt{u_1^2 + a^2}$ or $u_1 = \sqrt{u^2 + a^2}$ determine an isometry between these two surfaces.

9.8 Suppose that the functions e, f, and g are all zero for a surface S. Show that the surface is a portion of the plane.

9.9 Does there exist a surface with coordinate chart $x = x(u,v)$ and associated functions:

$$E = 1, F = 0, G = e^u, \quad e = e^u, f = 0, g = 1?$$

9.10 Show that the umbilic points form a closed set in a regular surface.

9.11† Suppose that $K(p) = 0$ for all p in a neighborhood W of a point in a surface S, and that the mean curvature $H(p) \neq 0$ for all p in W. Show that the unique asymptotic curve through each point $p \in W$ is a straight line in $\mathbb{R}^3$, and that the tangent plane is constant for all points along any such line.

10

Geodesics

It is as if he (al-Karābīsī) intended the notion which Archimedes expressed saying that it is the shortest distance which connects two points.

AL-NAYRĪZĪ (MS QOM 6526, 9TH–10TH CENTURY)

But if a particle is not forced to move upon a determinate curve, the curve which it describes possesses a singular property, which had been discovered by metaphysical considerations; but which is in fact nothing more that a remarkable result of the preceding differential equations. It consists in this, that the integral $\int v\,ds$, comprised between the two extreme points of the described curve, is less than on every other curve.

P. S. LAPLACE, MÉCANIQUE CÉLESTE (1799)

To generalize other familiar geometric notions to surfaces we need a notion of "line." Such a "line" should enjoy at least one of the elementary properties of straight lines in the plane, where lines are:

(1) the curves of shortest length joining two points (Archimedes);
(2) the curves of plane curvature identically zero (Huygens, Leibniz, Newton); and
(3) the curves whose unit tangent and its derivative are linearly dependent.

In order to have geometric significance, a defining notion for a "line" on a surface must be intrinsic, that is, independent of the choice of coordinates. We first adapt the condition of zero plane curvature. Let $\alpha\colon (-\epsilon,\epsilon) \to S$ be a curve on S, parametrized by arc length. The unit tangent vector is denoted by $T(s) = \alpha'(s)$. Consider $T'(s) = \alpha''(s)$; since $\alpha(s)$ is a unit speed curve, $T'(s)$ is perpendicular to $T(s)$. However, there are infinitely many directions perpendicular to $T(s)$. The directions most related to the surface lie in the tangent plane $T_{\alpha(s)}(S)$.

Definition 10.1. *The* **intrinsic normal** *to $\alpha(s)$, a unit speed curve on S, is the vector:*

$$n_\alpha(s) = N(\alpha(s)) \times T(s).$$

Since $[N(\alpha(s)), T(s), n_\alpha(s)]$ is an orthonormal frame and $T'(s)$ is perpendicular to $T(s)$, we have:

$$\alpha''(s) = T'(s) = \langle T'(s), N\rangle N + \langle T'(s), n_\alpha(s)\rangle n_\alpha(s).$$

From chapter 8 we know that the normal curvature of the surface at $\alpha(s)$ in the direction of $T(s)$, gotten by the intersection of the plane spanned by $[T(s), N]$ with the surface S, is given by:

$$k_n(T(s)) = \mathrm{II}_{\alpha(s)}(T(s)) = -\mathrm{I}_{\alpha(s)}(\frac{d}{ds}(N(\alpha(s)), T(s)) = \mathrm{I}_{\alpha(s)}(T'(s), N).$$

Thus, $\alpha''(s) = k_n(T(s))N + \mathrm{I}_{\alpha(s)}(T'(s), n_\alpha(s))n_\alpha(s)$ and the component in the direction $n_\alpha(s)$ lies in the tangent plane and is normal to the tangent vector. The coefficient of $n_\alpha(s)$ leads to the following definition.

Definition 10.2. *The* **geodesic curvature** *of a unit speed curve $\alpha(s)$ on a surface S is the function $k_g(s) = \mathrm{I}_{\alpha(s)}(\alpha''(s), n_\alpha(s)) = \langle \alpha''(s), n_\alpha(s) \rangle$.*

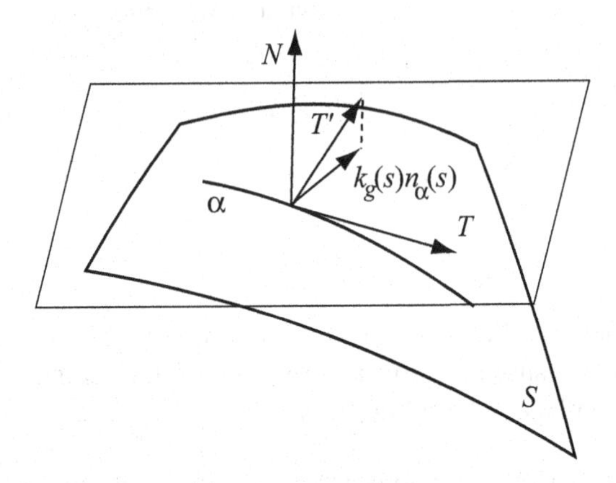

Geodesic curvature was introduced in 1830 by FERDINAND MINDING (1806–85). Minding was a student of Gauss in Göttingen and later professor of Mathematics at Dorpat (now Tartu in Estonia). This university was far removed from the research centers of mathematics of the day, but it was an outpost for work in differential geometry at the time, mostly through the efforts of a teacher of Gauss, MARTIN BARTELS (1769–1836) who came to Dorpat in 1821. Bartels was an active correspondent with Gauss and he was keenly interested in differential geometry (Lumiste 1997).

The quantity $\mathrm{I}_{\alpha(s)}(\alpha''(s), n_\alpha(s))$ measures the amount that $\alpha''(s)$ is "perceivable" in the tangent plane to S at $\alpha(s)$. In $\mathbb{R}^3$ a unit speed curve has $\alpha''(s) = \kappa(s)\bar{N}(s)$, where $\bar{N}(s) = \dfrac{\alpha''(s)}{\|\alpha''(s)\|}$. On a surface S in $\mathbb{R}^3$, since $\alpha''(s) = k_n(\alpha'(s))N(s) + k_g(s)n_\alpha(s)$, it follows that $(\kappa(s))^2 = (k_n(\alpha'(s)))^2 + (k_g(s))^2$.

Recall that a quantity is intrinsic if it is expressible in terms of the component functions of the metric and so it is preserved by isometries.

Theorem 10.3 (Minding 1830). *Geodesic curvature is intrinsic.*

PROOF: We calculate k_g for a unit speed curve lying in a coordinate chart $x\colon (U \subset \mathbb{R}^2) \to S$ given by $\alpha(s) = x(u(s), v(s))$. Then $T(s) = \alpha'(s) = u'x_u + v'x_v$, and from the proof of *Theorema Egregium* we can write:

$$\begin{aligned}T'(s) = \alpha''(s) &= u''x_u + v''x_v + (u')^2 x_{uu} + 2u'v'x_{uv} + (v')^2 x_{vv} \\ &= u''x_u + v''x_v + (u')^2(\Gamma_{11}^1 x_u + \Gamma_{11}^2 x_v + eN) \\ &\quad + 2u'v'(\Gamma_{12}^1 x_u + \Gamma_{12}^2 x_v + fN) + (v')^2(\Gamma_{22}^1 x_u + \Gamma_{22}^2 x_v + gN).\end{aligned}$$

Ignoring the component in the normal direction we have:

$$\begin{aligned}k_g(s)n_\alpha(s) = (u'' &+ (u')^2\Gamma_{11}^1 + 2u'v'\Gamma_{12}^1 + (v')^2\Gamma_{22}^1)x_u \\ &+ (v'' + (u')^2\Gamma_{11}^2 + 2u'v'\Gamma_{12}^2 + (v')^2\Gamma_{22}^2)x_v.\end{aligned}$$

Since $n_\alpha(s)$ has unit length,

$$k_g(s) = \langle k_g(s)n_\alpha(s), n_\alpha(s)\rangle = \langle k_g(s)n_\alpha(s), N(\alpha(s)) \times T(s)\rangle.$$

The cross product satisfies the relation $\langle \mathbf{u}, \mathbf{v} \times \mathbf{w}\rangle = \langle \mathbf{w} \times \mathbf{u}, \mathbf{v}\rangle$. Since $x_u \times x_v = \sqrt{EG - F^2}N$, we are led to the formula:

$$\begin{aligned}k_g(s) &= \langle T(s) \times k_g(s)n_\alpha(s), N\rangle \\ &= \langle (u'x_u + v'x_v) \times (u'' + (u')^2\Gamma_{11}^1 + 2u'v'\Gamma_{12}^1 + (v')^2\Gamma_{22}^1)x_u, N\rangle \\ &\quad + \langle (u'x_u + v'x_v) \times (v'' + (u')^2\Gamma_{11}^2 + 2u'v'\Gamma_{12}^2 + (v')^2\Gamma_{22}^2)x_v, N\rangle. \\ &= \sqrt{EG - F^2}(-v')(u'' + (u')^2\Gamma_{11}^1 + 2u'v'\Gamma_{12}^1 + (v')^2\Gamma_{22}^1) \\ &\quad + \sqrt{EG - F^2}(u')(v'' + (u')^2\Gamma_{11}^2 + 2u'v'\Gamma_{12}^2 + (v')^2\Gamma_{22}^2) \\ &= \det\begin{pmatrix} u' & u'' + (u')^2\Gamma_{11}^1 + 2u'v'\Gamma_{12}^1 + (v')^2\Gamma_{22}^1 \\ v' & v'' + (u')^2\Gamma_{11}^2 + 2u'v'\Gamma_{12}^2 + (v')^2\Gamma_{22}^2 \end{pmatrix}.\end{aligned}$$

If $\phi\colon S \to S'$ is an isometry, then at $\phi(x(u_0, v_0))$ the expression for k_g for $\alpha(s) = x(u(s), v(s))$ is the same for the curve $\phi(x(u(s), v(s)))$ since it depends only on the components of the metric, the Christoffel symbols, and $(u(s), v(s))$. ∎

When a curve $\beta(t)$ is not a unit speed curve, we can compute the geodesic curvature along $\beta(t)$ using properties of the cross product. We write the arc length parametrized curve associated to $\beta(t)$ as $\alpha(s) = \beta(t(s))$, where $t(s)$ is the inverse function of the arc length. Then, as usual,

$$T(s) = \alpha'(s) = \beta'(t(s))\frac{dt}{ds} \text{ and } T'(s) = \alpha''(s) = \beta''(t(s))\left(\frac{dt}{ds}\right)^2 + \beta'(t(s))\frac{d^2t}{ds^2}.$$

It follows that $T(s) \times T'(s) = [\beta'(t(s)) \times \beta''(t(s))]\left(\dfrac{dt}{ds}\right)^3$. Expressing everything in sight in terms of $t_0 = t(s_0)$ we find:

$$k_g|_{\beta(t_0)} = \langle T(s_0) \times T'(s_0), N\rangle = \frac{\langle \beta'(t_0) \times \beta''(t_0), N\rangle}{(ds/dt)^3} = \frac{\langle \beta'(t_0) \times \beta''(t_0), N\rangle}{\langle \beta'(t_0), \beta'(t_0)\rangle^{3/2}}.$$

Finally, if we write $\beta(t) = x(u(t), v(t))$ in a coordinate chart for the surface, we have $\beta'(t_0) = u'x_u + v'x_v$ and:

$$\begin{aligned}\beta''(t) &= u''x_u + v''x_v + (u')^2 x_{uu} + 2u'v'x_{uv} + (v')^2 x_{vv} \\ &= u''x_u + v''x_v + (u')^2(\Gamma^1_{11}x_u + \Gamma^2_{11}x_v + eN) \\ &\quad + 2u'v'(\Gamma^1_{12}x_u + \Gamma^2_{12}x_v + fN) + (v')^2(\Gamma^1_{22}x_u + \Gamma^2_{22}x_v + gN).\end{aligned}$$

From this expression we get the formula:

$$k_g|_{\beta(t_0)} = \frac{\sqrt{EG - F^2}}{\langle \beta'(t_0), \beta'(t_0)\rangle^{3/2}} \det \begin{pmatrix} u' & u'' + (u')^2\Gamma^1_{11} + 2u'v'\Gamma^1_{12} + (v')^2\Gamma^1_{22} \\ v' & v'' + (u')^2\Gamma^2_{11} + 2u'v'\Gamma^2_{12} + (v')^2\Gamma^2_{22} \end{pmatrix}.$$

If $T'(t)$ points entirely in the direction of the normal to the surface, then the intrinsic part of $T'(t)$ vanishes. An inhabitant of the surface would see $T'(t) = \mathbf{0}$, since such an observer only perceives acceleration in the tangent plane. This idea leads to the following class of curves on a surface.

Definition 10.4. *A* **geodesic** *on a surface S is a curve on S whose geodesic curvature is identically zero.*

On the unit sphere $S^2 \subset \mathbb{R}^3$ consider the curve given by the intersection of a plane Π with the sphere. This curve is planar and $T'(s)$ lies in Π. If the normal to the sphere at a point on such a curve does not lie in Π, then $T'(s)$ has a component in the tangent direction and $k_g \neq 0$. It follows that intersections of a plane with the sphere are geodesics if and only if they are great circles, that is, the plane Π contains the origin.

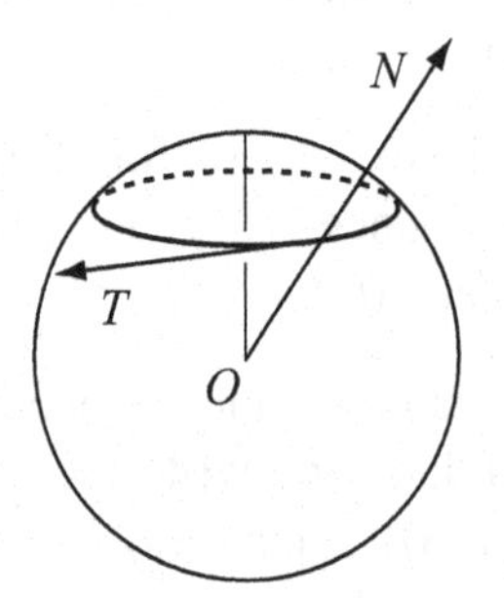

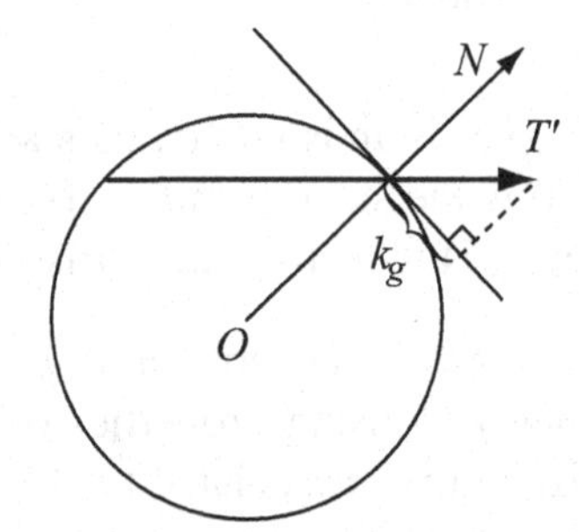

By a similar argument on a surface of revolution a parallel through a point of a graph is a geodesic if and only if it is over a critical point of the graph.

From the local formula for a parametrized curve and the linear independence of x_u and x_v, we see that k_g is identically zero when the following differential equations, the **geodesic equations**, hold:

$$\begin{cases} u'' + (u')^2\Gamma^1_{11} + 2u'v'\Gamma^1_{12} + (v')^2\Gamma^1_{22} = 0 \\ v'' + (u')^2\Gamma^2_{11} + 2u'v'\Gamma^2_{12} + (v')^2\Gamma^2_{22} = 0. \end{cases} \qquad (\spadesuit)$$

This analytic expression determines the geodesics in the ordinary plane. The line element is given by $ds^2 = du^2 + dv^2$ and so all of the Christoffel symbols vanish. The geodesic equations (♠) become $u'' = 0$, $v'' = 0$, which determine a line in the plane, as expected.

The geodesic equations depend on the choice of coordinates. For example, let's consider the plane, this time with the parametrization given by polar coordinates,

$$y\colon \mathbb{R}\times(-\pi,\pi)\to\mathbb{R}^2-\{(x,0)\mid x\leq 0\} \qquad y(\rho,\theta)=(\rho\cos\theta,\rho\sin\theta).$$

The coefficient functions of the metric are $E=1$, $F=0$, and $G=\rho^2$, which determine Christoffel symbols $\Gamma^1_{22}=-\rho$, $\Gamma^2_{12}=1/\rho$, and the others zero. The geodesic equations in this coordinate system are given by:

$$\rho''+(\rho')^2\Gamma^1_{11}+2\rho'\theta'\Gamma^1_{12}+(\theta')^2\Gamma^1_{22}=\rho''-(\theta')^2\rho=0,$$

$$\theta''+(\rho')^2\Gamma^2_{11}+2\rho'\theta'\Gamma^2_{12}+(\theta')^2\Gamma^2_{22}=\theta''+2\frac{\rho'\theta'}{\rho}=0.$$

Suppose a geodesic is given by a curve $\gamma : t\mapsto(\rho(t),t)$. Then $\theta'=1$ and $\theta''=0$, and the equations become:

$$\rho''-\rho=0, \text{ and } 2\frac{\rho'}{\rho}=0.$$

Angling for the expected answer, consider the differential equation:

$$\frac{1}{\rho}+\frac{d^2}{dt^2}\frac{1}{\rho}=\frac{1}{\rho}-\frac{\rho''}{\rho^2}+2\frac{(\rho')^2}{\rho^3}=\frac{\rho-\rho''}{\rho^2}+\frac{2}{\rho}\left(\frac{\rho'}{\rho}\right)^2=0.$$

Thus the function $1/\rho(t)$ satisfies $y+y''=0$ giving:

$$\frac{1}{\rho(t)}=a\cos t+b\sin t, \text{ and so } a\rho(t)\cos t+b\rho(t)\sin t=1,$$

which is the equation of the straight line $ax(t)+by(t)=1$.

The differential equations (♠) imply some of the important properties of geodesics.

Proposition 10.5. *Suppose $\beta\colon(-\epsilon,\epsilon)\to S$ is a curve on a surface S and $\beta(t)=x(u(t),v(t))$ in a coordinate chart on S. If β satisfies the geodesic equations, then:*

(1) β is a geodesic.
(2) β is parametrized by a multiple of arc length.
(3) For $r>0$ the curve $\bar\beta\colon(-\epsilon/r,\epsilon/r)\to S$ given by $\bar\beta(t)=\beta(rt)$ is also a geodesic.

PROOF: The geodesic equations imply $k_g(t)=0$ for all t and so β is a geodesic. If we write $\beta''(t)$ in terms of the frame $[x_u,x_v,N]$, the equations (♠) imply $\beta''(t)=lN$ and $k_g=0$. Furthermore,

$$\frac{d}{dt}\langle\beta'(t),\beta'(t)\rangle=2\langle\beta''(t),\beta'(t)\rangle=2\langle lN,\beta'(t)\rangle=0.$$

Thus, $\langle\beta'(t),\beta'(t)\rangle$ is a constant function and $\beta(t)$ is parametrized by a multiple of arc length.

Finally, (♠) is a system of degree two homogeneous differential equations and so the substitution $t \mapsto rt$ in $x(u(t),v(t))$ does not affect the conditions (♠). Thus, $\bar{\beta}(t)=\beta(rt)$ as defined is also a geodesic. ■

Since geodesic curvature is an intrinsic quantity, an isometry between surfaces preserves geodesics. A planar example with interesting features is given by a cone, given by the surface of revolution of the line $y = mx$ around the x-axis. This surface has coordinates:

$$x(u,v) = (u, mu\cos v, mu\sin v) \text{ for } 0 < u < \infty, -\pi < v < \pi.$$

A cone can be "rolled out" on a plane as follows. Consider the mapping:

$$X\colon (0,\infty)\times\left(\frac{-\pi m}{\sqrt{1+m^2}} < \theta < \frac{\pi m}{\sqrt{1+m^2}}\right) \to \mathbb{R}^3,$$

$$X(\rho,\theta) = \frac{1}{\sqrt{1+m^2}}\left(\rho, m\rho\cos\left(\frac{\sqrt{1+m^2}}{m}\theta\right), m\rho\sin\left(\frac{\sqrt{1+m^2}}{m}\theta\right)\right).$$

As a parametrization, the associated coefficients of the metric on the cone are $E = 1$, $F = 0$, and $G = \rho^2$, and so X is an isometry. The image of straight lines in the plane determines geodesics on the cone.

Lines in the plane have another property that transfers to geodesics on the cone. Let ψ denote the angle between a given line and a circle of radius R centered at the origin. Parametrize the line $ax+by=1$ by $t\mapsto(t,(1-at)/b)$ with tangent vector $(1,-a/b)$, and denote a point of intersection by $(R\cos\theta, R\sin\theta)$. Thus $aR\cos\theta + R\sin\theta = 1$ and the cosine of ψ is given by:

$$\cos\psi = \frac{(1,-a/b)\cdot(-R\sin\theta, R\cos\theta)}{\sqrt{1+(a/b)^2}\cdot R} = \frac{-bR\sin\theta - aR\cos\theta}{R\sqrt{a^2+b^2}} = \frac{-1}{R\sqrt{a^2+b^2}}.$$

It follows that $R\cos\psi = -1/\sqrt{a^2+b^2}$, a constant. On the cone, this observation implies that geodesics make an angle with the parallels, the images of the concentric circles $\rho =$ a constant, such that the radius of the circle times the cosine of the angle made with the circle is a constant.

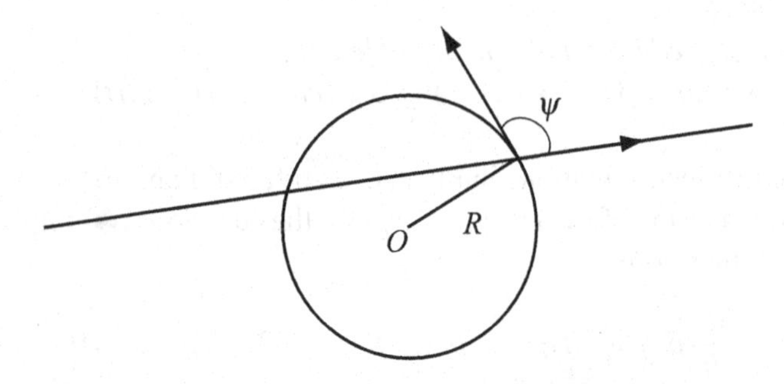

More generally, geodesics on a surface of revolution satisfy such a condition. Let $c\colon (a,b) \to \mathbb{R}^2$, $c(t) = (\lambda(t), \mu(t))$, be a one-to-one regular curve in the plane and:

$$x\colon (a,b) \times (-\pi, \pi) \to \mathbb{R}^3, \quad x(u,v) = (\lambda(u)\cos v, \lambda(u)\sin v, \mu(u)),$$

the coordinate chart for the surface of revolution S_c generated by the curve. The coefficient functions of the metric are $E = (\lambda'(u))^2 + (\mu'(u))^2$, $F = 0$, and $G = (\lambda(u))^2$. The associated Christoffel symbols are $\Gamma_{11}^1 = E_u/2E$, $\Gamma_{22}^1 = -G_u/2E$, $\Gamma_{12}^2 = G_u/2G$, and $\Gamma_{12}^1 = \Gamma_{11}^2 = \Gamma_{22}^2 = 0$. The meridian curves $t \mapsto x(t, v_0)$ are geodesics because $u' = 1$, $u'' = 0$, $v' = 0 = v''$ and so the determinant giving k_g vanishes.

Suppose $\alpha(t) = x(u(t), v(t))$ is any geodesic on S_c. Consider the angle ψ between the geodesic and a parallel $t \mapsto x(u_0, t)$: Since $F = 0$ and $\alpha(t)$ is parametrized by a multiple of arc length, ψ is determined by:

$$\cos\psi = \frac{\alpha'(t) \cdot x_v}{\|\alpha'(t)\| \cdot \|x_v\|} = \frac{(u'x_u + v'x_v) \cdot x_v}{\|\alpha'(t)\| \cdot \|x_v\|} = \frac{v'G}{C \cdot \sqrt{G}}.$$

Then $\lambda(u(t)) \cdot \cos\psi = \sqrt{G} \cdot \dfrac{v'G}{C \cdot \sqrt{G}}$. It follows that:

$$\begin{aligned}
\frac{d}{dt}(\lambda(u(t)) \cdot \cos\psi) &= \frac{1}{C}\frac{d}{dt}(v'(t)G) = \frac{1}{C}(v''G + v'(G_u u' + G_v v')) \\
&= \frac{G}{C}\left(v'' + 2u'v'\frac{G_u}{2G}\right) = \frac{G}{C}\left(v'' + 2u'v'\Gamma_{12}^2\right) \\
&= \frac{G}{C}\left(v'' + (u')^2\Gamma_{11}^2 + 2u'v'\Gamma_{12}^2 + (v')^2\Gamma_{22}^2\right) = 0,
\end{aligned}$$

because $\Gamma_{11}^2 = \Gamma_{22}^2 = 0$ in this case. Thus $\lambda(u(t)) \cdot \cos\psi$ is constant. Since the parallels and meridians are perpendicular, we have proved:

Proposition 10.6 (Clairaut's Relation). *On a surface of revolution, a geodesic $\alpha(t)$ makes an angle $\psi(t)$ with the meridians for which* $\sin\psi(t)$ *times the radius of the surface at $\alpha(t)$ is a constant.*

If we interpret a geodesic on a surface as a particle, freely moving in $\mathbb{R}^3$, but constrained to remain on the surface of revolution, then the angular momentum, radius times $\sin\psi$, of the particle is conserved.

Alexis Claude Clairaut (1713–1765) discovered many properties of curves as a young man, becoming a member of the French Academy for this work at age 16. He went on to contribute many significant results on issues of natural science, including the motion of the moon, the three-body problem, and the shape of the Earth.

We next consider the condition of Archimedes for being a line–a curve that locally minimizes distance. The geodesic equations provide the tool for analyzing this condition.

Theorem 10.7. *Suppose* $\alpha\colon(-\epsilon,\epsilon)\to S$ *is a unit speed curve on a surface S, and for any a, b in* $(-\epsilon,\epsilon)$ *we let* $\alpha(a)=p$, $\alpha(b)=q$. *If* $\alpha(s)$ *is the curve of shortest distance joining p to q in S, then* α *is a geodesic.*

PROOF: The proof is a page out of the calculus of variations, a subject that finds many applications in differential geometry (Dacorogna 2004). The goal is to show that $k_g(s)=0$ for $a<s<b$. Suppose that this condition fails and there is some $s_0\in(a,b)$ for which $k_g(s_0)\neq 0$. By the continuity of $k_g(s)$ there are values j, k, with $a<j<s_0<k<b$ with $k_g(s)\neq 0$ on $[j,k]$, and $\alpha([j,k])\subset x(U)$ for some coordinate chart x. Let $\lambda\colon[j,k]\to\mathbb{R}$ be any smooth function satisfying:

(1) $\lambda(j)=\lambda(k)=0$.
(2) $\lambda(s)k_g(s)>0$ for all $s\in(j,k)$.

Let $n_\alpha(s)=N\times\alpha'(s)$ be the intrinsic normal along the curve $\alpha(s)$. Define the functions $\bar{u}$, $\bar{v}\colon[j,k]\to\mathbb{R}$ by $\lambda(s)n_\alpha(s)=\bar{u}(s)x_u+\bar{v}(s)x_v$. Extend $\bar{u}$ and $\bar{v}$ to $[a,b]$ by $\bar{u}(s)=0=\bar{v}(s)$ for $a\leq s\leq j$ and $k\leq s\leq b$. Define:

$$C(r,s)=x(u(s)+r\bar{u}(s),v(s)+r\bar{v}(s)),$$

where $\alpha(s)=x(u(s),v(s))$ and $|r|$ is small enough for this expression to make sense. By the properties of the function λ, it follows that $C(r,a)=p$, $C(r,b)=q$, and $C(0,s)=\alpha(s)$. Thus r parametrizes a family of curves joining p and q. For a fixed value of r, we write $\alpha_r(s)=C(r,s)$.

The length of $\alpha_r(s)$, as it varies over the value of r, determines the function,

$$L(r)=\int_a^b\sqrt{\left\langle\frac{d\alpha_r}{ds},\frac{d\alpha_r}{ds}\right\rangle}\,ds.$$

By assumption, $L(r)$ has a minimum at $r=0$. Consider $L'(r)$:

$$L'(r)=\frac{d}{dr}\int_a^b\sqrt{\left\langle\frac{d\alpha_r}{ds},\frac{d\alpha_r}{ds}\right\rangle}\,ds=\int_a^b\frac{\partial}{\partial r}\sqrt{\left\langle\frac{d\alpha_r}{ds},\frac{d\alpha_r}{ds}\right\rangle}\,ds$$
$$=\int_a^b\left(\left\langle\frac{\partial^2 C}{\partial s\,\partial r},\frac{\partial C}{\partial s}\right\rangle\Big/\sqrt{\left\langle\frac{\partial C}{\partial s},\frac{\partial C}{\partial s}\right\rangle}\right)ds.$$

For $r = 0$, since $\alpha_0(s) = \alpha(s)$ is unit speed, $\left.\left\langle \frac{\partial C}{\partial s}, \frac{\partial C}{\partial s} \right\rangle\right|_{r=0} = 1$. Integrating we obtain:

$$\begin{aligned}
0 = L'(0) &= \int_a^b \left.\left\langle \frac{\partial^2 C}{\partial s \partial r}, \frac{\partial C}{\partial s} \right\rangle\right|_{r=0} ds \\
&= \int_a^b \left[\frac{d}{ds} \left.\left\langle \frac{\partial C}{\partial r}, \frac{\partial C}{\partial s} \right\rangle\right|_{r=0} - \left.\left\langle \frac{\partial C}{\partial r}, \frac{\partial^2 C}{\partial s^2} \right\rangle\right|_{r=0} \right] ds \\
&= \left.\left\langle \frac{\partial C}{\partial r}, \frac{\partial C}{\partial s} \right\rangle\right|_{r=0} - \int_a^b \left.\left\langle \frac{\partial C}{\partial r}, \frac{\partial^2 C}{\partial s^2} \right\rangle\right|_{r=0} ds.
\end{aligned}$$

By the definition of $C(r,s)$, on the interval $[j,k]$ we have:

$$\frac{\partial C}{\partial r} = \bar{u}(s)x_u + \bar{v}(s)x_v = \lambda(s)n_\alpha(s).$$

Since $n_\alpha(s)$ is perpendicular to $\frac{\partial C(0,s)}{\partial s} = \alpha'(s)$, we know that $\left.\left\langle \frac{\partial C}{\partial r}, \frac{\partial C}{\partial s} \right\rangle\right|_{r=0} = 0$. Hence,

$$\begin{aligned}
0 = L'(0) &= -\int_a^b \left.\left\langle \frac{\partial C}{\partial r}, \frac{\partial^2 C}{\partial s^2} \right\rangle\right|_{r=0} ds = -\int_a^b \langle \lambda(s)n_\alpha(s), \alpha''(s) \rangle \, ds \\
&= -\int_j^k \lambda(s)k_g(s)\, ds < 0.
\end{aligned}$$

This is a contradiction that followed from the assumption that $k_g(s_0) \neq 0$. Thus, we have shown that $k_g(s) = 0$ along $\alpha(s)$ and $\alpha(s)$ is a geodesic. ∎

The term *geodesic* comes from the science of geodesy, which is concerned with measurements of the Earth's surface. Many mathematicians contributed to the theoretical foundations of geodesy, most notably Gauss. The Königsberg mathematician F. W. Bessel (1784–1846) published an analysis of several geodetic surveys to determine more precisely the shape of the Earth as an ellipsoid of rotation. C. G. J. Jacobi (1804–51) made a study of the "shortest curves" on an ellipsoid of rotation that he referred to as "geodesic curves," motivated by such curves on the Earth's surface. Earlier writers such as Johann Bernoulli as well as Gauss had used the term *linea brevissima*–"shortest curves"–without worrying about ambiguities like the choice of direction along a great circle joining two points on a sphere. In 1844 the term "geodesic curve" replaced "shortest curve" in the influential work of J. Liouville (1809–82) who took the term from the work of Jacobi (Struik 1950). For a modern account of geodesy, see Meyer (2010).

The geodesic equations (♠) are grist for the mill in the theory of differential equations. The following general theorem leads to many useful geometric results. We refer the reader to Spivak (1970, Vol. 1) for a proof.

Theorem 10.8. *Let $F\colon \mathbb{R}^n \times \mathbb{R}^n \to \mathbb{R}^n$ be a smooth function and consider the differential equation:*

$$\frac{d^2c}{dt^2} = F\left(c, \frac{dc}{dt}\right).$$

Then, for all $(\mathbf{x}_0, \mathbf{v}_0) \in \mathbb{R}^n \times \mathbb{R}^n$, there is an open set $U \times V$ of $(\mathbf{x}_0, \mathbf{v}_0)$ and an $\epsilon > 0$ such that, for any $(\mathbf{x}, \mathbf{v}) \in U \times V$, the equation has a unique solution: $c_{\mathbf{v}}\colon (-\epsilon, \epsilon) \to \mathbb{R}^n$, satisfying the initial conditions, $c_{\mathbf{v}}(0) = \mathbf{x}$ and $c'_{\mathbf{v}}(0) = \mathbf{v}$. Moreover, the mapping $f\colon U \times V \times (-\epsilon, \epsilon) \to \mathbb{R}^n$ given by $f(\mathbf{x}, \mathbf{v}, t) = c_{\mathbf{v}}(t)$ is smooth.

The system of differential equations (♠) can be written to satisfy the conditions of the theorem and so we can apply it to guarantee the existence of geodesics. The restriction to an open set $U \times V$ in the theorem implies a restriction on the length of tangent vectors $\mathbf{v}$ to which the result applies. More precisely, define the **norm** of $\mathbf{v}$ in the tangent space $T_p(S)$ by $\|\mathbf{v}\| = \sqrt{\mathrm{I}_p(\mathbf{v}, \mathbf{v})}$.

Corollary 10.9. *If p is a point in a surface S, then there is an open set U of p and an $\epsilon > 0$ such that, if $q \in U$ and $\mathbf{v} \in T_q(S)$ with $\|\mathbf{v}\| < \epsilon$, then there is a unique geodesic $\gamma_{\mathbf{v}}\colon (-1, 1) \to S$ with $\gamma_{\mathbf{v}}(0) = q$ and $\gamma'_{\mathbf{v}}(0) = \mathbf{v}$. Moreover the mapping:*

$$C\colon \{(q, \mathbf{v}) \mid q \in U, \mathbf{v} \in T_q(S), \text{and } \|\mathbf{v}\| < \epsilon\} \times (-1, 1) \to S$$

given by $C(q, \mathbf{v}, t) = \gamma_{\mathbf{v}}(t)$ is smooth.

By applying the homogeneity property of the equations (♠) we can reparametrize the results of the corollary as follows:

If p is a point in a surface S, then there is an open set U of p and an $\epsilon > 0$ such that, if $q \in U$ and $\mathbf{v} \in T_q(S)$ with $\|\mathbf{v}\| = 1$, then there is a unique geodesic $\gamma_{\mathbf{v}}\colon (-\epsilon, \epsilon) \to S$ with $\gamma_{\mathbf{v}}(0) = q$ and $\gamma'_{\mathbf{v}}(0) = \mathbf{v}$. Moreover the mapping:

$$C\colon \{(q, \mathbf{v}) \mid q \in U, \mathbf{v} \in T_q(S), \text{and } \|\mathbf{v}\| = 1\} \times (-\epsilon, \epsilon) \to S$$

given by $C(q, \mathbf{v}, t) = \gamma_{\mathbf{v}}(t)$ is smooth.

The uniqueness part of the theorem allows us to finish a long-anticipated proof that great circle segments are *all* of the geodesics on the sphere. To wit, if $\gamma(s)$ is a unit speed geodesic through a point $p = \gamma(0)$ with $\mathbf{v} = \gamma'(0)$, then, by rotations of S^2 (isometries), we can take p to be on the Equator and $\mathbf{v}$ pointing north. We know that the meridian through this point is a great circle and hence a geodesic. Thus, the transformed γ lies along this great circle. The inverse rotations take great circles to great circles, and so γ lies along a great circle.

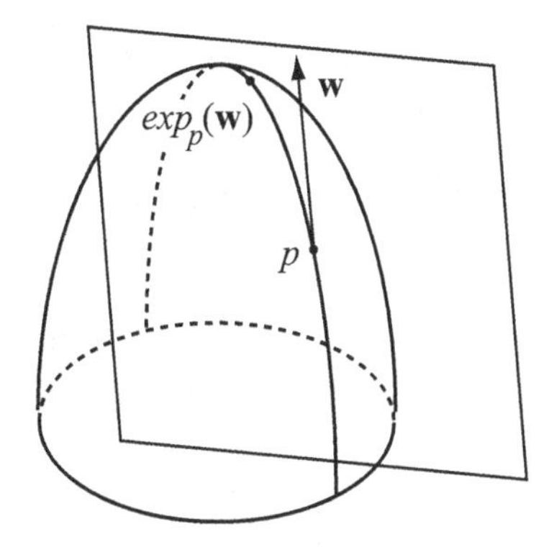

The existence of geodesics in every direction from a point p in S lead to a generalization of polar coordinates. In the normed linear space, $T_p(S)$ denote the *(open) ball of radius* δ by $B_\delta(p)$, and the *sphere of radius* δ by $S_\delta(p)$:

$$B_\delta(p) = \{\mathbf{w} \in T_p(S) \mid \|\mathbf{w}\| < \delta\}$$

$$\text{and } S_\delta(p) = \{\mathbf{w} \in T_p(S) \mid \|\mathbf{w}\| = \delta\}.$$

With this notation we can define the following important mapping.

Definition 10.10. *The* **exponential map** $\exp_p \colon B_\epsilon(p) \to S$ *is defined by:*

$$\exp_p(\mathbf{w}) = \gamma_{\mathbf{w}/\|\mathbf{w}\|}(\|\mathbf{w}\|),$$

that is, $\exp_p(\mathbf{w})$ *is the point in S gotten by traveling for length* $\|\mathbf{w}\|$ *along the unique geodesic through p determined by the direction* $\mathbf{w}/\|\mathbf{w}\|$.

The key properties of the exponential map are listed in the next theorem. A complete proof of these properties would require a deeper study of the differential equations involved. We refer the reader, once again, to Spivak, (1970, Vol. 1) for details.

Theorem 10.11. *To each* $p \in S$ *there is a value* ϵ_p *with* $0 < \epsilon_p \leq \infty$ *such that:*

(1) *The function* $\exp_p \colon B_{\epsilon_p}(p) \to S$ *maps* $B_{\epsilon_p}(p)$ *diffeomorphically onto a neighborhood of p. (Such a neighborhood of a point p is called a* **normal neighborhood***.)*
(2) *Any two points in* $\exp_p(B_{\epsilon_p}(p))$ *are joined by a unique geodesic of length less than* $2\epsilon_p$.

The exponential map gets its name from its form in a special higher-dimensional context where it is realized by exponentiation. Historically it is a little out of place in this discussion. However, it greatly simplifies the arguments to follow, and it organizes several notions from the classical theory of surfaces. In chapter 12 we will consider useful coordinates constructed from the exponential map. These coordinates appear in the work of Gauss (1825) and Riemann (1892), and so the exponential map may be traced back to Göttingen. The exponential map leads to analytic expressions of some synthetic notions.

Euclid revisited I: The Hopf–Rinow Theorem

The value ϵ_p in Theorem 10.11 tells us how far from the point p we can travel along any unit speed geodesic emanating from p. Euclid's Postulate II requires that any straight line segment be extendable in a straight line to any given length. In the language of surfaces, we interpret this postulate as the following property.

Definition 10.12. *A surface S is said to be* **geodesically complete** *if every geodesic $\gamma\colon [a,b] \to S$ can be extended to a geodesic $\gamma\colon \mathbb{R} \to S$.*

Equivalently, S is geodesically complete if, for any point $p \in S$ and any unit vector $\mathbf{v} \in T_p(S)$, the expression $\exp_p(r\mathbf{v})$ makes sense for any $r \in \mathbb{R}$. This leads to another reformulation of geodesic completeness.

Proposition 10.13. *A surface S is geodesically complete if and only if for every $p \in S$ the domain of $\exp_p$ is all of $T_p(S)$.*

Geodesic completeness is a global property of the surface S. Naïvely, to verify that it holds, we would need to check every geodesic emanating from each point. It is easy to construct a surface where geodesic completeness fails. For instance, let $S = S^2 - \{\text{South Pole}\}$. Then from any point in S the direction "due south" yields a geodesic of limited length.

We generalize Euclid's Postulate I to another global property of a surface S: For any pair of points p, $q \in S$, there exists a unique length-minimizing geodesic joining p to q. The term *length-minimizing* is important: Contrast the two geodesics joining New York to Chicago on the great circle through these cities. Furthermore, on the sphere there are infinitely many length-minimizing geodesics joining any pair of antipodal points.

Counterexamples to the existence part of Postulate I are also simple to construct. Consider $S = \mathbb{R}^2 - \{(0,0)\}$, the Euclidean plane with the origin removed. *No* geodesic joins $(-1,0)$ to $(1,0)$.

In what follows we examine the question of the existence of geodesics joining points, leaving the uniqueness to the next chapter. Our discussion leads to a condition on a surface that obtains the generalization of Euclid's Postulates I and II.

A minimal condition to realize Postulate I is that the surface S be connected. In fact, from the definition of a surface, connectedness implies that for any pair of points in S there is some piecewise regular curve joining the points. To a regular curve $\alpha\colon [t_0,t_1] \to S$ denote the **length** of $\alpha(t)$ by $L(\alpha) = \displaystyle\int_{t_0}^{t_1} \sqrt{\mathrm{I}_{\alpha(t)}(\alpha'(t),\alpha'(t))}\,dt$.

Definition 10.14. *The* **distance** *between two points p and q in a connected surface S is given by:*

$$d(p,q) = \inf\{L(\alpha) \mid \alpha \text{ is a piecewise regular curve in } S \text{ joining } p \text{ to } q\}.$$

Proposition 10.15. *The distance function $d\colon S \times S \to \mathbb{R}$ makes S into a metric space.*

PROOF: The triangle inequality $d(p,r)+d(r,q) \geq d(p,q)$ follows by joining curves and the definition of distance as an infimum. Certainly $d(p,q) = d(q,p)$, so it suffices to prove that $d(p,q) = 0$ implies that $p = q$.

Suppose $d(p,q) = 0$ and $p \neq q$. Let $\epsilon_p > 0$ be a radius for which the exponential map $\exp_p : B_{\epsilon_p}(p) \to S$ is a local diffeomorphism. Notice that if ϵ_p is small enough, all points q for which $d(p,q) < \epsilon_p$ lie in $\exp_p(B_{\epsilon_p}(p))$, that is, for $r \in \exp_p(B_{\epsilon_p}(p))$ there is a $\mathbf{u} \in B_{\epsilon_p}(p)$ such that $\|\mathbf{u}\| = d(p,q)$ and $r = \exp_p(\mathbf{u})$. Since $d(p,q) = 0 < \epsilon_p$, $q \in B_{\epsilon_p}(p)$ and so $q = \exp_p(\mathbf{v})$ for some $\mathbf{v} \in B_{\epsilon_p}(p)$. Since $p \neq q$ and $\exp_p$ is a one-to-one mapping, $\mathbf{v} \neq \mathbf{0}$. However, it then follows that $\|\mathbf{v}\| \neq 0$. The fact that $d(p,q) = \|\mathbf{v}\|$ yields a contradiction. Thus $p = q$. ∎

The surface $S = \mathbb{R}^2 - \{(0,0)\}$ lacks a certain property that relates to geodesic completeness: S is not a *complete* metric space. Recall that a sequence of points $\{a_n\}$ in a metric space (M,d) is a **Cauchy sequence** if, for $\epsilon > 0$, there is an index $N = N(\epsilon)$ with $d(a_k, a_l) < \epsilon$ whenever $k, l \geq N$. A metric space M is **complete** if every Cauchy sequence of points in M converges to a point in M. That is, given a Cauchy sequence $\{a_n\}$ of points in M there is a point $a \in M$ for which, given any $\epsilon > 0$, there is an index $J = J(\epsilon)$ for which $d(a, a_m) < \epsilon$ for all $m \geq J$. To see that $\mathbb{R}^2 - \{(0,0)\}$ is not complete, consider the Cauchy sequence $\{(1/n, 0)\}$.

In fact, metric completeness of a surface as a metric space (S,d) is the key to Postulates I and II.

Theorem 10.16 (Hopf–Rinow Theorem). *Let S be a connected surface. Then the following conditions are equivalent:*

(1) *(S,d) is a complete metric space.*
(2) *S is geodesically complete.*

Furthermore, each condition implies that for any p, $q \in S$ there is a geodesic $\gamma : [a,b] \to S$ with $\gamma(a) = p$, $\gamma(b) = q$, and $L(\gamma) = d(p,q)$.

PROOF: We first show that condition 2 implies the last statement of the theorem. Suppose that S is geodesically complete. Let $\rho = d(p,q)$. Consider $\epsilon_p > 0$ small enough so that $\exp_p : B_\epsilon(p) \to S$ is a local diffeomorphism and any pair of points in $\exp_p(B_{\epsilon_p}(p))$ are joined by a unique unit speed geodesic. For some $0 < \delta < \epsilon_p$, let $\Sigma = \exp_p(S_\delta(p))$; notice that $S_\delta(p)$ is compact and hence so is Σ. Let p_0 be the point in Σ satisfying $d(p_0, q) \leq d(s,q)$ for all $s \in \Sigma$.

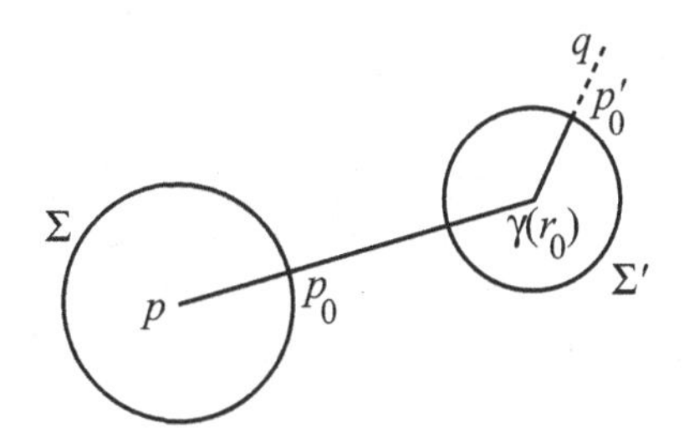

Such a point exists by the continuity of the function $x \mapsto d(x,q)$, a property of distance functions, and the compactness of Σ. By the choice of ϵ_p, $p_0 = \exp_p(\delta \mathbf{v}_0)$ for some unit vector $\mathbf{v}_0 \in S_1(p) \subset T_p(S)$. We claim that $\exp_p(\rho \mathbf{v}_0) = q$. To prove this claim we show that the geodesic $\gamma(t) = \exp_p(t\mathbf{v}_0)$, for $0 \leq t \leq \rho$, satisfies:

$$d(\gamma(t),q) = \rho - t.$$

By connectedness there are piecewise regular curves joining p to q. Any such curve must pass through Σ. Since distance is defined in terms of piecewise regular curves and distance is minimized by geodesics to points on Σ, at the point p_0 we can write:

$$\rho = d(p,q) = \min_{s \in \Sigma}\{d(p,s) + d(s,q)\} = d(p_0,q) + \delta.$$

Thus $d(\gamma(\delta),q) = \rho - \delta$.

Let $r_0 = \sup\{r \in [\delta,\rho] \mid d(\gamma(r),q) = \rho - r\}$. By the continuity of the metric $d(\gamma(r_0),q) = \rho - r_0$. Suppose that $r_0 < \rho$. Around $\gamma(r_0)$ there is a normal neighborhood given by $\exp_{\gamma(r_0)}(B_{\epsilon'}(\gamma(r_0)))$. Let $0 < \delta' < \epsilon'$. There is another circle $\Sigma' = \exp_{\gamma(r_0)}(S_{\delta'}(\gamma(r_0)))$ and a point $p_0' \in \Sigma'$ such that:

$$d(p_0',q) \leq d(s',q) \text{ for all } s' \in \Sigma'.$$

We write again:

$$d(\gamma(r_0),q) = \min_{s' \in \Sigma'}\{d(\gamma(r_0),s') + d(s',q)\} = \delta' + d(p_0',q),$$

and so $d(p_0',q) = (\rho - r_0) - \delta'$. This implies:

$$d(p,p_0') \geq d(p,q) - d(p_0',q) = r_0 + \delta'.$$

The piecewise geodesic curve gotten by going along $\gamma(s)$ to $\gamma(r_0)$ and then running the geodesic from $\gamma(r_0)$ to p_0' has length $r_0 + \delta'$. This curve achieves the minimal length. We show in a lemma and corollary to follow that a piecewise geodesic curve of minimal length must in fact be part of a single geodesic. That geodesic must be $\gamma(t)$ by uniqueness, and so $\gamma(r_0 + \delta') = p_0'$. But this violates the definition of r_0 since $d(\gamma(r_0 + \delta'),q) = \rho - (r_0 + \delta')$. Thus r_0 must be ρ and q is joined to p by a length-minimizing geodesic.

A consequence of this proof is that if $A \subset S$ is a subset of diameter ρ and $p \in A$, then $\exp_p \colon T_p(S) \to S$ maps the closed disk $B_\rho(p) \cup S_\rho(p)$ onto a compact set containing A. This implies that the closure of A is compact, and so, by an elementary result in point set topology, any Cauchy sequence of points in S converges to a point in A. Hence (S,d) is a complete metric space and condition 2 implies 1.

Suppose (S,d) is a complete metric space, and $\gamma\colon (a,b) \to S$ a unit speed geodesic. Then consider a Cauchy sequence in (a,b), say $\{t_n\}$, which converges to b. The unit speed assumption implies that the sequence $\{\gamma(t_n)\}$ is a Cauchy sequence in S and so it has a limit, which we write as $\gamma(b)$. By Corollary 10.9 (existence and uniqueness of geodesics), since $\gamma(b)$ is defined and $\gamma'(b)$ can be computed as a left limit, we can extend $\gamma(s)$ beyond b. Continuing the argument along in both directions we can extend the domain of $\gamma(s)$ to all of $\mathbb{R}$ and S is geodesically complete. ∎

Lemma 10.17. *If $\alpha\colon [a,b] \to S$ is a piecewise regular curve lying in the image of $B_{\epsilon_p}(p)$ under the exponential map and $\alpha(t) = \exp_p(u(t)X(t))$ for $0 < u(t) < \epsilon_p$ and $\|X(t)\| = 1$, then*

$$\int_a^b \sqrt{\mathrm{I}_{\alpha(t)}(\alpha'(t), \alpha'(t))}\,dt \geq |u(b) - u(a)|,$$

where equality holds if and only if $u(t)$ is monotonic and $X(t)$ is constant, that is, $\alpha(t)$ is a segment of a radial geodesic from p.

PROOF: Let $A(u,t) = \exp_p(uX(t))$ so that $\alpha(t) = A(u(t),t)$. It follows that:

$$\alpha'(t) = \frac{\partial A}{\partial u}\frac{du}{dt} + \frac{\partial A}{\partial t}.$$

Since $\|X(t)\| = 1$ for all t, $\|\partial A/\partial u\| = 1$ by the construction of the exponential map. Furthermore, $\left\langle \dfrac{\partial A}{\partial u}, \dfrac{\partial A}{\partial t} \right\rangle = 0$ by an argument similar to the proof of Theorem 10.7 (see also the proof of Lemma 12.2). It follows that:

$$\mathrm{I}_{\alpha(t)}(\alpha'(t), \alpha'(t)) = \left|\frac{du}{dt}\right|^2 + \left(\frac{\partial A}{\partial t}\right)^2 \geq \left|\frac{du}{dt}\right|^2,$$

and so:

$$\int_a^b \sqrt{\mathrm{I}_{\alpha(t)}(\alpha'(t), \alpha'(t))}\,dt \geq \int_a^b \left|\frac{du}{dt}\right| dt \geq \left|\int_a^b \frac{du}{dt}\,dt\right| \geq |u(b) - u(a)|.$$

To obtain equality we need $\partial A/\partial t = 0$, that is, $X'(t) = \mathbf{0}$, and $X(t)$ is constant. We also need that $u(t)$ is monotone to obtain $\displaystyle\int_a^b \left|\frac{du}{dt}\right| dt = |u(b) - u(a)|$. ∎

Corollary 10.18. *If q, $q' \in \exp_p(B_{\epsilon_p}(p))$ are joined by a unit speed geodesic $\gamma\colon [0,r] \to S$ and by a piecewise regular curve $\alpha\colon [0,r] \to S$, then $d(q,q') = L(\gamma) \leq L(\alpha)$, where equality occurs if and only if $\alpha(t)$ is a reparametrization of $\gamma(t)$.*

The metric conditions of the Hopf–Rinow Theorem imply analogs of Euclid's Postulates I and II from the point set topology of the surface S. For example, if a surface S is compact and connected, then it is a complete metric space and so it is geodesically complete. Global properties such as metric completeness or compactness cannot be determined merely from local information. However, they are important ingredients in the construction of models for geometries.

Exercises

10.1 Suppose S_1 and S_2 are surfaces and $\phi: S_1 \to S_2$ is an isometry. Suppose that $\gamma: (-\epsilon, \epsilon) \to S_1$ is a geodesic. Prove in detail that $\phi \circ \gamma: (-\epsilon, \epsilon) \to S_2$ is also a geodesic.

10.2† Let S be a surface in $\mathbb{R}^3$ and Π a plane that intersects S in a curve $\alpha(s)$. Show that $\alpha(s)$ is a geodesic if Π is a plane of symmetry of S, that is, the two sides of S are mirror images across Π. Apply this to the sphere and to an ellipsoid of revolution.

10.3 Let $\gamma(t)$ be a straight line segment in $\mathbb{R}^3$ contained in a surface S. Prove that $\gamma(t)$ is a geodesic.

10.4 Let $\alpha: [0,1] \to S^2$ be the curve $\alpha(t) = (\cos(e^t), \sin(e^t), 0)$. Show that α is a geodesic on S^2, but in the latitude-longitude parametrization of S^2, $\alpha(t)$ does not satisfy the differential equations (♠).

10.5 Let $\mathrm{II}(\mathbf{v},\mathbf{w}) = \frac{1}{2}(\mathrm{II}((v)+\mathbf{w}) - \mathrm{II}(\mathbf{v}) - \mathrm{II}(\mathbf{w}))$ denote the inner product determined by the second fundamental form. Prove the analogue of the Frenet-Serret Theorem for curves $\alpha(t)$ on S. Use the frame $[T, n_\alpha, N]$ and show:

$$T' = k_g n_\alpha + \mathrm{II}(T)N,$$
$$n'_\alpha = -k_g T + \mathrm{II}(T, n_\alpha)N,$$
$$N' = -\mathrm{II}(T)T - \mathrm{II}(T, n_\alpha)n_\alpha.$$

The function defined $\tau_g(t) = \mathrm{II}_{\alpha(t)}(T(t), n_\alpha(t))$ is called the **geodesic torsion** of α.

10.6† Prove the converse of Clairaut's Relation: If a curve on a surface of revolution shares none of its length with a parallel, and if such a curve satisfies the condition that the product of the radius at any point on the curve and the sine of the angle made by the curve with the meridian at that point is constant, then the curve is a geodesic.

10.7 Use the previous exercise to prove that the helices on a straight cylinder are geodesics on the cylinder.

10.8 Take the surface of revolution generated by the tractrix and investigate geodesics on it using Clairaut's Relation.

10.9 Suppose that S is geodesically complete. Show that for any point $p \in S$, the mapping $\exp_p$ has domain all of $T_p(S)$.

10.10 For the reader with some knowledge of topology: Show that the topologies on a surface S as a subset of $\mathbb{R}^3$ and as a metric space (S, d) are the same topology.

10.11† Suppose that a surface is geodesically complete and S satisfies the property that there is some $r > 0$ such that, for all $p, q \in S$, $d(p,q) < r$. Show that S is compact.

10.12 Suppose that S_1 is connected and geodesically complete and S_2 is connected and satisfies the property that every pair of points in S_2 is joined by a unique geodesic. If $\phi: S_1 \to S_2$ is a local isometry, show that, in fact, it is a global isometry.

10.13 Give a proof of Corollary 10.18.

Solutions to Exercises marked with a dagger appear in the appendix, pp. 335–336.

11

The Gauss–Bonnet Theorem

This theorem, if we mistake not, ought to be counted among the most elegant in the theory of curved surfaces ...

C.F. GAUSS (1827)

A triangle in the plane is determined by three line segments and the region they enclose; a triangle on a surface is determined by three geodesic segments that enclose a region. In *Disquisitiones generales circa superficies curvas* (Gauss 1825), Gauss deduced some of the basic properties of geodesic triangles. In particular he proved a general relation between area and angle sum that generalizes the properties of triangles on the sphere Proposition 1.4 and on the plane. In his paper of 1848, *Mémoire sur la Théorie générale des Surfaces* Bonnet (1865) introduced geodesic curvature $k_g(s)$ and applied Green's Theorem to prove a far-reaching generalization of Gauss's Theorem, this time for polygons whose sides are not necessarily geodesic segments. For compact, orientable surfaces the Gauss–Bonnet Theorem has remarkable topological implications.

In order to state the main theorem of the chapter we need to set out a handful of assumptions. Fix an orientable surface $S \subset \mathbb{R}^3$ and suppose $R \subset S$ is a region in S satisfying the following properties:

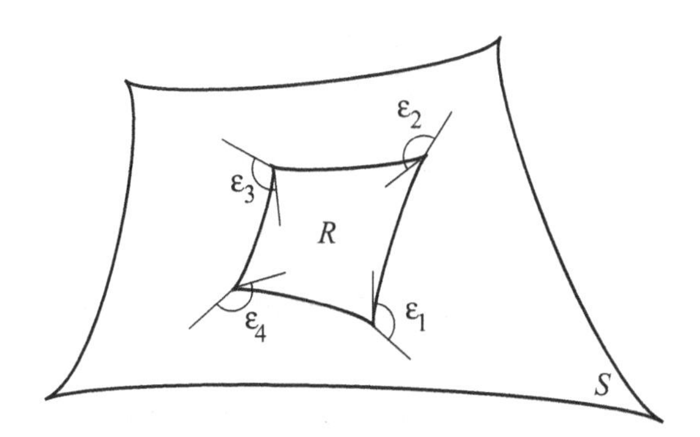

(1) The region R is simply connected in S. That is, any closed curve in R can be contracted to a point without leaving the region. Thus R is like a convex subset of S. Furthermore, the interior of R in S is connected. This condition excludes cases like the region enclosed by a figure-8 curve.

(2) The boundary of R is given by a curve $\alpha\colon [0,s_n] \to S$. Furthermore, α is one-one, piecewise differentiable, and closed; that is, it may be written:

$$\alpha = (\alpha_1, \alpha_2, \ldots, \alpha_n), \quad \alpha_i\colon [s_{i-1}, s_i] \to S,$$

where each α_i is unit speed and differentiable on its domain. Also $\alpha_i(s_i) = \alpha_{i+1}(s_i)$ for $i = 1, \ldots, n-1$, and $\alpha_n(s_n) = \alpha_1(0)$. The curves $\alpha_i(s)$ are the *sides* of a general sort of polygon lying on S. The points $\alpha(s_i)$ are the vertices of the polygons.

At each vertex, the tangent vectors $\alpha_i'(s_i)$ and $\alpha_{i+1}'(s_i)$ (along with $\alpha_1'(0)$ and $\alpha_n'(s_n)$) exist and the angle between them is well-defined. We call this angle ϵ_i, the *exterior angle* at the ith vertex, and it satisfies

$$\cos\epsilon_i = \mathrm{I}_{\alpha_i(s_i)}(\alpha_i'(s_i), \alpha_{i+1}'(s_i))$$

(and $\cos\epsilon_n = \mathrm{I}_{\alpha_n(s_n)}(\alpha_1'(0), \alpha_n'(s_n))$). If we assume $0 \leq |\epsilon_i| \leq \pi$, then the previous equation determines ϵ_i up to sign. A *cusp* is a vertex where $\alpha_i'(s_i) = -\alpha_{i+1}'(s_i)$ and so the exterior angle is $\pm\pi$. If a vertex is not a cusp, then we can fix the sign of ϵ_i by requiring the angle to lie between 0 and π and checking the orientation of the ordered basis $[\alpha_i'(s_i), \alpha_{i+1}'(s_i), N(\alpha_i(s_i))]$. If the basis is right-handed, the sign is positive, left-handed and the sign is negative.

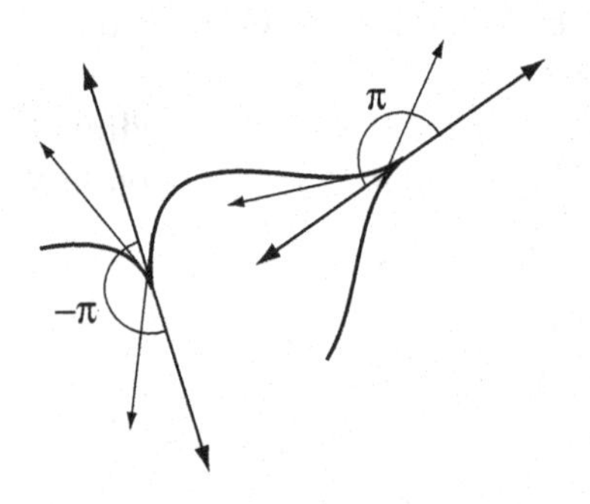

To fix the sign at a cusp, choose a nearby point on α_i, say $\alpha_i(s_i - \eta)$ for $\eta > 0$ and close to 0; choose a nearby point on α_{i+1}, say $\alpha_{i+1}(s_i + \eta)$, and consider the ordered basis:

$$[\alpha_i'(s_i - \eta), \alpha_{i+1}'(s_i + \eta), N(\alpha_i(s_i))].$$

Since the curves are one-to-one as functions and the surface is regular, this ordered basis is well-defined near the vertex and its handedness remains the same as η goes to zero. These choices determine the sign of π at a cusp.

(3) The region R and its boundary are contained in a coordinate chart $x\colon (U \subset \mathbb{R}^2) \to S$. Furthermore, we can take the component functions of the metric for x to satisfy $F = 0$, that is the coordinate curves are orthogonal (see Corollary 9.11).

At the heart of the proof of the Gauss–Bonnet Theorem is a result about piecewise differentiable curves on a surface like the boundary of the region R. Though out of place historically, a theorem of Heinz Hopf (1894–1971) has a natural place in our discussion and it was tacitly assumed by Gauss and Bonnet.

To state the result we introduce functions $\phi_i\colon [s_{i-1}, s_i] \to \mathbb{R}$ along curves α_i. These functions measure how the tangent $\alpha_i'(s)$ turns around the region R. Let $\phi_i(s)$ measure the angle from x_u to $\alpha_i'(s)$ at $\alpha_i(s)$. Such a function is determined up to multiples of 2π and it satisfies the formula:

$$\cos\phi_i(s) = \mathrm{I}_{\alpha_i(s)}\left(\alpha_i'(s), \frac{x_u}{\sqrt{E}}\right).$$

To determine the values $\phi_i(s)$ begin by choosing an angle to represent $\phi_1(0)$. Then extend $\phi_1(s)$ along $[0, s_1]$ by continuity. At $s = s_1$ we make a jump in order to measure the angle via the curve α_2. The size of the jump, and hence the value of $\phi_2(s_1)$ is determined by the equation:

$$\phi_2(s_1) = \epsilon_1 + \phi_1(s_1).$$

Next, extend $\phi_2(s)$ along $[s_1, s_2]$ by continuity, and so on. In this fashion we can construct all of the functions ϕ_i. A key result was proved in 1935 by Hopf for such curves on a surface.

Theorem 11.1 (Hopf's Umlaufsatz). *The tangent vector along a closed piecewise differentiable curve enclosing a simply connected region turns through 2π, that is, $\phi_n(s_n) + \epsilon_n = \phi_1(0) + 2\pi$.*

The proof of this innocent-sounding result is lengthy and involves ideas that really belong in a course on topology. In order to avoid such a long detour, we postpone a sketch of the proof to the end of the chapter. For now we conclude with a useful corollary that follows from the construction of the functions $\phi_i(s)$.

Corollary 11.2. $\displaystyle\sum_{j=1}^{n} \int_{\alpha_j} \phi_j'(s)\,ds = 2\pi - \sum_{j=1}^{n} \epsilon_j.$

In Exercise 9.3 you have proved that an orthogonal coordinate chart ($F = 0$) has Gaussian curvature that takes the form:

$$K = -\frac{1}{2\sqrt{EG}}\left(\frac{\partial}{\partial v}\left(\frac{E_v}{\sqrt{EG}}\right) + \frac{\partial}{\partial u}\left(\frac{G_u}{\sqrt{EG}}\right)\right).$$

Geodesic curvature can be expressed locally along each curve $\alpha_i(s)$ using the functions $\phi_i(s)$ and the fact that $F = 0$. This convenient expression is related to the Gaussian curvature via Green's Theorem.

Lemma 11.3. *Along the curve $\alpha_i(s)$, $\displaystyle k_g(s) = \frac{1}{2\sqrt{EG}}\left[G_u\frac{dv}{ds} - E_v\frac{du}{ds}\right] + \frac{d\phi_i}{ds}$.*

Proof: Let $\mathbf{e}_u = \dfrac{x_u}{\sqrt{E}}$ and $\mathbf{e}_v = \dfrac{x_v}{\sqrt{G}}$ be the unit length coordinate vectors associated to the patch x. We have assumed already that $\mathbf{e}_u$ is orthogonal to $\mathbf{e}_v$. Since each $\alpha_i(s)$ is unit speed, we obtain the following equations for the tangent and intrinsic normal:

$$\alpha_i'(s) = \cos\phi_i(s)\mathbf{e}_u + \sin\phi_i(s)\mathbf{e}_v, \quad n_{\alpha_i}(s) = -\sin\phi_i(s)\mathbf{e}_u + \cos\phi_i(s)\mathbf{e}_v.$$

Taking the derivative of the tangent vector we find:

$$\begin{aligned}\alpha_i''(s) &= -\sin\phi_i(s)\frac{d\phi_i}{ds}\mathbf{e}_u + \cos\phi_i(s)\mathbf{e}_u' + \cos\phi_i(s)\frac{d\phi_i}{ds}\mathbf{e}_v + \sin\phi_i(s)\mathbf{e}_v' \\ &= \frac{d\phi_i}{ds}n_{\alpha_i}(s) + \cos\phi_i(s)\mathbf{e}_u' + \sin\phi_i(s)\mathbf{e}_v'.\end{aligned}$$

Recall the formula for geodesic curvature: $k_g(s) = \mathrm{I}_{\alpha_i(s)}(\alpha_i''(s), n_{\alpha_i}(s))$:

$$\begin{aligned}k_g(s) &= \mathrm{I}_{\alpha_i(s)}\left(\frac{d\phi_i}{ds}n_{\alpha_i}(s) + \cos\phi_i(s)\mathbf{e}_u' + \sin\phi_i(s)\mathbf{e}_v', n_{\alpha_i}(s)\right) \\ &= \frac{d\phi_i}{ds} - \sin\phi_i(s)\cos\phi_i(s)\langle\mathbf{e}_u,\mathbf{e}_u'\rangle - \sin^2\phi_i(s)\langle\mathbf{e}_u,\mathbf{e}_v'\rangle \\ &\quad + \cos^2\phi_i(s)\langle\mathbf{e}_v,\mathbf{e}_u'\rangle + \sin\phi_i(s)\cos\phi_i(s)\langle\mathbf{e}_v,\mathbf{e}_v'\rangle.\end{aligned}$$

Since $\mathbf{e}_u$ and $\mathbf{e}_v$ are unit speed and perpendicular, $\langle\mathbf{e}_u,\mathbf{e}_u'\rangle = 0 = \langle\mathbf{e}_v,\mathbf{e}_v'\rangle$ and $\langle\mathbf{e}_u',\mathbf{e}_v\rangle = -\langle\mathbf{e}_v',\mathbf{e}_u\rangle$. Therefore the formula simplifies to:

$$k_g(s) = \langle\mathbf{e}_u',\mathbf{e}_v\rangle + \frac{d\phi_i}{ds}.$$

In the plane, signed curvature is obtained from $\frac{d\phi_i}{ds}$. The correction term $\langle\mathbf{e}_u',\mathbf{e}_v\rangle$ is the rate of rotation of the coordinate system as it travels along the curve. To compute $\langle\mathbf{e}_u',\mathbf{e}_v\rangle$ we first consider $\mathbf{e}_u'$:

$$\mathbf{e}_u' = \frac{d\mathbf{e}_u}{ds} = \frac{d}{ds}\left(\frac{x_u}{\sqrt{E}}\right) = \frac{1}{\sqrt{E}}\left(x_{uu}\frac{du}{ds} + x_{uv}\frac{dv}{ds}\right) - x_u\left(\frac{E_u}{2E^{3/2}}\frac{du}{ds} + \frac{E_v}{2E^{3/2}}\frac{dv}{ds}\right).$$

Since $F = 0$, $F_u = 0$ and so $\langle x_{uu}, x_v\rangle = -\langle x_{uv}, x_u\rangle = -(1/2)E_v$. It follows that

$$\langle\mathbf{e}_u',\mathbf{e}_v\rangle = \frac{1}{\sqrt{EG}}\langle x_{uu},x_v\rangle\frac{du}{ds} + \frac{1}{\sqrt{EG}}\langle x_{uv},x_v\rangle\frac{dv}{ds} = \frac{1}{2\sqrt{EG}}\left(G_u\frac{du}{ds} - E_v\frac{dv}{ds}\right).$$

This proves the lemma. ■

We call $\sum_{i=1}^{n}\int_{\alpha_i} k_g(s)\,ds$ the *total geodesic curvature* along α, which is the boundary of the region R. By Lemma 11.3 we have:

$$\sum_{i=1}^{n}\int_{\alpha_i} k_g(s)\,ds = \sum_{j=1}^{n}\int_{\alpha_i}\frac{1}{2\sqrt{EG}}\left(G_u\frac{dv}{ds} - E_v\frac{du}{ds}\right)ds + \sum_{j=1}^{n}\int_{\alpha_i}\frac{d\phi_i}{ds}\,ds.$$

Recall Green's Theorem for a region $W \subset \mathbb{R}^2$ with boundary ∂W, a simple closed curve:

$$\int_{\partial W} P\,du + Q\,dv = \iint_W \left(\frac{\partial Q}{\partial u} - \frac{\partial P}{\partial v}\right)du\,dv.$$

For the case at hand we have:

$$\frac{1}{2}\int_{\partial x^{-1}(R)} \frac{G_u}{\sqrt{EG}}\,dv - \frac{E_v}{\sqrt{EG}}\,du$$
$$= \frac{1}{2}\iint_{x^{-1}(R)} \left(\frac{\partial}{\partial v}\left(\frac{E_v}{\sqrt{EG}}\right) + \frac{\partial}{\partial u}\left(\frac{G_u}{\sqrt{EG}}\right)\right) du\,dv$$
$$= -\iint_{x^{-1}(R)} \frac{-1}{2\sqrt{EG}}\left(\frac{\partial}{\partial v}\left(\frac{E_v}{\sqrt{EG}}\right) + \frac{\partial}{\partial u}\left(\frac{G_u}{\sqrt{EG}}\right)\right)\sqrt{EG}\,du\,dv$$
$$= -\iint_R K\,dA.$$

This observation with Lemma 11.3 proves the main theorem of the chapter:

Theorem 11.4 (Gauss–Bonnet Theorem). *If R is a simply connected region in a regular surface S bounded by a piecewise differentiable curve $\alpha(s)$ making exterior angles $\epsilon_1, \epsilon_2, \ldots, \epsilon_n$ at the vertices of α, then:*

$$\sum_{j=1}^{n}\int_{\alpha_j} k_g(s)\,ds + \iint_R K\,dA = 2\pi - \sum_{j=1}^{n}\epsilon_j.$$

This formula is due to Bonnet and generalizes the result of Gauss's *Disquisitiones* for geodesic polygons. Bonnet's Formula contains geodesic curvature, Gaussian curvature, and exterior angles, mixing up curves, angles, and areas into a remarkable relation. Gauss referred to his version of the theorem as *Theorema Elegantissimum* (Gauss 1825, §20).

Euclid revisited II: Uniqueness of lines

Gauss's version of the Gauss–Bonnet Theorem is concerned with geodesic triangles. Let $R = \triangle ABC$ be such a triangle with sides given by geodesic segments. The interior angles are given by $\angle A = \pi - \varepsilon_A$, $\angle B = \pi - \varepsilon_B$ and $\angle C = \pi - \varepsilon_C$. Thus $2\pi - \varepsilon_A - \varepsilon_B - \varepsilon_C = \angle A + \angle B + \angle C - \pi$.

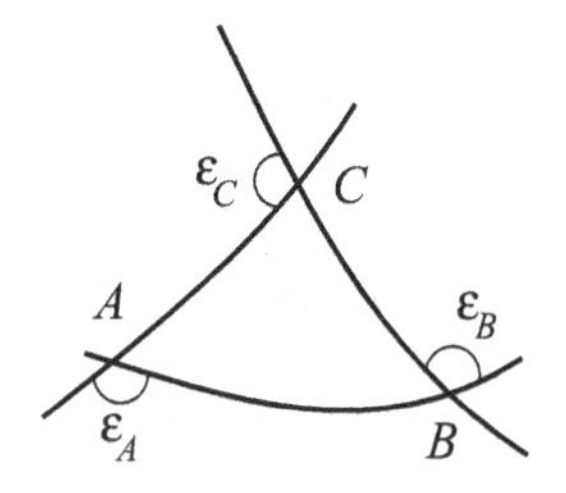

Since geodesic curvature vanishes on geodesics, we have proved Gauss's *Theorema Elegantissimum*:

Corollary 11.5. *For a geodesic triangle $\triangle ABC$ on a surfaces S,*

$$\iint_{\triangle ABC} K\,dA = \angle A + \angle B + \angle C - \pi.$$

Corollary 11.6.

(a) *The interior angle sum of a triangle in the Euclidean plane is π.*
(b) *On a surface of constant positive curvature, the area of a triangle is proportional to the angle excess.*
(c) *On a surface of constant negative curvature, the area of a triangle is proportional to the angle defect.*

PROOF: Statement (a) follows from the vanishing of Gaussian curvature on the plane. Statements (b) and (c) follow from Corollary 11.5, which for constant Gaussian curvature gives:

$$\iint_R K\,dA = K \iint_R dA = K \cdot \text{area}(R). \qquad \blacksquare$$

This corollary reproves an equivalent of Postulate V for the plane (Theorem 3.6) and it recovers Girard's formula for the area of a spherical triangle (Proposition 1.4). It also leads to a condition on a surface that models non-Euclidean geometry: In order to obtain the formula for area of a non-Euclidean triangle (Exercise 4.19), the surface must have constant negative curvature. In fact, more can be said. In chapter 10 we discussed Euclid's Postulate I and the existence of geodesics joining two distinct points in a surface. This observation leaves open the question of the uniqueness of geodesic segments joining two points. There are infinitely many geodesics joining the North and South Poles of the spheres. However, this phenomenon cannot happen on surfaces of nonpositive curvature.

Corollary 11.7. *On a surface of nonpositive curvature two geodesic segments γ_1, γ_2 meeting at a point p cannot meet at another point q so that the curves form the boundary of a simply connected region.*

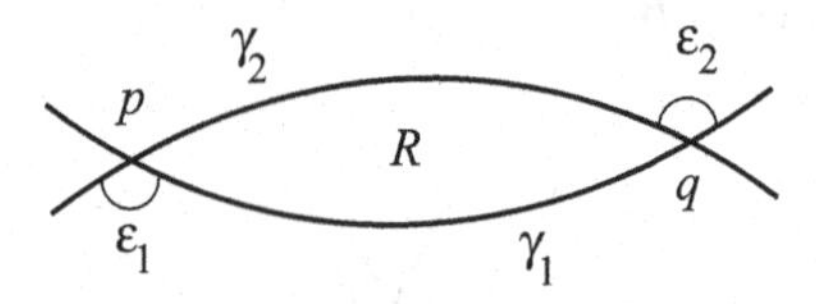

PROOF: Suppose such a figure can be formed. By the Gauss–Bonnet Theorem,

$$\iint_R K\,dA + \epsilon_1 + \epsilon_2 = 2\pi,$$

where ϵ_1 and ϵ_2 are the exterior angles at p and at q. Since geodesics in a given direction are unique, γ_1 and γ_2 cannot be mutually tangent, so ϵ_1 and ϵ_2 are both less than π. Since $K \leq 0$, the equation cannot be satisfied. ∎

Corollary 11.8. *On a complete, simply connected surface of nonpositive curvature, given two distinct points on such a surface, there is a unique geodesic joining them.*

Compact surfaces

Suppose that a surface S is compact, without boundary, and orientable. Because S is compact, for any continuous function $f\colon S \to \mathbb{R}$, the integral $\iint_S f\,dA$ is finite. In particular, we can integrate the Gaussian curvature over the entire surface S. The Gauss–Bonnet Theorem tells us the answer for a reasonable subset of S. In particular, we can interpret the formula of Gauss and Bonnet as an expression for the total geodesic curvature along the boundary of such a region.

Let us suppose further that the surface S has a **triangulation**,

$$\mathcal{P} = \{\Delta_i \subset S \mid i = 1,2,\dots,m\},$$

where each Δ_i is a generalized triangle on S, that is, $\Delta_i = x(\Delta A_i B_i C_i)$ and $\Delta A_i B_i C_i$ is a triangle in the domain $U \subset \mathbb{R}^2$ of a coordinate chart x. Furthermore, we assume that $S = \bigcup_i \Delta_i$ and $\Delta_i \cap \Delta_j$ is either empty, a vertex, or a shared side of each triangle.

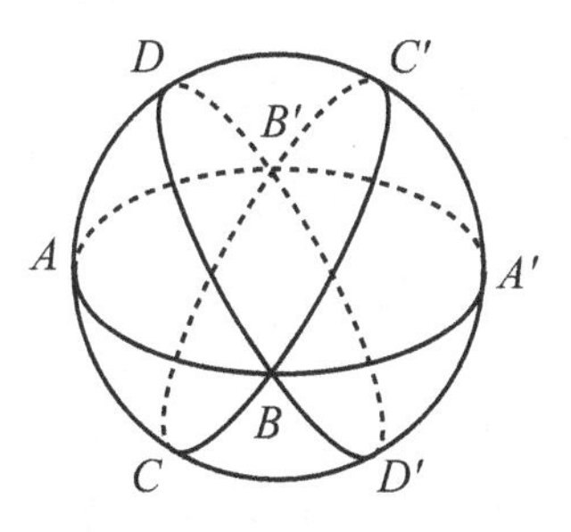

The existence of triangulations of surfaces was taken for granted in the nineteenth century. Gauss, as Director of the Göttingen Astronomical Observatory, had considerable experience doing geodetic surveys of the Kingdom of Hannover in which large geographical regions were broken up into geodesic triangles (see volume 9 of his collected works). For small enough regions of a surface, the geodetic approach of laying out geodesic triangles can lead to a triangulation. To prove that all compact surfaces have a triangulation requires the use of topological results such as the Jordan Curve Theorem.The first complete proof was due to Tibor Radó (1895–1965) in 1925. The interested reader may find a proof in Moise (1977), and Doyle and Moran (1968).

Using the orientation, we can take each triangle to be oriented as follows: Each boundary $\partial\Delta_i = \alpha_i$ is parametrizable as a unit speed, piecewise differentiable curve. Choose the direction so that $[\alpha_i'(s), n_{\alpha_i}(s), N(\alpha_i(s))]$ is a right-handed frame at each regular point on the curve. Orientability implies that we can make this choice coherently across all of S.

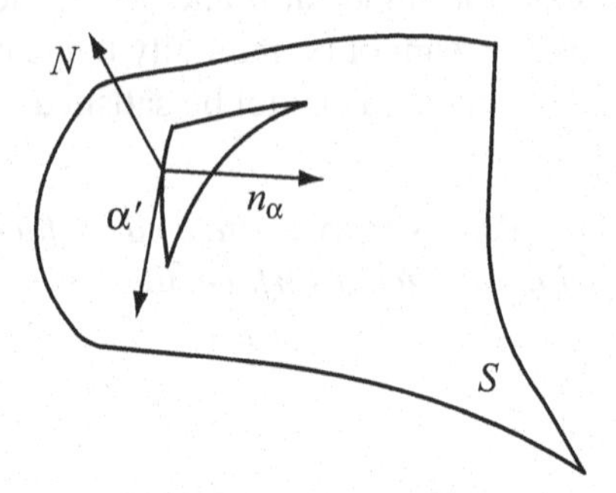

Associate to a triangulation $\mathcal{P}$ the numbers:

$\mathbf{F}$ = the number of faces (triangles) in the collection $\mathcal{P}$,

$\mathbf{E}$ = the number of edges in the collection $\mathcal{P}$,

$\mathbf{V}$ = the number of vertices in the collection $\mathcal{P}$.

Define the **Euler–Poincaré characteristic** of the triangulation $\mathcal{P}$ as the integer $\chi(S;\mathcal{P}) = \mathbf{V} - \mathbf{E} + \mathbf{F}$.

Proposition 11.9. *If S is a compact, orientable surface without boundary, then*

$$\iint_S K\,dA = 2\pi\,\chi(S;\mathcal{P}).$$

PROOF: From the definition of a triangulation, each triangle contributes three edges to $\mathbf{E}$. Because S has no boundary, each edge is shared by two triangles. Thus $3\mathbf{F} = 2\mathbf{E}$.

By the Gauss–Bonnet Theorem we have the following equation:

$$\iint_S K\,dA = \sum_{l=1}^{\mathbf{F}} \iint_{\Delta_i} K\,dA = \sum_{l=1}^{\mathbf{F}} \angle_{l1} + \angle_{l2} + \angle_{l3} - \pi + \sum_{l=1}^{\mathbf{F}} \int_{\alpha_l} k_g(s)\,ds,$$

where the $\angle_{lj}$ are the interior angles in each generalized triangle Δ_l.

Because each triangle is oriented, contributions by a given edge to the sum $\sum_{l=1}^{\mathbf{F}} \int_{\alpha_l} k_g(s)\,ds$ cancel for contiguous triangles. It follows from the absence of a boundary that the integral sum $\sum_l \int_{\alpha_l} k_g(s)\,ds$ vanishes. Thus, we have:

$$\iint_S K\,dA = \left(\sum_{l=1}^{\mathbf{F}} \angle_{l1} + \angle_{l2} + \angle_{l3}\right) - \pi\mathbf{F}.$$

At each vertex the interior angles sum to 2π so the sum of all the interior angles is $2\pi\mathbf{V}$. Also, because $3\mathbf{F} = 2\mathbf{E}$, we have $\pi(3\mathbf{F} - 2\mathbf{E}) = 0$ and so:

$$\begin{aligned}\iint_S K\,dA &= 2\pi\mathbf{V} - \pi\mathbf{F} = 2\pi\mathbf{V} - \pi\mathbf{F} + \pi(3\mathbf{F} - 2\mathbf{E})\\ &= 2\pi(\mathbf{V} - \mathbf{E} + \mathbf{F}) = 2\pi\,\chi(S;\mathcal{P}).\end{aligned}$$ ∎

Gauss referred to the integral $\iint_S K\,dA$ as the *total* or *integral curvature* of the surface S. Since the total curvature does not depend on the triangulation, we have proved that the Euler–Poincaré characteristic is independent of the choice of triangulation.

Corollary 11.10. *Given two triangulations $\mathcal{P}$ and $\mathcal{P}'$ of a compact, oriented surface S without boundary, $\chi(S;\mathcal{P}) = \chi(S;\mathcal{P}')$.*

We can write $\chi(S)$ for the value $\frac{1}{2\pi}\iint_S K\,dA$ for the value $\mathbf{V} - \mathbf{E} + \mathbf{F}$ associated to any triangulation of S. In 1758, Euler introduced the formula $\mathbf{V} - \mathbf{E} + \mathbf{F} = 2$ for polyhedra that are homeomorphic to a sphere (in our parlance). This formula leads to a proof that there are only five Platonic solids (Exercise 11.6). In 1639, however, René Descartes (1596–1650) had proved a result that anticipated Euler's formula and the Gauss–Bonnet Theorem. To state his result, consider a polyhedron that is homeomorphic to a sphere, such as the icosahedron or the cube. At each vertex the sum of the interior angles of the faces that meet at that vertex falls short of 2π by an amount we call the *defect* of the vertex. If one sums the defect of all of the vertices, then Descartes showed that the result is 4π. Descartes's proof uses spherical trigonometry. The paper in which Descartes proved this result only appeared in print in 1860 and so it had no direct influence on Euler or on the subsequent development. A general context for Euler's formula was found by J. Henri Poincaré (1854–1912) in 1895. Using his newly defined homology theory, Poincaré showed that the alternating sum $\mathbf{V} - \mathbf{E} + \mathbf{F}$ is a topological invariant of a surface. Thus the sum is independent of the choice of triangulation. In this context, the Gauss–Bonnet Theorem implies the topological invariance of the total curvature. Poincaré's work marks the beginning of the modern study of the interplay between geometry, topology, and analysis.

If we deform a surface without altering its topological type, then Poincaré's Theorem implies that there is no change in the Euler–Poincaré characteristic. The invariance may be interpreted as a kind of rigidity. For example, if you poke your finger into a balloon (idealized as S^2), it swells away from your finger to decrease the Gaussian curvature because curvature increased around your finger. However, if you require that the deformation maintain the condition of constant curvature, the class of deformations is severely restricted. In the case of S^2 the class of rigid deformations in $\mathbb{R}^3$ contains only translations, rotations, and reflections.

The Euler–Poincaré characteristic also restricts the ways a surface may lie in $\mathbb{R}^3$. For example, the torus T can be shown to have the Euler–Poincaré characteristic

$\chi(T) = 0$. Since T is compact, it has a point that is furthest from the origin. At that point, the Gaussian curvature is positive. This implies that there are points on the torus where the Gaussian curvature is negative since $\iint_S K\,dA = 0$. Thus, there is no way for the torus to be embedded in $\mathbb{R}^3$ with constant curvature.

A surprising application of the Gauss–Bonnet Theorem to the theory of curves in $\mathbb{R}^3$ is due to Jacobi.

Theorem 11.11 (Jacobi 1842). *Suppose $\alpha\colon [0,r] \to \mathbb{R}^3$ is a unit speed, differentiable, closed curve with $\alpha'(s)$ and $\alpha''(s)$ linearly independent for all $s \in [0,r]$. If the Frenet–Serret frame associated to $\alpha(s)$ is denoted $[T(s), N(s), B(s)]$, and the image of $N(s)$ on S^2 is without self-intersection, then the curve $N(s)$ divides S^2 into two sets of equal areas.*

PROOF: Let R denote one of the regions into which $N(s)$ divides S^2. Suppose $\alpha(s)$ is oriented so that $N(s)$ is the boundary of R with R lying to the left of $N(s)$. By the Gauss–Bonnet Theorem and the fact that $K \equiv 1$ on S^2,

$$\iint_R dA + \int_{\partial R} k_g(s)\,ds = 2\pi.$$

Here $k_g(s)$ is the geodesic curvature of the curve $N(s)$ on S^2. From this formula the theorem follows by showing $\int_{\partial R} k_g(s)\,ds = 0$.

Though $\alpha(s)$ is parametrized by arc length, $N(s)$ need not be. The formula in chapter 10 for the geodesic curvature gives:

$$k_g(s) = \mathrm{I}_{N(s)}(N'(s) \times N''(s), N_{S^2}(N(s))) \cdot (\|N'(s)\|^{-3}),$$

where $N_{S^2}(p) = p$ is the unit normal to the surface S^2. By the Frenet–Serret Theorem (Theorem 6.5), $N'(s) = -\kappa_N(s)T(s) + \tau_N(s)B(s)$ and so $\|N'(s)\| = \sqrt{\kappa_N^2(s) + \tau_N^2(s)}$. Also,

$$\begin{aligned} N''(s) &= -\kappa_N'(s)T(s) - \kappa_N(s)T'(s) + \tau_N'(s)B(s) + \tau_N(s)B'(s) \\ &= -\kappa_N'(s)T(s) - (\kappa_N^2(s) + \tau_N^2(s))N(s) + \tau_N'(s)B(s). \end{aligned}$$

Putting this into our expression for geodesic curvature we get:

$$\begin{aligned} k_g(s) &= \frac{1}{(\kappa_N^2 + \tau_N^2)^{3/2}} \langle (-\kappa_N T + \tau_N B) \times (-\kappa_N' T - (\kappa_N^2 + \tau_N^2)N + \tau_N' B), N \rangle \\ &= \frac{\kappa_N(s)\tau_N'(s) - \kappa_N'(s)\tau_N(s)}{(\kappa_N^2(s) + \tau_N^2(s))^{3/2}}. \end{aligned}$$

To compute the integral $\int_0^r k_g(s)\,ds$, we apply a change of variables and denote the arc length along $N(s)$ by σ. Then,

$$\sigma(s) = \int_0^s \|N'(u)\|\,du = \int_0^s \sqrt{\kappa_N^2(u) + \tau_N^2(u)}\,du, \quad \text{and so } \frac{ds}{d\sigma} = \frac{1}{\sqrt{\kappa_N^2(s) + \tau_N^2(s)}}.$$

It follows that:

$$\frac{d}{d\sigma} \arctan\left(\frac{\tau_N(s)}{\kappa_N(s)}\right) = \frac{\kappa_N(s)\tau_N'(s) - \kappa_N'(s)\tau_N(s)}{\kappa_N^2(s) + \tau_N^2(s)} \frac{ds}{d\sigma} = \frac{\kappa_N(s)\tau_N'(s) - \kappa_N'(s)\tau_N(s)}{(\kappa_N^2(s) + \tau_N^2(s))^{3/2}}.$$

Since $\alpha(s)$ is a closed curve,

$$\int_0^r k_g(s)\,ds = \int_{\sigma(0)}^{\sigma(r)} \frac{d}{d\sigma} \arctan\left(\frac{\tau_N(s)}{\kappa_N(s)}\right) d\sigma = 0.$$

This proves Jacobi's Theorem. ∎

Jacobi gave two proofs of this theorem, the second published after his first proof was criticized by THOMAS CLAUSEN (1801–85). Jacobi's method of proof was very different, using poles of points on a curve on S^2 as they related to area. The proofs harken back to Euler, and tacitly criticize Gauss's differential geometry. The dispute over the first proof and its context are discussed by McCleary (1994).

We return to curves in the plane and a sketch of the proof of Hopf's *Umlaufsatz*.

A digression on curves

Closed curves play an important role in the study of geometry and topology. In this digression we prove some of their key properties that are used to prove the Gauss–Bonnet Theorem. These properties rely upon certain topological features of curves.

Suppose $\alpha\colon [0,r] \to \mathbb{R}^2$ is a closed, differentiable, unit speed curve in the plane. Suppose further that $\alpha(s)$ is one-to-one, except for $\alpha(0) = \alpha(r)$, and that $\alpha'(0) = \alpha'(r)$. The Jordan Curve Theorem (McCleary 2006) implies that the image of $\alpha(s)$ encloses a bounded region. Assume that $\alpha(s)$ is oriented so that the enclosed region lies on the left; that is, the orthogonal basis at each point $[T(s), N(s)]$ satisfies the condition that $N(s)$ points *into* the bounded region. If not, replace $\alpha(s)$ with $\alpha(r-s)$.

The principal example of a closed curve is the unit circle $S^1 \subset \mathbb{R}^2$, parametrized by $p\colon [0,2\pi] \to \mathbb{R}^2$, $p(\theta) = (\cos\theta, \sin\theta)$. The parametrization may be extended to a mapping $p\colon \mathbb{R} \to S^1$, given by the same formula, $p(r) = (\cos(r), \sin(r))$. This mapping "winds" the line onto the circle. It has the further property that around each point $(\cos\theta_0, \sin\theta_0) \in S^1$ there is an open set $U \subset S^1$, with $(\cos\theta_0, \sin\theta_0) \in U$, such that $p^{-1}(U) \subset \mathbb{R}$ is a union of countably many identical (that is, diffeomorphic) copies of U. Such an open set is called an *elementary neighborhood of* $(\cos\theta_0, \sin\theta_0)$.

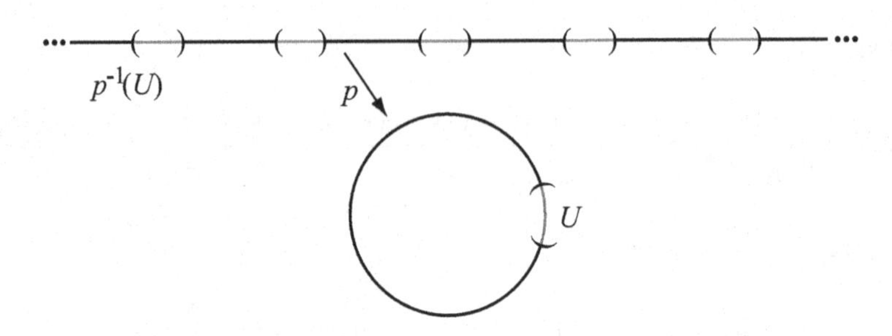

Each copy of U in $p^{-1}(U)$ may be identified with U via the mapping p. For example, the open set around the point $(1,0)$ given by the angles between $-\pi/6$ and $\pi/6$ has inverse image given by:

$$p^{-1}(\{(\cos\theta, \sin\theta) \mid -\pi/6 < \theta < \pi/6\}) = \bigcup_{k=-\infty}^{\infty} (2\pi k - (\pi/6), 2\pi k + (\pi/6)) \subset \mathbb{R}.$$

The identification via diffeomorphism on elementary neighborhoods allows us to work on the circle or on the line whenever we work in the small. Given a continuous function $f\colon [a,b] \to S^1$, we can construct a **lift** of f, that is, a continuous function $\hat{f}\colon [a,b] \to \mathbb{R}$, which satisfies $p \circ \hat{f} = f$. To construct the lift, choose a point in $\hat{a} \in \mathbb{R}$ with $p(\hat{a}) = f(a)$. Take an elementary neighborhood U which contains $f(a)$. Since $p(\hat{a}) = f(a)$, we have $\hat{a} \in p^{-1}(U)$. There is a diffeomorphism $\eta\colon U \to \hat{U} \subset p^{-1}(U)$, the piece of $p^{-1}(U)$ that contains $\hat{a}$. Consider the composite $\eta \circ f$ which is defined on $f^{-1}(U) \subset [a,b]$. Thus, we get a continuous function $\eta \circ f : f^{-1}(U) \to \mathbb{R}$ with $p \circ \eta \circ f = f$. Since $f^{-1}(U) \subset [a,b]$ is an open set containing a, it also contains $a + \epsilon$ for some $\epsilon > 0$. We can repeat the procedure of lifting to the value $f(a+\epsilon)$. Because $[a,b]$ is compact, the repetitions eventually construct a lift $\hat{f}\colon [a,b] \to \mathbb{R}$ with $p \circ \hat{f} = f$.

The tangent vector to the closed curve $\alpha\colon [0,r] \to \mathbb{R}^2$ determines a differentiable mapping $T\colon [0,r] \to S^1$. Furthermore, this mapping satisfies $T(0) = T(r)$. If we construct a lift of T, say $\widehat{T}\colon [0,r] \to \mathbb{R}$, then we have $p \circ \widehat{T}(0) = p \circ \widehat{T}(r)$ and so $\widehat{T}(r) - \widehat{T}(0)$ is a multiple of 2π.

Theorem 11.12. *If $\alpha\colon [0,r] \to \mathbb{R}^2$ is a closed, unit speed curve in the plane with $\alpha(s)$ one-to-one except at $\alpha(0) = \alpha(r)$, and $\alpha'(0) = \alpha'(r)$, and if $\alpha(s)$ is oriented so that the enclosed region is on the left, then $\widehat{T}(r) - \widehat{T}(0) = 2\pi$.*

PROOF: A SKETCH OF A PROOF: To prove the theorem, we introduce an auxiliary functions $H\colon S_r \to \mathbb{R}^2$ where $S_r = \{(a,b) \mid 0 \leq a \leq b \leq r\}$. Define H by:

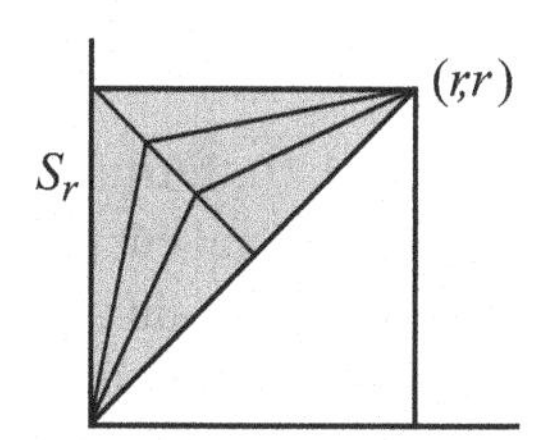

$$H(x,y) = \begin{cases} \dfrac{\alpha(y)-\alpha(x)}{\|\alpha(y)-\alpha(x)\|}, & \text{if } x<y, \text{ and } (x,y)\neq(0,r), \\ T(s), & \text{if } x=y=s, \\ -T(0), & \text{if } x=0 \text{ and } y=r. \end{cases}$$

We prove that the mapping H is continuous on S_r and $H(s,s) = T(s)$:

$$\lim_{y\to x}\frac{\alpha(y)-\alpha(x)}{\|\alpha(y)-\alpha(x)\|} = \lim_{y\to x}\frac{\alpha(y)-\alpha(x)}{y-x}\,\frac{y-x}{\|\alpha(y)-\alpha(x)\|} = \frac{T(x)}{\|T(x)\|} = T(x) = H(x,x).$$

At the corner point $(0,r)$, we have:

$$\lim_{x\to 0^+}\frac{\alpha(r)-\alpha(x)}{\|\alpha(r)-\alpha(x)\|} = -\lim_{x\to 0^+}\frac{\alpha(x)-\alpha(0)}{\|\alpha(x)-\alpha(0)\|} = -T(0)$$

$$\lim_{y\to r^-}\frac{\alpha(y)-\alpha(0)}{\|\alpha(y)-\alpha(0)\|} = -\lim_{y\to r^-}\frac{\alpha(r)-\alpha(y)}{\|\alpha(r-\alpha(y))\|} = -T(r),$$

and since $T(0) = T(r)$, H is continuous at $(0,r)$.

Consider the family of piecewise linear curves in S_r as pictured. Composing with H we get a continuous deformation of the mapping T to the mapping given by H restricted to the edges $\{0\}\times[0,r]\cup[0,r]\times\{r\}$. Explicitly, we define:

$$\lambda_t(s) = \begin{cases} H(0,2s), & \text{if } t=0 \text{ and } 0\leq s\leq r/2, \\ H(2s-r,r), & \text{if } t=0 \text{ and } r/2\leq s\leq r, \\ H\left(s, \left(\dfrac{2}{t}-1\right)s\right), & \text{if } 0<t\leq 1 \text{ and } 0\leq s\leq \dfrac{tr}{2}, \\ H\left(s, \dfrac{ts+2r(1-t)}{2-t}\right), & \text{if } 0<t\leq 1 \text{ and } \dfrac{tr}{2}\leq s\leq r. \end{cases}$$

Two facts finish the proof:

Fact 1. The difference of the values at the endpoints of a lift of a closed curve is unchanged under deformation.

By the definition of $\lambda_t(s)$, $\lambda_0(s)$ is H restricted to the edges $\{0\}\times[0,r]\cup[0,r]\times\{r\}$ and $\lambda_1(s)$ is $T(s)$. We can construct a lift of the entire family $\lambda_t(s)$ on S_r by an argument similar to the one given earlier to lift a curve. This gives a mapping $\widehat{\lambda_t(s)}\colon [0,r]\to\mathbb{R}$. The difference of the values at the endpoints of the lifts, that is, $\widehat{\lambda_t}(r) - \widehat{\lambda_t}(0)$, varies continuously over t. However, the difference is also an integer multiple of 2π. Since the set of integer multiples of 2π is discrete, the value $\widehat{\lambda}_t(r)-\widehat{\lambda}_t(0)$ is constant as a function of t. This proves Fact 1.

Fact 2. The difference of the values at the endpoints of a lift of the curve, obtained by restricting H to $\{0\}\times[0,r]\cup[0,r]\times\{r\}$, is 2π.

We prove the fact for curves $\alpha(s)$ with $\alpha(0)=(0,0)$ and $\alpha'(0)=(1,0)$. Furthermore, if we write $\alpha(s)=(x(s),y(s))$, then we want $y(s)\geq 0$. For more general curves we can always find a value $0\leq s_0\leq r$ for which the trace of $\alpha(s)$ lies entirely on one side of the tangent line to $\alpha(s)$ through $\alpha(s_0)$. Extend the domain of $\alpha(s)$ to all of $\mathbb{R}$ by $\alpha(s+r)=\alpha(s)$. We work with the one-to-one curve $\alpha\colon [s_0,s_0+r)\to\mathbb{R}^2$ and reparametrize it to have domain $[0,r)$. By a rotation and translation we can arrange the particular conditions desired. The parameter for the particular case may be rotated back to the original curve $\alpha(s)$ through a family of one-to-one curves, all with the same trace as $\alpha(s)$, and the argument given in Fact 1 can be applied to this family to show that the difference of the values at the endpoints of the lifts remains constant.

With these assumptions, $H(0,s)=\dfrac{\alpha(s)-\alpha(0)}{\|\alpha(s)-\alpha(0)\|}=\dfrac{\alpha(s)}{\|\alpha(s)\|}=\dfrac{(x(s),y(s))}{\sqrt{(x(s))^2+(y(s))^2}}$ lies above the x-axis for $0<s<r$. Furthermore, $H(0,0)=T(0)$ and $H(0,r)=-T(0)$ and so $\widehat{H}|_{\{0\}\times[0,r]}(r)-\widehat{H}|_{\{0\}\times[0,r]}(0)=\pi$. A similar argument for $\widehat{H}|_{[0,r]\times\{r\}}$ gives $\widehat{H}|_{[0,r]\times\{r\}}(r)-\widehat{H}|_{[0,r]\times\{r\}}(0)=\pi$. Putting the contributions of each part of the curve together we get $\widehat{H}|_{[0,r]\times\{r\}}(r)-\widehat{H}|_{\{0\}\times[0,r]}(0)=2\pi$. ∎

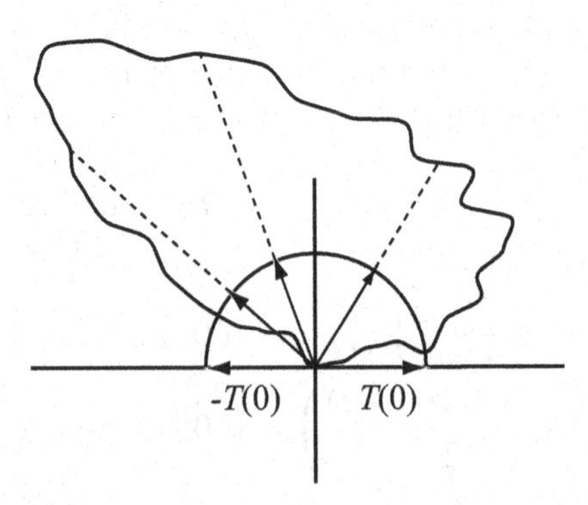

The general technical setting for results of this kind is the homotopy theory of closed curves. In that context we could speak in terms of winding numbers, covering spaces, and homotopy classes of curves. Since we will not need these ideas later, we refer the interested reader to Massey (1997) for a more thorough discussion.

If we write $T(s)=(\cos\widehat{T}(s),\sin\widehat{T}(s))$, then:

$$T'(s)=(-\widehat{T}'(s)\sin\widehat{T}(s),\widehat{T}'(s)\cos\widehat{T}(s))=\widehat{T}'(s)N(s),$$

and so the directed curvature of $\alpha(s)$ is given by $\kappa(s) = \widehat{T}'(s)$. The difference in values of the lift at the endpoints can be written as the integral:

$$\widehat{T}(r) - \widehat{T}(0) = \int_0^r \widehat{T}'(s)\,ds = \int_0^r \kappa(s)\,ds.$$

This value is called the *total curvature of the plane curve* $\alpha(s)$.

We next generalize Theorem 11.12 to a case more suitable to generalized polygons. Suppose that $\alpha(s)$ is a closed, piecewise differentiable curve in the plane. That is,

$$\alpha = (\alpha_1, \alpha_2, \ldots, \alpha_n) \colon [0, s_n] \to \mathbb{R}^2,$$

where $0 < s_1 < s_2 < \cdots < s_n$ and each $\alpha_i \colon [s_{i-1}, s_i] \to \mathbb{R}^2$ is a one-to-one, unit speed, differentiable curve. Then $\alpha_i(s_i) = \alpha_{i+1}(s_i)$ and $\alpha_n(s_n) = \alpha_1(0)$. Assume that $\alpha(s)$ is one-to-one, that is, no piece of $\alpha(s)$ crosses any other.

We call the trace of $\alpha(s)$ a **generalized polygon** in the plane, that is, a polygon made up of possibly curved sides. At each vertex, we define an *external angle* as we did for surfaces at the beginning of the chapter: Suppose $0 \leq |\epsilon_i| \leq \pi$ and:

$$\pm\epsilon_i = \arccos(\alpha_i'(s_i) \cdot \alpha_{i+1}'(s_i)),$$

($\pm\epsilon_n = \arccos(\alpha_n'(s_n) \cdot \alpha_1'(0))$). To give a sign to the angle, measure from $\alpha_i'(s)$ in a counterclockwise direction to $\alpha_{i+1}'(s_i)$; if the answer lies between 0 and π, the sign is positive; if the answer exceeds π by any amount, say η, then ϵ_i is given by $-(\eta + \pi/2)$. Cusps are handled the same way as for cusps on general surfaces.

We can associate a lift of the tangent directions along this piecewise differentiable curve as before: Choose an angle that represents $T_1(0) = \alpha_1'(0)$, say $\hat{a}$, and let $\widehat{T}_1(0) = \hat{a}$. By the same lifting argument, lift $T_1(s)$ to $\widehat{T}_1(s)$. To cross over to $\alpha_2(s)$, the angle of the tangent vector turns through the exterior angle ϵ_1. Define $\widehat{T}_2(s_1) = \widehat{T}_1(s_1) + \epsilon_1$. This determines the representative for $\widehat{T}_2(s_1)$ and so we can lift along $\alpha_2(s)$. Continuing in this manner, we can lift each $T_i(s)$ ending with $T_n(s_n)$.

Proposition 11.13. $\widehat{T}_n(s_n) = \epsilon_n = \widehat{T}_1(0) + 2\pi$.

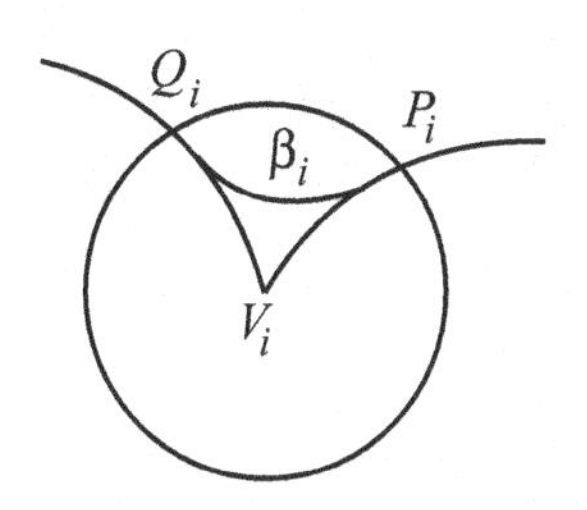

Proof: Construct a circle centered at each vertex $V_i = \alpha_i(s_i) = \alpha_{i+1}(s_i)$, small enough so that only portions of $\alpha_i(s)$ and $\alpha_{i+1}(s)$ lie in it. Denote the first point of entry into the circle along $\alpha_i(s)$ by $P_i = \alpha_i(t_i)$ and the point of departure along $\alpha_{i+1}(s)$ by $Q_i = \alpha_{i+1}(u_i)$. As the radius of the circle goes to zero, the angle $\angle P_iV_iQ_i$ converges to ϵ_i. Construct a curve β_i within the circle at V_i such that $\beta_i \colon [t_i, u_i] \to \mathbb{R}^2$ is differentiable and unit speed, and has $\beta'(t_i) = \alpha_i'(t_i)$ and $\beta_i'(u_i) = \alpha_{i+1}'(u_i)$. Lifting $T_{\beta_i}(s)$ so that $\widehat{T}_{\beta_i}(t_i) = \widehat{T}_i(t_i)$, we get that the total change along β_i of the tangent vector from t_i to u_i is close to ϵ_i.

By construction the curve made up of portions of the $\alpha_i(s)$ outside the circles and $\beta_i(s)$ in the circles is sufficiently differentiable to apply Theorem 11.12. and get a difference in the lift of the endpoints to be 2π. Since the contributions of the $\beta_i(s)$ is close to the sum of the exterior angles, the lift we constructed for the piecewise differentiable curve $\alpha(s)$ has a difference at the endpoints close to 2π minus the contribution of β_n. This proves the proposition. ■

Finally, we complete the proof of the Hopf *Umlaufsatz* (Theorem 11.1). We assumed that the region R is a subset of a regular, orientable surface S, and that R is contained in the image of a coordinate chart for S. Furthermore, the boundary of R is a generalized polygon on S. Therefore, we can express R as $x(W)$, where W is a subset of the domain $U \subset \mathbb{R}^2$ of the coordinate chart x with W satisfying all of the assumptions that R does.

The coordinate chart x induces a new inner product on the tangent space at each point of U that is different from the Euclidean inner product on $\mathbb{R}^2$. If $\eta_1(t) = (u_1(t), v_1(t))$ and $\eta_2(t) = (u_2(t), v_2(t))$ are curves in U that meet at $\eta_1(0) = \eta_2(0)$, then the inner product between the tangents to the curves is:

$$\langle \eta_1'(0), \eta_2'(0)\rangle_S = \mathrm{I}_p(u_1'(0)x_u + v_1'(0)x_v, u_2'(0)x_u + v_2'(0)x_v),$$

where $p = x(\eta_1(0))$. For example, the angle ψ between the line parallel to the u-axis and the line parallel to the v-axis through a point in U is determined by $\cos\psi = F(u_0, v_0)/\sqrt{E(u_0, v_0)G(u_0, v_0)}$. Denote the usual Euclidean inner product (the dot product) on $\mathbb{R}^2$ by $\langle\ ,\ \rangle_{\mathbf{E}}$. We can interpolate between the surface inner product and the Euclidean inner product by forming the sum:

$$\langle \eta_1'(0), \eta_2'(0)\rangle_t = (1-t)\langle \eta_1'(0), \eta_2'(0)\rangle_{\mathbf{E}} + t\langle \eta_1'(0), \eta_2'(0)\rangle_S.$$

The reader can check that, for all $t \in [0,1]$, the function $\langle\ ,\ \rangle_t$ is an inner product (Exercise 11.7). Suppose that the boundary of W is a piecewise differentiable and closed curve $\beta\colon [0, s_n] \to \mathbb{R}^2$. We can form the functions $\phi_i^t(s)$ that lift the tangents along the curve with respect to the inner product $\langle\ ,\ \rangle_t$. Notice that the angles between x_u and $\alpha_i'(s)$ may differ as t varies. When $t = 0$ the inner product is Euclidean and so our argument above gives the Hopf *Umlaufsatz* for the surface $\mathbb{R}^2$. As the inner product changes continuously, the functions $\phi_i^t(s)$ change continuously, and so the difference of the values of the lift at the beginning and endpoint of the curve plus the last angle gives a continuous mapping to multiples of 2π. Since this set is discrete, the value is constant. Thus, on the surface the equation

$$\phi_n(s_n) + \epsilon_n = \phi_1(0) + 2\pi$$

holds, that is, the tangent along a closed, piecewise differentiable curve enclosing a simply connected region in S turns through 2π.

The last part of the proof anticipates a more general notion of surface. For some values of t, the inner product $\langle\ ,\ \rangle_t$ may not correspond to the inner product at a point in some surface lying in $\mathbb{R}^3$. The pair $(U, \langle\ ,\ \rangle_t)$ for a fixed value of t is the coordinate chart for an *abstract surface*, a topic we will take up in chapter 13.

Exercises

11.1 Let T^2 be the torus in $\mathbb{R}^3$. Triangulate it and compute its Euler–Poincaré characteristic.

11.2 Show that the infinite cylinder $S^1 \times \mathbb{R} \subset \mathbb{R}^3$ has infinitely many geodesics joining two points (x,y,z) and (x',y',z') with $z \neq z'$. We know that $K = 0$ on the cylinder. Is this a contradiction to Corollary 11.7.

11.3 Let $R' \subset S$ be a subset of a surface with boundary $\partial R'$, a piecewise differentiable curve on S. Suppose further that R' can be triangulated so that each triangle is simply connected and lies in a coordinate chart. Show that the Gauss–Bonnet Theorem for R' takes the form:

$$\sum_j \epsilon_j + \int_{\partial R'} k_g(s)\,ds + \iint_{R'} K\,dA = 2\pi\chi(R').$$

11.4† On a surface S with Gaussian curvature nonpositive everywhere, show that a closed geodesic, if one exists, cannot be the boundary of a bounded region in S.

11.5 Begin with a sphere and add g *handles* to it. A handle is a cylinder that is joined to the sphere by removing two disks from the sphere and glueing the cylinder on at the holes, one end along one edge, the other end along the other edge. For example, when $g = 1$, one gets a surface that is homeomorphic to the torus. Show that this glueing procedure gives an orientable surface in $\mathbb{R}^3$ and that such a surface has Euler–Poincaré characteristic $2 - 2g$.

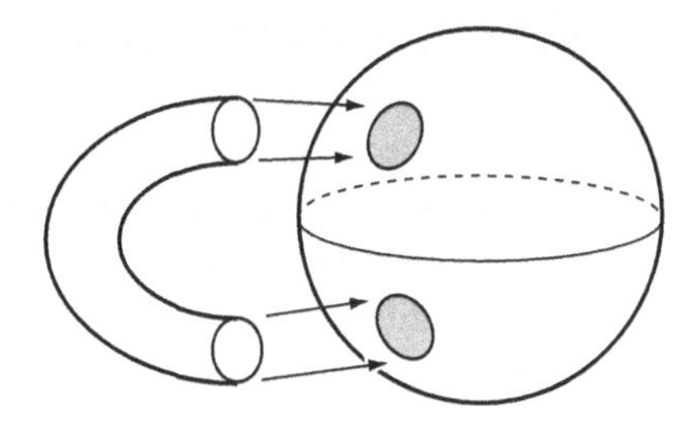

11.6 From Euler's formula prove that there are only five Platonic solids.

11.7 Given two inner products on a vector space V, say $\langle\ ,\ \rangle_1$ and $\langle\ ,\ \rangle_2$, show that the linear combination for $0 \leq t \leq 1$, $\langle v,w\rangle = (1-t)\langle v,w\rangle_1 + t\langle v,w\rangle_2$ is also an inner product on V.

11.8 The sphere S^2 can be triangulated by geodesic triangles, all of them congruent to each other. Is it possible to construct an analogous *tiling* of the sphere using congruent geodesic quadrangles so that four such quadrangles meet at each vertex.

11.9† The Jordan Curve Theorem states that a simple (no double points) closed curve in the plane is the boundary of a simply connected, bounded set. If we add that the curve is regular and has curvature always positive, then the bounded set is convex. Hadamard's Theorem generalizes this result to compact, closed surfaces in $\mathbb{R}^3$, that is, suppose S is a closed, compact surface in $\mathbb{R}^3$ with $K > 0$ at every point on the surface. Then S is the boundary of a convex subset of $\mathbb{R}^3$ and S is homeomorphic to S^2. To prove this, we need some topological facts: Any closed, compact, orientable surface in $\mathbb{R}^3$ with positive Euler characteristic is homeomorphic to S^2. Prove that, under our conditions, $\chi(S) > 0$. Then show that the Gauss map is a diffeomorphism $N\colon S \to S^2$, which is onto.

Solutions to Exercises marked with a dagger appear in the appendix, pp. 336–337.

12

Constant-curvature surfaces

I remark only that I have developed Astral Geometry so far that I can completely solve all its problems as soon as the constant k is given.

C. F. GAUSS (TO GERLING), (1819)

...we have sought, to the extent of our ability, to convince ourselves of the results of Lobachevskiĭ's doctrine; then, following the tradition of scientific research, we have tried to find a real substrate for this doctrine, ...

E. BELTRAMI (1868)

We finish the task of building a foundation for the geometry of Part I by developing the conditions that a surface must satisfy in order to provide a model of the geometries of interest. We take our cue from the metric space properties of surfaces found in chapter 10. In particular, it makes sense to speak of a *circle* in a surface S of radius r centered at a given point $p \in S$, that is, the set of points $q \in S$ such that $d(p,q) = r$. Of course, it is possible that this collection is an empty set for some radii. When such a circle determines a curve on S, then we may compute its circumference as a function of the radius.

Let $\mathrm{circum}_p(r)$ denote the circumference of a circle on S of radius r centered at p. In the Euclidean plane, the circumference of a circle of radius r is $2\pi r$. For the other geometries, the circumference takes the form:

$$\mathrm{circum}_p(r) = 2\pi R \sin\left(\frac{r}{R}\right) = 2\pi r - \frac{\pi r^3}{3R^2} + \cdots, \text{ on a sphere of radius } R \text{ (for } r < \pi R\text{)},$$

$$\mathrm{circum}_p(r) = 2\pi k \sinh\left(\frac{r}{k}\right) = 2\pi r + \frac{\pi r^3}{3k^2} + \cdots, \text{ in non-Euclidean geometry.}$$

The Taylor series for the spherical and non-Euclidean cases illustrate the deviation from the familiar Euclidean formula.

In Chapter 9 we developed special systems of coordinates that are directly related to the geometry of a surface, such as systems with coordinate curves given by lines of curvature or asymptotic lines. We add to this collection a pair of systems that generalize polar and rectangular coordinates in the Euclidean plane. In the plane the pencil of lines through the origin leads to polar coordinates. The pencil of lines perpendicular to a given line leads to rectangular coordinates.

Suppose p is a point in a surface S. Consider the pencil of geodesics through a given point. The exponential map may be used to define a coordinate chart as follows: Fix a vector $\mathbf{w}$ of unit length in $T_p(S)$. Let $a\colon (-\pi,\pi) \to T_p(S)$ be a smooth mapping satisfying the property that $\|a(\theta)\| = 1$ and $\mathrm{I}_p(\mathbf{w}, a(\theta)) = \cos\theta$ for all $\theta \in (-\pi,\pi)$. The image of $a\colon (-\pi,\pi) \to T_p(S)$ is the unit circle in $T_p(S)$ (minus one point) and $a(\theta)$ parametrizes the circle according to the angle made with the fixed vector $\mathbf{w}$. For some $\epsilon_p > 0$ the exponential map is a diffeomorphism of $B_{\epsilon_p}(p) \subset T_p(S)$ onto an open set containing $p \in S$ (Theorem 10.11). We use these data to construct a coordinate chart.

Definition 12.1. Geodesic polar coordinates *are given by the coordinate chart:*

$$x\colon (0,\epsilon_p) \times (-\pi,\pi) \to S, \text{ defined } x(r,\theta) = \exp_p(ra(\theta)).$$

A **geodesic circle** *of radius $r < \epsilon_p$ centered at p is the locus of points in S of distance r from p. Alternatively, it is the image of the set $\{\mathbf{u} \in T_p(S) \mid \|\mathbf{u}\| = r\}$ under the exponential map. For a fixed angle θ_0, the geodesic given by $\gamma\colon (0,r] \to S$, $\gamma(s) = \exp_p(sa(\theta_0))$ is a* **geodesic radius** *of the geodesic circle of radius r.*

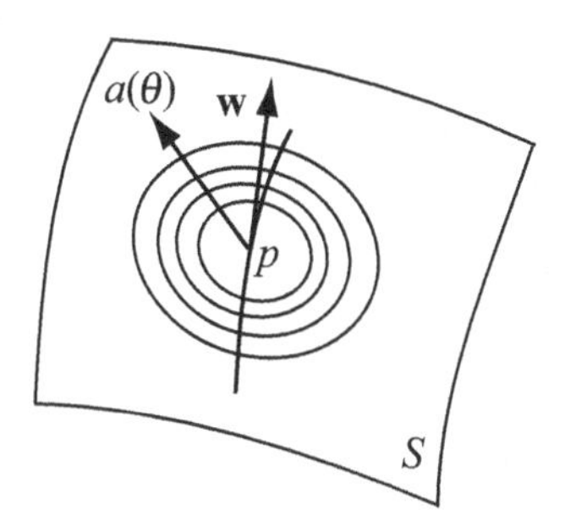

This definition contains the generalization of the familiar notions of circle, radius, and polar coordinates in the Euclidean plane. The restriction to radii of length less that ϵ_p allows us to apply the analytical properties of the exponential map. Euclid proved an important and useful property of circles and their radii in the plane (Book III, Proposition 18), which Gauss generalized to geodesic circles. Gauss gave two different proofs of the following lemma in §15 of the *Disquisitiones*.

Lemma 12.2 (Gauss's Lemma). *Geodesic circles are perpendicular to their geodesic radii.*

PROOF: Fix a point p in the surface S and by Theorem 10.11 there is an open ball $B_{\epsilon_p}(p) \subset T_p(S)$ on which the exponential map is a diffeomorphism. Suppose that $\alpha\colon (I \subset \mathbb{R}) \to T_p(S)$ is a curve in the tangent plane satisfying $\|\alpha(t)\| = r < \epsilon_p$ for all $t \in I$, an interval. The curve $\beta(t) = \exp_p(\alpha(t))$ is a portion of a geodesic circle of radius r. Define the family of curves:

$$\sigma_t(s) = \exp_p(s\alpha(t)), \quad 0 \le s \le 1.$$

Then $\sigma_t(1) = \beta(t)$, and for a fixed $t_0 \in I$, $\sigma_{t_0}(s)$ is a geodesic. Our goal is to prove that for each $t_0 \in I$,

$$\left\langle \frac{d}{dt}\sigma_t(1)\bigg|_{t=t_0}, \frac{d}{ds}\sigma_{t_0}(s)\bigg|_{s=1} \right\rangle = 0.$$

In this proof we consider the family of curves $\sigma_t(s)$ as a variation of a fixed geodesic $\sigma_{t_0}(s)$. When we proved that a shortest curve in a surface is a geodesic (Theorem 10.7), we considered a variation of the length function. Here we vary an *energy* function defined by:

$$E(t) = \int_0^1 \left\langle \frac{d\sigma_t}{ds}, \frac{d\sigma_t}{ds} \right\rangle ds.$$

The energy function is constant because it is the length r of the geodesic radius from p to $\beta(t)$. Taking the derivative with respect to t we obtain:

$$\begin{aligned}
0 = E'(t_0) &= \frac{d}{dt}\left(\int_0^1 \left\langle \frac{d\sigma_t}{ds}, \frac{d\sigma_t}{ds}\right\rangle ds\right)\Bigg|_{t=t_0} \\
&= \int_0^1 \frac{\partial}{\partial t}\left\langle \frac{d\sigma_t}{ds}, \frac{d\sigma_t}{ds}\right\rangle ds\Bigg|_{t=t_0} = \int_0^1 2\left\langle \frac{\partial^2\sigma_t}{\partial t\partial s}, \frac{d\sigma_t}{ds}\right\rangle ds\Bigg|_{t=t_0} \\
&= \int_0^1 \left[\frac{d}{ds}\left\langle \frac{d\sigma_t}{dt}, \frac{d\sigma_t}{ds}\right\rangle\Bigg|_{t=t_0} - \left\langle \frac{\partial\sigma_t}{\partial t}, \frac{\partial^2\sigma_t}{\partial s^2}\right\rangle\Bigg|_{t=t_0}\right] ds \\
&= 2\left\langle \frac{d\sigma_t}{dt}, \frac{d\sigma_t}{ds}\right\rangle\Bigg|_{t=t_0}\Bigg|_{s=0}^{s=1} - 2\int_0^1 \left\langle \frac{\partial\sigma_t}{\partial t}, \frac{\partial^2\sigma_t}{\partial s^2}\right\rangle\Bigg|_{t=t_0} ds.
\end{aligned}$$

The first term in this final expression can be evaluated:

$$\left\langle \frac{d\sigma_t}{dt}, \frac{d\sigma_t}{ds}\right\rangle\Bigg|_{t=t_0}\Bigg|_{s=0}^{s=1} = \left\langle \frac{d}{dt}\sigma_t(1)\Big|_{t=t_0}, \frac{d\sigma_{t_0}}{ds}(1)\right\rangle - \left\langle \frac{d}{dt}\sigma_t(0)\Big|_{t=t_0}, \frac{d\sigma_{t_0}}{ds}(0)\right\rangle.$$

Since $\sigma_t(0)$ is always the point p, the second term is 0. The first term is the inner product we wish to compute. It suffices then to show that the integral that is the last term in the expression for $E'(t_0)$ is zero.

The geodesic $\sigma_{t_0}(s)$ is constant speed and satisfies the property that $\dfrac{d^2\sigma_{t_0}}{ds^2} = lN$; that is, the acceleration of the curve $\sigma_{t_0}(s)$ points in a direction normal to the surface. Since $\dfrac{d\sigma_t}{dt}$ is the tangent vector to the curve $\sigma_t(s_0)$, it lies in the tangent plane to the surface and so we conclude that $\left\langle \dfrac{\partial\sigma_t}{\partial t}, \dfrac{\partial^2\sigma_t}{\partial s^2}\right\rangle\Bigg|_{t=t_0} = 0$. This proves the lemma. ∎

Consider the line element for geodesic polar coordinates. The coordinate curves are given by geodesic circles when r is constant, and geodesic rays emanating from p when θ is constant. By the definition of the exponential map, the curves with θ held constant are unit speed geodesics and so $E = 1$. By Gauss's Lemma, the coordinate curves meet orthogonally and so $F = 0$. Thus,

$$ds^2 = dr^2 + G(r,\theta)d\theta^2.$$

This is a particularly simple expression for the line element from which we can easily compute the associated apparatus of intrinsically defined functions.

The lone function $G(r,\theta)$ is not without restriction. In order to get at conditions on G, recall a particular property of polar coordinates for $\mathbb{R}^2$: The origin is a singular point for this coordinate chart. It is, however, a kind of false singularity as the change to rectangular coordinates has no singularity at the origin. If we consider the set $(0,\epsilon_p)\times(-\pi,\pi)$ as a subset of $\mathbb{R}^2$ in polar coordinates, then we can change variables from rectangular coordinates:

$$X\colon (B_{\epsilon_p}(p) - \{\mathbf{0}\} \subset \mathbb{R}^2) \longrightarrow (0,\epsilon_p)\times(-\pi,\pi)$$
$$X(x,y) = \left(\sqrt{x^2+y^2}, \arctan\left(\frac{y}{x}\right)\right),$$

the usual inverse to $(r,\theta) \mapsto (r\cos\theta, r\sin\theta)$. However, the origin is not a singular point for rectangular coordinates, and we can pass to the limit as $r \to 0^+$ and use rectangular coordinates to smooth out the singular point.

Since the change of coordinates given by X is an isometry, we can recompute ds^2 in x and y: Recall that $dr = \dfrac{x\,dx + y\,dy}{r}$ and $d\theta = \dfrac{x\,dy - y\,dx}{r^2}$.

$$\begin{aligned}
ds^2 &= dr^2 + G(r,\theta)\,d\theta^2 \\
&= \left(\frac{x\,dx + y\,dy}{r}\right)^2 + G(r,\theta)\left(\frac{x\,dy - y\,dx}{r^2}\right)^2 \\
&= \left(\frac{x^2}{r^2} + \frac{G(r,\theta)y^2}{r^4}\right)dx^2 + 2\left(1 - \frac{G(r,\theta)}{r^2}\right)\frac{xy}{r^2}\,dx\,dy + \left(\frac{y^2}{r^2} + \frac{G(r,\theta)x^2}{r^4}\right)dy^2 \\
&= \hat{E}dx^2 + 2\hat{F}dx\,dy + \hat{G}dy^2.
\end{aligned}$$

It follows from this expression and $x^2 + y^2 = r^2$ that:

$$\hat{E} - 1 = \frac{x^2}{r^2} + \frac{G(r,\theta)y^2}{r^4} - 1 = \frac{x^2}{r^2} - \frac{x^2+y^2}{r^2} + \frac{G(r,\theta)y^2}{r^4} = \frac{y^2}{r^2}\left(\frac{G(r,\theta)}{r^2} - 1\right),$$

and similarly $\hat{G} - 1 = \dfrac{x^2}{r^2}\left(\dfrac{G(r,\theta)}{r^2} - 1\right)$, and so $x^2(\hat{E}-1) = y^2(\hat{G}-1)$. Since this equation holds for any (x,y) in the region of definition, it follows that $\hat{E} - 1 = my^2 + \cdots$ and $\hat{G} - 1 = mx^2 + \cdots$. Substituting these relations into the equations for $\hat{E} - 1$ and $\hat{G} - 1$ we find that:

$$\frac{G(r,\theta)}{r^2} = 1 + mr^2 + \cdots \text{ and so } \sqrt{G(r,\theta)} = r + \frac{m}{2}r^3 + \cdots.$$

Thus we arrive at the following conditions on the function G:

Proposition 12.3. *For geodesic polar coordinates with $ds^2 = dr^2 + G(r,\theta)d\theta^2$, and for any fixed angle θ_0 with $-\pi < \theta_0 < \pi$, we have:*

$$\lim_{r\to 0^+} G(r,\theta_0) = 0, \quad \textit{and} \quad \lim_{r\to 0^+} \frac{\partial\sqrt{G(r,\theta_0)}}{\partial r} = 1.$$

In finding the analytic conditions on $G(r,\theta)$ we computed a bit of the Taylor series of $\sqrt{G(r,\theta)}$ around $r = 0$. The particular form of the line element leads to the following simple form of the Gaussian curvature in this coordinate chart (Exercise 9.3: Check this!):

$$K = \frac{-1}{\sqrt{G(r,\theta)}} \frac{\partial^2\sqrt{G(r,\theta)}}{\partial r^2}.$$

Substituting our results for the series associated to $\sqrt{G}$ we get the equation:

$$-K\left(r + \frac{m}{2}r^3 + \cdots\right) = -K\sqrt{G} = \frac{\partial^2\sqrt{G}}{\partial r^2} = 3mr + \cdots,$$

and so we conclude, near $r = 0$, that $3m = -K$ and:

$$\sqrt{G(r,\theta)} = r - K(p)\frac{r^3}{6} + \cdots,$$

where $K(p)$ is the Gaussian curvature at the point p. Notice that another choice of the unit vector $\mathbf{w}$ that determines $\theta = 0$ gives another form of the polar coordinates for which the variable r remains the same. Thus, these formulas do not depend on a choice of θ along which r goes to zero.

In geodesic polar coordinates, for $r_0 < \epsilon_p$, the geodesic circle of radius r_0 is the coordinate curve $x(r_0,\theta)$ for $-\pi \leq \theta \leq \pi$, and its circumference is given by:

$$\begin{aligned}\text{circum}_p(r_0) &= \int_{-\pi}^{\pi} \sqrt{G(r_0,\theta)}\,d\theta = \int_{-\pi}^{\pi} \left(r_0 - K(p)\frac{r_0^3}{6} + \cdots\right) d\theta \\ &= 2\pi r_0 - \frac{\pi K(p)}{3} r_0^3 + \cdots.\end{aligned}$$

In 1848, J. Bertrand (1822–1900) and V. Puiseux (1820–83) gave the following geometric interpretation of Gaussian curvature.

Proposition 12.4. *If $\text{circum}_p(r)$ is the circumference of a geodesic circle of radius r centered at the point p in a surface S, then:*

$$K(p) = \lim_{r\to 0^+} \frac{6\pi r - 3\,\text{circum}_p(r)}{\pi r^3}.$$

The proof follows from the Taylor series formula for the circumference of a circle.

Lemma 12.2 generalizes to other families of curves. For example, if we fix a unit speed curve $\alpha\colon (-\eta,\eta) \to S$, then in an open set of S containing $\alpha(0)$ we can define a family of curves:

$$\gamma_t(s) = \exp_{\alpha(t)}(sr(N \times \alpha'(t))) = \exp_{\alpha(t)}((sr)n_\alpha(t))$$

for $r < \epsilon$, the bound associated with $\alpha(0)$ by exp in this neighborhood. The curves $t \mapsto \gamma_t(s_0)$ are called *geodesic parallels* to the curve $\alpha(t)$ and they are a fixed distance from $\alpha(t)$. The family $\gamma_{t_0}(s)$ consists of geodesics that are all perpendicular to $\alpha(t)$, and following the proof of Gauss's lemma with little change, they can be shown to be perpendicular to the geodesic parallels. This leads to a version of rectangular coordinates in a neighborhood of $p = \alpha(0)$ with the curve $\alpha(t)$ acting as the x-axis and the normal geodesics acting as the lines $x = x_0$. The associated chart gives coordinates called **geodesic parallel coordinates** and the line element in this chart also takes the form $ds^2 = dx^2 + G(x,y)\,dy^2$. When $\alpha(t)$ is a geodesic, this procedure generalizes the pencil of lines perpendicular to a given line.

Euclid revisited III: Congruences

Geodesic circles allow us to interpret Euclid's Postulate III in the language of surfaces. On a surface S, suppose $p \in S$ is taken to be the center and $q \in S$ is joined to p by a geodesic segment taken to be the radius of a circle in S. When the exponential map is defined on $B_r(\mathbf{0}_p) \subset T_p(S)$, where $r > d(p,q)$, we can construct the geodesic circle centered at p with radius pq. In general, however, this construction is restricted to small radii by the general theory of differential equations on the surface (Corollary 10.9). This restriction can be overcome by assuming that the surface is geodesically complete, and the exponential map has all of the tangent plane $T_p(S)$ as its domain. When exp is one-to-one, the image of the set of points a fixed distance from $\mathbf{0}_p$ in $T_p(S)$ maps via exp to a geodesic circle of that radius.

In synthetic geometry, Euclid, and later Archimedes, Gauss, Lobachevskiĭ, and J. Bolyai, considered all circles with a given radius to be congruent. As Proposition 12.4 shows, on a general surface the circumference of a circle of radius r may depend on the Gaussian curvature at its center. This observation leads us to a further assumption about the surfaces that might be models of non-Euclidean geometry—such a surface must have constant Gaussian curvature.

From the formula for the circumference of a circle in non-Euclidean geometry (Theorem 4.32) and the formula of Bertrand and Puiseux (Proposition 12.4), we can finally give an interpretation for the mysterious constant k of non-Euclidean geometry found in the work of Gauss, Lobachevskiĭ, and J. Bolyai. To wit, comparing Taylor series for the circumference as a function of the radius we get:

$$2\pi k \sinh\left(\frac{r}{k}\right) = 2\pi r + \frac{\pi r^3}{3k^2} + \cdots = 2\pi r - \frac{\pi K(p)}{6} r^3 + \cdots,$$

and so $2/k^2 = -K(p) = -K$, a constant. Thus, a surface modeling non-Euclidean geometry must have constant negative curvature. This condition also appears in the formulas for area given by Corollary 11.5 and the relationship between area and angle defect in non-Euclidean geometry.

Using geodesic polar coordinates and the assumption of constant Gaussian curvature we can interpret Euclid's Postulate IV for surfaces. Given two right angles at points $P \neq Q$ in a surface, we can construct a local isometry taking open disks containing P and Q to each other with P going to Q and the right angles mapped onto one another. To do this, compose geodesic polar coordinates and their inverses, oriented to take the right angles to each other. The angles are congruent when this local isometry extends to an isometry of the whole surface to itself (chapter 9). Euclid's Postulate IV then assumes the existence of sufficiently many isometries of the surface to itself. If for each pair of points p and q in S and pair of unit tangent vectors $\mathbf{v} \in T_p(S)$, $\mathbf{w} \in T_q(S)$, there is an isometry $S \to S$ taking p to q and $\mathbf{v}$ to $\mathbf{w}$, then we call the surface **point transitive**. The development of these ideas leads to the theory of transformation groups. A point transitive surface is a surface of constant curvature.

The work of Minding

A general question discussed in chapter 9 is whether two surfaces can be compared geometrically; that is, given surfaces S_1 and S_2, is there an isometry between them? This problem can be posed locally and is called the *problem of Minding: Given S_1 and S_2 when is there a mapping $S_1 \to S_2$ for which $ds_1 = ds_2$ at corresponding points, where ds_i is the line element on S_i?* A necessary condition is the equality of the Gaussian curvature at corresponding points. The solution to this problem in the special case of constant curvature surfaces is due to Minding.

Theorem 12.5 (Minding 1839). *Two surfaces S_1 and S_2 with constant Gaussian curvatures K_1 and K_2 are locally isometric if and only if $K_1 = K_2$.*

PROOF: We have already shown (Corollary 9.5) that locally isometric surfaces have the same Gaussian curvature at corresponding points. To prove the converse, choose $p \in S_1$ and $q \in S_2$ and introduce geodesic polar coordinates at each point with the same coordinates (r, θ) in $(0, \epsilon) \times (-\pi, \pi)$. The line element takes the form:

$$ds_1^2 = dr^2 + G_1(r,\theta)d\theta^2 \text{ and } ds_2^2 = dr^2 + G_2(r,\theta)d\theta^2.$$

Furthermore, the functions G_1 and G_2 satisfy:

$$\frac{1}{\sqrt{G_1}} \frac{\partial^2 \sqrt{G_1}}{\partial r^2} = -K_1 = -K_2 = \frac{1}{\sqrt{G_2}} \frac{\partial^2 \sqrt{G_2}}{\partial r^2}.$$

This is a special case of the differential equation $\dfrac{\partial^2 f}{\partial r^2} + Kf = 0$ with the initial conditions $\lim_{r\to 0^+} f = 0$ and $\lim_{r\to 0^+} \dfrac{\partial f}{\partial r} = 1$, where $K = K_1 = K_2$, a constant. The uniqueness of solutions to differential equations with initial conditions implies $G_1 = G_2$. The desired local isometry is the composition of $\exp_p^{-1}$ with the coordinates around q. ∎

Minding's Theorem tells us that locally we can compare any two surfaces sharing the same constant curvature. Any constant curvature surface then is a local model for all surfaces of that curvature. One route to constructing model surfaces is the differential equation to be satisfied by $\sqrt{G}$, namely, for K constant,

$$\frac{\partial^2 f}{\partial r^2} + Kf = 0 \text{ such that } \lim_{r\to 0^+} f = 0 \text{ and } \lim_{r\to 0^+} \frac{\partial f}{\partial r} = 1.$$

We consider the cases $K = 0$, $K > 0$, and $K < 0$ separately.

Surfaces with $K = 0$

In this case, $\dfrac{\partial^2 \sqrt{G}}{\partial r^2} = 0$ and so $\sqrt{G(r,\theta)} = rf_1(\theta) + f_2(\theta)$. The initial conditions, which are independent of θ, imply that $f_1(\theta) = 1$ and $f_2(\theta) = 0$. Thus, $G(r,\theta) = r^2$ and $ds^2 = dr^2 + r^2\, d\theta^2$, the line element for polar coordinates on the Euclidean plane, which we can take as our model of constant zero curvature.

Surfaces with $K = 1/R^2$

The differential equation $\dfrac{\partial^2 f}{\partial r^2} + \dfrac{1}{R^2} f = 0$ has the general solution

$$f(r,\theta) = A(\theta)\cos(r/R) + B(\theta)\sin(r/R).$$

The initial conditions imply that the $f(r,\theta) = \sqrt{G(r,\theta)} = R\sin(r/R)$, and the line element becomes $ds^2 = dr^2 + R^2 \sin^2(r/R)\, d\theta^2$. In chapter 7bis we met this line element associated to the coordinates on the sphere given by

$$x(r,\theta) = (R\sin(r/R)\cos\theta, R\sin(r/R)\sin\theta, R\cos(r/R)),$$

polar coordinates on the sphere of radius R centered at the origin in $\mathbb{R}^3$, given by longitude and distance from the north pole, known as azimuthal projection. We already know that such a sphere is a surface of constant curvature $1/R^2$. By Minding's Theorem we may take it as the canonical model of such a surface.

A particular feature of the spheres in $\mathbb{R}^3$ is that they are compact surfaces on which geodesics are closed curves. If we look for surfaces in $\mathbb{R}^3$ of constant Gaussian curvature that are also compact, then we find that spheres are the only examples.

Theorem 12.6 (Liebmann 1899). *A compact, connected, regular surface S in $\mathbb{R}^3$ with constant Gaussian curvature and without singularities or boundary is a sphere.*

PROOF: We first observe that a compact surface has a point on it with positive curvature. To see this, consider the z-coordinate of points on the surface. This projection is a continuous function on a compact space and so it achieves a maximum at some point $p \in S$. At such a point of maximal height the tangent plane $T_p(S)$ bounds the surface, that is, S lies entirely on one side of this tangent plane. Thus $K(p)$ must be nonnegative by the characterization of the Gaussian curvature as the product of the principal curvatures. If the point were planar, then the surface would contain a line segment, and so, by completeness, a line, making the surface unbounded. This violates compactness (S is closed and bounded in $\mathbb{R}^3$). Thus the curvature at p is positive.

We next prove that every point on the surface S is umbilic. If a point p were not umbilic, then the principal curvatures at p, k_1 and k_2 would satisfy $k_1 > k_2$, giving the maximal and minimal normal curvatures at p. The following lemma, due to Hilbert, shows that this cannot happen.

Lemma 12.7 (Hilbert 1901). *Suppose p is a nonumbilic point in a regular surface S for which $k_1 > k_2$ are the principal curvatures on S. Suppose k_1 achieves a local maximum at p and k_2 achieves a local minimum at p. Then $K(p) \leq 0$.*

PROOF: Since p is an nonumbilic point of S, there is a coordinate chart around p with coordinate lines given by lines of curvature (Corollary 9.11). For such coordinates we have $k_1 = \dfrac{e}{E}$ and $k_2 = \dfrac{g}{G}$. Furthermore, the Mainardi–Codazzi Equations take the form:

$$e_v = \frac{E_v}{2}\left(\frac{e}{E} + \frac{g}{G}\right) = \frac{E_v}{2}(k_1 + k_2), \quad g_u = \frac{G_u}{2}\left(\frac{e}{E} + \frac{g}{G}\right) = \frac{G_u}{2}(k_1 + k_2).$$

Writing $e = k_1 E$ we take the partial derivative of both sides with respect to v and we get $e_v = \dfrac{\partial k_1}{\partial v}E + k_1 E_v$. Combining this relation with the Mainardi–Codazzi Equations we get:

$$\frac{\partial k_1}{\partial v} = \frac{E_v}{2E}(k_2 - k_1).$$

Similarly, writing $g = k_2 G$, we obtain the equation $\dfrac{\partial k_2}{\partial u} = \dfrac{G_u}{2G}(k_1 - k_2)$. Since k_1 is a local maximum and k_2 a local minimum at p, we have $\dfrac{\partial k_1}{\partial v} = 0 = \dfrac{\partial k_2}{\partial u}$, and so we have $E_v = G_u = 0$ at p.

Since the coordinate curves are perpendicular, the Gaussian curvature can be computed from the formula:

$$K = -\frac{1}{2\sqrt{EG}}\left(\frac{\partial}{\partial v}\left(\frac{E_v}{\sqrt{EG}}\right) + \frac{\partial}{\partial u}\left(\frac{G_u}{\sqrt{EG}}\right)\right).$$

Expanding the derivatives and simplifying the expression we get

$$K = -\frac{1}{2EG}(E_{vv} + G_{uu}) + \frac{1}{4EG}\left(\frac{E_v^2}{E} + \frac{E_v G_v}{G} + \frac{G_u^2}{G} + \frac{E_u G_u}{E}\right).$$

Thus, at p, the Gaussian curvature takes the form $K = \frac{-1}{2EG}(E_{vv} + G_{uu})$.

Since k_1 is a local maximum and k_2 a local minimum at p, the second partials of k_1 and k_2 satisfy

$$\frac{\partial^2 k_1}{\partial v^2} \leq 0 \text{ and } \frac{\partial^2 k_2}{\partial u^2} \geq 0.$$

We rewrite $E_v = \frac{2E}{k_2 - k_1}\frac{\partial k_1}{\partial v}$ and $G_u = \frac{2G}{k_2 - k_1}\frac{\partial k_2}{\partial u}$, then E_{vv} and G_{uu} at p relate to the principal curvatures by

$$E_{vv} = \frac{\partial^2 k_1}{\partial v^2}\frac{2E}{(k_2 - k_1)} \geq 0, \qquad G_{uu} = \frac{\partial^2 k_2}{\partial u^2}\frac{2G}{(k_1 - k_2)} \geq 0.$$

Substituting into our expression for the Gaussian curvature, we see that $K(p) \leq 0$. ■

To finish the proof of Liebmann's theorem, we show that if every point on S is an umbilic point, then S is a sphere. Suppose $x\colon (U \subset \mathbb{R}^2) \to S$ is a coordinate chart with $p \in x(U)$. Since every point is an umbilic point, every curve $\alpha(t)$ on S is a line of curvature. For lines of curvature we can apply Rodrigues's formula (Exercise 8.12), which gives, for some differentiable function $\lambda(t)$,

$$\frac{d}{dt}(N(\alpha(t))) = \lambda(\alpha(t))\alpha'(t).$$

In particular, for the coordinate curves we find

$$\frac{d}{du}(N(x(u,0))) = \lambda(u,0)x_u, \quad \frac{d}{dv}(N(x(0,v))) = \lambda(0,v)x_v,$$

that is, $N_u = \lambda x_u$ and $N_v = \lambda x_v$. Taking a second derivative we get the expresssions $N_{uv} = \lambda_v x_u + \lambda x_{uv}$ and $N_{vu} = \lambda_u x_v + \lambda x_{vu}$. Since $x_{uv} = x_{vu}$ and $N_{uv} = N_{vu}$ we obtain

$$-\lambda_u x_v + \lambda_v x_u = \mathbf{0}.$$

Since x_u and x_v are linearly independent, $\lambda_u = \lambda_v = 0$ and so λ is a constant. Integrating Rodrigues's formula we get, for any curve $\alpha(t)$ on S,

$$\alpha(t) = \mathbf{z} + \frac{1}{\lambda}N(\alpha(t)),$$

and this equation characterizes curves on a sphere of radius $1/\lambda$. By connectedness, all of S lies on a sphere. Since S is without boundary, S is a sphere and we have proved Liebmann's Theorem. ■

Surfaces with $K = -1/R^2$

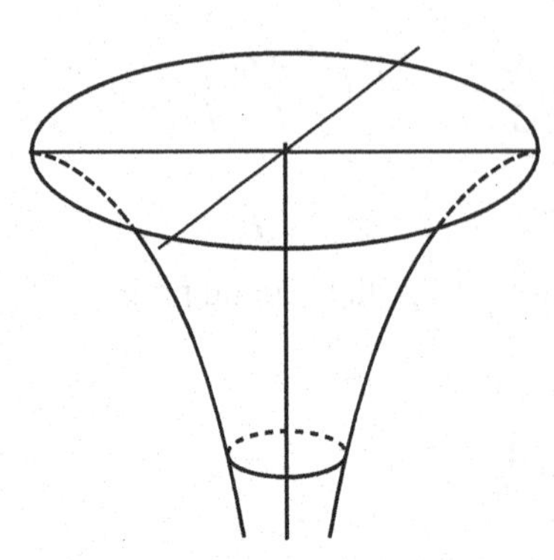

The pseudosphere

Minding's Theorem tells us that any model of a surface of constant negative curvature is locally isometric to any other surface of the same curvature, so we need only one model to study the geometry of such a surface near a point. For convenience, fix $R = 1$. In chapter 8 we considered the surface of revolution generated by the tractrix:

$$\hat{\Theta}(t) = (\cos(t), \ln|\sec(t) + \tan(t)| - \sin(t)).$$

This surface was called the **pseudosphere** by Beltrami (Coddington 1905). and it has Gaussian curvature $K \equiv -1$.

The differential equation that determines geodesic polar coordinates on a surface of constant negative curvature is $\frac{\partial^2 f}{\partial r^2} - \frac{1}{R^2} f = 0$. This equation has a general solution

$$f(r,\theta) = A(\theta)\cosh(r/R) + B(\theta)\sinh(r/R).$$

The initial conditions imply that:

$$f(r,\theta) = \sqrt{G(r,\theta)} = R\sinh(r/R),$$

and for $R = 1$ the line element becomes $ds^2 = dr^2 + \sinh^2(r)d\theta^2$. Thus, without constructing geodesic polar coordinates on the surface of revolution of the tractrix, we have obtained the line element that would result.

The local geometry of this surface will be seen to be non-Euclidean. Trigonometry on the surface is hyperbolic trigonometry, that is, the trigonometry of a sphere of radius $\sqrt{-1}$. It is also like the sphere of radius -1 because its surface area is 4π.

Is this the surface that we have been seeking as a model of non-Euclidean geometry? In fact, the pseudosphere falls short of our goal: *it is not complete*. Above the xy-plane we could continue the pseudosphere by reflection, but on the circle of points where $z = 0$ the surface is singular. Geodesics starting near this "equator" are unable to continue past it smoothly and so the surface is incomplete.

To correct this failure of the pseudosphere, we look for an example of a complete surface of constant negative curvature in $\mathbb{R}^3$. However, we arrive at a surprise ending discovered by Hilbert.

Suppose there is a surface $S \subset \mathbb{R}^3$ of constant negative Gaussian curvature, $K \equiv -C^2$. By Corollary 9.11, for each point p in S, there is an open set containing p with asymptotic lines as coordinate curves. We first show that these coordinates have some even nicer properties.

Proposition 12.8. *If p is a point in a surface $S \subset \mathbb{R}^3$ of constant Gaussian curvature $K \equiv -C^2$, then there is an open set of S containing p with all coordinate curves unit speed and asymptotic.*

PROOF: We apply the Mainardi-Codazzi equations. Suppose a coordinate chart $x\colon (-\epsilon,\epsilon) \times (-\eta,\eta) \to S$ with $p = x(0,0)$ has asymptotic lines as coordinate curves. At the end of chapter 9 we showed that:

$$f_u = \frac{f}{EG-F^2}\left(\frac{1}{2}(EG-F^2)_u + FE_v - EG_u\right),$$
$$f_v = \frac{f}{EG-F^2}\left(\frac{1}{2}(EG-F^2)_v + FG_u - GE_v\right).$$

Since $e = 0 = g$, the value of the Gaussian curvature is given by:

$$K = \frac{-f^2}{EG-F^2} = -C^2.$$

It follows that $f^2 = C^2(EG-F^2)$, and so $2f_uf = C^2(EG-F^2)_u$. Substituting the expression for f_u given above we get:

$$\begin{aligned}(EG-F^2)_u &= \frac{f^2}{C^2(EG-F^2)}[(EG-F^2)_u + 2FE_v - 2EG_u]\\ &= (EG-F^2)_u + 2FE_v - 2EG_u,\end{aligned}$$

which implies that $FE_v = EG_u$. Similarly, we obtain $FG_u = GE_v$ from the other Mainardi–Codazzi Equation. Multiplying each equation by the appropriate function we get:

$$F^2E_v = EFG_u = EGE_v \quad \text{and} \quad F^2G_u = FGE_v = EGG_u,$$

from which it follows that $(EG-F^2)E_v = 0 = (EG-F^2)G_u$. Since the surface is regular, $EG-F^2 \neq 0$ and so $E_v = 0 = G_u$. It follows that $E = E(u)$ and $G = G(v)$.

We make the change of coordinates given by:

$$u \mapsto \int_0^u E(t)^{1/2}dt \quad \text{and} \quad v \mapsto \int_0^v G(\tau)^{1/2}d\tau.$$

These are just the length functions on the coordinate curves. The inverse of this coordinate change composed with x has unit speed coordinate curves in asymptotic directions. ∎

An immediate consequence of the proposition is an open set in S containing the point p with line element given by:

$$ds^2 = du^2 + 2Fdudv + dv^2.$$

Any such system of coordinates is called a **Tchebychev net**. Such a parametrization was studied by P.L. TCHEBYCHEV (1821–94) in a paper of 1878 (Tchebychev). If $x\colon (a,b)\times(c,d)\to S$ is a Tchebychev net, one can view the domain as a piece of cloth with fibers parallel to the u- and v-axes, then x puts the cloth onto a patch of S without stretching the fibers. In a coordinate chart given by a Tchebychev net the Gaussian curvature takes on a simple form.

Corollary 12.9. *Let S be a surface in $\mathbb{R}^3$ and $x\colon (a,b)\times(c,d)\to S$ a Tchebychev net on S. Let $\omega(u,v)$ denote the angle between the coordinate curves x_u and x_v at (u,v). Then the Gaussian curvature satisfies $K=\dfrac{-1}{\sin\omega}\dfrac{\partial\omega}{\partial u\partial v}$.*

PROOF: First notice that $F=\cos\omega$ by the basic properties of the dot product and the unit speed coordinate curves. Since $E=G=1$, we have $E_u=E_v=0=G_u=G_v$, and $EG-F^2=1-\cos^2\omega=\sin^2\omega$. We apply the formula for curvature given in the proof of *Theorema Egregium* (chapter 9):

$$\begin{aligned}
K&=\frac{1}{(EG-F^2)^2}\det\begin{pmatrix}F_{uv}&0&F_u\\F_v&1&F\\0&F&1\end{pmatrix}\\
&=\frac{1}{(EG-F^2)^2}[F_{uv}(1-F^2)+FF_uF_v]\\
&=\frac{1}{\sin^4\omega}\left[(1-\cos^2\omega)\frac{\partial^2\cos\omega}{\partial u\partial v}+\cos\omega\frac{\partial\cos\omega}{\partial u}\frac{\partial\cos\omega}{\partial v}\right]\\
&=\frac{1}{\sin^4\omega}\left[-\sin^3\omega\frac{\partial^2\omega}{\partial u\partial v}-\sin^2\omega\cos\omega\frac{\partial\omega}{\partial u}\frac{\partial\omega}{\partial v}+\sin^2\omega\cos\omega\frac{\partial\omega}{\partial u}\frac{\partial\omega}{\partial v}\right]\\
&=\frac{-1}{\sin\omega}\frac{\partial^2\omega}{\partial u\partial v}.
\end{aligned}$$

■

Suppose S is geodesically complete; we extend a single special coordinate patch over the entire surface by a kind of analytic continuation.

Proposition 12.10. *Let S be a complete surface of constant negative curvature in $\mathbb{R}^3$. Then there is a Tchebychev net $x\colon \mathbb{R}^2\to S$.*

PROOF: Let p_0 be a point in S. Suppose $l\in\mathbb{R}$ is the least upper bound of the set:

$$\{r>0\mid \text{there exists an asymptotic Tchebychev net } x\colon(-r,r)\times(-r,r)\to S \text{ with } x(0,0)=p_0\}.$$

Since the Tchebychev net has coordinate curves parametrized by arc length, the images of x are open squares on S. Consider the union of all of these charts, which gives a coordinate chart $x\colon(-l,l)\times(-l,l)\to S$. Let R denote the image of x. Consider the boundary of the subset R, denoted $\operatorname{bdy}R$. It is closed, bounded, and hence compact, and by the metric completeness of S, $\operatorname{bdy}R\subset S$. Around each point

$q \in \text{bdy} R$ there is an asymptotic Tchebychev net $y_q \colon (-\epsilon_q, \epsilon_q) \times (-\epsilon_q, \epsilon_q) \to S$. The collection $\{y_q\}$ provides an open cover of bdy$R$, and by compactness, only finitely many of the charts, $y_{q_i}, i = 1, \ldots, n$, are needed to cover the boundary. Taking ϵ to be the minimum value of the ϵ_{q_i}, we can extend x to a square of side $l + \epsilon$, which contradicts the least upper bound property of l. Since we know at least one asymptotic Tchebychev net exists around each p_0, we find that the largest such coordinate chart has domain all of $\mathbb{R}^2$. ∎

Hilbert's Theorem

We come to a remarkable theorem of Hilbert (1901), the proof of which utilizes the machinery of covering spaces–a worthy but lengthy detour to our story. We present a proof by Holmgren from a paper that appeared in 1902.

Theorem 12.11. *There is no complete surface in $\mathbb{R}^3$ of constant negative curvature.*

PROOF: Combining Corollary 12.9 and Proposition 12.10, we know that a surface $S \subset \mathbb{R}^3$ of constant negative curvature $K = -C^2$ is equipped with a function $\omega \colon \mathbb{R}^2 \to \mathbb{R}$ giving the angle between x_u and x_v at each point $x(u,v)$, with $0 < \omega(u,v) < \pi$. Furthermore,

$$\frac{\partial^2 \omega}{\partial u \partial v} = C^2 \sin \omega. \qquad (\diamond)$$

Since the sine is positive over these values of ω, the differential equation satisfied by ω implies that $\partial\omega/\partial u$ is an increasing function of v, and so for all u we have

$$\frac{\partial \omega}{\partial u}(u,v) > \frac{\partial \omega}{\partial u}(u,0) \text{ for } v > 0.$$

Integration over the interval $[a,b]$ with respect to u gives the inequality

$$\int_a^b \frac{\partial \omega}{\partial u}(u,v)\,du > \int_a^b \frac{\partial \omega}{\partial u}(u,0)\,du, \text{ that is,}$$

$$\omega(b,v) - \omega(a,v) > \omega(b,0) - \omega(a,0), \text{ for } v > 0 \text{ and } a < b. \qquad (\dagger)$$

Integrating the differential equation $(\diamond)$ with respect to v implies $\dfrac{\partial \omega}{\partial u} \neq 0$ everywhere in the plane, and, in particular, $\dfrac{\partial \omega}{\partial u}(0,0) \neq 0$. The differential equation also enjoys a certain symmetry: If $\omega(u,v)$ is a solution, then so is $\omega(-u,-v)$. Switching solutions if necessary, we can assume that $\dfrac{\partial \omega}{\partial u}(0,0) > 0$.

Choose three values $0 < u_1 < u_2 < u_3$ such that $\dfrac{\partial\omega}{\partial u}(u,0) > 0$ for $0 \leq u \leq u_3$. Denote by ϵ the minimum value $\epsilon = \min\{\omega(u_3,0) - \omega(u_2,0), \omega(u_1,0) - \omega(0,0)\}$. By the inequality (†), for all $v \geq 0$ we have that

$$\omega(u_3,v) - \omega(u_2,v) > \epsilon \quad \text{and} \quad \omega(u_1,v) - \omega(0,v) > \epsilon.$$

Suppose $u_1 \leq u \leq u_2$ and $v \geq 0$. It follows that

$$\omega(u_1,v) \leq \omega(u,v) \leq \omega(u_2,v).$$

Also $\omega(u_1,v) \geq \omega(u_1,v) - \omega(0,v) > \epsilon$. Since $\omega(u_3,v) - \omega(u_2,v) > \epsilon$, we have $\omega(u_2,v) < \omega(u_3,v) - \epsilon < \pi - \epsilon$. Thus

$$\epsilon < \omega(u,v) < \pi - \epsilon.$$

It follows that $\sin\omega(u,v) > \sin\epsilon$ in the strip $[u_1,u_2] \times [0,\infty)$.

With these estimates we integrate $C^2 \sin\omega(u,v)$ over a rectangle $[u_1,u_2] \times [0,T]$:

$$\begin{aligned}\int_0^T \int_{u_1}^{u_2} C^2 \sin\omega(u,v)\,du\,dv &= \int_0^T \int_{u_1}^{u_2} \frac{\partial^2\omega}{\partial u \partial v}\,du\,dv \\ &= \omega(u_2,T) - \omega(u_1,T) - \omega(u_2,0) + \omega(u_1,0).\end{aligned}$$

Rearranging terms we get:

$$\begin{aligned}\pi > \omega(u_2,T) - \omega(u_1,T) &= \omega(u_2,0) - \omega(u_1,0) + \int_0^T \int_{u_1}^{u_2} C^2 \sin\omega(u,v)du dv \\ &\geq \omega(u_2,0) - \omega(u_1,0) + C^2 T(u_2 - u_1)\sin\epsilon.\end{aligned}$$

However, this cannot be true for large T. Therefore, there cannot be a function ω defined on all of $\mathbb{R}^2$ and satisfying the differential equation ($\diamond$). ■

This result puts the search for a model of non-Euclidean geometry into perspective. Our normal powers of visualization are restricted by the experience of our natural "space," $\mathbb{R}^3$. It was not for lack of effort that a model of non-Euclidean geometry was not discovered–it simply cannot be constructed in $\mathbb{R}^3$, except in a local manner.

One of the tools used in Hilbert's proof of Theorem 12.11 is a consequence proved in 1879 of the existence of an asymptotic Tchebychev net.

Proposition 12.12 (Hazzidakis 1879). *If S is a surface of constant negative curvature and $x\colon (a,b) \times (c,d) \to S$ is an asymptotic Tchebychev net, then any quadrilateral formed by coordinate curves in the image of x has area satisfying:*

$$\text{area} = -\frac{1}{K}\left(\sum\nolimits_{i=1}^{4} \alpha_i - 2\pi\right) < -\frac{2\pi}{K},$$

where the α_i denote the interior angles of the quadrilateral.

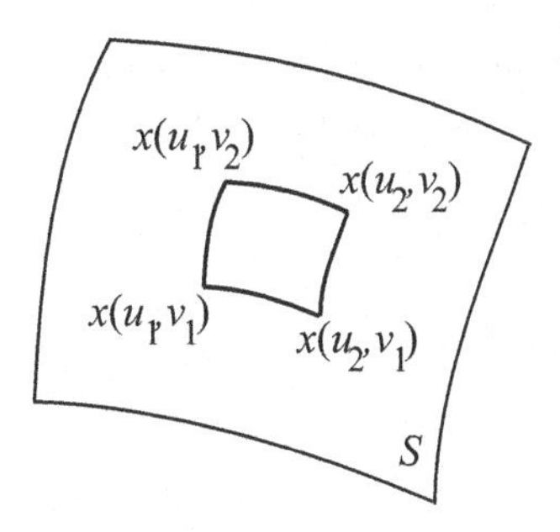

PROOF: Let the curvature be given by $K = -C^2$. A quadrilateral is determined by sides $x(u_1,v_1)$ to $x(u_2,v_1)$, $x(u_2,v_1)$ to $x(u_2,v_2)$, $x(u_2,v_2)$ to $x(u_1,v_2)$, and $x(u_1,v_2)$ to $x(u_1,v_1)$. The formula is proved by considering the Gaussian curvature on such a chart, $K = -\omega_{uv}/\sin\omega$, or equivalently, $C^2\sin\omega = \omega_{uv}$. Since we have a Tchebychev net, $E = G = 1$ and $F = \cos\omega$. It follows that $dA = \sqrt{EG-F^2}\,du\,dv = \sin\omega\,du\,dv$.

$$\begin{aligned}
C^2\text{area} &= \int_{v_1}^{v_2}\int_{u_1}^{u_2} C^2 dA = \int_{v_1}^{v_2}\int_{u_1}^{u_2} C^2\sin\omega\,du\,dv \\
&= \int_{v_1}^{v_2}\int_{u_1}^{u_2} \omega_{uv}\,du\,dv = \omega(u_2,v_2)-\omega(u_2,v_1)-\omega(u_1,v_2)+\omega(u_1,v_1) \\
&= \alpha_3-(\pi-\alpha_4)-(\pi-\alpha_2)+\alpha_1 = \sum_{i=1}^{4}\alpha_i - 2\pi.
\end{aligned}$$

Since the interior angles α_i are all less than π, it follows that $C^2\text{area} < 2\pi$. ∎

The *formula of Hazzidakis* may be applied to obtain another proof of Hilbert's Theorem by showing that the area of a complete surface of constant negative curvature is infinite. The formula puts the upper bound of $2\pi/C^2$ on the area of any quadrilateral of asymptotic sides, but this contradicts unbounded area.

In order to find a model of non-Euclidean geometry we next turn to a generalization of surfaces in $\mathbb{R}^3$.

Exercises

12.1 Suppose that a coodinate chart can be chosen for a portion of a surface such that the coordinate lines u =constant or v =constant are geodesics. Prove that the line element be brought into the form $ds^2 = du^2 + dv^2$, that is, the portion of the surface is developable on the plane near the point.

12.2† Prove the theorem Diguet's (1848), that:

$$K(p) = \lim_{r\to 0^+} \frac{12(\pi r^2 - \text{area}_p(r))}{\pi r^4},$$

where $\text{area}_p(r)$ is the area enclosed by a geodesic circle centered at p of radius r. (*Hint*: Use the formula for area given by Theorem 7.16 with geodesic polar coordinates.)

12.3 Prove that the tangent developable surface $x(s,t) = \alpha(s) + t\alpha'(s)$ defined in chapter 7 is locally isometric to the plane.

Solutions to Exercises marked with a dagger appear in the appendix, p. 337.

12.4 Prove one of the important properties of the exp mapping: Suppose that p and q are such that $q = \exp_p(\mathbf{w})$ with $\|\mathbf{w}\| < \epsilon_p$, and $\exp_p : B_{\epsilon_p}(p) \to S$ is a diffeomorphism onto an open disk in S. Show that the geodesic radius joining p to q is the shortest path in S. (*Hint*: Use the particular form of the metric in geodesic polar coordinates to show that a variation of the geodesic radius has longer length.)

12.5[†] A surface of revolution, $x(u,v) = (\lambda(u)\cos v, \lambda(u)\sin v, \mu(u))$, has constant curvature K if $(\lambda')^2 + (\mu')^2 = 1$ and $K = -\dfrac{\lambda''}{\lambda}$. For $K = 1$, the solutions to $\lambda'' = -\lambda$ take the form:

$$\lambda(u) = r_1 \cos u + r_2 \sin u = r\cos(u+b).$$

If we take $b = 0$, solve for $\mu(u)$ (you should get an elliptic integral). Describe the surfaces that arise if $r < 1$ (spindle-like surfaces) and if $r > 1$, (a bulge or *Wulsttyp*).

12.6 Prove the analog of the Gauss lemma for geodesic parallel coordinates, and obtain the associated line element $ds^2 = dx^2 + G(x,y)\,dy^2$. What form does $G(x,y)$ take on the sphere of radius R?

12.7 Show that the principal curvatures on a surface are differentiable as follows: Let H denote the mean curvature ($H = (k_1 + k_2)/2$) and K the Gaussian curvature ($K = k_1 k_2$) of a regular surface. Show that the principal curvatures k_i are computed by

$$k_i = H \pm \sqrt{H^2 - K}.$$

12.8 Determine the surface area of the pseudosphere. More generally, determine the surface area of the surface of revolution of a tractrix determined by dragging a weight at the end of a fixed length, say a.

PART C

Recapitulation and coda

13

Abstract surfaces

Basically the actual content of the entire argument belongs to a higher area of the general abstract study of quantity, independent of the spatial, whose object is the combinations of quantities that are connected through continuity, an area that is presently still little developed, and in which we also cannot move without a language borrowed from spatial pictures.

C.-F. GAUSS (1849)

The concept "two-dimensional manifold" or "surface" will not be associated with points in three-dimensional space; rather it will be a much more general abstract idea.

HERMANN WEYL (1913)

The development of differential geometry led us to a set of requirements for a model of non-Euclidean geometry–a geodesically complete surface of constant negative curvature. Hilbert's Theorem (Theorem 12.11) proves that there is no example of such a surface that is a regular surface in $\mathbb{R}^3$. To widen the search we turn to a more abstract notion of a surface. Other objects of geometric investigations in the 19th century suggested the need for a broader definition of surface. In 1808, ETIENNE LOUIS MALUS (1775–1812) had proven that if a system of light rays emanating from a point are reflected in an arbitrary surface, then the system of rays will remain orthogonal to the surface of points given by the wave front, the set of points reached at a fixed time from the origin. Such a surface is determined implicitly and its properties derived from the optics, not from an explicit description. WILLIAM ROWAN HAMILTON (1805–65) generalized Malus's Theorem in a context of higher dimensional spaces on which he based his analytical mechanics (Lützen 1995).

In his celebrated *Habilitationsvortrag* (1868), Riemann presented an abstraction of geometry which for surfaces led to new abstract objects on which the methods of differential geometry could be applied. Riemann's geometric development of complex function theory produced surfaces that are one-dimensional complex manifolds–Riemann surfaces–the properties of which were derived from arguments that interwove topology, analysis, and algebra. Separating these key properties in his talk, Riemann introduced a new approach to surfaces and their higher dimensional generalizations, his n-fold extended quantities (*n-fach ausgedehnte Grösse*). We take up the higher dimensional examples in chapter 15 and a translation of Riemann's *Habilitationsvortrag* follows that chapter. The development of Riemann's ideas between 1850 and 1900 is told in clear and abundant detail in the book and paper by Scholz (1980, 1999, respectively).

Surfaces in $\mathbb{R}^3$ as described in chapter 7 are concrete objects; their definition refers to a subset of $\mathbb{R}^3$, and the important geometric tools, such as the first and second fundamental forms, are derived from the Euclidean geometry of $\mathbb{R}^3$. The real workhorse for the intrinsic geometry of surfaces in $\mathbb{R}^3$, however, is the coordinate chart. To generalize the concept of surface we focus on these constructs.

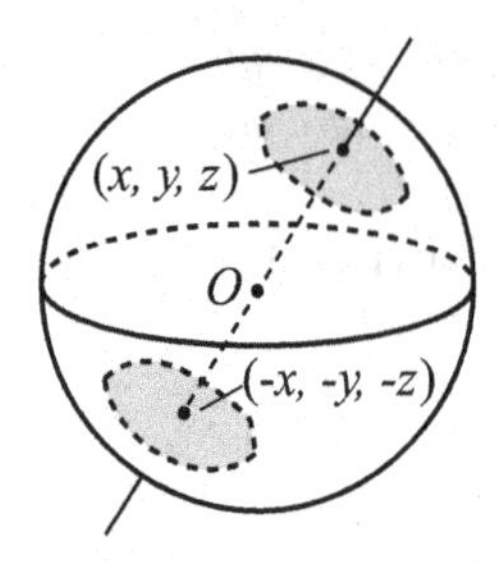

To motivate the definition, let us begin with an example. Let $\mathbb{R}P^2$ denote the set of lines that pass through the origin in $\mathbb{R}^3$. There are algebraic coordinates on $\mathbb{R}P^2$ defined by taking equivalence classes of nonzero triples, where $(x,y,z) \sim (rx,ry,rz)$ whenever $r \neq 0$. If an equivalence class $[x,y,z]$ has $z \neq 0$, then $[x,y,z] = [x/z, y/z, 1]$, and we find that a part of the plane, points (u,v) with u close to x/z and v close to y/z, determine points $[u,v,1] \in \mathbb{R}P^2$ with coordinates around $[x,y,z]$. Similarly, if $y \neq 0$ or $x \neq 0$ we can find planar coordinates around $[x,y,z]$. Thus $\mathbb{R}P^2$ is locally planar.

Alternatively, in a given equivalence class $[x,y,z] \in \mathbb{R}P^2$, there are two representatives satisfying $x^2+y^2+z^2=1$. If we take a coordinate chart on S^2, $x\colon (U \subset \mathbb{R}^2) \to S^2$ for which $x(U) \cap -x(U) = \emptyset$, then this defines a coordinate chart on $\mathbb{R}P^2$, $\bar{x}\colon (U \subset \mathbb{R}^2) \to \mathbb{R}P^2$, given by $\bar{x}(u,v) = [a(u,v), b(u,v), c(u,v)] = [x(u,v)]$. This set of coordinate charts obtains $\mathbb{R}P^2$ as a quotient of S^2 for which we identify pairs of antipodal points. Such a construction determines the open sets on $\mathbb{R}P^2$ as any subset whose inverse image under the quotient is open in S^2. The set $\mathbb{R}P^2$ with the collection of all possible coordinate charts as described is called the *real projective plane*.

Definition 13.1. *A smooth* **abstract surface** *(or smooth* **two-dimensional manifold***) is a set S equipped with a countable collection of one-to-one functions called* **coordinate charts** $\mathcal{A} = \{x_\alpha\colon (U_\alpha \subset \mathbb{R}^2) \to S \mid \alpha \in A\}$ *such that:*

(1) *for all* $\alpha \in A$, U_α *is an open subset of* $\mathbb{R}^2$.
(2) $\bigcup_\alpha x_\alpha(U_\alpha) = S$.
(3) *If* α *and* β *are in A and* $x_\alpha(U_\alpha) \cap x_\beta(U_\beta) = V_{\alpha\beta}$ *is nonempty, then the composite*

$$x_\alpha^{-1} \circ x_\beta \colon x_\beta^{-1}(V_{\alpha\beta}) \to x_\alpha^{-1}(V_{\alpha\beta})$$

is a smooth mapping (the **transition function***) between open subsets of* $\mathbb{R}^2$. *The collection* $\mathcal{A}$ *is called an* **atlas** *of charts on S if it is maximal with respect to this condition, that is, if* $x\colon U \to S$ *is another chart such that* $x_\alpha^{-1} \circ x$ *and* $x^{-1} \circ x_\alpha$ *are smooth for all* $\alpha \in A$, *then x is in* $\mathcal{A}$. *The choice of an atlas* $\mathcal{A}$ *is called a* **differentiable structure** *on S.*

There is a further more technical assumption about the set S with atlas $\mathcal{A}$:

(4) *The collection of subsets determined by the* $x_\alpha(U_\alpha) \subset S$ *for* x_α *in the atlas* $\mathcal{A}$ *is a subbasis for a topology on the set S, which is required to satisfy the Hausdorff condition. That is, if p and q are in S, then there are* α *and* β *such that* $p \in x_\alpha(U_\alpha)$, $q \in x_\beta(U_\beta)$, *and* $x_\alpha(U_\alpha) \cap x_\beta(U_\beta) = \emptyset$. *Furthermore, this topology is required to be second countable, that is, there is a countable collection of charts from the atlas* $\{x_n\} \subset \mathcal{A}$ *such that any open subset of S is a union of open sets taken from* $\{x_n(U_n)\}$.

The reader who is unacquainted with point set topology can find a discussion of the Hausdorff condition, second countability, and subbases for topologies in any good introductory topology book (for example, Munkres 2000). An atlas for a surface may be likened to a cartographic atlas—each coordinate chart is a page in the atlas, and on intersections of charts, there is a differentiable correspondence between corresponding points on the pages. The abstract surface, however, need not float in space like the Earth. It is not required to be a subspace of any particular $\mathbb{R}^n$.

EXAMPLES:

(1) The simplest surface is $\mathbb{R}^2$ with the atlas generated by the identity chart.
(2) All of the regular surfaces in $\mathbb{R}^3$ we have met in earlier chapters are two-dimensional manifolds.
(3) The real projective plane $\mathbb{R}P^2$ with the atlas generated by the small coordinate charts on S^2 is an abstract surface.

The atlas also provides the set S with a collection of special subsets, the open sets, that make it possible to define continuous functions to and from the surface. In fact, the charts provide the surface with a structure supporting the calculus.

Definition 13.2. *If S is a surface and* $f\colon S \to \mathbb{R}^n$ *a function, then we say that f is* **continuous** *if whenever* $V \subset \mathbb{R}^n$ *is an open set, then* $f^{-1}(V) \subset S$ *is an open set. We say that f is* **differentiable at a point** *p if, for any coordinate chart* $x_\alpha\colon (U_\alpha \subset \mathbb{R}^2) \to S$ *with* $p \in x_\alpha(U_\alpha)$, *the composite* $f \circ x_\alpha\colon U_\alpha \to \mathbb{R}^n$ *is differentiable at p. We say that f is* **differentiable** *if it is differentiable at all* $p \in S$. *Similarly a function* $g\colon \mathbb{R}^m \to S$ *is differentiable if composites with the inverses of coordinate charts* $x_\alpha^{-1} \circ g\colon \mathbb{R}^m \to U_\alpha$ *are differentiable.*

Two particularly important collections of differentiable functions are the class of smooth curves, $\lambda\colon (-\epsilon, \epsilon) \to S$, and the class of smooth, real-valued functions on S, $f\colon S \to \mathbb{R}$. We denote the set of all smooth, real-valued functions on S by $C^\infty(S)$. More locally, if $p \in S$, then let $C^\infty(p)$ denote the set of smooth, real-valued functions defined on some open set containing p in S. Both $C^\infty(S)$ and $C^\infty(p)$ are vector spaces with a multiplication. If $f, g \in C^\infty(p)$, then $f\colon U \to \mathbb{R}$ and $g\colon V \to \mathbb{R}$

and $p \in U \cap V$, also an open set. Let $q \in U \cap V$ and $r \in \mathbb{R}$, then:

$$(f+g)(q) = f(q) + g(q), \quad (rf)(q) = r(f(q)), \quad (fg)(q) = f(q)g(q).$$

Such a structure, a ring that is also a vector space, is called an *algebra* over $\mathbb{R}$; see Jacobson (2009) for more about the algebraic theory of algebras.

To define the length of a curve, the angle between curves, and the area of figures on a surface in $\mathbb{R}^3$, we used tangent vectors to curves and functions on the surface defined in terms of the such vectors. Vectors tangent to a surface are easily defined using the embedding of the surface in $\mathbb{R}^3$. For an abstract surface, we need to turn to a different interpretation of tangent vector, namely, the "direction" in which we take the derivative of a function on the surface.

Definition 13.3. *Given a smooth curve $\lambda\colon (-\epsilon,\epsilon) \to S$ through a point $p = \lambda(0)$ in S, define the* **tangent vector** *to $\lambda(t)$ at $t = 0$ as the linear mapping $\lambda'(0)\colon C^\infty(p) \to \mathbb{R}$, defined by*:

$$\lambda'(0)(f) = \frac{d}{dt}(f \circ \lambda(t))\bigg|_{t=0}.$$

The collection of all linear mappings associated to smooth curves through p is denoted by $T_p(S)$, the **tangent space** *of S at p.*

It is difficult to make geometric sense of such a definition for tangent vectors and the tangent space. In order to make some connection with our intuition, we prove the following:

Proposition 13.4. *For an abstract surface S, the tangent space at p $T_p(S)$ is a two-dimensional vector space. If X is a tangent vector, then $X(fg) = f(p)X(g) + g(p)X(f)$; that is, X satisfies the Leibniz Rule for the product of functions in $C^\infty(p)$.*

PROOF: We first introduce coordinates, prove our results for a single chart, and then show that our results are independent of the choice of chart. Suppose $x\colon (U \subset \mathbb{R}^2) \to S$ is a coordinate chart around $p = x(0,0)$. A curve $\lambda\colon (-\epsilon,\epsilon) \to S$ with $\lambda(0) = p$, passing through the chart, may be written as $\lambda(t) = x(u(t), v(t))$, where $(u(t), v(t)) = x^{-1}(\lambda(t)) \in U$. In this notation, the tangent vector determined by $X = \lambda'(0)$ acts as follows on a smooth function:

$$\begin{aligned} X(f) = \lambda'(0)(f) &= \frac{d}{dt} f(x(u(t), v(t)))\bigg|_{t=0} = \frac{\partial f}{\partial u}\frac{du}{dt}\bigg|_{t=0} + \frac{\partial f}{\partial v}\frac{dv}{dt}\bigg|_{t=0} \\ &= u'(0)\frac{\partial f}{\partial u} + v'(0)\frac{\partial f}{\partial u}. \end{aligned}$$

The analogue of the coordinate tangent vectors x_u and x_v to a surface in $\mathbb{R}^3$ are given by the tangent vectors determined by the coordinate curves $x(u, v_0)$ and $x(u_0, v)$. These are particular examples of smooth curves in S and their tangent vectors satisfy:

$$x(u, v_0)'(f) = \frac{\partial f}{\partial u} \text{ and } x(u_0, v)'(f) = \frac{\partial f}{\partial v}.$$

With this in mind it makes sense to denote these particular tangent vectors by $\dfrac{\partial}{\partial u}$ and $\dfrac{\partial}{\partial v}$.

In the case of a more general curve $\lambda(t)$, we can write:

$$\lambda'(0)(f) = \left[u'(0)\frac{\partial}{\partial u} + v'(0)\frac{\partial}{\partial v}\right](f),$$

and so tangent vectors may be written as linear combinations of the basis tangent vectors $\dfrac{\partial}{\partial u}$ and $\dfrac{\partial}{\partial v}$ at a point p. Furthermore, any linear combination, $a\dfrac{\partial}{\partial u} + b\dfrac{\partial}{\partial v}$ is realized as $\lambda'(0)$ for a curve $\lambda(t) = x(at, bt)$. The sum and scalar products of tangent vectors are clear from this representation and $T_p(S)$ is a two-dimensional vector space.

The Leibniz Rule follows directly from this local description—it is a property of partial derivatives.

Finally, if $y\colon (V \subset \mathbb{R}^2) \to S$ is some other coordinate chart around p, then the transition function $y^{-1} \circ x\colon U \cap x^{-1}(y(V)) \to V \cap y^{-1}(x(U))$ determines a change of basis for $T_p(S)$ as follows: Write $x(u, v)$ and $y(r, s)$ as coordinates. Then the coordinate curves for x become smooth curves in the chart determined by y and so:

$$\frac{\partial}{\partial u} = a\frac{\partial}{\partial r} + b\frac{\partial}{\partial s} \quad \text{and} \quad \frac{\partial}{\partial v} = c\frac{\partial}{\partial r} + d\frac{\partial}{\partial s}.$$

Because $y^{-1} \circ x$ is invertible, this linear transformation $\begin{pmatrix} a & b \\ c & d \end{pmatrix}$ is invertible and gives an isomorphism of $T_p(S)$ to itself. ∎

In fact, the matrix $\begin{pmatrix} a & b \\ c & d \end{pmatrix}$ is none other than the Jacobian of the transition function $y^{-1} \circ x$. This function may be written $y^{-1} \circ x(u, v) = (r(u, v), s(u, v))$ and so, given a smooth function f on S, we have:

$$\frac{\partial(f \circ x)}{\partial u} = \frac{\partial(f \circ y)}{\partial r}\frac{\partial r}{\partial u} + \frac{\partial(f \circ y)}{\partial s}\frac{\partial s}{\partial u}, \qquad \frac{\partial(f \circ x)}{\partial v} = \frac{\partial(f \circ y)}{\partial r}\frac{\partial r}{\partial v} + \frac{\partial(f \circ y)}{\partial s}\frac{\partial s}{\partial v}.$$

We denote this transformation in the notation of the tangent space by:

$$\begin{pmatrix} \dfrac{\partial}{\partial u} \\ \dfrac{\partial}{\partial v} \end{pmatrix} \mapsto \begin{pmatrix} \dfrac{\partial r}{\partial u} & \dfrac{\partial s}{\partial u} \\ \dfrac{\partial r}{\partial v} & \dfrac{\partial s}{\partial v} \end{pmatrix} \begin{pmatrix} \dfrac{\partial}{\partial r} \\ \dfrac{\partial}{\partial s} \end{pmatrix} = J(y^{-1} \circ x) \begin{pmatrix} \dfrac{\partial}{\partial r} \\ \dfrac{\partial}{\partial s} \end{pmatrix}.$$

Here $J(y^{-1} \circ x)$ denotes the Jacobian. Since $y^{-1} \circ x$ is also differentiable, we obtain the inverse of $J(y^{-1} \circ x)$, namely, $J(y^{-1} \circ x)^{-1} = J(x^{-1} \circ y)$.

Given two surfaces S and S', a function $\phi\colon S \to S'$ is **differentiable at a point** $p \in S$ if, for any coordinate charts $x_\alpha\colon (U_\alpha \subset \mathbb{R}^2) \to S$ with $p \in x_\alpha(U_\alpha)$ and $y_\beta\colon (V_\beta \subset \mathbb{R}^2) \to S'$ with $f(p) \in y_\beta(V_\beta)$, the composite $y_\beta^{-1} \circ \phi \circ x_\alpha\colon U_\alpha \to V_\beta$ is a differentiable mapping; that is, $y_\beta^{-1} \circ \phi \circ x_\alpha$ has continuous partial derivatives of all orders. A function $\phi\colon S \to S'$ is **differentiable** if it is differentiable at every point $p \in S$. A function $\phi\colon S \to S'$ is a **diffeomorphism** if ϕ is differentiable, one-to-one, and onto, and has a differentiable inverse function.

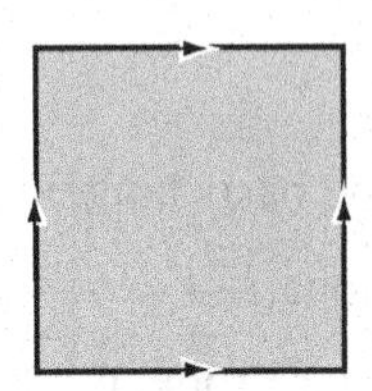

EXAMPLE: *The torus*: Let $I = [0,1]$ denote the closed unit interval in $\mathbb{R}$ and consider the equivalence relation on $I \times I$ given by $(0,v) \sim (1,v)$ and $(u,0) \sim (u,1)$. The quotient space of equivalence classes is a space T^2, homeomorphic to the usual torus, a cartesian product of circles, $S^1 \times S^1$. There is also a representation of the torus as a surface of revolution in $\mathbb{R}^3$ with coordinate chart:

$$x\colon (-\pi,\pi)\times(-\pi,\pi)\to\mathbb{R}^3, \quad x(u,v)=\left(\left(\frac{\cos v}{2}+1\right)\cos u, \left(\frac{\cos v}{2}+1\right)\sin u, \frac{\sin v}{2}\right).$$

The mapping ϕ that takes $(r,s) \in I \times I/{\sim}$ to $x(\pi r - \pi, \pi s - \pi)$ determines a diffeomorphism between the interior of the torus as $I \times I/{\sim}$ and the image of x in $\mathbb{R}^3$, the torus of revolution.

A differentiable function between abstract surfaces determines a differential mapping between tangent spaces.

Definition 13.5. *Given a differentiable mapping $\phi\colon S \to S'$ and a point p in S, the* **differential** *of ϕ at p,*

$$d\phi_p\colon T_p(S) \to T_{\phi(p)}(S'),$$

is given by $d\phi_p(\lambda'(0)) = \frac{d}{dt}(\phi \circ \lambda(t))\Big|_{t=0}$.

We leave it to the reader to prove that $d\phi_p$ is linear, and has the following local form: If $x\colon (U \subset \mathbb{R}^2) \to S$ is a coordinate chart around p, the curve $\lambda(t) = x(u(t), v(t))$ passes through $p = \lambda(0)$, and $y\colon (V \subset \mathbb{R}^2) \to S'$ is a coordinate chart around $\phi(p)$ with coordinates $y(r,s)$, then, in the bases $\left\{\frac{\partial}{\partial u}, \frac{\partial}{\partial v}\right\}$ for T_pS and $\left\{\frac{\partial}{\partial r}, \frac{\partial}{\partial s}\right\}$ for $T_{\phi(p)}(S')$, we have

$$\begin{aligned} d\phi_p(\lambda'(0)) &= u'(0)\left(\frac{\partial r}{\partial u}\frac{\partial}{\partial r} + \frac{\partial s}{\partial u}\frac{\partial}{\partial s}\right) + v'(0)\left(\frac{\partial r}{\partial v}\frac{\partial}{\partial r} + \frac{\partial s}{\partial v}\frac{\partial}{\partial s}\right) \\ &= \begin{pmatrix} \frac{\partial r}{\partial u} & \frac{\partial r}{\partial v} \\ \frac{\partial s}{\partial u} & \frac{\partial s}{\partial v} \end{pmatrix} \begin{pmatrix} u'(0) \\ v'(0) \end{pmatrix} = J(y^{-1} \circ \phi \circ x)(\lambda'(0)). \end{aligned}$$

The chain rule follows directly from this formula; that is, if $\phi\colon S \to S'$ and $\psi\colon S' \to S''$ are differentiable mappings, then $d(\psi \circ \phi)_p = d\psi_{\phi(p)} \circ d\phi_p$.

The Jacobian of the transition function between coordinate charts appears locally when we consider the identity mapping id: $S \to S$: In different coordinate charts around a point, the previous computation shows that $d(\mathrm{id})_p = J(y^{-1} \circ x)$ in local coordinates.

The last ingredient to add to an abstract surface is a line element. For surfaces in $\mathbb{R}^3$, ds^2 was induced by the embedding and the usual dot product for $\mathbb{R}^3$. Since tangent vectors were, in fact, vectors in $\mathbb{R}^3$, it was easy to see that the first fundamental form was well-defined and independent of the choice of coordinate chart. On an abstract surface we will have considerable freedom in choosing a line element, as long as the choice is made coherently with respect to changes of coordinates.

Definition 13.6. *A* **Riemannian metric** *on an abstract surface S is a choice for each $p \in S$ of a positive definite inner product $\langle\ ,\ \rangle_p$ on each tangent plane, $T_p(S)$, such that the choice varies smoothly from point to point.*

In detail, we require $\langle\ ,\ \rangle_p$ to satisfy, for all X, Y, and Z in $T_p(S)$, and $r \in \mathbb{R}$,

(1) $\langle rX + Y, Z\rangle_p = r\langle X, Z\rangle_p + \langle Y, Z\rangle_p$.
(2) $\langle X, Y\rangle_p = \langle Y, X\rangle_p$.
(3) $\langle X, X\rangle_p \geq 0$ and $\langle X, X\rangle_p = 0$ if and only if $X = 0$.

Smoothness is a local property, and so to "vary smoothly from point to point," we choose a coordinate chart $x\colon (U \subset \mathbb{R}^2) \to S$. By the linearity of $\langle\ ,\ \rangle_p$, the inner product is determined in the basis $\left\{\dfrac{\partial}{\partial u}, \dfrac{\partial}{\partial v}\right\}$ by the functions:

$$E = \left\langle \frac{\partial}{\partial u}, \frac{\partial}{\partial u}\right\rangle_p, \quad F = \left\langle \frac{\partial}{\partial u}, \frac{\partial}{\partial v}\right\rangle_p, \quad G = \left\langle \frac{\partial}{\partial v}, \frac{\partial}{\partial v}\right\rangle_p.$$

A Riemannian metric requires that these functions E, F, and G be smooth functions of u and v. Another requirement is that they "match" on the overlap of two charts. This condition is made precise by observing how $\{E, F, G\}$, the component functions of the metric, transform under transition functions.

Lemma 13.7. *If $x\colon (U \subset \mathbb{R}^2) \to S$ and $y\colon (V \subset \mathbb{R}^2) \to S$ are two coordinate charts for S and p is a point in $x(U) \cap y(V)$, then, at p:*

$$\begin{pmatrix} E & F \\ F & G \end{pmatrix} = J(y^{-1} \circ x)^t \begin{pmatrix} \hat{E} & \hat{F} \\ \hat{F} & \hat{G} \end{pmatrix} J(y^{-1} \circ x),$$

where $\{E, F, G\}$ are the component functions of the metric associated to x, $\{\hat{E}, \hat{F}, \hat{G}\}$ are those associated to y, and $J(y^{-1} \circ x)$ is the Jacobian matrix associated to the mapping $y^{-1} \circ x\colon x^{-1}(x(U) \cap y(V)) \to V$.

The proof is left to the reader. The transformational properties of the local description of a metric make it an instance of a "tensor field" on S. In chapter 15 we will consider more fully the analysis and algebra of tensors, which form the foundation for an approach to the local aspects of differential geometry of higher-dimensional manifolds.

Having defined a Riemannian metric, it is not immediate that for each abstract surface one can find a Riemannian metric for it. In particular, as a topological space, the abstract surface has a notion of open set. With a metric, the surface has another notion of open set coming from the metric space properties. We require that these notions coincide. The metric on a surface in $\mathbb{R}^3$ obtains the correct properties by inheriting them from the ambient space. Riemannian metrics for abstract surfaces do always exist, however, one must utilize all of the topological assumptions we made in the definition to prove this. We refer the interested reader to Warner (1983) for details.

EXAMPLES:

(1) *The flat torus*: We can endow the torus $T^2 = I \times I/\sim$ with a Riemannian metric given by the geometry of the unit square as a subset of $\mathbb{R}^2$. In particular, $E = G = 1$ and $F = 0$. This model of the torus will be shown to have constant zero curvature and so it differs significantly from the embedded torus in $\mathbb{R}^3$, which has points of positive and negative curvature.

(2) *Unrolling a surface of revolution*: Suppose we take the graph of a single-variable function $y = f(x) > 0$ over an open interval (a,b). Rotating this curve around the x-axis in $\mathbb{R}^3$ gives us a surface with coordinate charts:

$$\begin{cases} y_1\colon (a,b)\times(0,2\pi)\to S, \\ y_2\colon (a,b)\times(-\pi,\pi)\to S, \end{cases} \qquad y_1(u,v) = y_2(u,v) = (u, f(u)\cos v, f(u)\sin v).$$

The abstract surface we are interested in is given by $S' = (a,b)\times\mathbb{R}$ with a single chart given by the identity. The Riemannian metric is given by taking any curve in S' and composing it with y_1 or y_2 as extended by periodicity to S' and computing E, F, and G accordingly. The resulting component functions are determined by $f(x)$:

$$E(u,v) = 1 + (f'(u))^2, \qquad F(u,v) = 0, \qquad G(u,v) = (f(u))^2.$$

Thus we have "unrolled" the geometry of the surface of revolution onto the stripe $(a,b)\times\mathbb{R}$ in $\mathbb{R}^2$. This construction has the property that it preserves the local geometry of the surface of revolution, but it changes the topology; the abstract surface gotten from unrolling the surface of revolution is simply connected and the surface in $\mathbb{R}^3$ is not.

(3) The basis for our discussion of map projections in chapter 7bis was the longitude-latitude coordinates on the sphere. We can view the subset $(-\pi,\pi)\times(-\pi/2,\pi/2)\subset\mathbb{R}^2$ as an abstract surface modeling the sphere by assigning the Riemannian metric:

$$ds^2 = \cos^2\phi\, d\lambda^2 + d\phi^2.$$

Thus, the abstract surface $(-\pi,\pi)\times(-\pi/2,\pi/2)$, endowed with metric ds^2, is a representation of the unit sphere.

In the presence of an inner product and a notion of tangent vector to a curve, we can define the **arc length** of a curve $\alpha\colon (a,b)\to S$ by:

$$s(t)=\int_{a_0}^{t}\sqrt{\langle\alpha'(\tau),\alpha'(\tau)\rangle_{\alpha(\tau)}}d\tau$$

and the **angle** θ **between curves** at $\alpha_1(t)$ and $\alpha_2(t)$ is determined by:

$$\cos\theta=\frac{\langle\alpha_1'(t_p),\alpha_2'(t_p)\rangle_p}{\sqrt{\langle\alpha_1'(t_p),\alpha_1'(t_p)\rangle_p\cdot\langle\alpha_2'(t_p),\alpha_2'(t_p)\rangle_p}},$$

where $\alpha_1(t_p)=\alpha_2(t_p)=p$.

The notion of area generalizes to abstract surfaces as well. Suppose $R\subset S$ is a region in S lying entirely in a coordinate chart $x\colon (U\subset\mathbb{R}^2)\to S$. Then the area of R is given by the integral:

$$\iint_R dA=\iint_{x^{-1}(R)}\sqrt{EG-F^2}\,du\,dv.$$

The key geometric features of a surface in $\mathbb{R}^3$ discussed in chapters 7 through 12 were deduced from the associated first and second fundamental forms. In the case of an abstract surface, however, there is no normal direction defined on the surface and so the second fundamental form does not exist. The work of Gauss and, in particular, his *Theorema Egregium* free us to work without a normal direction.

Definition 13.8. *Given a coordinate chart $x\colon (U\subset\mathbb{R}^2)\to S$ for an abstract surface S with a Riemannian metric, define the* **Christoffel symbols** *as the functions $\Gamma_{ij}^k\colon (U\subset\mathbb{R}^2)\to\mathbb{R}$ satisfying the systems of equations:*

$$\begin{pmatrix}E & F\\ F & G\end{pmatrix}\begin{pmatrix}\Gamma_{11}^1\\ \Gamma_{11}^2\end{pmatrix}=\begin{pmatrix}\frac{1}{2}E_u\\ F_u-\frac{1}{2}E_v\end{pmatrix},\quad \begin{pmatrix}E & F\\ F & G\end{pmatrix}\begin{pmatrix}\Gamma_{12}^1\\ \Gamma_{12}^2\end{pmatrix}=\begin{pmatrix}\frac{1}{2}E_v\\ \frac{1}{2}G_u\end{pmatrix},$$

$$\begin{pmatrix}E & F\\ F & G\end{pmatrix}\begin{pmatrix}\Gamma_{22}^1\\ \Gamma_{22}^2\end{pmatrix}=\begin{pmatrix}F_v-\frac{1}{2}G_u\\ \frac{1}{2}G_v\end{pmatrix}.$$

The **Gauss-Riemann curvature** *is defined on the image of x as the function:*

$$K=\frac{1}{E}((\Gamma_{11}^2)_v-(\Gamma_{12}^2)_u+\Gamma_{11}^1\Gamma_{12}^2+\Gamma_{11}^2\Gamma_{22}^2-\Gamma_{12}^1\Gamma_{11}^2-\Gamma_{12}^2\Gamma_{12}^2).$$

From our proof of *Theorema Egregium* we could also have defined K as the expression:

$$K = \frac{1}{(EG-F^2)^2}\left[\det\begin{pmatrix} -\frac{1}{2}E_{vv}+F_{uv}-\frac{1}{2}G_{uu} & \frac{1}{2}E_u & F_u-\frac{1}{2}E_v \\ F_v-\frac{1}{2}G_u & E & F \\ \frac{1}{2}G_v & F & G \end{pmatrix} \right.$$
$$\left. -\det\begin{pmatrix} 0 & \frac{1}{2}E_v & \frac{1}{2}G_u \\ \frac{1}{2}E_v & E & G \\ \frac{1}{2}G_u & F & G \end{pmatrix}\right].$$

EXAMPLE: Suppose S is the plane $\mathbb{R}^2$ as an abstract surface. Choose $R > 0$ and endow the plane with the Riemannian metric given by:

$$E(u,v) = \frac{R^2(R^2+v^2)}{(R^2+u^2+v^2)^2},\ F(u,v) = \frac{-R^2uv}{(R^2+u^2+v^2)^2},\ G(u,v) = \frac{R^2(R^2+u^2)}{(R^2+u^2+v^2)^2}.$$

The origin of this choice of metric is told in chapter 14. Computing the Christoffel symbols leads to an algebraic morass that simplifies nicely, giving:

$$\Gamma_{11}^1 = \frac{-2u}{R^2+u^2+v^2},\quad \Gamma_{11}^2 = 0,\quad \Gamma_{12}^1 = \frac{-v}{R^2+u^2+v^2},$$
$$\Gamma_{12}^2 = \frac{-u}{R^2+u^2+v^2},\quad \Gamma_{22}^1 = 0,\quad \Gamma_{22}^2 = \frac{-2v}{R^2+u^2+v^2}.$$

The Gauss-Riemann curvature is calculated from the these functions:

$$\begin{aligned} K &= \frac{1}{E}((\Gamma_{11}^2)_v - (\Gamma_{12}^2)_u + \Gamma_{11}^1\Gamma_{12}^2 + \Gamma_{11}^2\Gamma_{22}^2 - \Gamma_{12}^1\Gamma_{11}^2 - \Gamma_{12}^2\Gamma_{12}^2) \\ &= \frac{(R^2+u^2+v^2)^2}{R^2(R^2+v^2)}\left(0 - \frac{\partial}{\partial u}\left(\frac{-u}{R^2+u^2+v^2}\right) + \frac{2u^2}{(R^2+u^2+v^2)^2}\right. \\ &\quad \left. +0-0-\frac{u^2}{(R^2+u^2+v^2)^2}\right) = \cdots = \frac{1}{R^2}. \end{aligned}$$

Thus the plane with this Riemannian metric has the same constant curvature as a sphere of radius R. Fixing the abstract surface and varying the metric is possible in this case; for example, if we let R go to infinity, the metric converges to $ds^2 = du^2 + dv^2$, the Euclidean metric on the plane with curvature zero.

To study the "lines" on an abstract surface we proceed by determining the curves that locally minimize distance, here given by arc length along the curve. As in chapter 10, this is a problem in the calculus of variations. We observe that the answer is local and so we can argue in a single coordinate chart. The argument carries over from the case of surfaces in $\mathbb{R}^3$ (Theorem 10.7) and we can make the following generalization.

Definition 13.9. *A curve $\gamma\colon(-r,r)\to S$ is a* **geodesic** *if, for each $-r<t<r$, in a coordinate chart around $\gamma(t)$, the following differential equation is satisfied:*

$$\begin{cases} u''+(u')^2\Gamma_{11}^1+2u'v'\Gamma_{12}^1+(v')^2\Gamma_{22}^1=0,\\ v''+(u')^2\Gamma_{11}^2+2u'v'\Gamma_{12}^2+(v')^2\Gamma_{22}^2=0.\end{cases}$$

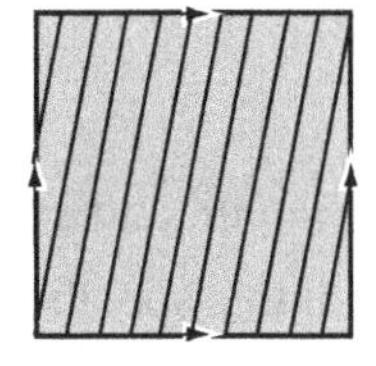

EXAMPLES: Consider the flat torus. In this case, the Christoffel symbols vanish and so the geodesic equations become $u''=0=v''$. In the chart for the torus, the geodesic equations are solved by a straight Euclidean line along the torus, taking the identifications into account. When the line has slope a rational number, notice that the geodesic eventually rejoins itself and is a closed curve. When the slope is irrational, the line has an image dense in the torus, that is, every point in the torus is arbitrarily close to the trajectory of the line.

When we unroll a surface of revolution onto a stripe in the plane, the Christoffel symbols are:

$$\Gamma_{11}^1=\frac{f'(u)f''(u)}{1+(f'(u))^2},\ \Gamma_{12}^2=\frac{f'(u)}{f(u)},\ \Gamma_{22}^1=\frac{-f'(u)f(u)}{1+(f'(u))^2},\ \Gamma_{11}^2=\Gamma_{12}^1=\Gamma_{22}^2=0.$$

The geodesic equations become:

$$\begin{cases} u''+(u')^2\dfrac{f'(u)f''(u)}{1+(f'(u))^2}-(v')^2\dfrac{f'(u)f(u)}{1+(f'(u))^2}=0,\\ v''+2u'v'\dfrac{f'(u)}{f(u)}=0.\end{cases}$$

Vertical lines, $t\mapsto(u_0,t)$, are geodesics whenever u_0 is a critical point of $f(u)$. By parametrizing the horizontal line:

$$t\mapsto(r(t),v_0)\text{ such that }r''=-(r')^2\frac{f'(r)f''(r)}{1+(f'(r))^2},$$

we also obtain a geodesic on this abstract surface.

On the plane with metric:

$$ds^2=\frac{R^2}{(R^2+u^2+v^2)^2}((R^2+v^2)du^2-2uv\,du\,dv+(R^2+u^2)dv^2),$$

the geodesic equations are:

$$\begin{cases} u''-(u')^2\dfrac{2u}{R^2+u^2+v^2}-2u'v'\dfrac{v}{R^2+u^2+v^2}=0\\ v''-2u'v'\dfrac{u}{R^2+u^2+v^2}-(v')^2\dfrac{2v}{R^2+u^2+v^2}=0.\end{cases}$$

Dividing the first by u' and the second by v' and rearranging we get:

$$\frac{u''}{u'} = \frac{2uu' + 2vv'}{R^2 + u^2 + v^2}, \quad \frac{v''}{v'} = \frac{2uu' + 2vv'}{R^2 + u^2 + v^2}.$$

These equations become:

$$\frac{d}{dt}(\ln(u')) = \frac{d}{dt}(\ln(R^2 + u^2 + v^2)) = \frac{d}{dt}(\ln(v')).$$

Thus, $\dfrac{d}{dt}(\ln(v'/u')) = 0$ and so $\ln(v'/u') = L$, a constant. Hence $v' = mu'$ and so $v(t) = mu(t) + b$, the equation of a line. The geodesics in the plane with this specific metric of constant curvature $K = 1/R^2$ have traces that are lines. We explore this abstract surface further in the next chapter.

Because a differential equation is the basis for the definition of a geodesic, we obtain all the familiar local results about them. For example, there is a unique geodesic in each direction from a point. The exponential mapping can be defined as in Chapter 10, and geodesic polar coordinates exist. The arc length provides the structure of a metric space on an abstract surface and the topology induced by this metric space structure coincides with the topology defined by the charts (Hicks, (1971) p. 80). Finally, the Hopf–Rinow Theorem (Theorem 10.16) holds for abstract surfaces.

Two abstract surfaces, each with a Riemannian metric, are considered equivalent if there is an *isometry* between the surfaces.

Definition 13.10. *A one-to-one, onto, differentiable function $\phi\colon S \to S'$ between abstract surfaces, each with a Riemannian metric, is an* **isometry** *if for all $p \in S$, and $V, W \in T_p(S)$, we have $\langle d\phi_p(V), d\phi_p(W)\rangle_{\phi(p)} = \langle V, W\rangle_p$.*

Since an isometry induces an isomorphism of tangent planes at each point, a generalized version of the Inverse Function Theorem for abstract manifolds tells us that an isometry is a diffeomorphism. It follows that we can use the charts from S to make charts on S' and so, at a point, there are charts on each surface such that the isometry sends the associated functions E, F, and G on S to identical functions on S'. It follows immediately that the Christoffel symbols and the Gauss–Riemann curvature are preserved by isometries.

For example, consider the flat torus T^2; is there a surface in $\mathbb{R}^3$ that is isometric to it? Since the flat torus is a compact space on which the curvature is constant, a surface in $\mathbb{R}^3$ that is isometric to the flat torus would be a compact surface with constant curvature. Theorem 12.6 of Liebmann shows that the compact, constant curvature surfaces in $\mathbb{R}^3$ are spheres and so there is no surface in $\mathbb{R}^3$ isometric to the flat torus. However, there is a mapping $\psi\colon T^2 \to \mathbb{R}^4$ given by:

$$(r, s) \mapsto (\cos r, \sin r, \cos s, \sin s).$$

We claim that this is an isometry from T^2 to the image of ψ contained in $S^1 \times S^1$. The coordinate charts on $S^1 \times S^1$ are images of open intervals $(a,b) \times (c,d)$ under the same mapping ψ and so ψ is certainly a diffeomorphism. The metric on $S^1 \times S^1$ is the one that is induced by the dot product on $\mathbb{R}^4$. We check the first of the component functions of the metric:

$$\begin{aligned} E &= \left\langle \frac{\partial}{\partial u}, \frac{\partial}{\partial u} \right\rangle_{u_0,v_0} \\ &= \frac{d}{dt}[(\cos t, \sin t, \cos v_0, \sin v_0) \cdot (\cos t, \sin t, \cos v_0, \sin v_0)] \\ &= (-\sin t, \cos t, 0, 0) \cdot (-\sin t, \cos t, 0, 0) = 1 = \hat{E}. \end{aligned}$$

Similarly, $F = \hat{F}$, and $G = \hat{G}$, and so ψ is an isometry.

This example of an isometry between an abstract surface and a subset of $\mathbb{R}^4$ generalizes the situation of the study of surfaces in $\mathbb{R}^3$. When there is a diffeomorphism between an abstract surface S and a subset of $\mathbb{R}^n$ for some n, then we say that S can be **embedded** in $\mathbb{R}^n$. Through the diffeomorphism S can inherit a Riemannian metric from the dot product on $\mathbb{R}^n$. If S has a Riemannian metric and the diffeomorphism is an isometry, we say that the surface S with the inherited Riemannian metric is **isometrically embedded** in $\mathbb{R}^n$. An embedding identifies the topology of the surface S with a surface embedded in $\mathbb{R}^n$. An isometric embedding identifies the geometry of the surface S with the geometry of a surface in $\mathbb{R}^n$. In particular, the Riemannian structure on S is identified via the isometry with the canonical Euclidean structure on subsets of $\mathbb{R}^n$.

By definition all of the classical surfaces in $\mathbb{R}^3$ of earlier chapters, if taken as abstract surfaces, have isometric embeddings in $\mathbb{R}^3$. The remarks above show that there is no isometric embedding of the flat torus T^2 in $\mathbb{R}^3$. Example 3,

$$((0,2\pi) \times (-\pi/2, \pi/2), ds^2 = \cos^2\phi \, d\lambda^2 + d\phi^2),$$

has an isometric embedding in $\mathbb{R}^3$ given by the unit sphere S^2. Hilbert's Theorem (Theorem 12.11) shows that there is no isometric embedding of a complete, constant negative curvature surface in $\mathbb{R}^3$. Getting a little ahead of ourselves, we mention that Blanuša (1955) exhibited an isometric embedding of a complete, constant negative curvature surface in $\mathbb{R}^6$. It is not known if it is possible to embed such a surface isometrically in $\mathbb{R}^5$. However, Rozendorn (1960) has proved that an immersion (a mapping that is not one-to-one in a nice way) into $\mathbb{R}^5$ is possible.

In the next chapter we reach the end of our search for a model of non-Euclidean geometry by constructing an abstract surface that is complete and has constant negative curvature.

Exercises

13.1 Suppose that $\mathcal{A}$ and $\mathcal{A}'$ are differentiable structures on an abstract surface S. Define the relation on differentiable structures by $\mathcal{A} \sim \mathcal{A}'$ if the union $\mathcal{A} \cup \mathcal{A}'$ defines a differentiable structure. Show that $\sim$ is an equivalence relation. Show that an atlas generated by $\mathcal{A}$ is the largest equivalence class under $\sim$ containing $\mathcal{A}$.

13.2 Let $V_i = \{\mathbf{x} = (x_1, x_2, x_3) \in \mathbb{R}^3 \mid x_i \neq 0\}$ for $i = 1, 2, 3$. Consider the mappings:

$$\phi_1(\mathbf{x}) = \left(\frac{x_2}{x_1}, \frac{x_3}{x_1}\right), \quad \phi_2(\mathbf{x}) = \left(\frac{x_1}{x_2}, \frac{x_3}{x_2}\right), \quad \phi_3(\mathbf{x}) = \left(\frac{x_1}{x_3}, \frac{x_2}{x_3}\right),$$

defined for $\phi_i \colon V_i \to \mathbb{R}^2$. The mappings ϕ_i satisfy the property $\phi_i(x, y, z) = \phi_i(rx, ry, rz)$ for $r \in \mathbb{R}, r \neq 0$, and so they determine mappings defined on $\mathbb{R}P^2 \to \mathbb{R}^2$. The inverses of the ϕ_i take the form:

$$\phi_1^{-1}(u, v) = [1, u, v], \quad \phi_2^{-1}(u, v) = [u, 1, v], \quad \phi_3^{-1}(u, v) = [u, v, 1].$$

Show that these mappings determine a differentiable structure on $\mathbb{R}P^2$.

13.3 Define what it means for an abstract surface to be *orientable*.

13.4 Prove Lemma 13.7.

13.5 Given a differentiable mapping $\phi \colon S \to S'$, prove that $d\phi_p$ is linear. Also, prove the chain rule, that is, if $\phi \colon S \to S'$ and $\psi \colon S' \to S''$ are differentiable mappings, then

$$d(\psi \circ \phi)_p = d\psi_{\phi(p)} \circ d\phi_p.$$

13.6 Show that the flat torus has constant zero Gauss-Riemann curvature.

13.7 Suppose that $\langle\, ,\, \rangle^1$ and $\langle\, ,\, \rangle^2$ are two Riemannian metrics on an abstract surface S. Show that a linear combination with positive coefficients of these inner products is also a Riemannian metric on S, that is, if $a, b > 0$, then $\langle V, W \rangle = a\langle V, W \rangle^1 + b\langle V, W \rangle^2$ is a positive definite inner product.

13.8† Consider the Riemannian metric $ds^2 = du^2 + (f(u, v))^2\, dv^2$ on a portion of the plane. Show that the curves $v = c$, a constant are geodesics on this abstract surface.

13.9 On the upper half-plane $\mathbb{H} = \{(u, v) \in \mathbb{R}^2 \mid$ Riemannian metric $ds^2 = \dfrac{du^2 + dv^2}{v^2}$. Compute the Gauss-Riemann curvature of this abstract surface. On all of $\mathbb{R}^2$ define the Riemannian metric $ds^2 = du^2 + e^{2v}\, dv^2$. Compute the Gauss-Riemann curvature of this abstract surface.

13.10† In $\mathbb{C}^2$ consider the set of all complex lines through the origin. Denote this set by $\mathbb{C}P^1$. Show that $\mathbb{C}P^1$ is an abstract surface that is diffeomorphic to S^2. (Hint: The points on a complex line through the origin take the form $(z, \alpha z)$ for some $\alpha \in \mathbb{C}$, except for the line $(0, z)$.)

Solutions to Exercises marked with a dagger appear in the appendix, p. 338.

14

Modeling the non-Euclidean plane

In recent times the mathematical public has begun to occupy itself with some new concepts which seem to be destined, in the case they prevail. to profoundly change the entire order of classical geometry.

E. BELTRAMI (1868)

If we adopt these definitions, the theorems of Lobachevskiĭ are true, that is to say, that all of the theorems of ordinary geometry apply to these new quantities, except those that are a consequence of the parallel postulate of Euclid.

H. POINCARÉ (1882)

The notion of an abstract surface frees us to seek models of non-Euclidean geometry without the restriction of finding a subset of Euclidean space. A set, not necessarily a subset of some $\mathbb{R}^n$, with coordinate charts and a Riemannian metric determines a geometric surface. With this new freedom we achieve our goal of constructing realizations of the geometry of Lobachevskiĭ, Bolyai, and Gauss and the well-known models of non-Euclidean geometry due to E. BELTRAMI (1835–1906) and to Poincaré (1908).

This chapter contains many computational details like a lot of nineteenth-century mathematics. The foundations for these calculations lie in the previous chapters. It will be the small details that open up new landscapes.

In an 1865 paper, Beltrami posed a natural geometric question: He sought local conditions on a pair of surfaces, S_1 and S_2, that guarantee that there is a local diffeomorphism of $S_1 \to S_2$ for which the geodesics on S_1 are carried to geodesics on S_2. Such a mapping is called a **geodesic mapping**. Beltrami (1865) solved the problem when the target surface is the Euclidean plane. He gave conditions for the existence of a mapping taking geodesics on a surface S to straight lines, the geodesics of the plane.

Theorem 14.1 (Beltrami 1865). *If there is a geodesic mapping from a surface S to the Euclidean plane, then the Gaussian curvature of the surface S is constant.*

PROOF: Suppose $f\colon (W \subset S) \to \mathbb{R}^2$ is a geodesic mapping, a local diffeomorphism defined from an open set $W \subset S$ to the plane. Let $U \subset \mathbb{R}^2$ be an open set lying in the image of f and take $x\colon (U \subset \mathbb{R}^2) \to S$ to be the chart given by $x = f^{-1}$. Since f is a geodesic mapping, straight lines in U go to geodesics in S. Write $ds^2 = Edu^2 + 2Fdudv + Gdv^2$ for the metric associated to x. Suppose $\gamma(t) = (u(t), v(t))$ maps to a geodesic $x \circ \gamma(t) = f^{-1} \circ \gamma(t)$ on S. By assumption $\gamma(t)$ is part of a Euclidean line, that is,

$$au(t) + bv(t) + c = 0,$$

where a and b are not both zero. Since $au' + bv' = 0$ and $au'' + bv'' = 0$, we have

$$\begin{pmatrix} u' & v' \\ u'' & v'' \end{pmatrix} \begin{pmatrix} a \\ b \end{pmatrix} = \begin{pmatrix} 0 \\ 0 \end{pmatrix}, \text{ and so } u'v'' - u''v' = 0.$$

A curve is a geodesic when the geodesic curvature vanishes, that is, when the following differential equation holds:

$$\begin{aligned} 0 &= \det \begin{pmatrix} u' & u'' + (u')^2\Gamma^1_{11} + 2u'v'\Gamma^1_{12} + (v')^2\Gamma^1_{22} \\ v' & v'' + (u')^2\Gamma^2_{11} + 2u'v'\Gamma^2_{12} + (v')^2\Gamma^2_{22} \end{pmatrix} \\ &= \Gamma^2_{11}(u')^3 + (2\Gamma^2_{12} - \Gamma^1_{11})(u')^2v' + (\Gamma^2_{22} - 2\Gamma^1_{12})u'(v')^2 - \Gamma^1_{22}(v')^3. \end{aligned}$$

Since U is an open subset of $\mathbb{R}^2$, there is an open disk of some radius centered at each point in U and entirely contained in U. Thus there are lines of every slope through each point. In particular, there is a geodesic $\gamma_1(t) = (u_1(t), v_1)$ with $u' \neq 0$ and $v' = 0$; another geodesic $\gamma_2(t) = (u_2, v_2(t))$ satisfying $u' = 0$ and $v' \neq 0$. Plugging $\gamma_1(t)$ into the geodesic equation, we find $\Gamma^1_{22} = 0$. From $\gamma_2(t)$ we find $\Gamma^2_{11} = 0$. Using a pair of directions $(u(t), v(t))$ for which $2u(t) - v(t) + c = 0$ and $u_0(t) - v_0(t) + c + 0 = 0$, we can apply the equations with $2u' = v'$ and $u_0' = v_0$, which leads to the relation $\Gamma^2_{22} - 2\Gamma^1_{12} = 0$. Similar choices give $2\Gamma^2_{12} - \Gamma^2_{11} = 0$. Thus the following relations hold:

$$\Gamma^2_{11} = 0 = \Gamma^1_{22}, \quad 2\Gamma^2_{12} = \Gamma^1_{11}, \quad \text{and} \quad 2\Gamma^1_{12} = \Gamma^2_{22}.$$

The Gauss Equations (chapter 9) relate the Christoffel symbols and Gaussian curvature:

$$\begin{aligned} &(\Gamma^1_{11})_v - (\Gamma^1_{12})_u + \Gamma^1_{11}\Gamma^2_{12} + \Gamma^2_{11}\Gamma^1_{22} - \Gamma^1_{12}\Gamma^2_{11} - \Gamma^2_{12}\Gamma^2_{12} = EK, && \text{(a)} \\ &(\Gamma^1_{12})_u - (\Gamma^1_{11})_v + \Gamma^2_{12}\Gamma^1_{12} - \Gamma^2_{11}\Gamma^1_{22} = FK, && \text{(b)} \\ &(\Gamma^1_{22})_u - (\Gamma^1_{12})_v + \Gamma^1_{22}\Gamma^1_{11} + \Gamma^2_{22}\Gamma^1_{12} - \Gamma^1_{12}\Gamma^1_{12} - \Gamma^2_{12}\Gamma^1_{22} = GK, && \text{(c)} \\ &(\Gamma^2_{12})_v - (\Gamma^2_{22})_u + \Gamma^1_{12}\Gamma^2_{12} - \Gamma^1_{22}\Gamma^2_{11} = FK. && \text{(d)} \end{aligned}$$

The relations associated to a geodesic mapping to the plane imply some simplications:

(a) $EK = \Gamma^2_{12}\Gamma^2_{12} - (\Gamma^2_{12})_u$,
(b) $FK = (\Gamma^1_{12})_u - (2\Gamma^2_{12})_v + \Gamma^1_{12}\Gamma^2_{12}$,
(c) $GK = \Gamma^1_{12}\Gamma^1_{12} - (\Gamma^1_{12})_u$,
(d) $FK = (\Gamma^2_{12})_v - (2\Gamma^1_{12})_u + \Gamma^1_{12}\Gamma^2_{12}$,

Subtracting (d) from (b) we get $(\Gamma^1_{12})_u = (\Gamma^2_{12})_v$ and so we can rewrite (b) and (d) as:

(b) $FK = \Gamma^1_{12}\Gamma^2_{12} - (\Gamma^2_{12})_v$,
(d) $FK = \Gamma^1_{12}\Gamma^2_{12} - (\Gamma^1_{12})_u$.

By smoothness we have the equations $(\Gamma^1_{12})_{uv} = (\Gamma^1_{12})_{vu}$ and $(\Gamma^2_{12})_{uv} = (\Gamma^2_{12})_{vu}$. Taking the partial derivative of line (a) with respect to v, and of line (b) with respect to u, we get:

$$\frac{\partial(EK)}{\partial v} = E_v K + EK_v = 2\Gamma^2_{12}(\Gamma^2_{12})_v - (\Gamma^2_{12})_{uv},$$
$$\frac{\partial(FK)}{\partial u} = F_u K + FK_u = \Gamma^1_{12}(\Gamma^2_{12})_u + \Gamma^2_{12}(\Gamma^1_{12})_u - (\Gamma^2_{12})_{vu}.$$

Combining these relations, we find:

$$EK_v - FK_u + K(E_v - F_u) = 2\Gamma^2_{12}(\Gamma^2_{12})_v - \Gamma^1_{12}(\Gamma^2_{12})_u - \Gamma^2_{12}(\Gamma^1_{12})_u.$$

Substituting from (a), (b), and (d) gives:

$$\begin{aligned} EK_v - FK_u &= -K(E_v - F_u) + 2\Gamma^2_{12}(\Gamma^1_{12}\Gamma^2_{12} - FK) \\ &\quad - \Gamma^1_{12}(\Gamma^2_{12}\Gamma^2_{12} - EK) - \Gamma^2_{12}(\Gamma^1_{12}\Gamma^2_{12} - FK) \\ &= -K(E_v - F_u) + K(\Gamma^1_{12}E - \Gamma^2_{12}F). \end{aligned}$$

From the definition of the Christoffel symbols, we have $\Gamma^1_{12}E + \Gamma^2_{12}F = \frac{1}{2}E_v$ and $\Gamma^1_{11}F + \Gamma^2_{11}G = F_u - \frac{1}{2}E_v$. The relations $\Gamma^2_{11} = 0$ and $\Gamma^1_{11} = 2\Gamma^2_{12}$ imply:

$$\begin{aligned} \Gamma^1_{12}E - \Gamma^2_{12}F &= \Gamma^1_{12}E + \Gamma^2_{12}F - 2\Gamma^2_{12}F - \Gamma^2_{11}G \\ &= \Gamma^1_{12}E + \Gamma^2_{12}F - (\Gamma^1_{11}F + \Gamma^2_{11}G) \\ &= \frac{1}{2}E_v - \left(F_u - \frac{1}{2}E_v\right) = E_v - F_u. \end{aligned}$$

Thus $EK_v - FK_u = 0$. Carrying out the same derivation for $\dfrac{\partial(FK)}{\partial v} - \dfrac{\partial(GK)}{\partial u}$, we obtain $FK_v - GK_u = 0$. In matrix form these equations become:

$$\begin{pmatrix} E & F \\ F & G \end{pmatrix}\begin{pmatrix} K_v \\ -K_u \end{pmatrix} = \begin{pmatrix} 0 \\ 0 \end{pmatrix}.$$

Since $EG - F^2 \neq 0$, it follows that $K_u = K_v = 0$, and so K is constant. ∎

In fact, there is a geodesic mapping of each of our local models of surfaces of constant curvature to the plane. The case of constant zero curvature is achieved by the identity mapping on a plane. For the surface of revolution of the tractrix, we leave the details to the reader. In the case of the sphere we have already seen in chapter 7bis that central projection takes great circles to straight lines. Let us describe this mapping more analytically.

Proposition 14.2. *The inverse of central projection of the lower hemisphere of a sphere of radius R centered at the origin to the plane tangent to the South Pole $(0,0,-R)$ has the form $x\colon \mathbb{R}^2 \to S$,*

$$x(u,v) = \frac{R}{\sqrt{R^2+u^2+v^2}}(u,v,-R).$$

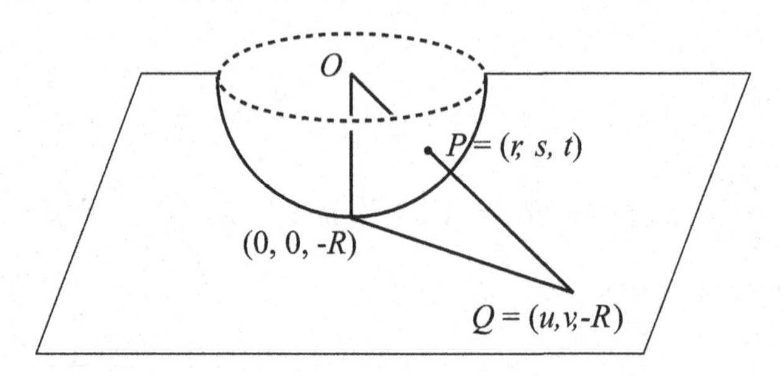

PROOF: Let $Q = (u,v,-R)$ denote a point in the plane tangent to the South Pole of S^2_R and consider the line segment in $\mathbb{R}^3$ joining Q to the origin. This passes through a point P on the sphere. Write the coordinates of the point $x(u,v) = P = (r,s,t)$. The linear dependence of OP and OQ implies:

$$(0,0,0) = OP \times OQ = (-Rs - tv, Rr + tu, rv - su).$$

From this equation it follows that $r = -\frac{tu}{R}$ and $s = -\frac{tv}{R}$. The condition $r^2+s^2+t^2 = R^2$ implies that $t = \frac{-R^2}{\sqrt{R^2+u^2+v^2}}$ and the proposition follows. ∎

From the inverse of central projection we can endow the plane with the geometry of the sphere S^2_R by inducing a Riemannian metric on $\mathbb{R}^2$ via the mapping $x\colon \mathbb{R}^2 \to S^2_R$. Since the sphere is a surface in $\mathbb{R}^3$ we compute directly:

$$x_u = \frac{R}{(R^2+u^2+v^2)^{3/2}}(R^2+v^2, -uv, Ru),$$

$$x_v = \frac{R}{(R^2+u^2+v^2)^{3/2}}(-uv, R^2+u^2, Rv).$$

These coordinate vectors determine the following line element on the plane induced by the geometry of the sphere:

$$ds^2 = R^2 \frac{(R^2+v^2)\,du^2 - 2uv\,du\,dv + (R^2+u^2)\,dv^2}{(R^2+u^2+v^2)^2}.$$

We studied the abstract surface $(\mathbb{R}^2, ds^2)$ in chapter 13 and found that it has constant curvature $K = 1/R^2$ and geodesics given by straight lines. Thus central projection is a geodesic mapping from the (open hemi-)sphere and the plane.

The Beltrami disk

In the calculation of the Gauss–Riemann curvature for $(\mathbb{R}^2, ds^2)$ with ds^2 induced by central projection, every expression involved depended on R^2, but not on R. From this observation, Beltrami (1867) made the leap that brings us to the first model of non-Euclidean geometry. He replaced R^2 with $-R^2$ to obtain the line element:

$$\begin{aligned} ds^2 &= -R^2 \frac{(-R^2+v^2)\,du^2 - 2uv\,du\,dv + (-R^2+u^2)\,dv^2}{(-R^2+u^2+v^2)^2} \\ &= R^2 \frac{(R^2-v^2)\,du^2 + 2uv\,du\,dv + (R^2-u^2)\,dv^2}{(R^2-u^2-v^2)^2}. \end{aligned}$$

This formula determines a Riemannian metric on the abstract surface given by the interior of the disk of radius R in $\mathbb{R}^2$ centered at $(0,0)$. The same computation from chapter 13 shows that the curvature is constant and equal to $-1/R^2$. Furthermore, since we have only changed the constant R^2, the relations $\Gamma_{11}^2 = 0 = \Gamma_{22}^1$, $2\Gamma_{12}^2 = \Gamma_{11}^1$ and $2\Gamma_{12}^1 = \Gamma_{22}^2$ continue to hold. Thus the geodesics on this abstract surface are Euclidean line segments.

For concreteness fix the value of R to be 1. Our abstract surface then is the interior of the unit disk in the plane, which we denote by:

$$\mathbb{D} = \{(u,v) \in \mathbb{R}^2 \mid u^2 + v^2 < 1\}.$$

We decorate the metric with a B (for Beltrami):

$$ds_B^2 = \frac{(1-v^2)\,du^2 + 2uv\,du\,dv + (1-u^2)\,dv^2}{(1-u^2-v^2)^2}.$$

Let $\mathbb{D}_B = (\mathbb{D}, ds_B)$ denote the **Beltrami model**. Let us investigate how the geometry of Lobachevskiĭ, Bolyai, and Gauss is realized in $\mathbb{D}_B$.

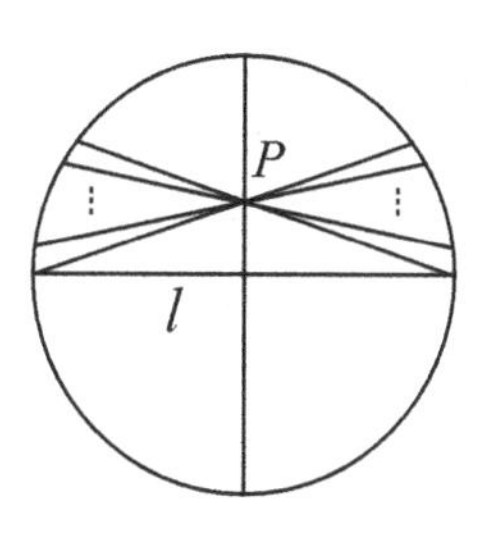

We can see immediately that Euclid's Postulate V fails on $\mathbb{D}_B$. Consider the accompanying diagram, where we have a point P not on a line l and an infinite family of lines (that is, geodesics) through P, each of which does not intersect the given line. We can see parallels through P to l as well: Parallels are the lines joining P to points on the unit circle where l meets it. Points on the unit circle are not in $\mathbb{D}_B$, so these lines do not meet l, but any line through P inside $\angle AP\Omega$ crosses l.

Postulate I holds because $\mathbb{D}$ is a convex subset of $\mathbb{R}^2$. To discuss Postulates II and IV we need to consider congruences. It is convenient to have polar coordinates at our disposal where $\mathbb{D} = \{(r,\theta) \mid 0 \leq r < 1\}$. To convert the metric let $u = r\cos\theta$, $v = r\sin\theta$, giving $du = \cos\theta\,dr - r\sin\theta\,d\theta$ and $dv = \sin\theta\,dr + r\cos\theta\,d\theta$. The

Beltrami metric can be rewritten in the following convenient form:

$$ds_B^2 = \frac{(1-u^2-v^2)(du^2+dv^2)+(u\,du+v\,dv)^2}{(1-u^2-v^2)^2}$$

$$= \frac{(1-r^2)(dr^2+r^2d\theta^2)+r^2dr^2}{(1-r^2)^2} = \frac{dr^2}{(1-r^2)^2}+\frac{r^2d\theta^2}{1-r^2}.$$

In this form, rotations, $r \mapsto r$, $\theta \mapsto \theta+\theta_0$, are clearly isometries, and reflection across the u-axis, $r \mapsto r$, $\theta \mapsto -\theta$, is an isometry. Reflections across other lines may be determined using their synthetic definition. First we need to construct the pencil of perpendiculars to a given line, that is, a geodesic in $\mathbb{D}_B$. If l is the line, we can rotate it so that it is perpendicular to the u-axis, passing through the point $P=(r_0,0)$. Through the points where l meets the boundary circle, construct the tangents to the circle that meet on the u-axis at a point Pol(l), the *pole* of the line l.

Proposition 14.3. *If A is any point on the line l and m is the line determined by the intersection of the line joining A and* Pol(l) *in $\mathbb{D}$, then m is perpendicular to l at A in $\mathbb{D}_B$.*

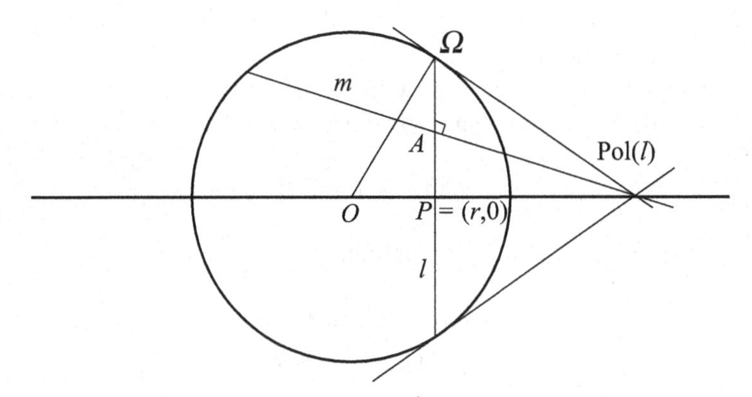

PROOF: Suppose $A=(r_0,b)$. We can parametrize the line l as $(u_0(t),v_0(t)) = (r_0,b+t)$ with $(u_0',v_0')=(0,1)$. The pole associated to l satisfies $\cos\angle\Omega O\mathrm{Pol}(l) = r_0 = 1/O\mathrm{Pol}(l)$, since $\angle O\Omega\mathrm{Pol}(l)$ is a right angle, and so the pole has coordinates $\mathrm{Pol}(l)=(1/r_0,0)$. We can parametrize the line m through A and Pol(l) with $(u',v')= \left(1,\dfrac{-r_0b}{1-r_0^2}\right)$, because $\dfrac{-r_0b}{1-r_0^2}$ is the slope of m. The cosine of the angle in $\mathbb{D}_B$ between l and m has its numerator given, up to a nonzero factor, by:

$$[u_0',v_0']\begin{bmatrix}E & F\\ F & G\end{bmatrix}\begin{bmatrix}u'\\ v'\end{bmatrix} = [0,1]\begin{bmatrix}1-b^2 & r_0b\\ r_0b & 1-r_0^2\end{bmatrix}\begin{bmatrix}1\\ \dfrac{-r_0b}{1-r_0^2}\end{bmatrix} = r_0b-r_0b=0.$$

Hence m is perpendicular to l. ■

After constructing perpendiculars to a given line l, we can define reflection across l: A point Q not on l lies on a line m perpendicular to l meeting l at A.

The point of reflection Q' lies on the other side of l from Q, on m, and satisfies $QA \cong AQ'$. We choose not to define reflection across l analytically, and so we will have to assume that the definition of reflection is an isometry. Later, when we consider other isometric models for the hyperbolic plane, it will be easier to give an analytic expression for reflection that can then be seen to be an isometry.

We can apply features of non-Euclidean geometry to make a synthetic construction of reflection, learned from (Greenberg) and using a property of parallels and the angle of parallelism to find Q'.

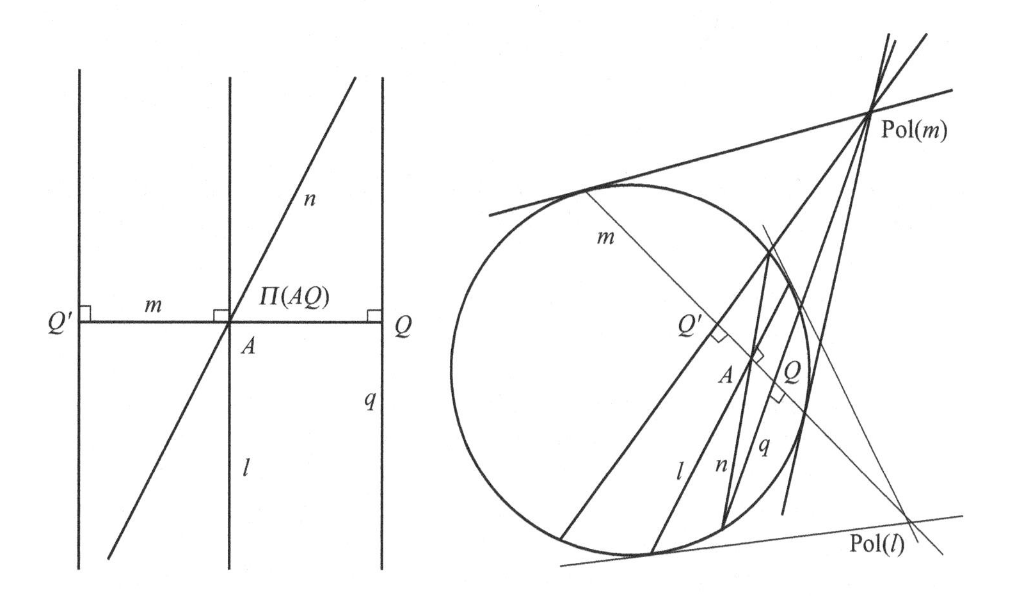

The construction on the left is pictured in the non-Euclidean plane and in the Beltrami plane on the right, where parallels are constructible from lines that meet on the boundary circle, and perpendiculars are constructed using poles. We erect a perpendicular m to l joining Q to A on l, and a perpendicular q to m through Q. There is a line n passing through A that is parallel to q. On the other side of l, there is a line parallel to n and perpendicular to m. It passes through the point Q' on m and since the angle of parallelism for AQ' is congruent to $\Pi(AQ)$, $AQ \cong AQ'$. Hence Q' is the reflection of Q across l. For the Beltrami disk the parallels and perpendiculars to a line can be constructed using points not in $\mathbb{D}$, in this case, the line joining the pole of m and the intersection of n with the boundary circle on the side of l opposite Q. We get the perpendicular to m passing through Q' on m.

Assuming that reflection is an isometry, we can introduce a useful notion that makes arguments in the model more convenient.

Proposition 14.4. *Let P be a point in $\mathbb{D}$ and l a line in the Beltrami disk passing through P. Then there is an isometry taking P to $O = (0,0)$ and l to a vertical line.*

PROOF: First rotate $\mathbb{D}$ to move P to P' on the u-axis. Let m be the line perpendicular to the u-axis through the midpoint between O and P'. Reflect across m which exchanges O and P' denoted P'' and O', respectively. The line l through P moved to l' through P' and finally l'' through $P'' = O$. Rotate $\mathbb{D}$ again to make l vertical. ■

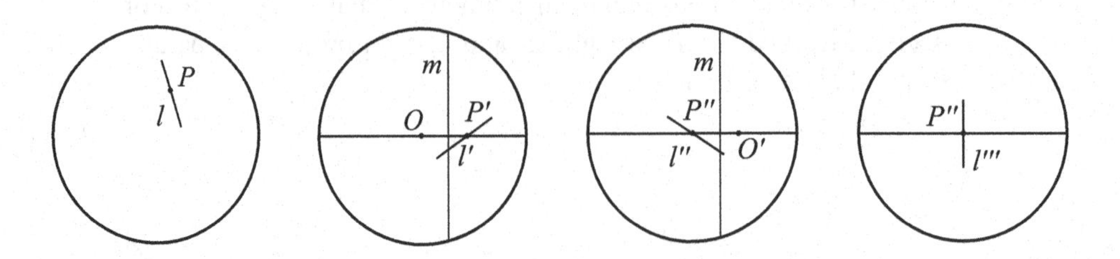

Corollary 14.5. *Given two points P and Q and a line l through P and a line m through Q, there is an isometry of the Beltrami disk taking P to Q and l to m.*

PROOF: First send P to O and l to the v-axis. Then reverse the isometry that takes Q to O and m to the v-axis. ■

There are many simplifications possible at the center of the Beltrami disk that make it a convenient point for computations and constructions: The u-axis and v-axis are given by $u = 0$ or $v = 0$ and so the middle term F of the metric vanishes. It follows that the geodesics $u \mapsto (u, v_0)$ and $v \mapsto (u_0, v)$ are perpendicular to the v-axis or u-axis, respectively. Also at O the metric takes the form $du^2 + dv^2$, and angles with vertex O are the same in $\mathbb{D}_B$ as they are in the Euclidean plane. With the corollary, we can make a construction at the point O and know that there is an isometry that moves the construction to any other point.

To discuss distance from O and circles centered at O we use polar coordinates on $\mathbb{D}$ to simplify computations. A line through the origin in $\mathbb{D}_B$ has polar equation $\theta = \theta_0$, a constant, and may be parametrized by $\alpha(t) = (t, \theta_0)$ for $0 \leq t \leq r_0$. If we write $P = (r_0 \cos\theta_0, r_0 \sin\theta_0)$, then the distance in $\mathbb{D}_B$, denoted by $d_B(O,P)$, is given by:

$$d_B(O,P) = \int_\alpha ds = \int_0^{r_0} \frac{dt}{1-t^2} = \frac{1}{2}\ln\left(\frac{1+r_0}{1-r_0}\right).$$

Notice that as r_0 approaches 1 the distance goes to infinity. Thus, lines through the origin (and hence all lines) are infinite in length. A Cauchy sequence of points in $(\mathbb{D}, ds_B)$, thought of as a metric space, is known to lie in a bounded closed disk of Euclidean radius $r_0 < 1$ centered at O. Since this disk is a compact set topologically, the Cauchy sequence converges to a point in the disk, and so as a metric space $(\mathbb{D}, ds_B)$ is complete. By the Hopf–Rinow Theorem, the abstract surface $(\mathbb{D}, ds_B)$ is geodesically complete, and hence a model of non-Euclidean geometry.

Solving for r_0 in terms of $d_B(O,P) = \rho$ we find $\rho = \frac{1}{2}\ln\left(\frac{1+r_0}{1-r_0}\right)$ implies $\frac{1+r_0}{1-r_0} = e^{2\rho}$, which gives $1+r_0 = e^{2\rho} - r_0e^{2\rho}$ and $r_0 = \frac{e^{2\rho}-1}{e^{2\rho}+1} = \frac{e^{\rho}-e^{-\rho}}{e^{\rho}+e^{-\rho}} = \tanh\rho$.

When we fix $r = \tanh(\rho)$ and vary θ we get a circle of radius ρ in $\mathbb{D}_B$. Its circumference is given by the integral:

$$\text{circum}(\rho) = \int_0^{2\pi} \frac{r\,d\theta}{\sqrt{1-r^2}} = \frac{2\pi r}{\sqrt{1-r^2}} = \frac{2\pi\tanh\rho}{\sqrt{1-\tanh^2\rho}} = 2\pi\sinh\rho.$$

This establishes Gauss's formula (Theorem 4.32 with $k = 1$).

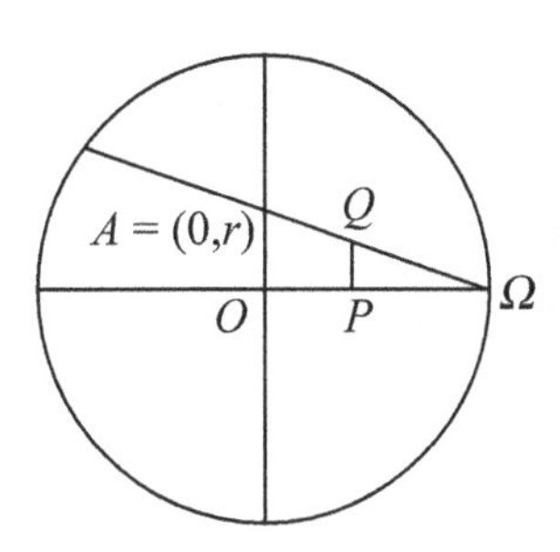

Saccheri (Theorem 3.13) established that non-Euclidean parallels are asymptotic. We have seen how points moving toward the boundary circle are at an unbounded distance from the center. To see how Saccheri's observation works in the Beltrami disk, consider the point $A = (0, r_0)$ and the parallel to the u-axis through A given by the line segment $A\Omega$ with $\Omega = (1,0)$. Choose a point $P = (a,0)$ on the u-axis and consider the line segment joining P to the line $A\Omega$, meeting at the point $Q = (a, r_0 - r_0a)$. The parametrization of this segment given by $C(t) = (a, r_0(1-a)t)$, for $0 \leq t \leq 1$, allows us to compute the length of PQ. Since $u' = 0$ and $v' = r_0(1-a)$ for the curve C, the arc length of PQ is given by the integral:

$$\int_0^1 \sqrt{G(C(t))(v')^2}\,dt = \int_0^1 r_0(1-a)\sqrt{\frac{1-a^2}{(1-a^2-r_0^2(1-a)^2t^2)^2}}\,dt$$
$$= r_0\sqrt{1-a^2}\int_0^1 \frac{dt}{(1+a)-r_0^2(1-a)t^2} = \frac{1}{2}\ln\frac{\sqrt{1+a}+r_0\sqrt{1-a}}{\sqrt{1+a}-r_0\sqrt{1-a}}.$$

As a goes to 1, the limit of this expression is zero, that is, the length of PQ goes to zero and parallels are asymptotic.

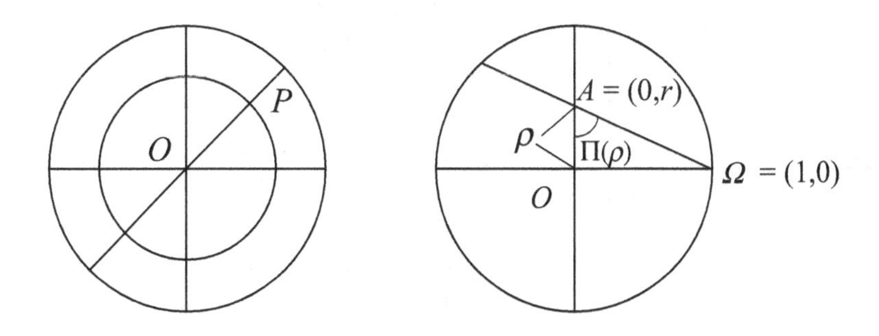

We next prove the Lobachevskiĭ–Bolyai Theorem (Theorem 4.29) on the angle of parallelism. Choose O as the point of intersection of the perpendicular geodesics given by the u- and v-axes. Let $A = (0, r_0)$ be a point a distance $\rho = \operatorname{arctanh}(r_0)$ from the origin. The parallel through A to the right of the v-axis passes through the point $\Omega = (1,0)$ on the limit circle, and can be parametrized by:

$$C_1(t) = (u_1(t), v_1(t)) = (t, r_0 - r_0 t), \text{ for } 0 \leq t \leq 1.$$

This gives $u_1' = 1$ and $v_1' = -r_0$. The line from A to the origin along the v-axis is parametrized by $C_2(t) = (u_2(t), v_2(t)) = (0, r_0 - t)$ with $0 \leq t \leq r_0$ for which $u_2' = 0$ and $v_2' = -1$. We put these data into the Beltrami metric to compute the cosine of the angle between the curves C_1 and C_2 at A, that is, the angle of parallelism:

$$\begin{aligned}\cos \Pi(\rho) &= \frac{E(0,r_0)u_1'u_2' + F(0,r_0)(u_1'v_2' + u_2'v_1') + G(0,r_0)v_1'v_2'}{\sqrt{\langle C_1'(0), C_1'(0)\rangle_{(0,r_0)}}\sqrt{\langle C_2'(0), C_2'(0)\rangle_{(0,r_0)}}} \\ &= \frac{r_0/(1-r_0^2)^2}{\sqrt{(1/(1-r_0^2)) + (r_0^2/(1-r_0^2)^2)}\sqrt{1/(1-r_0^2)^2}} \\ &= \frac{r_0}{1 - r_0^2 + r_0^2} = r_0 = \tanh \rho.\end{aligned}$$

The Lobachevskiĭ-Bolyai theorem follows from the computation:

$$\begin{aligned}\tan\left(\frac{\Pi(\rho)}{2}\right) &= \frac{1-\cos\Pi(\rho)}{\sin\Pi(\rho)} = \frac{1-\tanh(\rho)}{\sqrt{1-\tanh^2(\rho)}} = \frac{1-(e^\rho - e^{-\rho}/e^\rho + e^{-\rho})}{\sqrt{\dfrac{(e^\rho+e^{-\rho})^2 - (e^\rho - e^{-\rho})^2}{(e^\rho + e^{-\rho})^2}}} \\ &= \frac{2e^{-\rho}}{\sqrt{(e^\rho+e^{-\rho})^2 - (e^\rho - e^{-\rho})^2}} = e^{-\rho}.\end{aligned}$$

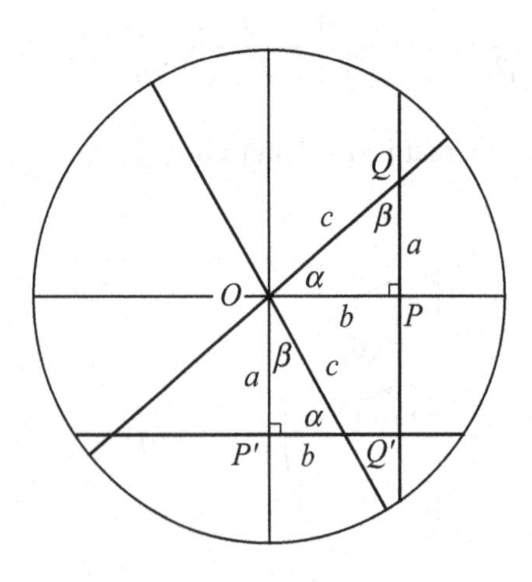

We close our discussion of the Beltrami model with area. In particular, we prove Gauss's Theorem (Exercise 4.19 and Corollary 11.5) that the area of a non-Euclidean triangle is proportional to the angle defect. We consider a right triangle $\triangle OPQ$ with one vertex at the convenient point O. Set out another copy of the right triangle $\triangle OP'Q'$ with P along the u-axis and P' along the v-axis. Suppose further that $OP \cong P'Q'$, $PQ \cong OP'$ and $OQ \cong OQ'$, and that the right angles are at P and P'. At the center of the Beltrami disk the measure of non-Euclidean angles coincides with their Euclidean measure since

$ds_B^2 = du^2 + dv^2$ at O. Let α denote the angle measure of angles $\angle POQ \cong \angle P'Q'O$, and let β denote the angle measure of $\angle PQO \cong \angle P'OQ'$.

Suppose the Euclidean coordinates of P and Q are given by $P = (x, 0)$ and $Q = (x, y)$. By parametrizing the segment PQ as the curve $\gamma(t) = (x, t)$ with $0 \leq t \leq y$, we can compute:

$$a = d_B(P,Q) = \int_\gamma ds = \int_0^y \sqrt{\frac{(1-t^2)\cdot 0 + 2xt\cdot 0\cdot 1 + (1-x^2)\cdot 1}{1-x^2-t^2}}\,dt$$
$$= \int_0^y \frac{\sqrt{1-x^2}}{1-x^2-t^2}\,dt = \frac{1}{2}\ln\left(\frac{\sqrt{1-x^2}+y}{\sqrt{1-x^2}-y}\right).$$

We also have:

$$b = d_B(O,P) = \frac{1}{2}\ln\left(\frac{1+x}{1-x}\right) \text{ and } c = d_B(O,Q) = \frac{1}{2}\ln\left(\frac{1+\sqrt{x^2+y^2}}{1-\sqrt{x^2+y^2}}\right).$$

These relations between coordinates and lengths can be inverted to give:

$$x = \tanh b, \quad \sqrt{x^2+y^2} = \tanh c, \quad \text{and} \quad \frac{y}{\sqrt{1-x^2}} = \tanh a.$$

Since $\operatorname{sech}^2 b = 1 - \tanh^2 b$, we can write $\cosh b = \dfrac{1}{\sqrt{1-x^2}}$ and so $\tanh a = y\cosh b$.

In the triangle $\triangle OP'Q'$, the point $P' = (0, y')$ is determined by $a = d_B(O, P') = \dfrac{1}{2}\ln\left(\dfrac{1+y'}{1-y'}\right)$, so $y' = \tanh a = y\cosh b$. Since the non-Euclidean and Euclidean measures of angles agree at O, we can compute sines and cosines from the Euclidean right triangles:

$$\sin\alpha = \frac{y}{\sqrt{x^2+y^2}}, \qquad \cos\beta = \frac{y'}{\sqrt{x^2+y^2}} = \frac{y\cosh b}{\sqrt{x^2+y^2}} = \cosh b \sin\alpha.$$

In Euclidean polar coordinates the triangle $\triangle OPQ$ can be parametrized by $0 \leq \theta \leq \alpha$ and $0 \leq r \leq x\sec\theta = \tanh b\sec\theta$. The polar line element for $\mathbb{D}_B$ gives:

$$E = \frac{1}{(1-r^2)^2}, \; F = 0, \text{ and } G = \frac{r^2}{1-r^2}.$$

Following chapter 13, to compute the area of $\triangle OPQ$, we integrate:

$$\text{area}(\triangle OPQ) = \iint_{\triangle OPQ} \sqrt{EG-F^2}\,dr\,d\theta = \int_0^\alpha \int_0^{\tanh b\sec\theta} \frac{r}{(1-r^2)^{3/2}}\,dr\,d\theta$$
$$= \int_0^\alpha \left(\frac{1}{\sqrt{1-(\tanh^2 b)(\sec^2\theta)}} - 1\right) d\theta; \qquad \text{let } t = \sin\theta,$$
$$= -\alpha + \int_0^{\sin\alpha} \frac{dt}{\sqrt{(1-\tanh^2 b)-t^2}} = -\alpha + \arcsin\left(\frac{\sin\alpha}{\sqrt{1-\tanh^2 b}}\right)$$

$$= \arcsin(\cosh b \sin\alpha) - \alpha = \arcsin(\cos\beta) - \alpha$$
$$= \frac{\pi}{2} - \beta - \alpha = \pi - \left(\alpha + \beta + \frac{\pi}{2}\right)$$

and this is the angle defect of $\triangle OPQ$.

We postpone a discussion of horocycles in the Beltrami disk until we have developed the isometric models of non-Euclidean geometry that bear Poincaré's name.

The Poincaré disk

One drawback of the Beltrami disk is the representation of angles–the rays forming the angle may be Euclidean line segments, but the angle measure can differ significantly from the Euclidean measure of the angle. For example, by Proposition 14.3, the pencil of perpendiculars to a given line l is determined by the pole of l and so generally only one line in the pencil is perpendicular to l in $\mathbb{D}_B$ and in the Euclidean sense.

We next apply some remarkable transformations to obtain a second model of non-Euclidean geometry that also has the interior of a Euclidean disk as its underlying abstract surface. In this model, however, angle measurement agrees with its Euclidean representation. Geodesics will fail to remain Euclidean straight lines. However, they are transformed into reasonable classical curves.

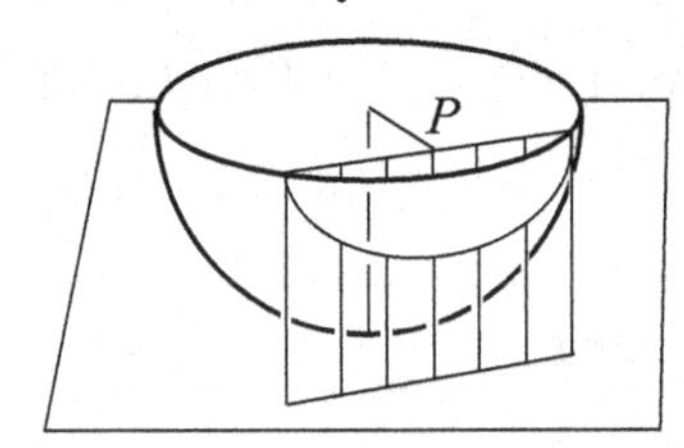

The construction requires an intermediate visit to the sphere. In chapter 7 we introduced the coordinate chart on S^2 given by orthographic projection:

$$x(u,v) = (u, v, -\sqrt{1-u^2-v^2}).$$

This projection maps $\mathbb{D}$ to the lower hemisphere of S^2 and the metric on $\mathbb{D}$ resulting from x is given by:

$$ds^2 = \frac{(1-v^2)du^2 + 2uv\,du\,dv + (1-u^2)dv^2}{1-u^2-v^2} = (1-u^2-v^2)ds_B^2.$$

By Proposition $7^{bis}.3$, the mapping given by x is conformal, and we have proved:

Proposition 14.6. *Orthographic projection of the Beltrami disk $\mathbb{D}_B$ to the lower hemisphere of the sphere S^2 is a conformal mapping.*

Orthographic projection takes the geodesics in $\mathbb{D}_B$, Euclidean line segments, to semicircles on the sphere that meet the equator at right angles. Follow orthographic projection by stereographic projection from the North Pole to the plane that is tangent to the South Pole. The lower hemisphere maps to the disk of radius 2, centered at the origin,

$$\mathbb{D}_2 = \{(x,y) | x^2 + y^2 < 4\}.$$

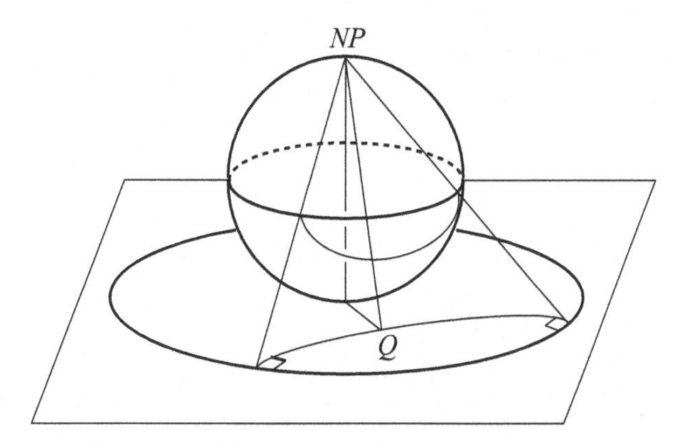

We induce a Riemannian metric on $\mathbb{D}_2$ by transferring ds_B from $\mathbb{D}_B$ via the diffeomorphism given by the composite of orthographic and stereographic projections. Parametrize the lower hemisphere of the unit sphere as $(u,v) \mapsto (u,v,-\sqrt{1-u^2-v^2}) = (u,v,w)$. Following chapter 7bis, stereographic projection sends such a point to $(x,y,-1)$, given by:

$$x = \frac{2u}{1+\sqrt{1-u^2-v^2}} = \frac{2u}{1-w}, \quad y = \frac{2v}{1+\sqrt{1-u^2-v^2}} = \frac{2v}{1-w}.$$

Since $w = -\sqrt{1-u^2-v^2}$, it follows that $dw = \dfrac{u\,du+v\,dv}{\sqrt{1-u^2-v^2}}$ and $\dfrac{dw^2}{w^2} = \dfrac{(u\,du+v\,dv)^2}{(1-u^2-v^2)^2}$. We can rewrite the Beltrami metric as:

$$ds_B^2 = \frac{(1-u^2-v^2)(du^2+dv^2)+(u\,du+v\,dv)^2}{(1-u^2-v^2)^2} = \frac{1}{w^2}(du^2+dv^2+dw^2).$$

From the analytic expression for stereographic projection we obtain the relations $2u = (1-w)x$ and $2v = (1-w)y$, from which we get:

$$2\,du = (1-w)\,dx - x\,dw, \text{ and } 2\,dv = (1-w)\,dy - y\,dw.$$

It follows that:

$$\begin{aligned} 4\,du^2+4\,dv^2+4\,dw^2 &= (1-w)^2(dx^2+dy^2) - 2(1-w)(x\,dx+y\,dy)\,dw \\ &\quad + (x^2+y^2+4)\,dw^2. \end{aligned}$$

From the relations between variables we deduce:

$$\begin{aligned} x^2+y^2 &= \frac{4u^2}{(1-w)^2} + \frac{4v^2}{(1-w)^2} = \frac{4}{(1-w)^2}(u^2+v^2) \\ &= \frac{4}{(1-w)^2}(1-w^2) = 4\frac{1+w}{1-w} = 4 + \frac{8w}{1-w}. \end{aligned}$$

Taking derivatives we have: $2(x\,dx + y\,dy) = \dfrac{8\,dw}{(1-w)^2}$. Rearranging we have

$$2(1-w)(x\,dx + y\,dy)\,dw = \frac{8}{1-w}\,dw^2.$$

However, $\dfrac{8}{1-w} = \dfrac{8(1-w)+8w}{1-w} = 8 + \dfrac{8w}{1-w} = x^2 + y^2 + 4$, and so the last two summands of $4\,du^2 + 4\,dv^2 + 4\,dw^2$ cancel. Returning to the metric we have shown:

$$ds_B^2 = \frac{1}{w^2}(du^2 + dv^2 + dw^2) = \frac{(1-w)^2}{4w^2}(dx^2 + dy^2) = \frac{dx^2 + dy^2}{\left(1 - \dfrac{x^2+y^2}{4}\right)^2} = ds_R^2.$$

The subscript R refers to Riemann; a metric of this form appears in the middle of his famous *Habilitation* lecture as part of a general consideration of higher-dimensional manifolds with constant curvature, discussed in the next chapter. We write $\mathbb{D}_P$ to denote the abstract surface with Riemannian metric $(\mathbb{D}_2, ds_R)$, which is called the **Poincaré disk**. Because $\mathbb{D}_P$ is isometric to $\mathbb{D}_B$, it too is a model of non-Euclidean geometry. Since stereographic projection is conformal and takes circles on the sphere to circles in the plane (Proposition $7^{bis}.7$), the geodesics in $\mathbb{D}_P$ are the images of the geodesics in $\mathbb{D}_B$, that is, circle segments which intersect the boundary circle in right angles.

The advantage of $\mathbb{D}_P$ over $\mathbb{D}_B$ is the conformal representation of angles. The metric is a multiple of the Euclidean metric and so angles between curves share the same measure as their Euclidean representation. Hence, we can visualize properties of angles more directly. For example, the sum of the interior angles of a triangle in $\mathbb{D}_P$ is less than π, because it is always less than the angle sum of the triangle with straight line edges in the Euclidean plane.

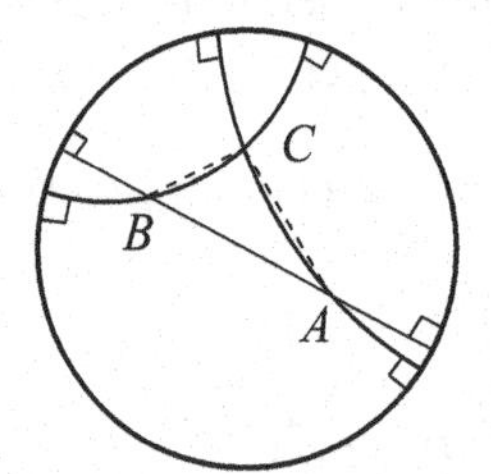

We can also prove that there is a regular octagon in $\mathbb{D}_P$ with all right angles: From the center of the disk $\mathbb{D}_2$, we send out eight rays each $\pi/4$ apart. At the boundary circle, the angle between the line parallel to an adjacent pair of rays and each ray is zero. Consider the family of (Euclidean) circles concentric to the limit circle. For each such circle, we construct the regular octagon with the intersection of the circle with the eight rays as vertices. As the circles decrease in radius, we obtain a regular octagon with increasing angles. The metric converges to the Euclidean metric at the origin, and, therefore, near the origin, the octagons are nearly Euclidean, that is, the angles are close to $3\pi/4$. By continuity there is a regular octagon for some concentric circle with all right angles.

Poincaré arrived at this model of non-Euclidean geometry from an entirely different direction than Beltrami. His investigations of complex differential equations, based on the work of L. I. FUCHS (1833–1902), led him to consider figures in a disk made up of arcs of circles meeting the boundary at right angles. The trick of sending such circles to straight line segments in the plane by composing stereographic projection with orthographic projection led him to a result he needed. He did not make the connection with the Beltrami disk until later in one of the most well-known accounts of mathematical discovery:

At that moment I left Caen where I then lived, to take part in a geological expedition organized by the École des Mines. The circumstances of the journey made me forget my mathematical work. When we arrived at Coutances we boarded an omnibus for I don't know what journey. At the moment when I put my foot on the step the idea came to me, without anything in my previous thoughts having prepared me for it; that the transformations I had made use of to define the Fuchsian functions were identical with those of non-Euclidean geometry. I did not verify this, I did not have time for it, since scarcely had I sat down in the bus than I resumed the conversation already begun, but I was entirely certain at once. On returning to Caen I verified the result at leisure to salve my conscience.

This story is often quoted in discussions of mathematical thought and the psychology of mathematical invention. For a detailed discussion of the related mathematics see Gray (1985). In fact, Beltrami had presented this model in his 1869 paper *Teoria fundamentale degli spazii di curvatura constante* (Beltrami 1869). That paper was concerned with extending the results of the *Saggio* (Beltrami 1868) to higher dimensions in the context of Riemann's theory of manifolds. Poincaré established another context, complex function theory, where the non-Euclidean plane appeared naturally and fruitfully. The popular name of the Poincaré disk model is due to Poincaré's efforts to bring the new geometry into the mainstream of nineteenth-century mathematics. His inspiration, around the year 1880, continued after the discovery of the disk model. Further studies in complex analysis led him to another realization of the non-Euclidean plane, which we discuss next.

The Poincaré half-plane

In a ground-breaking 1882 paper for many areas of mathematics, Poincaré, then 28 years old, applied orthograhic and stereographic projections to introduce another conformal model of the non-Euclidean plane.

To construct the model, project the Beltrami disk $\mathbb{D}_B \subset \mathbb{R}^2 \times \{0\} \subset \mathbb{R}^3$ orthographically to the lower hemisphere of the sphere S^2 as we did to construct $\mathbb{D}_P$. Rotate the sphere around the x-axis through a right angle to put the lower hemisphere in the half-space $y > 0$ of $\mathbb{R}^3$. This composite is given explictly by:

$$(u,v) \mapsto (u,v,-\sqrt{1-u^2-v^2}) \mapsto (u,\sqrt{1-u^2-v^2},v).$$

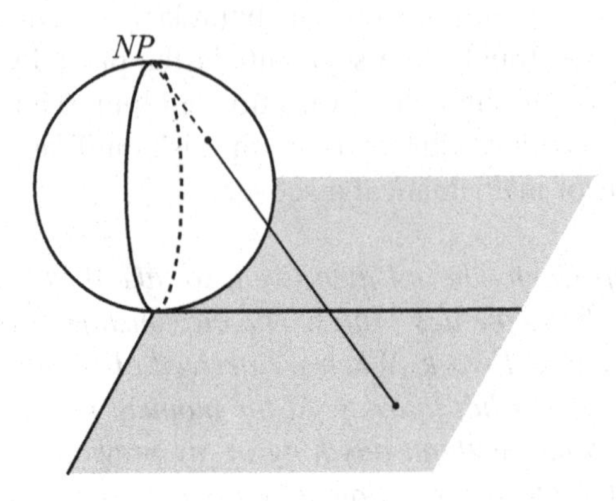

Finally apply stereographic projection from the North Pole. The "right hemisphere" of the sphere ($y > 0$) maps to the upper half-plane in $\mathbb{R}^2$ given by $\mathbb{H} = \{(u,v) \mid v > 0\}$. The composite mapping $\mathbb{D}_B \to \mathbb{H}$ is seen to be:

$$(u,v) \mapsto \left(\frac{2u}{1-v}, \frac{2\sqrt{1-u^2-v^2}}{1-v} \right).$$

What is the metric on $\mathbb{H}$ induced by this mapping from the metric on $\mathbb{D}_B$? Let $w = -\sqrt{1-u^2-v^2}$ so that $-w\,dw = u\,du + v\,dv$. Write $x = \dfrac{2u}{1-v}$ and $y = \dfrac{-2w}{1-v}$ to get $2u = x(1-v)$ and $-2w = y(1-v)$ and so:

$$2du = (1-v)\,dx - x\,dv, \qquad -2dw = (1-v)\,dy - y\,dv.$$

Squaring and summing we have an analogous formula to the case of $\mathbb{D}_P$:

$$4(du^2+dv^2+dw^2) = (1-v)^2(dx^2+dy^2) - 2(1-v)(x\,dx+y\,dy)\,dv + (x^2+y^2+4)\,dv^2.$$

Since $x^2+y^2 = \left(\dfrac{2u}{1-v}\right)^2 + \left(\dfrac{-2w}{1-v}\right)^2$, and $u^2+w^2 = 1-v^2$, we get:

$$x^2+y^2 = \frac{4}{(1-v)^2}(1-v^2) = \frac{4(1+v)}{(1-v)} = 4 + \frac{8v}{1-v}.$$

Thus $x^2 + y^2 + 4 = (8/1 - v)$. Taking the derivative we find:

$$2x\,dx + 2y\,dy = \frac{8}{(1-v)^2}\,dv,$$

from which it follows that $2(1-v)(x\,dx + y\,dy)\,dv = \dfrac{8}{(1-v)}\,dv^2 = (x^2+y^2+4)\,dv^2$. Therefore, we can write:

$$\begin{aligned} ds_B^2 &= \frac{1}{w^2}(du^2 + dv^2 + dw^2) = \frac{(1-v)^2}{4w^2}(dx^2 + dy^2) \\ &= \left(\frac{1-v}{-2w}\right)^2 (dx^2 + dy^2) = \frac{dx^2 + dy^2}{y^2} = ds_P^2. \end{aligned}$$

Let $\mathbb{H}$ denote the abstract surface with Riemannian metric $(\mathbb{H}, ds_P)$, which is called the **Poincaré half-plane**. This surface is isometric to $\mathbb{D}_B$ and so models the geometry of Lobachevskiĭ and Bolyai. By the properties of stereographic projection, the geodesics are the images of geodesics in $\mathbb{D}_B$, that is, either vertical lines, which correspond to straight lines in $\mathbb{D}_B$ that pass through the North Pole, or semicircles, which meet the x-axis at right angles. Like the Poincaré disk, the Poincaré half-plane is a conformal model of non-Euclidean geometry.

The advantage of $\mathbb{H}$ over the other models is its connection with complex variables. As Poincaré understood, by thinking of $\mathbb{H}$ as the upper half of the complex plane, well-known methods of group theory and complex analysis can be brought to bear. Let's take advantage of this serendipity. We do not assume that the reader has much familiarity with complex analysis–acquaintance with the elementary properties of the complex numbers will suffice. Similarly, the groups are represented as 2×2 matrices with the usual matrix multiplication.

The following transformations play a central role in the development of elliptic functions and modular forms and, consequently, all of complex analysis.

Definition 14.7. *A 2×2 matrix A with complex entries and nonzero determinant determines a* **linear fractional transformation** *(or Möbius transformation) given by:*

$$A = \begin{pmatrix} a & b \\ c & d \end{pmatrix} \quad \Rightarrow \quad z \mapsto T_A(z) = \frac{az+b}{cz+d}.$$

The most basic properties of linear fractional transformations are listed next, proofs of which are left to the reader. For many more details, the interested reader can consult Ahlfors (1979) or Needham (1999). Notice that a matrix A and its negative $-A$ both determine the same linear fractional transformation: $\dfrac{ax+b}{cz+d} = \dfrac{-az-b}{-cz-d}$.

The following geometric mappings are linear fractional transformations, and their associated matrices are given:

$$\begin{pmatrix} 1 & b \\ 0 & 1 \end{pmatrix} \longleftrightarrow z \mapsto z+b \text{ translation,}$$

$$\begin{pmatrix} a & 0 \\ 0 & 1 \end{pmatrix} \longleftrightarrow z \mapsto az, \text{ for } a > 0, \text{ similarity ,}$$

$$\begin{pmatrix} a & 0 \\ 0 & 1 \end{pmatrix} \longleftrightarrow z \mapsto az, \text{ for } |a| = 1, \text{ rotation,}$$

$$\begin{pmatrix} 0 & 1 \\ 1 & 0 \end{pmatrix} \longleftrightarrow z \mapsto \frac{1}{z} \text{ inversion.}$$

Proposition 14.8. *If A and B are 2×2 complex matrices, then*

$$T_{AB} = T_A \circ T_B. \textit{ In particular, } (T_A)^{-1} = T_{A^{-1}}.$$

Given a linear fractional transformation $T_A(z)$, in most cases we can rewrite it as follows:

$$\frac{az+b}{cz+d} = \frac{a}{c}\frac{acz+bc}{acz+ad} = \frac{a}{c}\frac{acz+ad+(bc-ad)}{acz+ad} = \frac{a}{c}\left(1 - \frac{\det A}{acz+ad}\right).$$

It follows that the geometric transformations–translations, similarities, rotations, and inversion–generate linear fractional transformations by composition. Hence, to prove that a property holds for all linear fractional transformations, we check that it holds for the geometric transformations and that the property is preserved under composition.

The domain of a linear fractional transformation is the set of all complex numbers except $z = -d/c$. By moving our discussion to the *Riemann sphere*, $\widehat{\mathbb{C}} = \mathbb{C} \cup \{\infty\}$, we can extend the domain to all complex numbers. The Riemann sphere is the result of mapping the complex numbers to the sphere by stereographic projection and adding the North Pole as infinity. We extend a linear fractional transformation to a mapping of the Riemann sphere to itself by letting $T_A(-d/c) = \infty$ and $T_A(\infty) = a/c$. Because $(T_A)^{-1}$ exists, $T_A : \widehat{\mathbb{C}} \to \widehat{\mathbb{C}}$ is one-one and onto.

In this formulation a line in the complex plane is a circle on the Riemann sphere passing through ∞ and so we can talk of both circles and lines as circles.

Proposition 14.9. *Linear fractional transformations take circles to circles.*

PROOF: The equation for a circle in the complex plane centered at p with radius r is $(z-p)(\bar{z}-\bar{p}) = r^2$, where $\bar{z}$ denotes the *complex conjugate* of z, $\overline{(a+bi)} = a-bi$. Multiplying this equation out we get a more useful form:

$$A|z|^2 + B\bar{z} + \bar{B}z + C = 0,$$

where A and C are real number and $|B|^2 \geq AC$. Notice that this form includes a line as the case $A = 0$.

To prove the proposition we check the statement on the geometric transformations. Clearly, translation takes circles to circles. If $T_A(z) = az$ for $a = pe^{i\theta}$, then we can multiply the equation for a circle by $|a|^2$ to get

$$\begin{aligned} 0 &= A|z|^2|a|^2 + B\bar{z}|a|^2 + \bar{B}z|a|^2 + C|a|^2 \\ &= A|az|^2 + Ba\overline{az} + \overline{Ba}(az) + C|a|^2 \end{aligned}$$

and $|Ba|^2 = |B|^2|a|^2 \geq AC|a|^2$. Thus, rotations and similarities take circles to circles. Finally, we check inversion: Divide the equation of a circle by $|z|^2$ to get:

$$C\left|\frac{1}{z}\right|^2 + Bd\frac{1}{z} + \bar{B}\frac{1}{\bar{z}} + A = 0.$$

Thus, the equation for a circle holds with A and C reversed for the image of inversion.

The statement of the proposition is closed under compositions, and so the proposition is proved. ∎

An important property of linear fractional transformations relates them to projective geometry and gives the foundation for another approach to non-Euclidean geometry that was developed by FELIX KLEIN (1849–1925). To state the property we need the following notion.

Definition 14.10. *Given distinct complex numbers z_1, z_2, z_3, and z_4, their* **cross ratio** *is given by*

$$(z_1, z_2; z_3, z_4) = \frac{z_1 - z_2}{z_1 - z_3} \cdot \frac{z_4 - z_3}{z_4 - z_2}.$$

The cross ratio depends on the ordering of the complex numbers z_i. We leave to the reader to determine the effect on the value of the cross ratio of a permutation of the values z_i. We extend the cross ratio to the Riemann sphere by allowing one of the z_i to equal ∞:

$$(z_1, z_2; z_3, \infty) = \frac{z_1 - z_2}{z_1 - z_3}.$$

Proposition 14.11. *Given a linear fractional transformation T_A and four distinct points on the Riemann sphere z_1, z_2, z_3, and z_4,*

$$(z_1, z_2; z_3, z_4) = (T_A(z_1), T_A(z_2); T_A(z_3), T_A(z_4)).$$

PROOF: We leave the details to the reader; the proposition follows from the simple computation:

$$T_A(z) - T_A(w) = \frac{az + b}{cz + d} - \frac{aw + b}{cw + d} = \frac{(ad - bc)(z - w)}{(cz + d)(cw + d)}.$$ ∎

The cross ratio determines a linear fractional transformation as follows: Given three distinct points on the Riemann sphere, z_2, z_3, and z_4, define:

$$T_{z_2,z_3,z_4}(z) = (z, z_2; z_3, z_4) = \frac{z - z_2}{z - z_3} \cdot \frac{z_4 - z_3}{z_4 - z_2}.$$

In this form we can see that the linear fractional transformation T_{z_2,z_3,z_4} takes z_2 to 0, z_3 to ∞, and z_4 to 1. Thus, we can take any circle (in the extended sense) to any other circle by choosing three points on each and applying the transformation $(T_{w_2,w_3,w_4})^{-1} \circ T_{z_2,z_3,z_4}$. We also know that, when four points lie on a circle or line, if we take three of the points to be z_2, z_3, and z_4, then the transformation T_{z_2,z_3,z_4} takes these three points to $0, \infty$, and 1, respectively. These points are on the extended circle that is the real line in the complex plane. The inverse of this transformation takes the real line to the circle determined by z_2, z_3, and z_4. If any other point z_1 lies on this circle, then $T_{z_2,z_3,z_4}(z_1)$ is real. Therefore, four distinct points on the Riemann sphere lie on a circle if and only if their cross ratio is real.

We return to the Poincaré half-plane $\mathbb{H}$. The main object of study in Poincaré's 1882 paper is the group of real 2×2 matrices $SL_2(\mathbb{R})$ with determinant one. These matrices determine the linear fractional transformations of the form:

$$T_A(z) = \frac{az + b}{cz + d}, \text{ where } A = \begin{pmatrix} a & b \\ c & d \end{pmatrix}, a, b, c, d \in \mathbb{R}, \text{ and } ad - bc = 1.$$

Any real 2×2 matrix with positive determinant determines an element in $SL_2(\mathbb{R})$ by rescaling all the entries by the square root of the determinant. By an argument analogous to the case for all linear fractional transformations, the group $SL_2(\mathbb{R})$ is generated by *translations*, $z \mapsto z + s$ with $s \in \mathbb{R}$, *similarities*, where for $a > 0$, $\begin{pmatrix} a & 0 \\ 0 & 1 \end{pmatrix}$ and $\begin{pmatrix} \sqrt{a} & 0 \\ 0 & 1/\sqrt{a} \end{pmatrix}$ determine the same linear fractional transformation), and *negative inversion*, that is, $z \mapsto -1/z$ determined by the matrix $\begin{pmatrix} 0 & -1 \\ 1 & 0 \end{pmatrix}$. The important property of these transformations is given by the following result.

Theorem 14.12. *A real linear fractional transformation T_A from $A \in SL_2(\mathbb{R})$ is an isometry of the Poincaré half-plane $\mathbb{H}$. Furthermore, any isometry of $\mathbb{H}$ is given by $z \mapsto T_A(z)$ or $z \mapsto T_A(-\bar{z})$.*

PROOF: We first check that real linear fractional transformations in $SL_2(\mathbb{R})$ take $\mathbb{H}$ to itself. If A is in $SL_2(\mathbb{R})$, $z = u + iv$, then $-\bar{z} = -u + iv$ and:

$$\begin{aligned} T_A((\pm u) + iv) &= \frac{a(\pm u) + aiv + b}{c(\pm u) + civ + d} = \frac{(a(\pm u) + b) + aiv}{(c(\pm u) + d) + civ} \cdot \frac{(c(\pm u) + d) - civ}{(c(\pm u) + d) - civ} \\ &= \frac{(a(\pm u) + b)(c(\pm u) + d) + acv^2}{(c(\pm u) + d)^2 + (cv)^2} + \frac{(ad - bc)vi}{(c(\pm u) + d)^2 + (cv)^2}. \end{aligned}$$

If $v > 0$ and $ad - bc = 1$, then $T_A(z)$ and $T_A(-\bar{z})$ are in $\mathbb{H}$. Because a linear fractional transformation takes an extended circle to an extended circle, T_A also takes the u-axis (the real line), the boundary of $\mathbb{H}$, to itself. In this case, the singular point of T_A, $z = -d/c$, is a real number, and the image of ∞ is also real. Finally, T_A has an inverse and so it is a one-to-one and onto transformation on $\mathbb{H}$.

To show that T_A is an isometry, we write $w = T_A(z)$, and compare the metric ds_P at corresponding points. The comparison is made especially easy by changing coordinates from (x,y) to $(z,\bar{z})$ where $z = x+iy$ and $\bar{z} = x - iy$. Then $dz = dx + i\,dy$ and $d\bar{z} = dx - i\,dy$ and it follows that $dx^2 + dy^2 = dz\,d\bar{z}$. Solving for y we get $y = \dfrac{z-\bar{z}}{2i}$ and so:

$$ds_P^2 = \frac{dx^2+dy^2}{y^2} = -4\frac{dz\,d\bar{z}}{(z-\bar{z})^2}.$$

At $w = T_A(z)$, the change of coordinates takes the form:

$$dw = \frac{a(cz+d)dz - c(az+b)dz}{(cz+d)^2} = \frac{(ad-bc)dz}{(cz+d)^2} = \frac{dz}{(cz+d)^2}.$$

Since a, b, c, and d are real, $\bar{w} = \overline{T_A(z)} = T_A(\bar{z})$ and so $d\bar{w} = \dfrac{\bar{z}}{(c\bar{z}+d)^2}$. We compute:

$$\frac{dw\,d\bar{w}}{(w-\bar{w})^2} = \frac{dz/(cz+d)^2\,d\bar{z}/(c\bar{z}+d)^2}{\left[\dfrac{az+b}{cz+d} - \dfrac{a\bar{z}+b}{c\bar{z}+d}\right]^2} = \frac{dz\,d\bar{z}}{[(az+b)(c\bar{z}+d)-(a\bar{z}+b)(cz+d)]^2}$$
$$= \frac{dz\,d\bar{z}}{[(ad-bc)z-(ad-bc)\bar{z}]^2} = \frac{dz\,d\bar{z}}{(z-\bar{z})^2}.$$

Thus, T_A is an isometry, globally defined on $\mathbb{H}$. Reversing the roles of z and $-\bar{z}$, the transformation $w = T_A(-\bar{z})$ satisfies $\bar{w} = \overline{T_A(-\bar{z})} = T_A(-z)$ and so the same calculation proves that $z \mapsto T_A(-\bar{z})$ is an isometry on $\mathbb{H}$.

To show that every congruence of the Poincaré half-plane can be expressed in terms of transformations based on $SL_2(\mathbb{R})$, we show that any pair of points P and Q in $\mathbb{H}$ and any pair of unit speed geodesics, γ_1, passing through P, γ_2 through Q, can be mapped by a real linear fractional transformation taking P to Q and γ_1 onto γ_2. The existence and uniqueness of geodesics through a given point and the fact that isometries take geodesics to geodesics imply that we have constructed all of the congruences. Geodesics in $\mathbb{H}$ may be vertical lines or semicircles with centers on the real axis. We treat the case of a pair of semicircles and leave the other cases to the reader.

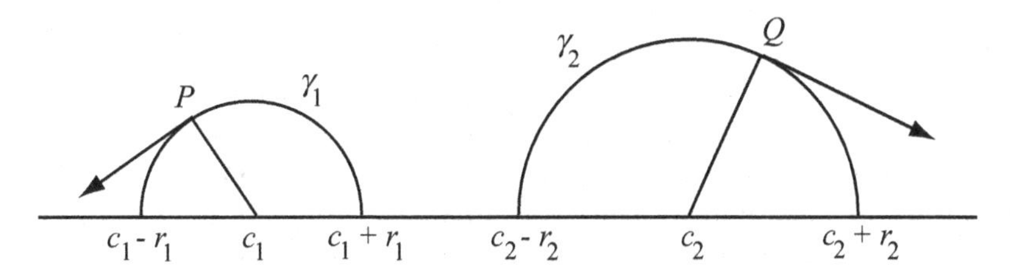

It suffices to show that there is a sequence taking a unit speed geodesic γ and a point P on γ, together with the unit tangent at P, $\gamma'(t_P)$, to the vertical geodesic given by the y-axis in such a way that P maps to i, and $\gamma'(t_P)$ maps to the vertical unit tangent at i pointing upward. Suppose γ is a semicircle centered at $c \in \mathbb{R}$ with radius r. We carry this construction out by steps:

(1) Translate $z \mapsto z - c$ taking γ to $\hat{\gamma}$ a semicircle centered at the origin with radius r. The point P goes to $P' = P - c$.
(2) Apply the similarity $z \mapsto (1/r)z$ which takes $\hat{\gamma}$ to τ, the semicircle centered at the origin of radius one.
(3) Apply the linear fractional transformation $T_A \colon z \mapsto \dfrac{z-1}{z+1}$, determined by the matrix $\begin{pmatrix} 1/\sqrt{2} & -1/\sqrt{2} \\ 1/\sqrt{2} & 1/\sqrt{2} \end{pmatrix}$, that takes 1 to 0 and -1 to ∞. Evaluating at i we see:
$$\frac{i-1}{i+1} = \frac{-(i-1)^2}{(1+i)(1-i)} = \frac{-(-1-2i+1)}{2} = i.$$
Thus τ is mapped to the y-axis. The point P'' maps to a point iv_P with $v_P > 0$.
(4) Apply the similarity determined by the matrix $\begin{pmatrix} 1/\sqrt{v_P} & 0 \\ 0 & \sqrt{v_P} \end{pmatrix}$ and iv_P maps to i.

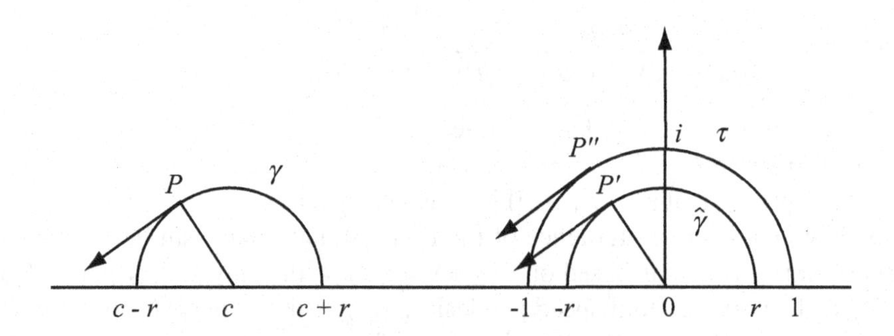

Following $\gamma'(t_P)$ throughout this sequence of isometries, we end at a unit tangent vector $\mathbf{u}_P$ at i along the y-axis. If $\mathbf{u}_P$ points upward we are done. If $\mathbf{u}_P$ points down, we apply the linear fractional transformation $z \mapsto -1/z$. This mapping preserves the y-axis and $i \mapsto i$, but it reverses the orientation taking $\mathbf{u}_P$ to $-\mathbf{u}_P$. Thus, for any geodesic γ together with a point P on γ and the unit tangent vector to γ at P, there is a linear fractional transformation taking γ to the y-axis, P to i, and the unit tangent vector at P to the upward pointing unit tangent to the y-axis at i.

Finally, given any pair of points P and Q in $\mathbb{H}$ and any pair of unit speed geodesics, γ_1 passing through P, γ_2 through Q, we apply the steps to each pair of point and geodesic. Reversing the steps for (Q, γ_2), we can begin at (P, γ_1) and apply linear fractional transformations from $SL_2(\mathbb{R})$ to map P to Q and γ_1 to γ_2 in such a way that their orientations map from one onto the other. ∎

The consequences of this theorem are many and substantial. We remark on those most relevant to our goals.

(1) The simplicity of the metric on $\mathbb{H}$ and the explicit group of isometries permit us to find a formula for distance in the Poincaré half-plane. First we derive the distance between two points on a vertical geodesic, $P = (a, y_1)$ and $Q = (a, y_2)$. The function $C(t) = (a, y_1 + t(y_2 - y_1))$ for $0 \leq t \leq 1$ provides a parametrization of the vertical geodesic with $C'(t) = (0, y_2 - y_1)$. The distance between the points P and Q in $\mathbb{H}$ is denoted $d_{\mathbb{H}}(P,Q)$ and is given by:

$$d_{\mathbb{H}}(P,Q) = \left| \int_C \sqrt{\frac{dx^2 + dy^2}{y^2}} \right| = \left| \int_0^1 \frac{(y_2 - y_1)\,dt}{y_1 + t(y_2 - y_1)} \right|$$
$$= \left| \ \ln(y_1 + t(y_2 - y_1)) \Big|_0^1 \ \right| = \left| \ln\left(\frac{y_2}{y_1}\right) \right|.$$

Holding P fixed and varying Q, it is immediate from this formula that vertical geodesics are infinite in length.

Given two points that do not lie on a vertical line, say P' and Q', we determine the semicircle with its center on the real line that passes through P' and Q' by a Euclidean procedure: Construct the perpendicular bisector of the Euclidean line segment $P'Q'$. The intersection of this perpendicular bisector and the real line is the center of the semicircle. Suppose that semicircle has center c and radius r so that it intersects the real line at points $c - r$ and $c + r$. Consider the following sequence of transformations that takes P' and Q' to a vertical line:

(a) Translate c to 0 by $z \mapsto z - c$.
(b) Dilate by the radius to bring the semicircle to the unit semicircle, $z \mapsto (1/r)z$.
(c) Map the unit semicircle to the y-axis by $z \mapsto \dfrac{z-1}{z+1}$.

The composition of these operations is given by:

$$T(z) = \frac{z - (c+r)}{z - (c+r)}.$$

Let $P' = z_1$ and $Q' = z_2$, then $T(z_1) = iy_1$ and $T(z_2) = iy_2$. By the formula for distance along a vertical line we find:

$$d_{\mathbb{H}}(P',Q') = \left| \ln\left(\frac{y_2}{y_1}\right) \right| = \left| \ln\left(\frac{(z_2 - (c-r))/(z_2 - (c+r))}{(z_1 - (c-r))/(z_1 - (c+r))} \right) \right|$$
$$= \left| \ln\left(\frac{(z_2 - (c-r))}{(z_2 - (c+r))} \cdot \frac{(z_1 - (c+r))}{(z_1 - (c-r))} \right) \right|.$$

Notice that the innermost expression in the distance formula is a cross ratio,

$$\frac{(z_2-(c-r))}{(z_2-(c+r))}\cdot\frac{(z_1-(c+r))}{(z_1-(c-r))}=\frac{((c+r)-z_1)}{((c+r)-z_2)}\cdot\frac{((c-r)-z_2)}{((c-r)-z_1)}=(c+r,z_1;z_2,c-r).$$

The cross ratio $(c+r,z_1;z_2,c-r)$ is a real number (since the points lie on a circle). This number times its conjugate is simply the cross ratio squared, but the complex expression becomes:

$$\frac{|z_2-(c-r)|^2}{|z_2-(c+r)|^2}\cdot\frac{|z_1-(c+r)|^2}{|z_1-(c-r)|^2}=\left(\frac{d_E(z_2,(c-r))}{d_E(z_2,(c+r))}\cdot\frac{d_E(z_1,(c+r))}{d_E(z_1,(c-r))}\right)^2,$$

where $d_E(z,w)$ is the Euclidean distance between the points z and w as elements in $\mathbb{R}^2$. Thus,

$$d_{\mathbb{H}}(P',Q')=\left|\ln\left(\frac{d_E(z_2,(c-r))}{d_E(z_2,(c+r))}\cdot\frac{d_E(z_1,(c+r))}{d_E(z_1,(c-r))}\right)\right|.$$

(2) The other models of the non-Euclidean plane are isometric to $\mathbb{H}$ and so we can relate the distance function on $\mathbb{H}$ to the distance functions on the Beltrami disk $\mathbb{D}_B$ and the Poincaré disk $\mathbb{D}_P$. We treat the case of $\mathbb{D}_B$ here and leave $\mathbb{D}_P$ to the reader.

Suppose P and Q are points in $\mathbb{D}_B$. Let the distance function on $\mathbb{D}_B$ be denoted by $d_B(P,Q)$. Because we have an isometry $\phi\colon \mathbb{D}_B\to\mathbb{H}$, we can compute distance by:

$$d_B(P,Q)=d_{\mathbb{H}}(\phi(P),\phi(Q)).$$

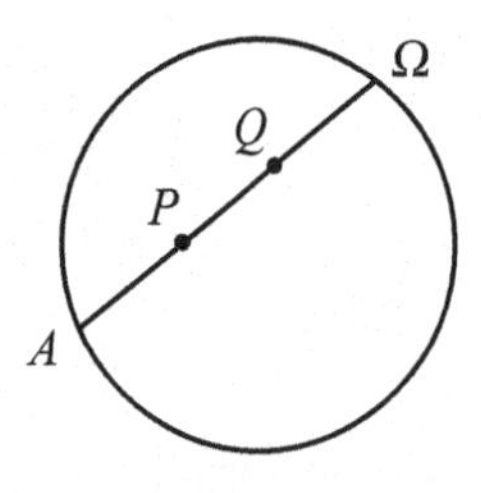

Extend the line segment PQ in $\mathbb{D}_B$ until it meets the limit circle at points A and Ω. The points A and Ω go to the intersection of the geodesic in $\mathbb{H}$ through $\phi(P)$ and $\phi(Q)$ and the real line. The formula for distance in $\mathbb{H}$ in terms of Euclidean distances yields:

$$d_B(P,Q)=\left|\ln\left(\frac{d_E(\phi(Q),\phi(A))}{d_E(\phi(Q),\phi(\Omega))}\cdot\frac{d_E(\phi(P),\phi(\Omega))}{d_E(\phi(P),\phi(A))}\right)\right|.$$

Recall that the isometry ϕ is given by $\phi(u,v)=\left(\dfrac{2u}{1-v},\dfrac{2\sqrt{1-u^2-v^2}}{1-v}\right)$. Write $P=(u_1,v_1),Q=(u_2,v_2),A=(u_0,v_0)$, and $\Omega=(u_\infty,v_\infty)$. To compute $d_B(P,Q)$, we need to simplify the following horrendous expression involving the squares of the distances:

$$\frac{\left(\frac{2u_2}{1-v_2}-\frac{2u_0}{1-v_0}\right)^2+\left(\frac{2\sqrt{1-u_2^2-v_2^2}}{1-v_2}\right)^2}{\left(\frac{2u_2}{1-v_2}-\frac{2u_\infty}{1-v_\infty}\right)^2+\left(\frac{2\sqrt{1-u_2^2-v_2^2}}{1-v_2}\right)^2}$$

$$\cdot\frac{\left(\frac{2u_1}{1-v_1}-\frac{2u_\infty}{1-v_\infty}\right)^2+\left(\frac{2\sqrt{1-u_1^2-v_1^2}}{1-v_1}\right)^2}{\left(\frac{2u_1}{1-v_1}-\frac{2u_0}{1-v_0}\right)^2+\left(\frac{2\sqrt{1-u_1^2-v_1^2}}{1-v_1}\right)^2}.$$

We can use $u_0^2+v_0^2=1=u_\infty^2+v_\infty^2$ and the fact that P, Q, A, and Ω lie on a line, say $v=mu+b$, to significantly simplify this expression (a vertical line in $\mathbb{D}_B$ needs another simple argument). After some fuss, it becomes, in fact,

$$d_B(P,Q)=\frac{1}{2}\left|\ln\left(\frac{|u_2-u_0|}{|u_2-u_\infty|}\cdot\frac{|u_1-u_\infty|}{|u_1-u_0|}\right)\right|.$$

With the Euclidean metric on $\mathbb{D}_B$, notice that:

$$d_E(Q,A)=\sqrt{(u_2-u_0)^2+(v_2-v_0)^2}=\sqrt{(m^2+1)(u_2-u_0)^2}.$$

Similar expressions hold for $d_E(Q,\Omega), d_E(P,A)$, and $d_E(P,\Omega)$, so we can form the expression:

$$\frac{d_E(Q,A)}{d_E(Q,\Omega)}\cdot\frac{d_E(P,\Omega)}{d_E(P,A)}=\frac{|u_2-u_0|}{|u_2-u_\infty|}\cdot\frac{|u_1-u_\infty|}{|u_1-u_0|}.$$

Substituting this into the formula for distances in $\mathbb{D}_B$ we get:

$$d_B(P,Q)=\frac{1}{2}\left|\ln\left(\frac{d_E(Q,A)}{d_E(Q,\Omega)}\cdot\frac{d_E(P,\Omega)}{d_E(P,A)}\right)\right|.$$

This generalizes the case of lines through the origin that we computed earlier.

If we treat the four points P,Q,A, and Ω as complex numbers, then we can use the fact that they lie on a line to substitute the cross ratio $(Q,A;\Omega,P)$ for the expression in Euclidean distances. Therefore, we can write:

$$d_B(P,Q)=\frac{1}{2}|\ln((Q,A;\Omega,P))|.$$

The presence of the cross ratio and the representation of geodesics as lines strongly suggest connections to projective geometry. These connections were made by Felix Klein in his 1871 paper, *Über die sogenannte Nicht-Euclidische Geometrie.* He based his work on an idea of SIR ARTHUR CAYLEY (1821–91) whose 1859 paper, *A sixth memoir upon quantics*, contained a discussion of distance on subsets of projective space. Today the Beltrami model is most often referred to as the *Klein model*, or the *Klein–Beltrami model*, or even the *Cayley–Klein–Beltrami model*. The modern term for non-Euclidean geometry, **hyperbolic geometry**, was coined by Klein in this paper. Our emphasis on the differential geometric approach, and the dates of publication of the relevant papers by Beltrami and Klein led to our choice of nomenclature. For an exposition of the projective viewpoint see Coxeter (1947) or Klein (1928).

(3) The distance formula of **(1)** and the conformal nature of the Poincaré half-plane lead to a simple proof of the Lobachevskiĭ–Bolyai formula for the angle of parallelism (Theorem 4.29). Since there are sufficient isometries to go anywhere in the plane, we argue at a convenient place with convenient lines. Let l denote the y-axis and τ the unit semicircle centered at 0. Then l and τ are perpendicular. Suppose P is a point on τ to the right of l. We can express $P = \cos\theta + i\sin\theta$ where θ is the angle made by the segment $0P$ with the real axis. The perpendicular bisector of $0P$ meets the real axis at a point C, which is the center of the semicircle m through P and 0, which is the line in $\mathbb{H}$ parallel to l through P.

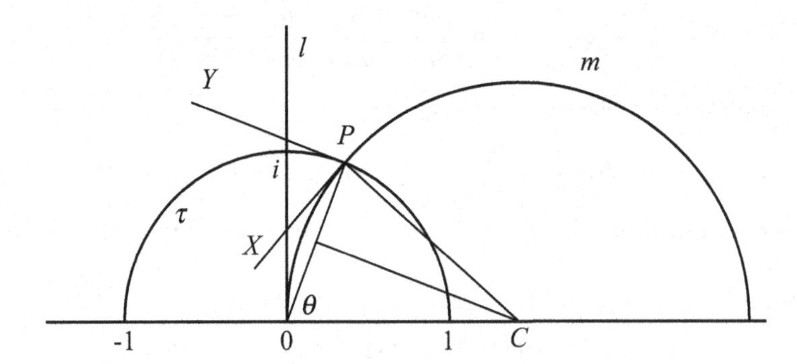

We can make some observations from the Euclidean point of view. Let PY denote the tangent to τ at P. Then $\angle 0PY$ is a right angle. Let PX denote the tangent to m at P, and so $\angle CPX$ is a right angle. Since $0C$ and CP are radii of m, $\angle 0PC = \theta$. Then $\angle 0PX = \pi/2 - \theta$. Hence $\angle XPY = \angle 0PY - \angle 0PX = \pi/2 - (\pi/2 - \theta) = \theta$. Thus the angle of parallelism associated to the length of the line segment on τ between i and P is θ.

Computing the distance from i to P we get:

$$\begin{aligned} d_{\mathbb{H}}(i,P) &= |\ln(1, \cos\theta + i\sin\theta; i, -1)| \\ &= \left|\ln\left(\frac{1-(\cos\theta + i\sin\theta)}{1-i} \cdot \frac{-1-i}{-1-(\cos\theta + i\sin\theta)}\right)\right| = \left|\ln\left(\frac{\sin\theta}{1+\cos\theta}\right)\right|. \end{aligned}$$

We know that $\dfrac{\sin\theta}{1+\cos\theta} = \tan(\theta/2)$, and so $d_{\mathbb{H}}(i,P) = |\ln(\tan(\theta/2)|$. Since $0 < \theta < \pi/2$, $0 < \tan(\theta/2) < 1$, and so we can write:

$$\rho = d_{\mathbb{H}}(i,P) = -\ln(\tan(\theta/2)),$$

that is, $e^{-\rho} = \tan\left(\dfrac{\Pi(\rho)}{2}\right)$, the Lobachevskiĭ–Bolyai formula.

(4) From the convenient center $O = (0,0)$ of the Beltrami model we were able to determine the circumference of a circle of radius ρ with polar coordinates. Such a circle in $\mathbb{D}_B$ is also a Euclidean circle (a special circumstance for this center) and, furthermore, orthographic projection maps such circles to circles on the sphere. One of the special properties of stereographic projection is that it takes circles on the sphere to circles (in the extended sense) in the plane (Proposition $7^{bis}.7$). Thus, these circles of any radius centered at O in $\mathbb{D}_B$ map isometrically to non-Euclidean circles in $\mathbb{D}_P$ and $\mathbb{H}$, which take the form of Euclidean circles. By applying an isometry of $\mathbb{H}$ we can move any given circle to pass through any point of $\mathbb{H}$. Since linear fractional transformations map Euclidean circles to Euclidean circles, we have shown that all non-Euclidean circles in $\mathbb{H}$ are curves given by Euclidean circles.

However, the center of such a non-Euclidean circle will differ from its Euclidean center. Let $(x-a)^2 + (y-b)^2 = r^2$ (here $b > r > 0$) be the Euclidean equation of a circle in $\mathbb{H}$. The Euclidean center, (a,b), is not the non-Euclidean center. To determine the center in $\mathbb{H}$ we can choose a diameter and halve it. The most convenient diameter is the vertical one, the line $\{(a,y)|y > 0\}$.

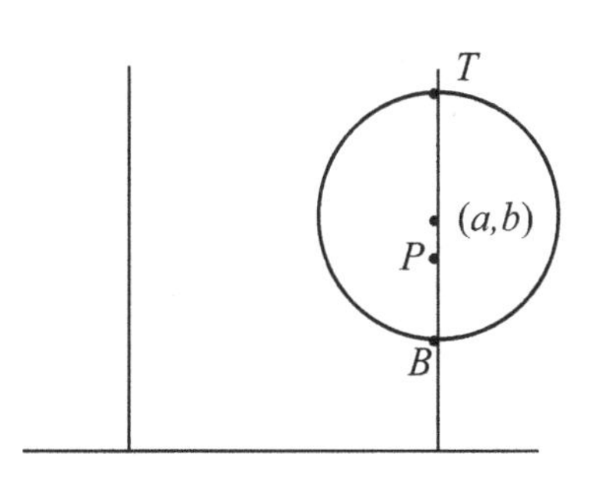

Let $B = (a, b-r)$ and $T = (a, b+r)$ denote the bottom and top, respectively, of the circle. The line segment BT is a diameter and so the midpoint, $P = (a,p)$, is the center of the circle as a circle in $\mathbb{H}$.

To compute the radius we use the distance formula:

$$d_{\mathbb{H}}(P,B) = \ln\left(\frac{p}{b-r}\right) = \ln\left(\frac{b+r}{p}\right) = d_{\mathbb{H}}(P,T),$$

which implies that $p = \sqrt{b^2 - r^2}$ and the center is found. The radius becomes $\rho = \ln\left(\dfrac{\sqrt{b^2-r^2}}{b-r}\right) = \ln\left(\sqrt{\dfrac{b+r}{b-r}}\right)$.

These equations are easily reversed to determine the non-Euclidean circle with center $P = (a,p)$ and radius ρ. We leave it to the reader to prove that it is the Euclidean circle of radius $p\sinh(\rho)$ and center $(a, p\cosh(\rho))$.

Similar results hold for the circles in $\mathbb{D}_P$. The Beltrami model is more difficult–circles with centers away from $(0,0)$ are *not* Euclidean circles. In fact, non-Euclidean circles in $\mathbb{D}_B$ are represented by Euclidean ellipses. Their shape can be deduced from orthographic projection of arbitrary circles on the sphere to the disk. The appearance of conics in this setting is further clarified in the projective formulation of Klein (1928).

Having determined the circles in $\mathbb{H}$, we have also determined the open balls of any radius in $\mathbb{H}$ that give the metric topology on $\mathbb{H}$. With a small adjustment for a different center, it is evident the $\mathbb{H}$ is homeomorphic to $\mathbb{R}^2$ with the usual metric, and since $\mathbb{R}^2$ with the Euclidean metric is complete, so is $\mathbb{H}$. Once again we see from the Hopf–Rinow Theorem that $\mathbb{H}$ is geodesically complete.

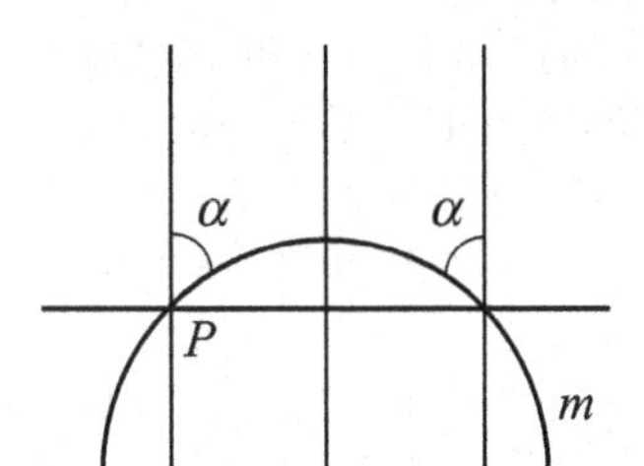

(5) In the non-Euclidean plane the horocycle is a new geometric object. Recall that it can be thought of as a "circle with center at infinity." One of the beauties of our explicit models is that "infinity" is a visible place–the boundary circle for $\mathbb{D}_B$ or $\mathbb{D}_P$, and the real axis (with ∞) for $\mathbb{H}$.

The simplest pencil of parallel lines to visualize in $\mathbb{H}$ is the set $\mathbf{P}_l$ of vertical lines that are all parallel to l, the y-axis, in the "up" direction. If we choose a point P in $\mathbb{H}$ and any semicircular geodesic m through P, m forms an angle α with the vertical line through P. To find the other point on m that corresponds to P (Definition 4.5), we look for the vertical line that makes an angle α on the opposite side of the vertical as at P. Since the model is conformal, we can solve this question with Euclidean tools. We reflect $\mathbb{H}$ across a vertical line that is perpendicular to m (through the "uppermost" point of m). The reflection takes m to itself and the vertical through P to a vertical through the corresponding point. Thus, the points that correspond to P all share the same y coordinate, and so a horocycle with respect to the pencil of vertical lines $\mathbf{P}_l$ is a horizontal Euclidean line.

In this context we can prove another important theorem about the non-Euclidean plane: The ratio of lengths of concentric arcs of two horocycles intercepted between two lines in the pencil of parallels satisfy $\text{length}(\widehat{AB}) = \text{length}(\widehat{A'B'})e^{l(AA')}$ (Theorem 4.9).

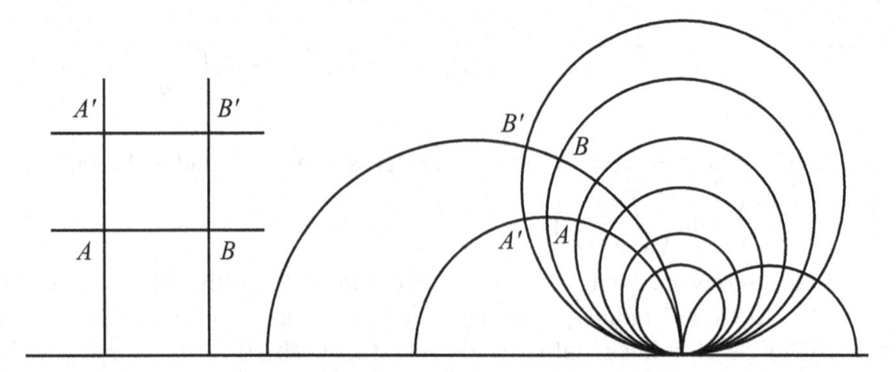

Working with the pencil of lines parallel to the y-axis, the vertical lines, we can compute the length of arcs of a horocycle between two parallels by parametrizing the horizontal segment, say $\widehat{AB}$ and using the metric: Suppose $A = a + iy_1$ and $B = b + iy_1$, $A' = a + iy_2$ and $B' = b + iy_2$. Then we parametrize $\widehat{AB}$ by $C(t) = bt + a(1-t) + y_1$ with $C'(t) = b - a$:

$$\text{length}(\widehat{AB}) = \int_C \sqrt{\frac{dz\,d\bar{z}}{(z-\bar{z})^2}} = \int_0^1 \left|\frac{b-a}{y_1}\right| dt = \frac{b-a}{y_1}.$$

From **(1)** we know that $l(AA') = d_{\mathbb{H}}(A, A') = d_{\mathbb{H}}(B, B') = |\ln(y_2/y_1)|$. Hence,

$$\text{length}(\widehat{AB}) = \frac{b-a}{y_1} = \frac{b-a}{y_2} \cdot \frac{y_2}{y_1} = \text{length}(\widehat{A'B'})e^{l(AA')}.$$

Every other horocycle can be constructed by applying the isometries of $\mathbb{H}$. A real linear fractional transformation takes a horizontal line to a circle with one point on the real axis (the image of ∞) and the pencil of vertical lines to a pencil of semicircles with centers on the real axis, all passing through a fixed point, again the image of ∞.

The picture for $\mathbb{D}_P$ is the same–horocycles are Euclidean circles with one point on the boundary circle. In the case of $\mathbb{D}_B$, horocycles are Euclidean ellipses in general.

At this point, we have completed the story of Postulate V. The existence of a geodesically complete abstract surface of constant negative curvature implies that Postulate V cannot be proved from Postulates I through IV. Euclid is vindicated in his choice to call Postulate V an axiom. A search stretching over two millennia is ended, not with a proof, but with a new geometric universe to study.

We can prove many other results of non-Euclidean geometry through the analytic methods developed here. The methods of differential geometry provide the foundation on which Euclidean and non-Euclidean geometry are special cases of a vastly richer set of geometric objects, each with its own story to tell and connected in new ways to other parts of mathematics.

In my paper (McCleary 2002), I present a version of the Beltrami metric with a parameter based on the curvature. I use the metric to define trigonometric functions with a parameter that specializes to spherical geometry when positive and hyperbolic geometry when negative. In this way uniform arguments for both trigonometries can be given. An unexpected consequence is a proof of the irrationality of π and of e depending on choice of parameter.

Exercises

14.1 Suppose there is a geodesic mapping of a surface S to the plane. Taking the inverse of the mapping as coordinates on S. show that the induced metric on the plane satisfies the differential equations:

$$\begin{aligned} 2EF_u - EE_v - FE_u &= 0, \\ 2GF_v - GG_u - FG_v &= 0, \\ GE_u - 2EG_u - 2FF_u + 3FE_v &= 0, \\ EG_v - 2GE_v - 2FF_v + 3FG_u &= 0. \end{aligned}$$

14.2 By a change of scale on the coordinates u and v, show that a solution to the differential equations given in Exercise 14.1 is the following set of metric functions where R and a are constants:

$$E = R^2 \frac{a^2 + v^2}{(a^2 + u^2 + v^2)^2}, \quad F = -R^2 \frac{uv}{(a^2 + u^2 + v^2)^2}, \quad G = R^2 \frac{(a^2 + u^2)}{(a^2 + u^2 + v^2)^2}.$$

14.3 Construct a geodesic mapping from the surface of revolution of the tractrix to the plane.

14.4 In the Beltrami disk $\mathbb{D}_B$, let $A = (0, r)$ and $O = (0, 0)$. Take the v-axis as a line through O not containing A. Suppose l is a line through A that does not intersect the v-axis but is not one of the two parallels through A to the v-axis. Show that l and the u-axis share a perpendicular.

14.5 Prove the assertion for abstract surfaces made by Riemann in his celebrated *Habilitationsvortrag* that the metric:

$$ds^2 = \frac{du^2 + dv^2}{\left(1 + \frac{\alpha}{4}(u^2 + v^2)\right)^2}$$

determines a surface of constant curvature given by the value α.

14.6 Show that the transformation of the complex plane $z \mapsto \dfrac{z+i}{i+iz}$ is an isometry between the abstract surfaces $\mathbb{D}_P \to \mathbb{H}$.

14.7 Prove Proposition 14.8. Prove Proposition 14.11.

14.8 Show that the set of complex number's satisfying the equation:

$$A|z|^2 + B\bar{z} + \bar{B}z + C = 0,$$

where A and C are real and $|B|^2 \geq AC$, forms a circle if $A \neq 0$ and a line if $A = 0$.

Solutions to Exercises marked with a dagger appear in the appendix, pp. 338–339.

14.9 Derive the formula for distance in $\mathbb{H}$ directly as follows: We can parametrize the geodesics that are not vertical lines conveniently. Recall that $\cosh^2 t - \sinh^2 t = 1$. Dividing by $\cosh^2 t$ and rearranging terms we get $\tanh^2 t + \operatorname{sech}^2 t = 1$. Since $\tanh t > 0$ for all real t, the following is a parametrization of the semicircle with center c on the real axis and radius r (the so-called *Weierstrass coordinates*):

$$t \mapsto (r\tanh t + c, r\operatorname{sech} t).$$

Use this parametrization and the Poincaré metric to obtain the formula for distance between two points on a semicircular geodesic.

14.10 Let z_1, z_2, z_3, and z_4 be distinct complex numbers, and let $w = (z_1, z_2; z_3, z_4)$ denote their cross ratio. Determine, in terms of w, all of the values taken on by the cross ratio when the four given distinct values z_1, z_2, z_3, and z_4 are permuted.

14.11 The distance function in the Poincaré half-plane is related to the cross ratio. By manipulating the expressions a little, prove that when $P = x_1 + iy_1$ and $Q = x_2 + iy_2$ and P and Q lie on a semicircle centered at $c \in \mathbb{R}$ of radius r, that

$$d_{\mathbb{H}}(P,Q) = \left| \ln\left(\frac{x_1 - c + r}{x_2 - c + r} \cdot \frac{y_2}{y_1} \right) \right|.$$

14.12 Show that the area of a hyperbolic square is bounded by a constant, while the area of a hyperbolic circle is not bounded.

14.13 Show directly that a circle of radius $0 < r < 1$ in $\mathbb{D}_B$ centered at $(0,0)$ maps via the isometry $\phi\colon \mathbb{D}_B \to \mathbb{H}$ to a Euclidean circle. Which one?

14.14 Show that the non-Euclidean circle in $\mathbb{H}$ with center $P = (a,p)$ and radius ρ is the Euclidean circle of radius $p\sinh(\rho)$ and center $(a, p\cosh(\rho))$. What is the area of this circle?

14.15 We know that a non-Euclidean circle in $\mathbb{D}_P$ is a Euclidean circle and determine the center and radius from the Euclidean data, $(x-a)^2 + (y-b)^2 = r^2$ for $0 < r < 1 - (a^2 + b^2)$.

14.16 Given a line l in $\mathbb{H}$ give a construction for the pencil of lines in $\mathbb{H}$ all of which are perpendicular to l.

14.17† Prove that a Euclidean circle of radius $p\sinh(\rho)$ with center $(a, p\cosh(\rho))$ corresponds to a non-Euclidean circle with center $P = (a,p)$ and radius ρ in $\mathbb{H}$.

14.18† Use the analytic methods developed so far to prove the following important results in non-Euclidean trigonometry—Bolyai's Hyperbolic Law of Sines (Theorem 4.31):

$$\frac{\sin A}{\sinh a} = \frac{\sin B}{\sinh b} = \frac{\sin C}{\sinh c},$$

and the hyperbolic Pythagorean Theorem (Theorem 4.30):

$$\cosh C = \cosh a \cosh b.$$

15

Epilogue: Where from here?

Alongside Euclid, who still holds pride of place, Lobachevskiĭ and Riemann have played a particularly important role in research on the foundation of geometry, but Riemann comes before the others.

SOPHUS LIE (1893)

... the nature of the relation between the quantities is frequently rendered more intelligible by regarding them (if only two or three in number) as the co-ordinates of a point in a plane or in space: for more than three quantities there is, from the greater complexity of the case, the greater need of such a representation; ...

ARTHUR CAYLEY (1869)

During the 19th century the development of analysis and mechanics led mathematicians to the use of expressions with many degrees of freedom. The formulation of notions like kinetic energy as a quadratic form and the principle of least action as a variational problem raised the possibility of applying geometric methods to mechanical problems in higher dimensions. As has been argued by Scholz (1980) and Lützen (1995), the acceptance of non-Euclidean geometry in the second half of the 19th century, together with the development of higher dimensional phenomena in mechanics and projective geometry, set the stage for the introduction of higher dimensional geometry as a framework to represent and solve naturally arising problems. By freeing geometry from its Euclidean origins it was possible to imagine a study of geometric objects in any dimension on which complex mechanical systems moved according to well known principles.

On the tenth day of June 1854, G. F. BERNHARD RIEMANN (1826–60) delivered a lecture to the Philosophical Faculty of the University in Göttingen to fulfill the requirements of promotion to *Privatdozent*, a position allowing him to receive fees for his lectures. As was customary, Riemann offered three possible topics for his lecture. The first two dealt with electricity, and the third with the foundations of geometry. Against usual practice, Gauss, a member of his committee, chose the third topic and Riemann offered the lecture *On the hypotheses that lie at the foundations of geometry*. There he launched the next stage of development of differential geometry. A translation of Riemann's words follows this chapter.

The text of the lecture did not appear in Riemann's lifetime. Its posthumous publication in 1868 brought a nearly immediate response from the community of mathematicians working on geometry. Coincidentally in 1868, Beltrami's *Essay on the interpretation of non-Euclidean geometry* appeared, in which he introduced the disk model of chapter 14. He focused on the non-Euclidean plane, leaving unresolved the problem of realizing the three dimensional non-Euclidean space of Lobachevskiĭ and Bolyai. In fact, he thought that such a space was impossible: *"We believe we have attained this goal for the planar part of the doctrine, but we believe that it is impossible to proceed further"* (Beltrami 1868). Shortly after Riemann's lecture appeared, Beltrami (1868b) published an analysis of spaces of constant curvature in many dimensions based on Riemann's ideas that includes a detailed version of non-Euclidean three dimensional space. See Stillwell (1996) for an English translation of and commentary on these papers of Beltrami and others.

The reader will find that the mathematical content of Riemann's lecture is sparse and condensed. However, there was enough detail to give direction to subsequent generations of geometers. One of the key points is the separation of the point set properties of a higher-dimensional space, its topology, from the possible metric properties. As in the case of an abstract surface, many metrics may be defined on a particular point set. When the point set is the space of our universe, then a choice of a metric is a physical problem because there is no "natural" choice of the geometry of the physical world. This insight dispels the philosophical confusion over the primacy of Euclidean or non-Euclidean geometry. Riemann had steeped himself in Gauss's theory of surfaces and fully appreciated the intrinsic viewpoint; it is the cornerstone of his approach. Further discussions of these historical developments may be found in Scholz (1980), Gray (2007), Spivak (1979, Vol.2), and Torretti (1978).

We begin our discussion of Riemann's ideas with his "concept of space," that is, the appropriate notion of a point set for higher dimensions.

Manifolds (differential topology)

What is a many-dimensional space? Riemann proposed the concept of "multiply extended quantities," which was refined for the next fifty years and is given today by the notion of a manifold.

Definition 15.1. *An **n-dimensional manifold** is a set M (often denoted M^n) equipped with a countable collection of one-to-one functions called* **coordinate charts***:*

$$\mathcal{C} = \{x_\alpha : (U_\alpha \subset \mathbb{R}^n) \to M \mid \alpha \in A\},$$

such that:

(1) *Each U_α is an open set in $\mathbb{R}^n$.*
(2) $\bigcup_{\alpha \in A} x_\alpha(U_\alpha) = M$.
(3) *For x_α and x_β in $\mathcal{C}$, if $x_\alpha(U_\alpha) \cap x_\beta(U_\beta) = V_{\alpha\beta} \neq \emptyset$, then the composite:*

$$x_\alpha^{-1} \circ x_\beta : x_\beta^{-1}(V_{\alpha\beta}) \to x_\alpha^{-1}(V_{\alpha\beta})$$

is a smooth mapping (the **transition function***) between open subsets of $\mathbb{R}^n$. The collection $\mathcal{C}$ generates a maximal set of such charts $\mathcal{A}$ called an* **atlas** *of charts on M. An atlas is maximal in the following sense: If $x\colon U \to M$ is another chart for which $x_\alpha^{-1} \circ x$ and $x^{-1} \circ x_\alpha$ are smooth for all $x_\alpha \in \mathcal{C}$, then x is in the atlas $\mathcal{A}$. The atlas generated by $\mathcal{C}$ is called a* **differentiable structure** *on M.*

(4) *The collection of subsets $\{x(U)\}$ for x in the atlas determined by $\mathcal{A}$ is a subbasis for a topology on the set M, which is required to satisfy the Hausdorff condition. That is, if p and q are in M, then there are coordinate charts in the atlas $y\colon V \to M$ and $z\colon W \to M$ such that $p \in y(V)$, $q \in z(W)$, and $y(V) \cap z(W) = \emptyset$. Finally, this topology is required to be second countable.*

This definition generalizes the definition of an abstract surface, which is a two dimensional manifold.

If $p \in M^n$ is a point in the image of a coordinate chart $x\colon (U \subset \mathbb{R}^n) \to M$, then we denote $x^{-1}(p) = (x^1, \ldots, x^n)$. The superscript convention is chosen to agree with the algebraic formalism of tensor analysis on which the modem local theory of manifolds is based. More on this later.

EXAMPLES:

(1) The simplest n-dimensional manifold is $\mathbb{R}^n$ with the atlas generated by the identity chart. More generally, the atlas containing the identity chart makes any open subset of $\mathbb{R}^n$ an n-dimensional manifold by restricting to that subset.
(2) All of the abstract surfaces of chapter 13, and hence all of the classical surfaces in $\mathbb{R}^3$ of chapter 7, are two-dimensional manifolds.
(3) The unit sphere centered at the origin in $\mathbb{R}^n$ is the set:

$$S^{n-1} = \{(x^1, \ldots, x^n) \in \mathbb{R}^n \mid \textstyle\sum_j (y^j)^2 = 1\}.$$

We can define coordinate charts for the sphere with domain the open set in $\mathbb{R}^{n-1}$:

$$E = \{(y^1, \ldots, y^{n-1}) \mid \textstyle\sum_j (y^j)^2 < 1\}.$$

For each i, with $1 \le i \le n$, define two mappings $x_i^{\pm}\colon E \to S^{n-1}$ given by:

$$x_i^{\pm}(y^1, \ldots, y^{n-1}) = (y^1, \ldots, y^{i-1}, \pm\sqrt{1 - \textstyle\sum_j (y^j)^2}, y^i, \ldots, y^{n-1}).$$

The set of $2n$ charts $\{x_i^{\pm} \mid 1 \le i \le n\}$ covers S^{n-1} and generates an atlas making the sphere an $(n-1)$-dimensional manifold.
(4) Let $\mathbb{RP}^n$ denote the set of all lines through the origin in $\mathbb{R}^{n+1}$. There are algebraic coordinates $[x^0, \ldots, x^n]$ for points $\mathbb{RP}^n$ defined as the equivalence classes under the relation:

$$(x^0, \ldots, x^n) \sim (rx^0, \ldots, rx^n) \text{ whenever } r \neq 0.$$

In each equivalence class there are exactly two representatives for which $\sum_j (x^j)^2 = 1$, where the given line meets S^n. Coordinate charts for S^n that satisfy $x(U) \cap (-x(U)) = \emptyset$ determine coordinate charts on $\mathbb{RP}^n$. The n-dimensional manifold $\mathbb{RP}^n$ with the atlas generated by these charts is called the *real projective n-space.*

Differentiable mappings between manifolds are defined as in the case of mappings to and from abstract surfaces (Definition 13.2).

Definition 15.2. *Given two manifolds M and M', of dimensions n and n', respectively, a function $\phi\colon M \to M'$ is differentiable at a point $p \in M$ if, for any coordinate charts, $x_\alpha\colon (U_\alpha \subset \mathbb{R}^n) \to M$ around p and $y_\beta\colon (V_\beta \subset \mathbb{R}^{n'}) \to M'$ around $\phi(p)$, the composite $y_\beta^{-1} \circ \phi \circ x_\alpha$ is a differentiable mapping $U_\alpha \to V_\beta$. A function $\phi\colon M \to M'$ is* **differentiable** *if it is differentiable at every point $p \in M$. A differentiable function $\phi\colon M \to M'$ is a* **diffeomorphism** *if ϕ is one-to-one and onto, and has a differentiable inverse function.*

For example, the mapping $\pi : S^n \to \mathbb{RP}^n$ given by $(x^0, \ldots, x^n) \mapsto [x^0, \ldots, x^n]$ is a differentiable mapping.

The definition of tangent vectors and the tangent space at a point in manifold is a generalization of the definition for abstract surfaces. Denote the set of all smooth, real-valued functions on M by $C^\infty(M)$. Locally, at $p \in M$, let $C^\infty(p)$ denote the set of smooth, real-valued functions defined on an open subset of M containing p. Both $C^\infty(M)$ and $C^\infty(p)$ are real algebras: If q is in an open set containing p on which f and g are defined, then $(f+g)(q) = f(q) + g(q)$, $(rf)(q) = r(f(q))$, and $(fg)(q) = f(q)g(q)$.

Definition 15.3. *Given a curve $\lambda\colon (-\epsilon, \epsilon) \to M$ through a point $p = \lambda(0)$ in M, define the* **tangent vector** *to λ at $t = 0$ as the linear mapping:*

$$\lambda'(0)\colon C^\infty(p) \to \mathbb{R}, \quad \lambda'(0)(f) = \frac{d}{dt}(f \circ \lambda(t))\bigg|_{t=0}.$$

The collection of all such linear mappings for all smooth curves through p is denoted by $T_p(M)$, the **tangent space** *of M at p.*

Proposition 15.4. *$T_p(M)$ is an n-dimensional vector space. Furthermore, if $X \in T_p(M)$ and f, $g \in C^\infty(p)$, then $X(fg) = f(p)X(g) + g(p)X(f)$, that is, X satisfies the Leibniz rule for the product of functions in $C^\infty(p)$.*

The coordinate curves associated to a chart, $t \mapsto x(x^1, \ldots, x^{i-1}, t, x^{i+1}, \ldots, x^n)$, have tangent vectors that form a basis for the tangent space $T_p(M)$ denoted $\left\{\dfrac{\partial}{\partial u^i}, i = 1, \ldots, n\right\}$.

Definition 15.5. *Given a differentiable mapping* $\phi\colon M \to M'$ *and a point* p *in* M, *the* **differential** *of* ϕ *at* p,

$$d\phi_p\colon T_p(M) \longrightarrow T_{\phi(p)}(M'),$$

is given by $d\phi_p(\lambda'(0)) = \dfrac{d}{dt}(\phi\circ\lambda(t))\Big|_{t=0}$.

As in the case of surfaces, $d\phi_p$ is linear. If $x\colon (U\subset\mathbb{R}^n)\to M$ is a coordinate patch containing p, then $\lambda(t) = x(x^1(t),\ldots,x^n(t))$; for $y\colon (V\subset\mathbb{R}^n)\to M'$ a coordinate patch containing $\phi(p)$ with coordinates $y(y^1,\ldots,y^{n'})$, we have a local expression for $d\phi_p$:

$$d\phi_p(\lambda'(0)) = \sum\nolimits_{i,j}(x^i)'(0)\frac{\partial(y^{-1}\circ\phi\circ x)^j}{\partial x^i}\frac{\partial}{\partial y^j}.$$

It follows from the linearity and invertibility of the differential associated to a diffeomorphism $\phi\colon M\to M'$ that $\dim M = \dim M'$.

EXAMPLES:

(5) A source of manifolds with beautiful properties is the family of groups of matrices. In particular, we may consider the collection of all $n\times n$ matrices with real entries as the space $\mathbb{R}^{n^2}$. The determinant mapping $\det\colon \mathbb{R}^{n^2}\to\mathbb{R}$ is smooth and the inverse image of the regular value zero is a closed subset of $\mathbb{R}^{n^2}$. The complement of $\det^{-1}(\{0\})$ is the n^2-dimensional manifold, $\mathrm{GL}_n(\mathbb{R})$ of invertible $n\times n$ matrices.

The symmetric $n\times n$ matrices, that is, those that satisfy $A = A^t$ (the transpose of A), form a subset of $\mathbb{R}^{n^2}$ of dimension $n(n+1)/2$ obtained by mapping $\mathbb{R}^{n(n+1)/2}$ to $\mathbb{R}^{n^2}$ as the entries along the diagonal and above the diagonal of a symmetric matrix. The intersection of the symmetric matrices with $\mathrm{GL}_n(\mathbb{R})$ is the subspace of the manifold $\mathrm{GL}_n(\mathbb{R})$ of invertible symmetric matrices. We define the mapping $F\colon \mathrm{GL}_n(\mathbb{R})\to\mathrm{GL}_n(\mathbb{R})$ by $F(A) = AA^t$. This mapping is smooth, and it makes a matrix symmetric because $(A^t)^t = A$ and $(AB)^t = B^tA^t$. In the discussion at the end of chapter 6, we proved that the group $O(n)$ of rigid motions of $\mathbb{R}^n$ is made up of invertible $n\times n$ matrices satisfying $A^{-1} = A^t$ (Proposition 6.15). This condition means that $O(n) = F^{-1}(\mathrm{id})$, where id is the identity matrix. To determine the dimension of $O(n)$, we look at the differential $dF_A\colon T_A(\mathrm{GL}_n(\mathbb{R}))\to T_{\mathrm{id}}(\mathrm{GL}_n(\mathbb{R}))$ for $A\in O(n)$. Since F is constant along the $O(n)$ directions, $T_A(O(n))$ lies in the kernel of dF_A. The image $F(\mathrm{GL}_n(\mathbb{R}))$ lies in the symmetric matrices, and so the image of the tangent space has dimension at most $n(n+1)/2$. By applying a version of the Implicit Function Theorem (chapter 7) one can prove that the dimension of $O(n)$ is as large as it can be, that is, $n(n-1)/2$, which is $n^2 - n(n+1)/2$, the dimension of the kernel of dF_A. Thus, $O(n)$ is an $n(n-1)/2$-dimensional manifold.

(6) Given a manifold M^n we can associate to it a $2n$-dimensional manifold by "glueing together" all of the tangent spaces of M: Let $TM = \bigcup_{p\in M} T_p(M)$. A

point in TM has the form (p,X), where $p \in M$ and $X \in T_p(M)$. If $x\colon (U \subset \mathbb{R}^n) \to M$ is a coordinate chart containing p, then we can define a chart $\bar{x}\colon (U \times \mathbb{R}^n \subset \mathbb{R}^{2n}) \to TM$ around (p,X) by letting:

$$\bar{x}(x^1,\ldots,x^n,z^1,\ldots,z^n) = \left(x(x^1,\ldots,x^n), \sum_{i=1}^{n} z^i \frac{\partial}{\partial x^i} \right).$$

To prove that we have a manifold, we check the transition functions between charts. If $y\colon (V \subset \mathbb{R}^n) \to M$ is another chart containing p, then we can write $p = x(x^1,\ldots,x^n) = y(y^1,\ldots,y^n)$ and $x^i = x^i(y^1,\ldots,y^n)$, so $\dfrac{\partial}{\partial x^i} = \sum_{j=1}^{n} \dfrac{\partial y^j}{\partial x^i}\dfrac{\partial}{\partial y^j}$.
The effect on the coordinates for the tangent space is given by:

$$\bar{x}^{-1} \circ \bar{y}(y^1,\ldots,y^n,z^1,\ldots,z^n) = \left(x^1,\ldots,x^n, \sum_{i=1}^{n} z^i \frac{\partial x^1}{\partial y^i},\ldots,\sum_{i=1}^{n} z^i \frac{\partial x^n}{\partial y^i} \right).$$

The Jacobian of this transformation is given by:

$$J(\bar{x}^{-1} \circ \bar{y}) = \begin{pmatrix} J(x^{-1} \circ y) & 0 \\ H & J(x^{-1} \circ y) \end{pmatrix},$$

where H denotes the matrix $(h_{jk}) = \left(\sum_i \dfrac{\partial^2 x^j}{\partial y^i \partial y^k} \right)$. The determinant of the Jacobian $J(\bar{x}^{-1} \circ \bar{y})$ is $(\det J(x^{-1} \circ y))^2$, which is nonzero, and so the transition functions are smooth and nondegenerate. Thus TM is a $2n$-dimensional manifold.

The manifold TM is equipped with a smooth mapping $\pi\colon TM \to M$ given by $\pi(p,X) = p$. The mapping π is called the **tangent bundle** of M. The manifold TM is also called the *total space* of the tangent bundle.

Let $M = \mathbb{R}^n$ and consider the tangent bundle $T\mathbb{R}^n$. Because $\mathbb{R}^n$ has an atlas generated by a single chart, $T\mathbb{R}^n$ may be expressed as $\mathbb{R}^n \times \mathbb{R}^n$, that is, a product. In general $TM \neq M \times \mathbb{R}^n$ for most n-dimensional manifolds. The failure is often a consequence of the topology of the manifold. The study of the topological properties of manifolds is the focus of differential topology, which developed into an important area of mathematics in the century after Riemann's lecture appeared; see Dubrovin et al. (1985) or Guillemin and Pollack (1974).

Vector and tensor fields

With the tangent bundle, $\pi\colon TM \to M$, we can define vector fields in a simple manner.

Definition 15.6. *A differentiable* **vector field** *on a manifold M is a smooth mapping $X\colon M \to TM$ satisfying $\pi \circ X = \mathrm{id}_M$, that is, $X(p) = (p,v)$ for some $v \in T_p(M)$.*

The economy of language is clear–a vector field is a differentiable mapping of manifolds with a certain property. In order to describe a vector field locally, we choose a coordinate patch $x\colon (U \subset \mathbb{R}^n) \to M$ and write $X(p) = (p, v(p))$. Then, $v(p) = \sum_{i=1}^{n} a^i(p) \frac{\partial}{\partial x^i}$ for differentiable functions a^i defined on U. Equivalently, if we choose a set of functions $\{a^i_U\colon U \to \mathbb{R} \mid 1 \le i \le n\}$ on each coordinate chart $x\colon U \to M$ and require on the overlap of charts that the sets of functions transform into one another coherently, then we have determined a vector field. The transformation rule for coherence is derived as follows: For two coordinate patches $x\colon U \to M$ and $y\colon V \to M$ write:

$$X(p) = \sum_i a^i_U(p) \frac{\partial}{\partial x^i} = \sum_j a^j_V(p) \frac{\partial}{\partial y^j}.$$

By the chain rule we have $\frac{\partial}{\partial x^i} = \sum_j \frac{\partial y^j}{\partial x^i} \frac{\partial}{\partial y^j}$, and so, replacing and recombining we have, for all j:

$$a^j_V = \sum_i a^i_U \frac{\partial y^j}{\partial x^i}.$$

This expression is the desired transformation rule and it is the defining property for a $\binom{1}{0}$-tensor, that is, the sums $\sum_i a^i_U(p) \frac{\partial}{\partial x^i}$, one for each coordinate chart $x\colon (U \subset \mathbb{R}^n) \to M$, form the $\binom{1}{0}$-tensor field (or vector field). Notice that it is the analytic fact, the chain rule, that determines the transformation rule, *not* the geometry.

The coordinate charts provide a locally Euclidean structure on a manifold that allows us to locally define derivatives, vector fields, and integrals. In a given coordinate chart, such expressions may be manipulated as if we are working in an open subset of $\mathbb{R}^n$, but without coherence, they may not extend to all of the manifold. The problem we consider next is how to decide when a collection of analytic expressions in coordinates is not dependent on the choice of coordinates. First, some examples.

Vector fields are globally defined as smooth mappings $X\colon M \to TM$. Dual to vector fields are the *1-forms*, expressions that appear in line integrals. Locally, we write $\sum_i \theta_i dx^i$. A change of variables leads to a transformation rule:

$$\sum_i \theta^U_i dx^i = \sum_j \theta^V_j dy^j \text{ implies } \theta^V_j = \sum_i \theta^U_i \frac{\partial x^i}{\partial y^j}.$$

This equation follows from the familiar rule $dx^i = \sum_k \frac{\partial x^i}{\partial y^k} dy^k$. A collection of sums $\sum_i \theta^U_i dx^i$, one for each coordinate chart, is called a $\binom{0}{1}$-**tensor field** (or a

1-form) on M if the coefficient functions on overlapping charts are related by the transformation rule.

In the hierarchy of tensors, a differentiable function $f\colon M \to \mathbb{R}$ is a $\binom{0}{0}$-tensor. The partial derivatives of f with respect to different charts transform according to the rule governing a $\binom{0}{1}$-tensor,

$$\frac{\partial f}{\partial y^j} = \sum\nolimits_i \frac{\partial f}{\partial x^i}\frac{\partial x^i}{\partial y^j},$$

where $f(y^1,\ldots,y^n) = f(x^1(y^1,\ldots,y^n),\ldots,x^n(y^1,\ldots,y^n))$. To recover f from its derivatives, we apply Stokes's Theorem (Spivak 1965) and integrate the 1-form:

$$df = \sum_{i=1}^{n} \frac{\partial f}{\partial x^i} dx^i = \sum_{j=1}^{n} \frac{\partial f}{\partial y^j} dy^j.$$

We note that the "derivative" df of the $\binom{0}{0}$-tensor f is a $\binom{0}{1}$-tensor.

Metrical relations (Riemannian manifolds)

In chapter 13 we considered the case of a Riemannian metric on an abstract surface S. To set the stage for generalization, we introduce a more flexible notation for the component functions of a Riemannian metric on a surface:

$$g_{11} = E, \quad g_{12} = g_{21} = F, \quad \text{and} \quad g_{22} = G.$$

The line element is an aggregate of four functions (g_{ij}), and an example of a $\binom{0}{2}$-tensor field on the surface S. That is, $\{g_{ij}\}$ satisfies the transformation rule on overlapping charts:

$$ds^2 = \sum\nolimits_{r,s} g_{rs} dx^r dx^s = \sum\nolimits_{i,j} \bar{g}_{ij} dy^i dy^j.$$

Lemma 13.7 implies that:

$$\bar{g}_{ij} = \sum\nolimits_{r,s} g_{rs} \frac{\partial x^r}{\partial y^i}\frac{\partial x^s}{\partial y^j}.$$

In the second part of his lecture, Riemann turned to the problem of making elementary geometric constructions possible on a manifold for which one requires lines and circles. To define these special metric curves, we need a notion of length, that is, a line element on the manifold.

Definition 15.7. *A* **Riemannian metric** *on a manifold M is a choice, for each $p \in M$, of a positive-definite inner product $\langle\ ,\ \rangle_p$ on each tangent space $T_p(M)$, such that the choice varies smoothly from point to point. A manifold with a choice of Riemannian metric is called a* **Riemannian manifold**.

For a particular coordinate chart $x\colon (U \subset \mathbb{R}^n) \to M$, the inner product may be represented by a symmetric matrix of smooth functions $(g_{ij}(x^1,\ldots,x^n))$. If X and Y are tangent vectors at $p = x(u^1,\ldots,u^n)$, then $X = \sum_i a^i \frac{\partial}{\partial x^i}$ and $Y = \sum_i b^i \frac{\partial}{\partial x^i}$, and:

$$\langle X, Y\rangle_p = (a^1,\ldots,a^n)(g_{ij}(x^1,\ldots,x^n))\begin{pmatrix} b^1 \\ \vdots \\ b^n \end{pmatrix}.$$

Independence of the metric on the choice of a coordinate chart requires that the functions g_{ij} form a $\binom{0}{2}$-tensor.

Lemma 15.8. *If $x\colon (U \subset \mathbb{R}^n) \to M$ and $y\colon (V \subset \mathbb{R}^n) \to M$ are two coordinate charts for M, and p is a point in $x(U) \cap y(V)$, then:*

$$(\bar{g}_{rs}) = J(x^{-1} \circ y)^t (g_{ij}) J(x^{-1} \circ y),$$

where (g_{ij}) is the matrix of component functions of the metric associated to x, $(\bar{g}_{rs})$ is the matrix associated to y, and $J(x^{-1} \circ y)$ is the Jacobian matrix associated to the mapping $x^{-1} \circ y\colon (y^{-1}(x(U) \cap y(V)) \subset \mathbb{R}^n) \to (x^{-1}(x(U) \cap y(V)) \subset \mathbb{R}^n)$.

The proof is left to the reader. Component-wise, this matrix equation gives us:

$$\bar{g}_{rs} = \sum_{i,j} g_{ij} \frac{\partial x^i}{\partial y^r} \frac{\partial x^j}{\partial y^s},$$

which is the defining property of a $\binom{0}{2}$-tensor field on M. The fact that the metric is positive-definite implies that the matrix (g_{ij}) is invertible; we denote its inverse by (g^{kl}). The Jacobian of $x^{-1} \circ y$ is also an invertible matrix:

$$J(x^{-1} \circ y) = \left(\frac{\partial x^i}{\partial y^j}\right), \text{ and so } J(x^{-1} \circ y)^{-1} = J(y^{-1} \circ x) = \left(\frac{\partial y^i}{\partial x^j}\right).$$

These relations imply the identities $\sum_k \frac{\partial y^r}{\partial x^k} \frac{\partial x^k}{\partial y^s} = \delta^r_s$ and $\sum_k g_{ik} g^{kj} = \delta^j_i$, where δ^m_n is the Kronecker delta, $\delta^m_m = 1$, and, if $n \neq m$, $\delta^m_n = 0$. These rules lead to new tensor fields.

Vector fields, differentials for line integrals, and Riemannian metrics are all geometrically important examples of tensors, which satisfy appropriate transformation rules on overlapping charts. More generally, we can define more complex tensors.

Definition 15.9. *For each coordinate chart* $x\colon (U \subset \mathbb{R}^n) \to M$*, an aggregate of functions:*

$$\{T^{i_1\cdots i_s}_{j_1\cdots j_r}\colon U \to \mathbb{R} \mid 1 \leq i_1,\ldots,i_s,j_1,\ldots,j_r \leq n\},$$

constitutes an $\binom{s}{r}$**-tensor field** *on* M *if, on* $y\colon (V \subset \mathbb{R}^n) \to M$*, an overlapping chart, the functions* $\bar{T}$ *associated to the coordinate chart* y *satisfy the transformation rule:*

$$\bar{T}^{k_1\cdots k_s}_{l_1\cdots l_r} = \sum_{i_1,\ldots,i_s,j_1,\ldots,j_r} T^{i_1\cdots i_s}_{j_1\cdots j_r} \frac{\partial y^{k_1}}{\partial x^{i_1}} \cdots \frac{\partial y^{k_s}}{\partial x^{i_s}} \frac{\partial x^{j_1}}{\partial y^{l_1}} \cdots \frac{\partial x^{j_r}}{\partial y^{l_r}},$$

In order to build a natural algebraic home for tensors, we would need a sizeable detour through the linear algebra of tensor products and the dual and tensor product bundles associated to the tangent bundle; see Warner (1983) or Lee (2003) for a thorough foundation. Riemann and his immediate successors did not have these algebraic notions, though they manipulated the complex expressions associated to these constructions.

The key feature of the definition of a tensor field is the transformation rule that implies that the tensor field does not depend on a choice of coordinates. The founder of this algebraic structure is G. RICCI-CURBASTRO (1853–1925). He and his student T. LEVI-CIVITA (1873–1941) presented a unified treatment of tensors and their calculus in a paper (Ricci, Levi-Civita 1901) on the *absolute differential calculus*, which refers to the independence of choice of coordinates. The absolute differential calculus was renamed tensor analysis by Einstein in 1916.

The relation between a Riemannian metric and its inverse leads to operations on tensors.

Proposition 15.10. *Suppose* $\{T^{i_1\cdots i_s}_{j_1\cdots j_r}\}$ *is an* $\binom{s}{r}$*-tensor field on* M*, an* n*-dimensional Riemannian manifold. Then the following expressions:*

$$U^{i_1\cdots\widehat{i_p}\cdots i_s}_{\lambda j_1\cdots j_r} = \sum_{m=1}^{n} g_{\lambda m} T^{i_1\cdots i_{p-1} m i_{p+1}\cdots i_s}_{j_1\cdots j_r}, \text{ and } V^{\lambda i_1\cdots i_s}_{j_1\cdots\widehat{j_q}\cdots j_r} = \sum_{m=1}^{n} g^{\lambda m} T^{i_1\cdots i_s}_{j_1\cdots j_{q-1} m j_{q+1}\cdots j_r}$$

constitute an $\binom{s-1}{r+1}$*- and an* $\binom{s+1}{r-1}$*-tensor field, respectively. The operation of* **contraction** *of indices on a tensor field is defined by:*

$$W^{i_1\cdots\widehat{i_p}\cdots i_s}_{j_1\cdots\widehat{j_q}\cdots j_r} = \sum_{m=1}^{n} T^{i_1\cdots i_{p-1} m i_{p+1}\cdots i_s}_{j_1\cdots j_{q-1} m j_{q+1}\cdots j_r}.$$

The contraction of an $\binom{s}{r}$*-tensor field is an* $\binom{s-1}{r-1}$*-tensor field.*

Here, $\widehat{\ }$ means to omit the index marked; for example, $\widehat{i_1} i_2 \cdots i_n = i_2 \cdots i_n$.

PROOF: It suffices to demonstrate the transformation rules. We prove one case and leave the others to the reader.

$$\bar{U}^{k_1\cdots\widehat{k_p}\cdots k_s}_{\lambda l_1\cdots l_r} = \sum_{m=1}^{n} \bar{g}_{\lambda m}\bar{T}^{k_1\cdots k_{p-1}mk_{p+1}\cdots k_s}_{l_1\cdots l_r}$$

$$= \sum_{\substack{u,v,m,i_1,\ldots,i_s\\ j_1,\ldots,j_r}} g_{uv}\frac{\partial x^u}{\partial y^\lambda}\frac{\partial x^v}{\partial y^m}T^{i_1\cdots i_s}_{j_1\cdots j_r}\frac{\partial y^{k_1}}{\partial x^{i_1}}\cdots\frac{\partial y^{k_{p-1}}}{\partial x^{i_{p-1}}}\frac{\partial y^m}{\partial x^{i_p}}\frac{\partial y^{k_{p+1}}}{\partial x^{i_{p+1}}}\cdots\frac{\partial y^{k_s}}{\partial x^{i_s}}\frac{\partial x^{j_1}}{\partial y^{l_1}}\cdots\frac{\partial x^{j_r}}{\partial y^{l_r}}$$

$$= \sum_{\substack{u,i_1,\ldots,i_s\\ j_1,\ldots,j_r}} g_{ui_p}T^{i_1\cdots i_s}_{j_1\cdots j_r}\frac{\partial y^{k_1}}{\partial x^{i_1}}\cdots\frac{\partial y^{k_{p-1}}}{\partial x^{i_{p-1}}}\frac{\partial y^{k_{p+1}}}{\partial x^{i_{p+1}}}\cdots\frac{\partial y^{k_s}}{\partial x^{i_s}}\frac{\partial x^u}{\partial y^\lambda}\frac{\partial x^{j_1}}{\partial y^{l_1}}\cdots\frac{\partial x^{j_r}}{\partial y^{l_r}}$$

$$= \sum_{\substack{i_1,\ldots,\widehat{i_p},\ldots,i_s\\ u,j_1,\ldots,j_r}} U^{i_1\cdots\widehat{i_p}\cdots i_s}_{uj_1\cdots j_r}\frac{\partial x^u}{\partial y^\lambda}\frac{\partial y^{k_1}}{\partial x^{i_1}}\cdots\frac{\partial y^{k_{p-1}}}{\partial x^{i_{p-1}}}\frac{\partial y^{k_{p+1}}}{\partial x^{i_{p+1}}}\cdots\frac{\partial y^{k_s}}{\partial x^{i_s}}\frac{\partial x^u}{\partial y^\lambda}\frac{\partial x^{j_1}}{\partial y^{l_1}}\cdots\frac{\partial x^{j_r}}{\partial y^{l_r}}.$$

Here we used the relation $\sum_{m=1}^{n}\frac{\partial x^v}{\partial y^m}\frac{\partial y^m}{\partial x^{i_p}} = \delta^v_{i_p}$ and the definition of U. ■

These procedures of "raising," "lowering," and contracting indices allow us to change the types of tensors, sometimes arriving at geometrically significant expressions from other such tensors (see Proposition 15.14).

Returning to geometry, we define the **arc length** along a curve $\lambda(t)$ in a Riemannian manifold $(M,\langle\ ,\ \rangle)$ by:

$$s(t) = \int_{a_0}^{t}\sqrt{\langle\lambda'(\tau),\lambda'(\tau)\rangle_{\lambda(\tau)}}\,d\tau = \int_{a_0}^{t}\sqrt{\sum_{i,j}g_{ij}(dx^i/d\tau)(dx^j/d\tau)}\,d\tau,$$

where $\lambda(t) = x(x^1(t),\ldots,x^n(t))$. The angle θ between curves $\lambda_1(t)$ and $\lambda_2(t)$ is determined by:

$$\cos\theta = \frac{\langle\lambda_1'(t_p),\lambda_2'(t_p)\rangle_p}{\sqrt{\langle\lambda_1'(t_p),\lambda_1'(t_p)\rangle_p\cdot\langle\lambda_2'(t_p),\lambda_2'(t_p)\rangle_p}},$$

where $\lambda_i(t_p) = p$ for $i = 1, 2$. The **line element** is given by the local expression:

$$ds^2 = \sum_{i,j}g_{ij}\,dx^i\,dx^j.$$

The transformation rule (based on the chain rule) $dy^k = \sum_j\frac{\partial y^k}{\partial x^j}dx^j$, gives:

$$ds^2 = \sum_{i,j}g_{ij}\,dx^i\,dx^j = \sum_{r,s}\bar{g}_{rs}\,dy^r\,dy^s,$$

that is, arc length is independent of choice of coordinates.

By separating the manifold from its metrical properties, Riemann made the topological and the geometric aspects of a geometric object independent. For example, let $M^2 = \mathbb{R}^2$ and contrast the Riemannian manifolds:

$$(\mathbb{R}^2, ds^2 = dx^2 + dy^2) \text{ and } (\mathbb{R}^2, ds^2 = dx^2 + e^{2x}dy^2).$$

In the first case, we have the Euclidean plane and in the second case, another model for the geometry of Bolyai and Lobachevskiĭ. Riemann attributed the confusion over Postulate V to the assumption that the plane, as a two-dimensional manifold, has a unique geometry, that is, a unique choice of metric. By separating the metric relations from the space itself, he sought to eliminate the confusion. For three-dimensional space $\mathbb{R}^3$, Riemann showed that there are many metrics and so geometry, as the science of the space in which we live, must depend on physical experience.

One direction of generalization was prompted by "experience," that is, physics. In 1905, ALBERT EINSTEIN (1879–1955) introduced his special Theory of Relativity. In 1908, HERMANN MINKOWSKI (1864–1909) lectured to the 80th Assembly of the German Scientific Union in Köln on *Space and Time*. The text of the lecture begins:

> *The views of space and time which I wish to lay before you have sprung from the soil of experimental physics, and therein lies their strength. They are radical. Henceforth space by itself, and time by itself, are doomed to fade away into mere shadows, and only a kind of union of the two will preserve an independent reality.*

The union of space and time that lies at the basis of special relativity may be described by a slight generalization of the Riemannian metric. We drop the condition that the inner product be positive definite and instead require the weaker condition that it be nondegenerate. Thus, for $p \in M$, for X, Y, and Z in $T_p(M)$, and $r \in \mathbb{R}$,

(1) $\langle rX + Y, Z\rangle_p = r\langle X, Z\rangle_p + \langle Y, Z\rangle_p$.
(2) $\langle X, Y\rangle_p = \langle Y, X\rangle_p$.
(3) If $\langle X, Y\rangle_p = 0$ for all $Y \in T_p(M)$, then $X = 0$ (*nondegeneracy*).

If we represent such a metric locally by a matrix (g_{ij}), then Sylvester's Law of Inertia (Lang 2010) implies that, at a particular point $p \in M$, there is a change of coordinates in which the line element takes the form:

$$ds^2 = \sum_{i,j} g_{ij}\, dx^i\, dx^j = \sum_{k=1}^{l} dy^k\, dy^k - \sum_{m=l+1}^{n} dy^m\, dy^m.$$

In this case we say that the **index** of (g_{ij}) is $n - l$. A *pseudo-Riemannian manifold* is a manifold with a nondegenerate metric of constant index. A *Lorentz manifold* is a pseudo-Riemannian manifold with a metric of index 1.

The simplest Lorentz manifold was introduced in Minkowski's lecture. *Minkowski space* is the manifold $\mathbb{R}^4$ endowed with the Lorentz metric:

$$ds^2 = dx^2 + dy^2 + dz^2 - c^2\,dt^2,$$

where c denotes the speed of light. Tangent vectors in Minkowski space fall into three classes:

$$\begin{aligned} \textit{spacelike:} &\quad \text{if } \langle X,X\rangle_p > 0 \text{ or } X = 0, \\ \textit{null:} &\quad \text{if } \langle X,X\rangle_p = 0, \\ \textit{timelike:} &\quad \text{if } \langle X,X\rangle_p < 0. \end{aligned}$$

This classification models causality and the physical assumption that the speed of light is constant in every frame of reference. A curve or *world line* of a physically meaningful object must have its tangent vector everywhere timelike or null in order to preserve this assumption.

The study of differential geometry and, in particular, pseudo-Riemannian manifolds was spurred on by the development of relativistic physics. The interested reader may consult O'Neill (1983) for a comprehensive introduction to pseudo-Riemannian geometry and the many books on general relativity in the physics library.

Curvature

When can two Riemannian metrics be transformed into one another? This natural question for Riemannian geometry is called the *local equivalence problem*. Several constructions may be reduced to this problem; for example, on the overlap of two coordinate charts, the metrics in the coordinates must transform one to the other, and so this condition requires a solution to the local equivalence problem in the definition of a Riemannian metric. If two manifolds are to be isometric, then at corresponding points we must have a solution to the local equivalence problem. Riemann reduced certain analytic questions from the study of the heat equation to this problem. On this question he offered a counting argument that was a key step.

A metric is determined by $n(n+1)/2$ functions on each coordinate chart, namely, the component functions $\{g_{ij}\}$. A change of coordinates is given by the expression:

$$x(x^1,\ldots,x^n) = y(y^1(x^1,\ldots,x^n),\ldots,y^n(x^1,\ldots,x^n)).$$

This introduces n functions changing at most n of the functions $\{g_{ij}\}$, and leaving $n(n-1)/2$ choices. Riemann argued that there must be some set of $n(n-1)/2$ functions that determine the metric. When $n = 2$, there are three functions, E, F, and G, that make up the components of the metric. Two may be fixed by a change of coordinates (say, to geodesic polar coordinates), leaving one function to determine the metric completely. That one function is the Gauss–Riemann curvature. The work of

Minding (chapter 12) bears this out. The general case is answered by the Riemann curvature tensor, which is determined by a set of $n(n-1)/2$ functions (Proposition 15.15). The theorem that these functions fix a metric uniquely was proved in 1951 by the modern geometer E. CARTAN (1869–1951). The special cases of $n=2$ and $n=3$ were solved by G. DARBOUX (1842–1917) and L. BIANCHI (1856–1928), respectively.

The Riemann curvature tensor is formalized in the work of E. B. CHRISTOFFEL (1829–1900), in particular, in his 1869 paper in Crelle's Journal, *Über die Transformation homogener Differentialausdrücke zweiten Grades*. Though it appeared after Riemann's death, this work is a natural extension of Riemann's ideas. Christoffel acknowledged his debt at the end of the paper, where he thanked Dedekind for making some of the unpublished papers of Riemann available. The Christoffel symbols (chapter 9) were introduced in this paper. Christoffel's computations form the basis for the invariant approach to Riemannian geometry, which characterized its next stage of development.

Suppose that a given metric has component functions $\{g_{ij}\}$ in one set of coordinates, and $\{\bar{g}_{ij}\}$ in another set of coordinates. Recall the transformation rule for the metric,

$$\bar{g}_{rs} = \sum_{i,j} g_{ij} \frac{\partial x^i}{\partial y^r} \frac{\partial x^j}{\partial y^s}.$$

This expression may be differentiated to give the equation:

$$\frac{\partial \bar{g}_{rs}}{\partial y^t} = \sum_{i,j,k} \frac{\partial g_{ij}}{\partial x^k} \frac{\partial x^k}{\partial y^t} \frac{\partial x^i}{\partial y^r} \frac{\partial x^j}{\partial y^s} + \sum_{i,j} g_{ij} \left(\frac{\partial^2 x^i}{\partial y^r \partial y^t} \frac{\partial x^j}{\partial y^s} + \frac{\partial x^i}{\partial y^r} \frac{\partial^2 x^j}{\partial y^s \partial y^t} \right).$$

Notice that the derivatives of the metric coefficients do *not* satisfy the property of being a tensor. The second derivative terms bar the expression from being a tensor. We turn this to an advantage later.

These derivatives may be permuted and summed with signs to obtain the **Christoffel symbols of the first kind** associated to $\{g_{ij}\}$:

$$[jk,i] = \frac{1}{2} \left(\frac{\partial g_{ij}}{\partial x^k} + \frac{\partial g_{ik}}{\partial x^j} - \frac{\partial g_{jk}}{\partial x^i} \right).$$

These functions are sometimes denoted $\Gamma_{jk,i}$. Let $\overline{[st,r]}$ denote the Christoffel symbols of the first kind associated to the metric $\{\bar{g}_{rs}\}$. Substituting the analogous expressions for the $\overline{[st,r]}$ into the definition of the $[jk,i]$ and simplifying we get:

$$\begin{aligned} \overline{[st,r]} &= \frac{1}{2} \left(\frac{\partial \bar{g}_{rs}}{\partial y^t} + \frac{\partial \bar{g}_{rt}}{\partial y^s} - \frac{\partial \bar{g}_{st}}{\partial y^r} \right) \\ &= \sum\nolimits_{i,j,k} [jk,i] \frac{\partial x^i}{\partial y^r} \frac{\partial x^j}{\partial y^s} \frac{\partial x^k}{\partial y^t} + \sum\nolimits_{i,j} g_{ij} \frac{\partial x^i}{\partial y^r} \frac{\partial^2 x^j}{\partial y^s \partial y^t}. \end{aligned}$$

The **Christoffel symbols of the second kind** associated to $\{g_{ij}\}$ are defined as:

$$\Gamma^i_{jk} = \sum\nolimits_{l=1}^{n} g^{il}[jk,l].$$

(Christoffel's original notation was $[jk,i] = \begin{bmatrix} jk \\ i \end{bmatrix}$ and $\Gamma^i_{jk} = \begin{Bmatrix} jk \\ i \end{Bmatrix}$. In Riemann's work, $p_{ijk} = 2[jk,i]$.) Substituting the expression for $[jk,i]$ and switching the roles of the charts, we get:

$$\begin{aligned}
\Gamma^i_{jk} &= \sum_{r,s,t,l} g^{il}\overline{[st,r]}\frac{\partial y^r}{\partial x^l}\frac{\partial y^s}{\partial x^j}\frac{\partial y^t}{\partial x^k} + \sum_{r,s,l} g^{il}\bar{g}_{rs}\frac{\partial y^r}{\partial x^l}\frac{\partial^2 y^s}{\partial x^j \partial x^k} \\
&= \sum_{r,s,t,u,v,l} \bar{g}^{uv}\frac{\partial x^i}{\partial y^u}\frac{\partial x^l}{\partial y^v}\overline{[st,r]}\frac{\partial y^r}{\partial x^l}\frac{\partial y^s}{\partial x^j}\frac{\partial y^t}{\partial x^k} + \sum_{r,s,u,v,l} \bar{g}^{uv}\bar{g}_{rs}\frac{\partial x^i}{\partial y^u}\frac{\partial x^l}{\partial y^v}\frac{\partial y^r}{\partial x^l}\frac{\partial^2 y^s}{\partial x^j \partial x^k} \\
&= \sum_{r,s,t,u} \bar{g}^{ur}\overline{[st,r]}\frac{\partial y^s}{\partial x^j}\frac{\partial y^t}{\partial x^k}\frac{\partial x^i}{\partial y^u} + \sum_{r,s,u} \bar{g}^{ur}\bar{g}_{rs}\frac{\partial x^i}{\partial y^u}\frac{\partial^2 y^s}{\partial x^j \partial x^k} \\
&= \sum_{s,t,u} \bar{\Gamma}^u_{st}\frac{\partial y^s}{\partial x^j}\frac{\partial y^t}{\partial x^k}\frac{\partial x^i}{\partial y^u} + \sum_{s} \frac{\partial x^i}{\partial y^s}\frac{\partial^2 y^s}{\partial x^j \partial x^k},
\end{aligned}$$

where we have applied the relations $\sum_i \frac{\partial x^i}{\partial y^v}\frac{\partial y^r}{\partial x^l} = \delta^r_v$ and $\sum_r \bar{g}^{ur}\bar{g}_{rs} = \delta^u_s$. Fix an index $s = \alpha$. We use the previous equations to isolate a second partial derivative:

$$\begin{aligned}
\frac{\partial^2 y^\alpha}{\partial x^j \partial x^k} &= \sum\nolimits_s \delta^\alpha_s \frac{\partial^2 y^s}{\partial x^j \partial x^k} = \sum\nolimits_{i,s} \frac{\partial y^\alpha}{\partial x^i}\left(\frac{\partial x^i}{\partial y^s}\frac{\partial^2 y^s}{\partial x^j \partial x^k}\right) \\
&= \sum\nolimits_i \left(\Gamma^i_{jk} - \sum\nolimits_{s,t,u} \bar{\Gamma}^u_{st}\frac{\partial y^s}{\partial x^j}\frac{\partial y^t}{\partial x^k}\frac{\partial x^i}{\partial y^u}\right)\frac{\partial y^\alpha}{\partial x^i} \\
&= \sum\nolimits_i \Gamma^i_{jk}\frac{\partial y^\alpha}{\partial x^i} - \sum\nolimits_{s,t} \bar{\Gamma}^\alpha_{st}\frac{\partial y^s}{\partial x^j}\frac{\partial y^t}{\partial x^k}.
\end{aligned}$$

Lemma 15.11 (Christoffel 1869). *Let Γ^i_{jk} denote the Christoffel symbols of the second kind associated to the metric (g_{ij}) and coordinates $(x^1,\ldots,x^n)$, and $\bar{\Gamma}^u_{st}$ those associated with $(\bar{g}_{rs})$ and $(y^1,\ldots,y^n)$. With respect to a change of coordinates:*

$$(x^1,\ldots,x^n) \xrightarrow{x} M \xrightarrow{y^{-1}} (y^1(x^1,\ldots,x^n),\ldots,y^n(x^1,\ldots,x^n)),$$

taking the metric (g_{ij}) to $(\bar{g}_{rs})$, we have, for all $1 \leq \alpha, j, k \leq n$,

$$\frac{\partial^2 y^\alpha}{\partial x^j \partial x^k} = \sum\nolimits_i \Gamma^i_{jk}\frac{\partial y^\alpha}{\partial x^i} - \sum\nolimits_{s,t} \bar{\Gamma}^\alpha_{st}\frac{\partial y^s}{\partial x^j}\frac{\partial y^t}{\partial x^k}.$$

This equation shows that the second mixed partials of the change of coordinates are expressible in terms of first partials and Christoffel symbols. It follows that this is true of all higher iterated partial derivatives. By his 1864 work on invariant theory, Christoffel reduced the local equivalence problem to an algebraic problem that he had already solved.

To discover the Riemann curvature we take another derivative:

$$\frac{\partial^3 y^\alpha}{\partial x^j \partial x^k \partial x^l} = \sum_m \left(\frac{\partial \Gamma^m_{jk}}{\partial x^l} \frac{\partial y^\alpha}{\partial x^m} + \Gamma^m_{jk} \frac{\partial^2 y^\alpha}{\partial x^m \partial x^l} \right) - \sum_{s,t,u} \frac{\partial \bar{\Gamma}^\alpha_{st}}{\partial y^u} \frac{\partial y^u}{\partial x^l} \frac{\partial y^s}{\partial x^j} \frac{\partial y^t}{\partial x^k}$$

$$- \sum_{s,t} \bar{\Gamma}^\alpha_{st} \left(\frac{\partial^2 y^s}{\partial x^j \partial x^l} \frac{\partial y^t}{\partial x^k} + \frac{\partial y^s}{\partial x^j} \frac{\partial^2 y^t}{\partial x^k \partial x^l} \right).$$

Expanding the second partial derivatives according to Lemma 15.11, we obtain the (somewhat horrifying) expression:

$$\begin{aligned} \frac{\partial^3 y^\alpha}{\partial x^j \partial x^k \partial x^l} &= \sum_m \frac{\partial \Gamma^m_{jk}}{\partial x^l} \frac{\partial y^\alpha}{\partial x^m} + \sum_{m,r} \Gamma^m_{jk} \Gamma^r_{ml} \frac{\partial y^\alpha}{\partial x^r} \\ &- \sum_{m,s,t} \Gamma^m_{jk} \bar{\Gamma}^\alpha_{st} \frac{\partial y^s}{\partial x^m} \frac{\partial y^t}{\partial x^l} - \sum_{s,t,u} \frac{\partial \bar{\Gamma}^\alpha_{st}}{\partial y^u} \frac{\partial y^u}{\partial x^l} \frac{\partial y^s}{\partial x^j} \frac{\partial y^t}{\partial x^k} \\ &- \sum_{r,s,t} \bar{\Gamma}^\alpha_{st} \Gamma^r_{jl} \frac{\partial y^s}{\partial x^r} \frac{\partial y^t}{\partial x^k} + \sum_{s,t,u,v} \bar{\Gamma}^\alpha_{st} \bar{\Gamma}^s_{uv} \frac{\partial y^u}{\partial x^j} \frac{\partial y^v}{\partial x^l} \frac{\partial y^t}{\partial x^k} \\ &- \sum_{r,s,t} \bar{\Gamma}^\alpha_{st} \Gamma^r_{kl} \frac{\partial y^t}{\partial x^r} \frac{\partial y^s}{\partial x^j} + \sum_{s,t,u,v} \bar{\Gamma}^\alpha_{st} \bar{\Gamma}^t_{uv} \frac{\partial y^u}{\partial x^k} \frac{\partial y^v}{\partial x^l} \frac{\partial y^s}{\partial x^j}. \end{aligned}$$

The point of all this computation is to compare what we get when we equate the mixed partials $\dfrac{\partial^3 y^\alpha}{\partial x^j \partial x^k \partial x^l} = \dfrac{\partial^3 y^\alpha}{\partial x^j \partial x^l \partial x^k}$. Recall that the Gauss equations leading to *Theorema Egregium* arose from $x_{uuv} = x_{uvu}$. Setting on the left everything that is made up of expressions related to the coordinate chart x, and on the right everything related to y, we obtain:

$$\sum_m \left[\left(\frac{\partial \Gamma^m_{jk}}{\partial x^l} - \frac{\partial \Gamma^m_{jl}}{\partial x^k} \right) + \sum_r (\Gamma^r_{jk} \Gamma^m_{rl} - \Gamma^r_{jl} \Gamma^m_{rk}) \right] \frac{\partial y^\alpha}{\partial x^m}$$

$$= \sum_{s,t,u} \left[\left(\frac{\partial \bar{\Gamma}^\alpha_{st}}{\partial y^u} - \frac{\partial \bar{\Gamma}^\alpha_{su}}{\partial y^t} \right) + \sum_r (\bar{\Gamma}^\alpha_{uv} \bar{\Gamma}^v_{st} - \bar{\Gamma}^\alpha_{tv} \bar{\Gamma}^v_{su}) \right] \frac{\partial y^s}{\partial x^j} \frac{\partial y^t}{\partial x^k} \frac{\partial y^u}{\partial x^l}.$$

If we multiply both sides by $\dfrac{\partial x^m}{\partial y^\alpha}$ and sum over all α, we obtain the equation:

$$\left(\frac{\partial \Gamma^m_{jk}}{\partial x^l} - \frac{\partial \Gamma^m_{jl}}{\partial x^k} \right) + \sum_r (\Gamma^r_{jk} \Gamma^m_{rl} - \Gamma^r_{jl} \Gamma^m_{rk})$$

$$= \sum_{s,t,u,\alpha} \left[\left(\frac{\partial \bar{\Gamma}^\alpha_{st}}{\partial y^u} - \frac{\partial \bar{\Gamma}^\alpha_{su}}{\partial y^t} \right) + \sum_v (\bar{\Gamma}^v_{st} \bar{\Gamma}^\alpha_{uv} - \bar{\Gamma}^v_{su} \bar{\Gamma}^\alpha_{tv}) \right] \frac{\partial y^s}{\partial x^j} \frac{\partial y^t}{\partial x^k} \frac{\partial y^u}{\partial x^l} \frac{\partial x^m}{\partial y^\alpha}.$$

Definition 15.12. *The* **Riemann curvature tensor** *is the* $\binom{1}{3}$*-tensor field given by* $\{R^m_{jlk} \mid 1 \leq m, j, k, l \leq n\}$*, where:*

$$R^m_{jlk} = \left[\left(\frac{\partial \Gamma^m_{jk}}{\partial x^l} - \frac{\partial \Gamma^m_{jl}}{\partial x^k} \right) + \sum\nolimits_r (\Gamma^r_{jk} \Gamma^m_{rl} - \Gamma^r_{jl} \Gamma^m_{rk}) \right].$$

The previous derivation gives the transformation rule going from y to x and hence a $\binom{1}{3}$-tensor: $R^m_{jlk} = \sum\nolimits_{s,t,u,\alpha} \bar{R}^\alpha_{sut} \frac{\partial y^s}{\partial x^j} \frac{\partial y^t}{\partial x^k} \frac{\partial y^u}{\partial x^l} \frac{\partial x^m}{\partial y^\alpha}$.

A necessary condition for a solution to the local equivalence problem emerges from the transformation rule: In order for two Riemannian metrics to be transformable, one to another by a change of coordinates, this equation must hold for all choices of m, j, k, and l. Since the ingredients that go into computing the functions R^m_{jlk} are derived from the component functions of the metric and their derivatives, we can test a pair of metrics for equivalence directly.

A special case of the local equivalence problem was treated in one of Riemann's unpublished works, his Paris Academy paper (Riemann 1861). The problem set by the Academy in 1858 concerned heat conduction and systems of isothermal curves. Riemann recast the problem in geometric terms that a given metric be transformed into the standard Euclidean metric. His essay lacked certain details and the Academy decided not to award the prize until the essay was revised. Riemann's failing health prevented him from completing the essay and the prize was withdrawn in 1868.

If a metric takes the form of the standard Euclidean metric in a coordinate system $(y^1, \ldots, y^n)$, then the Christoffel symbols $\bar{\Gamma}^\alpha_{st} = 0$ for all α, s, and t. Thus, a necessary condition for a metric $\{g_{ij}\}$ to be transformable to the Euclidean metric is the vanishing of the Riemann curvature tensor, that is, $R^m_{jlk} = 0$ for all m, j, k, and l.

In fact, this condition is also sufficient. To see this, we rewrite the problem in Riemann's formulation. If $(x^1, \ldots, x^n)$ are the coordinates for the given metric $\{g_{ij}\}$, then let $(y^1, \ldots, y^n)$ be the desired coordinates for the standard Euclidean metric $\{g_{ij} = \delta_{ij}\}$. Riemann derived the particular case of Lemma 15.11, namely:

$$\sum\nolimits_{r=1}^n \Gamma^r_{jk} \frac{\partial y^\beta}{\partial x^r} = \frac{\partial^2 y^\beta}{\partial x^j \partial x^k} = \frac{\partial}{\partial x^k} \left(\frac{\partial y^\beta}{\partial x^j} \right).$$

We can write this condition in a convenient matrix form as the equation:

$$\begin{pmatrix} \Gamma^1_{1k} & \Gamma^2_{1k} & \cdots & \Gamma^n_{1k} \\ \Gamma^1_{2k} & \Gamma^2_{2k} & \cdots & \Gamma^n_{2k} \\ \vdots & \vdots & & \vdots \\ \Gamma^1_{nk} & \Gamma^2_{nk} & \cdots & \Gamma^n_{nk} \end{pmatrix} \begin{pmatrix} \partial y^\beta / \partial x^1 \\ \partial y^\beta / \partial x^2 \\ \vdots \\ \partial y^\beta / \partial x^n \end{pmatrix} = \frac{\partial}{\partial x^k} \begin{pmatrix} \partial y^\beta / \partial x^1 \\ \partial y^\beta / \partial x^2 \\ \vdots \\ \partial y^\beta / \partial x^n \end{pmatrix}$$

or $F_k(\mathbf{x})(\nabla y^\beta) = \frac{\partial}{\partial x^k}(\nabla y^\beta)$, where $F_k(\mathbf{x})$ denotes the matrix of Christoffel symbols $\Gamma^{\text{column}}_{\text{row},k}(\mathbf{x})$ and ∇ denotes the gradient. A necessary condition for the existence of the new coordinates $(y^1, \ldots, y^n)$ is given by the equation of mixed partials:

$$\frac{\partial}{\partial x^l}\frac{\partial}{\partial x^k}(\nabla y^\beta) = \frac{\partial}{\partial x^k}\frac{\partial}{\partial x^l}(\nabla y^\beta).$$

To prove sufficiency recall a familiar integrability result from two dimensions:

FACT: If f, g: $\mathbb{R}^2 \to \mathbb{R}$ satisfy $\frac{\partial f}{\partial y} = \frac{\partial g}{\partial x}$ in an open set containing $(0,0)$, then, for $z_0 \in \mathbb{R}$, there is a function F: $(U \subset \mathbb{R}^2) \to \mathbb{R}$ with U an open set containing $(0,0)$ such that $F(0,0) = z_0$, and $\nabla F(x,y) = (f(x,y), g(x,y))$.

In n dimensions, the integrability conditions for the differential equations satisfied by the y^is are those that follow from $\frac{\partial}{\partial x^l}F_k(\mathbf{x})(\nabla y^\beta) = \frac{\partial}{\partial x^k}F_l(\mathbf{x})(\nabla y^\beta)$ (Spivak, 1970, vol. 2, p. 4D-6). This condition is equivalent to:

$$\frac{\partial^3 y^\beta}{\partial x^l \partial x^j \partial x^k} = \frac{\partial}{\partial x^l}\sum\nolimits_{r=1}^n \Gamma^r_{jk}\frac{\partial y^\beta}{\partial x^r} = \frac{\partial^3 y^\beta}{\partial x^k \partial x^j \partial x^l} = \frac{\partial}{\partial x^k}\sum\nolimits_{r=1}^n \Gamma^r_{jl}\frac{\partial y^\beta}{\partial x^r},$$

which follows from $R^m_{jlk} = 0$ for all m, j, k, and l.

The equivalence shows the effectiveness of Riemann's notion of curvature. There are many other formulations of the Riemann curvature tensor; we refer the reader to Spivak (1970, Vol. 2) for a choice of seven proofs of the following local equivalence theorem.

Theorem 15.13 (Riemann 1861). *An n-dimensional Riemannian manifold for which all the functions making up the Riemann curvature tensor vanish, that is, $R^i_{jkl} = 0$ for all choices of i, j, k, and l, is locally isometric to $\mathbb{R}^n$ with the standard Euclidean metric $ds^2 = \sum_i dx^i\,dx^i$.*

A Riemannian manifold that satisfies $R^i_{jkl} = 0$ for all i, j, k, and l is called **flat**. The Riemann curvature tensor measures the deviation from flatness of a manifold.

Associated to the Riemann curvature tensor is a $\binom{0}{4}$-tensor field obtained by lowering an index:

$$R_{ijkl} = \sum\nolimits_{r=1}^n g_{ir}R^r_{jkl}.$$

Notice that if $R^r_{jkl} = 0$ for all choices of indices, then $R_{ijkl} = 0$ as well. We will tease out a geometric interpretation of a subset of the R_{ijkl} in what follows.

The relation between the Riemann curvature tensor and the Gauss–Riemann curvature (Definition 13.8) is established by considering surfaces through a point that lie in a given manifold. Suppose X and Y are linearly independent tangent vectors at a point p in an n-dimensional Riemannian manifold M. We may choose the vectors X and Y as the coordinate vectors $\frac{\partial}{\partial x^1}$ and $\frac{\partial}{\partial x^2}$ to a surface S_{XY} gotten by fixing the rest of the coordinates of a chart $x: (U \subset \mathbb{R}^n) \to M$ with $x(0,\ldots,0) = p$. The surface is the image of x applied to that part of the x^1x^2-plane in $\mathbb{R}^n$ lying in U. By applying linear isomorphisms to $\mathbb{R}^n$, we can always arrange for any given coordinate chart around p to have such properties. The rest of the charts for M transform smoothly when restricted to the appropriate subspaces of their domains, and so we obtain an abstract surface S_{XY} containing p. Since the tangent plane of S_{XY} is a linear subspace of the tangent space of M at p, we may restrict the metric on M to a Riemannian metric on S_{XY}. We compute the Gauss–Riemann curvature of this surface.

Proposition 15.14. *Given linearly independent tangent vectors X and Y at a point p in an n-dimensional Riemannian manifold M, the* **sectional curvature** *$K_{XY}(p)$ is the Gauss–Riemann curvature at p of the surface S_{XY}, given by:*

$$K_{XY}(p) = \frac{R_{1212}}{EG - F^2},$$

where R_{1212} is a particular component of the lowered Riemann curvature tensor restricted to S_{XY} and $E = \left\langle \frac{\partial}{\partial x^1}, \frac{\partial}{\partial x^1} \right\rangle_p$, $F = \left\langle \frac{\partial}{\partial x^1}, \frac{\partial}{\partial x^2} \right\rangle_p$ and $G = \left\langle \frac{\partial}{\partial x^2}, \frac{\partial}{\partial x^2} \right\rangle_p$.

PROOF: The proposition is a restatement of the formulas from *Theorema Egregium* as they apply to abstract surfaces. To wit we have:

$$R^1_{212} = \frac{\partial \Gamma^1_{22}}{\partial x^1} - \frac{\partial \Gamma^1_{21}}{\partial x^2} + (\Gamma^1_{22}\Gamma^1_{11} - \Gamma^1_{12}\Gamma^1_{12} + \Gamma^2_{22}\Gamma^1_{12} - \Gamma^2_{12}\Gamma^1_{22}) = GK$$

$$R^2_{212} = \frac{\partial \Gamma^2_{22}}{\partial x^1} - \frac{\partial \Gamma^2_{12}}{\partial x^2} + (\Gamma^1_{22}\Gamma^2_{11} - \Gamma^1_{12}\Gamma^2_{12}) = -FK.$$

It follows that $R_{1212} = ER^1_{212} + FR^2_{212} = (EG - F^2)K$. ∎

Not all of the component functions R^r_{jkl} or R_{ijkl} of the forms of the Riemann curvature tensor are independent. In fact, there are many symmetries indicating that a deeper algebraic structure must be enjoyed by this tensor.

Proposition 15.15. *The following equations hold:*

(1) $R^i_{jkl} = -R^i_{jlk}$, *and* $R_{ijkl} = -R_{jikl}$.
(2) $R^i_{jkl} + R^i_{klj} + R^i_{ljk} = 0$.
(3) $R_{ijkl} = R_{klij}$.

By exploiting the symmetries of the curvature, one proves that the $\frac{n(n-1)}{2} = \binom{n}{2}$ functions R_{ijij} determine R_{ijkl} and hence R^r_{jkl}. These functions are the missing data that determine a metric as conjectured by Riemann.

A Riemannian manifold is said to have **constant curvature** if the sectional curvatures $K_{XY} = c$, a constant, for all linearly independent X and Y. Constant Gauss–Riemann curvature was a requirement of an abstract surface in order for it to have enough congruences. In his lecture Riemann proposed the condition that *"lines have a length independent of their configuration, so that every line can be measured by every other."* The local equivalence problem and a version of the exponential mapping imply that such a manifold M has constant curvature. Riemann also proposed examples of manifolds of constant curvature.

Proposition 15.16. *The subset* $\{(x^1,\ldots,x^n) \mid 1 + \frac{\alpha}{4}\sum_i (x^i)^2 > 0\}$ *of* $\mathbb{R}^n$ *with the metric*

$$ds^2 = \frac{\sum_i dx^i\,dx^i}{\left(1 + \frac{\alpha}{4}\sum_i (x^i)^2\right)^2}$$

is an n-dimensional Riemannian manifold of constant curvature.

PROOF: The component functions of this metric may be written $g_{ij} = \dfrac{\delta_{ij}}{F^2}$, where $F = 1 + \frac{\alpha}{4}\sum_i (x^i)^2$. If we write $f = \ln F$, then we have:

$$\frac{\partial g_{ij}}{\partial x^k} = -\delta_{ij}\left(\frac{2}{F^3}\frac{\partial F}{\partial x^k}\right) = \frac{-2\delta_{ij}}{F^2}\frac{\partial f}{\partial x^k}.$$

This equation leads to the following expression for the Christoffel symbols of the first kind:

$$[jk,l] = -\frac{1}{F^2}\left(\delta_{ij}\frac{\partial f}{\partial x^k} + \delta_{ik}\frac{\partial f}{\partial x^j} - \delta_{jk}\frac{\partial f}{\partial x^i}\right).$$

The Christoffel symbols of the second kind are given by:

$$\Gamma^i_{jk} = \sum_{m=1}^n g^{im}[jk,m] = \sum_{m=1}^n F^2\delta^{im}[jk,m] = F^2[jk,i].$$

Similarly, we can relate the lowered form of the Riemann curvature tensor with its usual form by:

$$R_{ijij} = \sum_{m=1}^n g_{im}R^m_{jij} = \sum_{m=1}^n \frac{\delta_{im}}{F^2}R^m_{jij} = \frac{1}{F^2}R^i_{jij}.$$

To compute the sectional curvature we use the fact that $K_{XY}(p)$ is linear in X and Y, and so it suffices to compute $K_{\frac{\partial}{\partial x^i},\frac{\partial}{\partial x^j}}$, where $i \neq j$.

$$\begin{aligned} R^i_{jij} &= \frac{\partial \Gamma^i_{jj}}{\partial x^i} - \frac{\partial \Gamma^i_{ji}}{\partial x^j} + \sum\nolimits_m \left(\Gamma^m_{jj}\Gamma^i_{mi} - \Gamma^m_{ji}\Gamma^i_{mj}\right) \\ &= \frac{\partial^2 f}{\partial x^i \partial x^i} + \frac{\partial^2 f}{\partial x^j \partial x^j} + \sum\nolimits_{m=1}^n ([jj,m][mi,i] - [ji,m][mj,i])F^4 \\ &= \frac{\partial^2 f}{\partial x^i \partial x^i} + \frac{\partial^2 f}{\partial x^j \partial x^j} + \left(\frac{\partial r}{\partial x^i}\right)^2 + \left(\frac{\partial f}{\partial x^j}\right)^2 - \sum\nolimits_{s=1}^n \left(\frac{\partial f}{\partial x^s}\right)^2. \end{aligned}$$

In our case, $f(x^1,\ldots,x^n) = \ln\left(1 + \frac{\alpha}{4}\sum\nolimits_i (x^i)^2\right)$ from which it follows that:

$$\frac{\partial f}{\partial x^s} = \frac{\alpha x^s}{2F}, \text{ and } \frac{\partial^2 f}{\partial x^s \partial x^s} = \frac{1}{4F^2}(2\alpha F - (\alpha x^s)^2).$$

Substituting this into the preceding expression we obtain $R_{ijij} = \alpha/F^4$, and since $g_{jj} = 1/F^2$,

$$K_{\frac{\partial}{\partial x^i},\frac{\partial}{\partial x^j}} = \frac{R_{ijij}}{g_{ii}g_{jj}} = \alpha.$$ ∎

When $n = 3$ and $\alpha = -1$, we obtain a model for Bolyai's and Lobachevskiĭ's non-Euclidean space. In particular, the model is represented by the interior of the ball of radius 2 in $\mathbb{R}^3$ and so, topologically, it is simply connected. Furthermore, the boundary of the model lies an infinite distance from any point in the space and so it is complete. The arguments of Bolyai and Lobachevskiĭ are realized in this manifold by restricting to planar sections of the ball; these are non-Euclidean planes that are isometric to the Beltrami disk. These remarks were established by Beltrami in his paper on spaces of constant curvature (Beltrami 1869). He also presented new examples of constant-curvature manifolds by applying the higher-dimensional analogs of stereographic and orthographic projection. In particular, he identified the upper half-space model:

$$\left(H^n = \{(x^1,\ldots,x^n) \in \mathbb{R}^n \mid x^n > 0\}, \quad ds^2 = \frac{(dx^1)^2 + \cdots + (dx^n)^2}{(x^n)^2}\right)$$

as an n-dimensional Riemannian manifold with constant negative curvature. In H^3 the planes given by $\{(x^1,x^2,c) \mid c$, a positive constant$\}$ are horospheres and it is immediate from the form of the line element that the induced geometry is Euclidean. Thus, Beltrami showed analytically that the Euclidean plane lies inside a three-dimensional non-Euclidean space (Theorem 4.28), a result shown synthetically by Wachter, Bolyai, and Lobachevskiĭ.

Covariant differentiation

The quantities making up a tensor field on M are part of the analytic data associated to a manifold. To extend our understanding of these analytic expressions, we introduce derivatives of the functions making up a tensor field. First consider a vector field, given locally on coordinate charts $x\colon (U \subset \mathbb{R}^n) \to M$ and $y\colon (V \subset \mathbb{R}^n) \to M$ by:

$$X = \sum_i a_U^i \frac{\partial}{\partial x^i} = \sum_j a_V^j \frac{\partial}{\partial y^j}.$$

Suppose we take the partial derivative of a_V^r with respect to a coordinate direction y^q. This operation determines a transformation rule as follows: Since $a_V^r = \sum_i a_U^i \dfrac{\partial y^r}{\partial x^i}$,

$$\frac{\partial a_V^r}{\partial y^q} = \sum_{i,j} \left(\frac{\partial a_U^i}{\partial x^j} \frac{\partial x^j}{\partial y^q} \frac{\partial y^r}{\partial x^i} + a_U^i \frac{\partial^2 y^r}{\partial x^i \partial x^j} \frac{\partial x^j}{\partial y^q} \right).$$

The presence of the second-order partial derivative shows us that we do **not** have a tensor. If the change of coordinates were linear, then the second derivative would vanish and the expression $\dfrac{\partial a_V^r}{\partial y^q}$ would transform as a $\binom{1}{1}$-tensor. In the preliminary stages of his development of general relativity, Einstein struggled with this problem, and for a brief time he postulated that the only physically interesting changes of coordinates were linear (Einstein 1913).

The restriction to linear changes of coordinates is unreasonable for both geometry and physics and so another approach is needed. Let us expand the previous expression with the help of Lemma 15.11.

$$\begin{aligned}
\frac{\partial a_V^r}{\partial y^q} &= \sum_{i,j} \left(\frac{\partial a_U^i}{\partial x^j} \frac{\partial x^j}{\partial y^q} \frac{\partial y^r}{\partial x^i} + a_U^i \frac{\partial^2 y^r}{\partial x^i \partial x^j} \frac{\partial x^j}{\partial y^q} \right) \\
&= \sum_{i,j} \frac{\partial a_U^i}{\partial x^j} \frac{\partial x^j}{\partial y^q} \frac{\partial y^r}{\partial x^i} + \sum_{i,j} a_U^i \left(\sum_l \Gamma_{ij}^l \frac{\partial y^r}{\partial x^l} - \sum_{s,t} \bar{\Gamma}_{st}^r \frac{\partial y^s}{\partial x^i} \frac{\partial y^t}{\partial x^j} \right) \frac{\partial x^j}{\partial y^q} \\
&= \sum_{i,j} \left(\frac{\partial a_U^i}{\partial x^j} + \sum_l a_U^l \Gamma_{lj}^i \right) \frac{\partial y^r}{\partial x^i} \frac{\partial x^j}{\partial y^q} - \sum_s a_V^s \bar{\Gamma}_{sq}^r,
\end{aligned}$$

where we have used $\sum_i a_U^i \dfrac{\partial y^s}{\partial x^i} = a_V^s$. Rearranging the terms we find:

$$\frac{\partial a_V^r}{\partial y^q} + \sum_s a_V^s \bar{\Gamma}_{sq}^r = \sum_{i,j} \left(\frac{\partial a_U^i}{\partial x^j} + \sum_l a_U^l \Gamma_{lj}^i \right) \frac{\partial y^r}{\partial x^i} \frac{\partial x^j}{\partial y^q},$$

that is, the expression $\dfrac{\partial a_V^r}{\partial y^q} + \sum_s a_V^s \bar{\Gamma}_{sq}^r$ transforms as a $\binom{1}{1}$-tensor.

Definition 15.17. *The* **covariant derivative** *of a* $\binom{1}{0}$*-tensor* $\{a^i\}$ *is given by:*

$$a^i_{\ ;j} = \frac{\partial a^i}{\partial x^j} + \sum_l a^l \Gamma^i_{lj}.$$

The set of functions $\{a^i_{\ ;j}\}$ *constitutes a* $\binom{1}{1}$*-tensor.*

The case of a $\binom{0}{1}$-tensor $\{\theta_i\}$ is similar. The analogous calculation leads to the expression:

$$\theta_{i;j} = \frac{\partial \theta_i}{\partial x^j} - \sum_l \theta_l \Gamma^l_{ij}.$$

The covariant derivative of a $\binom{0}{1}$-tensor is a $\binom{0}{2}$-tensor.

Let us consider what we have discovered in these formulas. First of all, notice that for the flat Euclidean metric the covariant derivative reduces to the ordinary partial derivative because $\Gamma^i_{jk} = 0$ for all i, j, k. Second, the new expression is a tensor again; that is, we have taken an aggregate of functions that behaves as a well-defined tensor on a manifold, and the covariant derivative once again determines a well-defined tensor on the manifold.

Finally, we can give the definition of the covariant derivative for a general tensor field. It is made up of parts depending on the upper and lower indices and behaves like a partial derivative following the Leibniz rule with respect to multiplication by a function on the manifold.

Definition 15.18. *The* **covariant derivative** *with respect to* x^k *of an* $\binom{s}{r}$*-tensor* $\{T^{i_1\cdots i_s}_{j_1\cdots j_r}\}$ *is given by:*

$$T^{i_1\cdots i_s}_{j_1\cdots j_r;k} = \frac{\partial T^{i_1\cdots i_s}_{j_1\cdots j_r}}{\partial x^k} + \sum_{l=1}^{s} \sum_{m=1}^{n} T^{i_1\cdots i_{l-1} m i_{l+1}\cdots i_s}_{j_1\cdots j_r} \Gamma^{i_l}_{km}$$
$$- \sum_{\lambda=1}^{r} \sum_{\mu=1}^{n} T^{i_1\cdots i_s}_{j_1\cdots j_{\lambda-1}\mu j_{\lambda+1}\cdots j_r} \Gamma^{\mu}_{kj_\lambda}.$$

The functions $\{T^{i_1\cdots i_s}_{j_1\cdots j_r;k}\}$ *constitute an* $\binom{s}{r+1}$*-tensor.*

So far we have presented a formalism to generalize the partial derivative and stay within the space of tensors. To put some geometry back into the discussion, let us consider a special case of interest, geodesics on an n-dimensional manifold M. We begin by recalling the Euclidean case. We can characterize a geodesic in $\mathbb{R}^n$ in three equivalent ways:

(1) It is the shortest curve joining two points along it.
(2) The second derivatives of the components vanish.
(3) The tangent vectors along the curve are all parallel to each other.

The second property suggests generalizing the operator $\frac{d^2}{dt^2}$ in the context of tensor analysis. Suppose that $\alpha\colon(-r,r)\to M$ is a regular curve. Then the curve may be expressed in local coordinates by:

$$\alpha(t)=x(x^1(t),\dots,x^n(t))=y(y^1(t),\dots,y^n(t))$$

for overlapping charts $x\colon(U\subset\mathbb{R}^n)\to M$ and $y\colon(V\subset\mathbb{R}^n)\to M$. The tangent vector may be written:

$$\alpha'(t)=\sum_i\frac{dx^i}{dt}\frac{\partial}{\partial x^i}=\sum_j\frac{dy^j}{dt}\frac{\partial}{\partial y^j},$$

and the ordinary rules of differentiation imply $\frac{dy^r}{dt}=\sum_i\frac{dx^i}{dt}\frac{\partial y^r}{\partial x^i}$, that is, the tangent vector along $\alpha(t)$ transforms as a $\binom{1}{0}$-tensor field.

Taking another derivative of the transformation rule we have:

$$\begin{aligned}\frac{d^2y^r}{dt^2}&=\sum_i\frac{d^2x^i}{dt^2}\frac{\partial y^r}{\partial x^i}+\sum_{i,j}\frac{dx^i}{dt}\frac{dx^j}{dt}\frac{\partial^2y^r}{\partial x^i\partial x^j}\\&=\sum_i\frac{d^2x^i}{dt^2}\frac{\partial y^r}{\partial x^i}+\sum_{i,j}\frac{dx^i}{dt}\frac{dx^j}{dt}\left(\sum_k\Gamma_{ij}^k\frac{\partial y^r}{\partial x^k}-\sum_{s,t}\bar\Gamma_{st}^r\frac{\partial y^s}{\partial x^i}\frac{\partial y^t}{\partial x^j}\right).\end{aligned}$$

From this expression we obtain:

$$\frac{d^2y^r}{dt^2}+\sum_{s,t}\bar\Gamma_{st}^r\frac{dy^s}{dt}\frac{dy^t}{dt}=\sum_k\left(\frac{d^2x^k}{dt^2}+\sum_{i,j}\Gamma_{ij}^k\frac{dx^i}{dt}\frac{dx^j}{dt}\right)\frac{\partial y^r}{\partial x^k}.$$

This equation is the defining property for covariant differentiation with respect to t; it takes a particular vector field, the field of tangents to the curve, and gives back a vector field of second derivatives. We denote this operation by D/dt:

$$\frac{D\alpha'}{dt}=\sum_l\left(\frac{d^2x^l}{dt^2}+\sum_{i,j}\Gamma_{ij}^l\frac{dx^i}{dt}\frac{dx^j}{dt}\right)\frac{\partial}{\partial x^l}.$$

More generally, we define the covariant derivative of any vector field defined along $\alpha(t)$.

Definition 15.19. *Given a regular curve on a manifold M with coordinates $\alpha(t)=x(x^1(t),\dots,x^n(t))$ and a vector field $X=\sum_{i=1}^n a^i\frac{\partial}{\partial x^i}$ defined on an open set containing the trace of $\alpha(t)$, the* **covariant derivative** *of X along $\alpha(t)$ is given by the expression:*

$$\frac{DX}{dt}=\sum_l\left(\frac{da^l}{dt}+\sum_{j,k}\Gamma_{jk}^l a^j\frac{dx^k}{dt}\right)\frac{\partial}{\partial x^l}.$$

The previous calculation generalizes to show that $\dfrac{DX}{dt}$ is also a vector field along $\alpha(t)$, that is, it has the correct transformation rule. In order to appreciate the importance of this formula we prove a somewhat technical lemma. This result lies at the heart of the derivation of the equation satisfied by a geodesic.

Lemma 15.20. *If M is an n-dimensional Riemannian manifold with metric denoted by $\langle\ ,\ \rangle_p$, then:*

$$\frac{d}{dt}\langle\alpha'(t),\alpha'(t)\rangle_{\alpha(t)} = 2\left\langle\frac{D\alpha'}{dt},\alpha'(t)\right\rangle_{\alpha(t)}.$$

PROOF: We simple compute:

$$\begin{aligned}
\frac{1}{2}\frac{d}{dt}\langle\alpha'(t),\alpha'(t)\rangle &= \frac{1}{2}\frac{d}{dt}\sum_{i,j} g_{ij}\frac{dx^i}{dt}\frac{dx^j}{dt} \\
&= \sum_{j,k,l}\frac{1}{2}\frac{\partial g_{jk}}{\partial x^k}\frac{dx^j}{dt}\frac{dx^k}{dt}\frac{dx^l}{dt} + \sum_{i,j} g_{ij}\frac{d^2x^i}{dt^2}\frac{dx^j}{dt} \\
&= \sum_{j,k,l}[jk,l]\frac{dx^j}{dt}\frac{dx^k}{dt}\frac{dx^l}{dt} + \sum_{i,j} g_{ij}\frac{d^2x^i}{dt^2}\frac{dx^j}{dt} \\
&= \sum_{i,j,k,l} g_{ij}\Gamma^i_{kl}\frac{dx^j}{dt}\frac{dx^k}{dt}\frac{dx^l}{dt} + \sum_{i,j} g_{ij}\frac{d^2x^i}{dt^2}\frac{dx^j}{dt} \\
&= \sum_{i,j} g_{ij}\left(\frac{d^2x^i}{dt^2} + \sum_{k,l}\Gamma^i_{kl}\frac{dx^k}{dt}\frac{dx^l}{dt}\right)\frac{dx^j}{dt} \\
&= \left\langle\frac{D\alpha'}{dt},\alpha'(t)\right\rangle.
\end{aligned}$$

■

If we compute the arc length along a curve $\alpha(t)$ that satisfies $\dfrac{D\alpha'}{dt} = 0$, we see from the lemma that $\langle\alpha'(t),\alpha'(t)\rangle$ is a constant and so arc length is proportional to the parameter. This property ties curves with $\dfrac{D\alpha'}{dt} = 0$ to arc length. Analysis similar to the proof of Theorem 10.7 shows that such curves in fact provide the shortest distance between points along them. We are led to the following definition.

Definition 15.21. *A curve $\gamma\colon(-r,r)\to M$ is a* **geodesic** *if $\dfrac{D\gamma'}{dt} = 0$, that is, for each $-r < t < r$, in a coordinate chart around $\gamma(t) = x(x^1(t),\ldots,x^n(t))$, the following differential equations are satisfied for $i = 1,\ldots,n$:*

$$\frac{d^2x^i}{dt^2} + \sum_{j,k}\Gamma^i_{jk}\frac{dx^j}{dt}\frac{dx^k}{dt} = 0.$$

All of the standard results about existence and uniqueness of geodesics follow from the theory of second-order differential equations. Furthermore, we can define the exponential mapping and the special coordinate systems based on exp.

The geodesic equations bear a formal identity with the equations arrived at through the geodesic curvature definition for geodesics on surfaces in $\mathbb{R}^3$. Is there a sense in which these equations and covariant differentiation are "geometric"? The last stage of the development of differential geometry that we will discuss in this book begins with an answer to this question. For this we recall the situation of surfaces in $\mathbb{R}^3$.

Consider a plane Π in $\mathbb{R}^3$ through the origin. The tangent plane to Π at each point is simply a copy of Π and we may identify every $T_p(\Pi)$ as Π. This identification makes taking the derivative of a vector field simple. If $X\colon \Pi \to T\Pi$ is a vector field and $x\colon (U \subset \mathbb{R}^2) \to \Pi$ a coordinate chart, then $X(x(a,b))$ and $X(x(a+t,b))$ can be thought of as being in the same plane. Thus,

$$\lim_{t\to 0} \frac{X(x(a+t,b)) - X(x(a,b))}{t} = \frac{\partial X}{\partial x^1}(x(a,b))$$

makes sense.

When we consider a more general surface $S \subset \mathbb{R}^3$, the vectors $X(x(a+t,b))$ and $X(x(a,b))$ are in *different* tangent planes that are usually not parallel. Hence, the difference quotient with limit definition of the derivative does not make sense. Every tangent plane to the surface S, however, is abstractly isomorphic to every other. The problem is to decide how to choose an isomorphism between $T_p(S)$ and $T_q(S)$ for each p and q in S. In Levi-Civita (1917), a formalism is introduced for making these choices of isomorphisms, solving the problem of giving a geometric interpretation of covariant differentiation by generalizing the parallelism enjoyed in Euclidean space. A similar approach was arrived at independently by J. A. SCHOUTEN (1883–1971) around the same time.

Definition 15.22. *Given a curve $\alpha\colon (-r,r) \to M$, where M is an n-dimensional Riemannian manifold and X a vector field defined along $\alpha(t)$, we say X is* **parallel** *along $\alpha(t)$ if $\dfrac{DX}{dt}(\alpha(t)) = \mathbf{0}$ for all $-r < t < r$.*

In these terms, a curve $\alpha(t)$ is a geodesic if the tangent $\alpha'(t)$ to the curve is parallel along the curve. In a flat Euclidean space the covariant derivative coincides with an ordinary derivative, and a vector field is parallel along a curve if it is constant along it. To see how this leads to an interpretation of covariant differentiation, we need a technical result:

Theorem 15.23. *Given a regular curve $\alpha\colon (-r,r) \to M$ and a vector field X defined on an open set containing the trace of $\alpha(t)$, there exists a unique vector field $\hat{X}$ defined along $\alpha(t)$ with $\hat{X}(\alpha(0)) = X(\alpha(0))$ and $\hat{X}$ parallel along α.*

The proof relies upon the properties of the differential equations for parallelism (Spivak, 1970, Vol. 2). We note that the equations are linear and so if $X = rX_1 + X_2$, then $\hat{X} = r\hat{X}_1 + \hat{X}_2$. We use this property to define a mapping from $T_{\alpha(0)}(M)$ to $T_{\alpha(s)}(M)$ for $-r < s < r$. Let $\tau_s(X(\alpha(0))) = \hat{X}(\alpha(s))$. For any tangent vector $Y \in T_{\alpha(0)}(M)$, there is a vector field X_Y defined on an open set containing $\alpha(0)$ with $X_Y(\alpha(0)) = Y$. Thus, our definition makes sense. The mapping τ_s is called **parallel transport** along $\alpha(t)$ and it is a linear isomorphism. The inverse of τ_s is defined by reversing the curve and linearity follows from the differential equations.

This notion leads to a geometric interpretation of covariant differentiation.

Theorem 15.24. *Let $x\colon (U \subset \mathbb{R}^n) \to M$ be a coordinate chart for an n-dimensional Riemannian manifold M around a point $p \in M$ with $p = x(0,\ldots,0)$. Suppose $c_k\colon (-r,r) \to M$ is the regular curve on M given by the coordinate curve:*

$$c_k(t) = x(0,\ldots,0,t,0,\ldots,0), \quad t \text{ in the } k\text{th place.}$$

Let $\tau_s\colon T_p(M) \to T_{c_k(s)}(M)$ for $-r < s < r$ denote the isomorphism given by parallel transport along c_k. Then, for any vector field $X = \sum_l a^l \frac{\partial}{\partial x^l}$ defined on an open set of M containing p, we have $\sum_l a^l{}_{;k} \frac{\partial}{\partial x^l} = \lim_{h \to 0} \frac{\tau_h^{-1}(X(c_k(h))) - X(p)}{h}$.

PROOF: Denote the tangent vector $\tau_h^{-1}(X(c_k(h)))$ by $\hat{X}_h(0)$. For each h near 0, we have a (possibly different) parallel vector field along $c_k(t) = (c_k^1(t),\ldots,c_k^n(t))$ given locally by:

$$\hat{X}_h(t) = \sum_l b^l(h,t)\frac{\partial}{\partial x^l}, \text{ at } c_k(t), \text{ with } b^l(h,h) = a^l(c_k(t)).$$

Since $\hat{X}_h$ is a parallel vector field, the following differential equations are satisfied:

$$\frac{db^l(h,t)}{dt} = \sum_{i,j} b^j(h,t)\frac{dc_k^i}{dt}\Gamma_{ij}^l(c_k(t)) = 0.$$

Recall that the derivative with respect to t along $c_k(t)$ is the partial derivative with respect to x^k so this differential equation becomes:

$$\frac{\partial b^l(h,t)}{\partial x^k} = \sum_j b^j(h,t)\Gamma_{kj}^l(c_k(t)) = 0.$$

By the mean value theorem we may write:

$$b^l(h,h) = b^l(h,0) + h\left.\frac{db^l(h,t)}{dt}\right|_{t=u_l}, \text{ for some } u_l \in (0,h).$$

With this in mind, we compute:

$$\lim_{h\to 0}\frac{\tau_h^{-1}(X(c_k(h)))-X(c_k(0))}{h}=\lim_{h\to 0}\frac{1}{h}\sum\nolimits_l\left(b^l(h,0)-a^l(c_k(0))\right)\frac{\partial}{\partial x^l}$$

$$=\lim_{h\to 0}\frac{1}{h}\sum\nolimits_l\left(b^l(h,h)-h\left.\frac{db^l(h,t)}{dt}\right|_{t=u_l}-a^l(c_k(0))\right)\frac{\partial}{\partial x^l}$$

$$=\sum\nolimits_l\left(\frac{\partial a^l}{\partial x^k}+\lim_{h\to 0}\sum_j b^l(h,u_l)\Gamma^l_{kj}(c_k(0))\right)\frac{\partial}{\partial x^l}$$

$$=\sum\nolimits_l\left(\frac{\partial a^l}{\partial x^k}+\sum_j a^l\Gamma^l_{kj}\right)\frac{\partial}{\partial x^l}$$

$$=\sum\nolimits_l a^l{}_{;k}\frac{\partial}{\partial x^l}.$$

■

The parallelism introduced by Levi-Civita (1917) has further properties that are explored in the exercises. Not only does it give an interpretation of the covariant derivative, but it uncovers a new foundation for differential geometry. At the heart of our flurry of indices is Christoffel's lemma (Lemma 15.11) that gives a kind of transformation rule for the Γ^i_{jk}:

$$\frac{\partial^2 y^\alpha}{\partial x^j\partial x^k}=\sum_i\Gamma^i_{jk}\frac{\partial y^\alpha}{\partial x^i}-\sum_{s,t}\bar{\Gamma}^\alpha_{st}\frac{\partial y^s}{\partial x^j}\frac{\partial y^t}{\partial x^k}.$$

Suppose $\{G^i_{jk}\mid 1\le i,j,k\le n\}$ is a collection of n^3 functions that satisfy the same transformation rule as the Γ^i_{jk}. Using the theory of differential equations as a tool, one can develop a notion of covariant differentiation and parallel transport by following the formal development given earlier, but based on the functions G^i_{jk}. Such a choice of functions $\{G^i_{jk}\}$ is called a (classical) **connection** for the notion of parallel transport it determines. In particular, this choice of structure is prior to the choice of Riemannian metric and belongs to the analytic machinery associated to the manifold. This level of generality was introduced by H. WEYL (1885–1955) in (Weyl 1918), which is motivated by Weyl's attempt to unify relativity and electromagnetism.

Among the possible connections associated to a Riemannian manifold, the one based on a choice of Riemannian metric and its associated Christoffel symbols is unique.

Theorem 15.25 (Fundamental Theorem of Riemannian Geometry). *Given a Riemannian manifold M there is one and only one connection $\{G^i_{jk}\}$ on M for which parallel transport is an isometry and the connection is symmetric, that is, $G^i_{jk}=G^i_{kj}$ for all i, j, and k.*

The connection singled out by this theorem is called the *Levi–Civita connection* or the *canonical connection* on $(M, \langle\ ,\ \rangle)$.

The coordinate viewpoint that we have presented in this chapter has given way in the twentieth century to coordinate-free constructions that do the work of the tensor calculus globally. The basic ideas for these developments lie in Levi–Civita's work on parallelism. Developments by E. Cartan have led to a general framework with a connection as foundation, generalizing geometric notions widely (Sharpe 1997, Kobayashi-Nomizu 1963).

We close with Gauss's motto from his Copenhagen *Preisschrift*, 1822, a paper in which he solves "the problem of mapping the parts of a given surface onto another such that the image and the mapped part are similar in the smallest parts":

Ab his via sternitur ad maiora ...
From here the way to greater (accomplishments) is smoothed.

It is the author's hope that this book has smoothed the way for the reader to study the current formulation of differential geometry with a sense of how that subject originated in the rest of geometry and with a deeper intuition about and appreciation for its inner workings, formalism, and potential.

Exercises

15.1 Prove that a connected 1-manifold is diffeomorphic to either the real line or a circle.

15.2 Given two manifolds M^m and N^n, construct the direct product manifold $M \times N$ of dimension $m+n$. Explain the charts in detail.

15.3 Prove Lemma 15.8.

15.4 Show directly that the upper half-space,

$$\left(H^n = \{(x^1,\ldots,x^n) \in \mathbb{R}^n \mid x^n > 0\}, \quad ds^2 = \frac{(dx^1)^2 + \cdots + (dx^n)^2}{(x^n)^2} \right)$$

is a Riemannian n-manifold of constant curvature.

15.5 Use the transformation rule for the Christoffel symbols to show that Γ^i_{jk} is **not** a tensor of any order.

15.6 Show that the sectional curvature K_{XY} is linear in each argument, that is, $K_{aX+bY,Z} = aK_{XZ} + bK_{YZ}$, and $K_{X,aY+bZ} = aK_{XY} + bK_{XZ}$.

15.7 Prove Ricci's lemma, that is, for a Riemannian metric $\{g_{ij}\}$ on a manifold M we have:

$$g_{ij;k} = 0 = g^{ij}{}_{;k}.$$

15.8† Prove the Bianchi identity for the Riemann curvature tensor:

$$R^m_{ijk;l} + R^m_{ikl;j} + R^m_{ilj;k} = 0.$$

Solutions to Exercises marked with a dagger appear in the appendix, p. 340.

15.9 Suppose that $\{a^i\}$ denotes a $\binom{1}{0}$-tensor field on a Riemannian manifold M. Prove the following identity due to Ricci:

$$a^i_{\ ;j;k} - a^i_{\ ;k;j} = -\sum_{l=1}^{n} a^l R^i_{ljk}.$$

15.10 Suppose $\{\theta_i\}$ is $\binom{0}{1}$-tensor. Show that the covariant derivative $\{\theta_{i;j}\}$ is a $\binom{0}{2}$-tensor.

15.11 Let $\Gamma(TM)$ denote the set of vector fields on a manifold M. Show that $\Gamma(TM)$ is a vector space.

15.12† Suppose that $\{G^i_{jk}\}$ is a collection of n^3 functions associated to each coordinate chart that satisfy the transformation rules for a classical connection. We seek a function $\nabla\colon \Gamma(TM) \times \Gamma(TM) \to \Gamma(TM)$ for which (1) ∇ is linear in both variables; (2) $\nabla(fX,Y) = f\nabla(X,Y)$, for $f \in C^\infty(M)$; (3) $\nabla(X,fY) = f\nabla(X,Y) + X(f)Y$. Such a function is called a *Koszul connection*. Locally, suppose:

$$\nabla\left(\frac{\partial}{\partial x^i}, \frac{\partial}{\partial x^j}\right) = \sum_k G^k_{ij} \frac{\partial}{\partial x^k}.$$

Show that the covariant derivative:

$$a^i_{\ ;j} = \frac{\partial a^i}{\partial x^j} + \sum\nolimits_l a^l G^i_{lj}$$

extends to a Koszul connection. Show further that, for the choice of the canonical connection $\{\Gamma^i_{jk}\}$ on a Riemannian manifold, the covariant derivative of a vector field X along a curve $\alpha(t)$ is given by:

$$\frac{DX}{dt} = \nabla(\alpha'(t), X).$$

Riemann's *Habilitationsvortrag*: On the hypotheses which lie at the foundations of geometry[1]

Plan of the investigation

As is well known, geometry presupposes as given both the concept of space and the basic principles for constructions in space. It gives only nominal definitions of these things, while their essential specifications appear in the form of axioms. The relationship between these presuppositions is left in the dark; one does not see whether, or to what extent, any connection between them is necessary, or a priori whether any connection between them is possible.

From Euclid to Legendre, to name the most famous of the modern reformers of geometry, this darkness has been dispelled neither by the mathematicians nor by the philosophers who have concerned themselves with it. This is undoubtedly because the general concept of multiply extended quantities, which includes spatial quantities, remains completely unexplored. I have therefore first set myself the task of constructing the concept of a multiply extended quantity from general notions of quantity. It will be shown that a multiply extended quantity is susceptible of various metric relations, so that Space constitutes only a special case of a triply extended quantity. From this however it is a necessary consequence that the theorems of geometry cannot be deduced from general notions of quantity, but that those properties that distinguish Space from other conceivable triply extended quantities can only be inferred from experience. Thus arises the problem of seeking out the simplest data from which the metric relations of Space can be determined, a problem which by its very nature is not completely determined, for there may be several systems of simple data that suffice to determine the metric relations of Space; for the present purposes, the most important system is that laid down as a foundation of geometry by Euclid. These data are—like all data—not necessary, but only of empirical certainty, they are hypotheses; one can therefore investigate their likelihood, which is certainly very great within the bounds of observation, and afterwards judge the legitimacy of extending them beyond the bounds of observation, both in the direction of the immeasurably large, and in the direction of the immeasurably small.

I. Concept of an n-fold extended quantity

In attempting to solve the first of these problems, the development of the concept of multiply extended quantity, I feel particularly entitled to request an indulgent

[1] Based on a translation by Michael Spivak; used with permission.

criticism, as I am little trained in these tasks of a philosophical nature where the difficulties lie more in the concepts than in the construction, and because I could not make use of any previous studies, except for some very brief hints that Privy Councillor Gauss has given in his second memoir on biquadratic residues, in the Göttingen Gelehrte Anzeige in his Jubilee-book, and some philosophical researches of Herbart.

1

Notions of quantity are possible only where there already exists a general concept which allows different realizations. Depending on whether or not a continuous transition of instances can be found between any two of them, these realizations form either a continuous or a discrete manifold; individual instances in the first case are called points and in the latter case elements of the manifold. Concepts whose instances form a discrete manifold are so numerous that some concept can always be found, at least in the more highly developed languages, under which any given collection of things can be comprehended (and in the study of discrete quantities, mathematicians could unhesitatingly proceed from the principle that given objects are to be regarded as all of one kind). On the other hand, reasons for creating concepts whose instances form a continuous manifold occur so seldom in everyday life that the position of sensible objects and colors are perhaps the only simple concepts whose instances form a multiply extended manifold. More numerous reasons for the generation and development of these concepts first occur in higher mathematics.

Distinct portions of a manifold, distinguished by a mark or by a boundary, are called quanta. Their quantitative comparison is effected in the case of discrete quantities by counting, in the case of continuous quantities by measurement. Measuring involves the super-position of the quantities to be compared; it therefore requires a means of transporting one quantity to be used as a standard for the others. Otherwise, one can compare two quantities only when one is a part of the other, and then only to decide "more" or "less," not "how much." The investigations about this which can be carried out in this case form a general division of the science of quantity, independent of measurement, where quantities are regarded, not as existing independent of position and not as expressible in terms of a unit, but as regions in a manifold. Such investigations have become a necessity for several parts of mathematics, for the treatment of many-valued analytic functions, and the dearth of such studies is one of the principal reasons why the celebrated Theorem of Abel and the contributions of Lagrange, Pfaff and Jacobi to the general theory of differential equations have remained unfruitful for so long. From this general part of the science of extended quantity which assumes nothing further than what is already contained in the same concept, it suffices for the present purposes to emphasize two points, the first of which will make clear the generation of the concept of a multiply extended manifold, the second reducing position fixing in a given manifold to numerical determinations and will make clear the essential character of an n-fold extension.

2

In a concept whose instances form a continuous manifold, if one passes from one instance to another in a well-determined way, the instances through which one has passed form a simply extended manifold, whose essential characteristic is that from a point in it a continuous movement is possible in only two directions, forwards and backwards. If one now imagines that this manifold passes to another, completely different one, and once again in a well-determined way, that is, so that every point passes to a well-determined point of the other, then the instances form similarly a doubly extended manifold. In a similar way, one obtains a triply extended manifold if one imagines that a doubly extended one passes in a well-determined way to a completely different one, and it is easy to see how one can continue this construction. If one considers the process as one in which the objects vary, instead of regarding the concept as fixed, then this construction can be characterized as a synthesis of a variability of $n+1$ dimensions from a variability of n dimensions and a variability of one dimension.

3

I will now show, conversely, how one can break up a variability, whose boundary is given, into a variability of one dimension and a variability of lower dimension. To this end one considers an arbitrary piece of a manifold of one dimension – with a fixed origin, so that points of it may be compared with one another – varying so that for every point of the given manifold it has a definite value, continuously changing with this point, or in other words, one takes within the given manifold a continuous function of position, and, moreover, one such function that is not constant along any part of the manifold. Every system of points where the function has a constant value then forms a continuous manifold of fewer dimensions than the given one. These manifolds pass continuously from one to another as the function changes; one can therefore assume that they all emanate from one of them, and generally speaking this will occur in such a way that every point of the first passes to a definite point of any other; the exceptional cases, whose investigation is important, need not be considered here. In this way, the determination of position in the given manifold is reduced to a numerical determination and to the determination of position in a manifold of fewer dimensions. It is now easy to show that this manifold has $n-1$ dimensions, if the given manifold is an n-fold extension. By an n-fold repetition of this process, the determination of position in an n-fold extended manifold is reduced to n numerical determinations, and therefore the determination of position in a given manifold is reduced, whenever this is possible, to a finite number of numerical determinations. There are, however, also manifolds in which the fixing of position requires not a finite number, but either an infinite sequence or a continuous manifold of numerical measurements. Such manifolds form, for example, the possibilities for a function in a given region, the possible shapes of a solid figure, and so forth.

II. Metric relations of which a manifold of n dimensions is susceptible, on the assumption that lines have a length independent of their configuration, so that every line can be measured by every other

Now that the concept of an n-fold extended manifold has been constructed, and its essential characteristic has been found in the fact that position fixing in the manifold can be reduced to n numerical determinations, there follows, as the second of the problems proposed above, an investigation of the metric relations of which such a manifold is susceptible, and of the conditions that suffice to determine them. These metric relations can be investigated only in abstract terms, and represented in context only through formulas; under certain assumptions, however, one can resolve them into relations which are individually capable of geometric representation, and in this way it becomes possible to express the results of calculation geometrically. Thus, to put this work on solid ground, although an abstract investigation with formulas certainly cannot be avoided, the results can be presented in geometric garb. The foundations of both parts of the question are contained in the celebrated treatise of Privy Councillor Gauss on curved surfaces.

1

Measurement requires an independence of quantity from position, which can occur in more than one way; the hypothesis that next presents itself, and which I shall develop here, is just this, that the length of lines be independent of their configuration, so that every line can be measured by every other. If fixing of position is reduced to numerical determinations, so that the position of a point in the given n-fold extended manifold is expressed by n varying quantities x_1, x_2, x_3, and so forth up to x_n, then specifying a line amounts to giving the quantities x as functions of one variable. The problem then is to set up a mathematical expression for the length of a line, for which purpose the quantities x must be thought of as expressible in units. I will treat this problem only under certain restrictions, and I first limit myself to lines in which the ratios of the quantities dx – the increments in the quantities x—vary continuously; one can then regard the lines as broken up into elements within which the ratios of the quantities dx may be considered to be constant and the problem then reduces to setting up at every point a general expression for the line element ds, which will therefore contain the quantities x and the quantities dx. I assume, secondly, that the length of the line element remains unchanged, up to first order, when all the points of this line element suffer the same infinitesimal displacement, whereby I simply mean that if all the quantities dx increase in the same ratio, the line element changes by the same ratio. Under these assumptions, the line element can be an arbitrary homogeneous function of the first degree in the quantities dx that remains the same when all the quantities dx change sign, and in which the arbitrary constants are functions of the quantities x. To find the simplest cases, I next seek an expression for the $(n-1)$-fold extended manifolds that are everywhere equidistant from the origin of the line element, that is, I seek a continuous function of position that distinguishes them from one another. This must either decrease or increase in all directions from the origin; I want to assume that it increases in all directions and therefore has a minimum at the origin. Then if its first and second differential quotients are finite, the first-order differential must vanish and the second-order differential cannot be negative; I assume that it is always positive. This differential expression of the second order remains constant if ds remains constant and increases quadratically if the quantities dx, and thus also ds, all increase in the same ratio; it is therefore equal to a constant times ds^2, and consequently ds equals the square root of an everywhere positive homogeneous function of the second degree in the quantities dx, in which the coefficients are continuous functions of the quantities x. In Space, if one expresses the location of a point by rectilinear coordinates, then $ds = \sqrt{\Sigma(dx)^2}$; Space is therefore included in this simplest case. The next simplest case would perhaps include the manifolds in which the line element may be expressed as the fourth root of a differential expression of the fourth degree. Investigation of this more general class would actually require no essentially different principles, but it would be rather time consuming and throw proportionally little new light on the study of Space, especially if the results cannot be expressed geometrically; I consequently restrict myself to those manifolds where the line element can be expressed by the square root of a differential expression of the second degree. One can transform such an expression into another similar one by substituting for the n independent variables, functions of n new independent variables. However, one cannot transform any expression into any other in this way; for the expression contains $n(n+1)/2$ coefficients which are arbitrary functions of the independent variables; by the introduction of new variables one can satisfy only n conditions, and can therefore make only n of the coefficients equal to given quantities. There remain $n(n-1)/2$ others, already completely determined by the nature of the manifold to be represented, and consequently $n(n-1)/2$ functions of position are required to determine its metric relations. Manifolds, such as the Plane and Space, in which the line element can be brought into the form $\sqrt{\Sigma(dx)^2}$ thus constitute only a special case of the manifolds to be investigated here; they clearly deserve a special name, and consequently, these manifolds, in which the square of the line element can be expressed as the sum of the squares of independent differentials, I propose to call flat. In order to survey the essential differences of all the manifolds representable in the assumed form, it is necessary to eliminate the features depending on the mode of presentation, which is accomplished by choosing the variable quantities according to a definite principle.

2

For this purpose, suppose one constructs the system of shortest lines emanating from a given point; the position of an arbitrary point can then be determined by the direction of the shortest line in which it lies, and by its distance along this line from the initial point, and it can therefore be expressed by the ratios of the quantities dx^0, that is, the quantities dx at the origin of this shortest line, and by the length s of this line. In place of the dx^0 one now introduces linear expressions $d\alpha$ formed from them in such a way that the initial value of the square of the line element will be equal to the sum of the squares of these expressions, so that the independent variables are: the quantity s and the ratio of the quantities $d\alpha$; finally, in place of the $d\alpha$ choose quantities $x_1, x_2, \ldots, x_n$ proportional to them, such that the sum of their squares equals s^2. If one introduces these quantities, then for infinitesimal values of x the square of the line element equals $\Sigma(dx)^2$, but the next order term in its expansion equals a homogeneous expression of the second degree in the $n(n-1)/2$ quantities $(x_1dx_2 - x_2dx_1), (x_1dx_3 - x_3dx_1), \ldots$, and is consequently an infinitely small quantity of the fourth order, so that one obtains a finite quantity if one divides it by the area of the infinitely small triangle at whose vertices variables have the values $(0,0,0,\ldots), (x_1,x_2,x_3,\ldots), (dx_1,dx_2,dx_3,\ldots)$. This quantity remains the same as long as the quantities x and dx are contained in the same binary linear forms, or as long as the two shortest lines from the initial point to x and from the initial point to dx remain in the same surface element, and therefore depends only on the position and direction of that element. It obviously equals zero if the manifold in question is flat, that is, if the square of the line element is reducible to $\Sigma(dx)^2$, and can therefore be regarded as the measure of deviation from flatness in this surface direction at this point. When multiplied by $-3/4$ it becomes equal to the quantity that Privy Councillor Gauss has called the curvature of a surface. Previously, $n(n-1)/2$ functions of position were found necessary in order to determine the metric relations of an n-fold extended manifold representable in the assumed form; hence if the curvature is given in $n(n-1)/2$ surface directions at every point, then the metric relations of the manifold may be determined, provided only that no identical relations can be found between these values, and indeed in general this does not occur. The metric relations of these manifolds, in which the line element can be represented as the square root of a differential expression of the second degree, can thus be expressed in a way completely independent of the choice of the varying quantities. A similar path to the same goal could also be taken in those manifolds in which the line element is expressed in a less simple way, for example, by the fourth root of a differential expression of the fourth degree. The line element in this more general case would not be reducible to the square root of a quadratic sum of differential expressions, and therefore in the expression for the square of the line element the deviation from flatness would be an infinitely small quantity of the second dimension, whereas for the other manifolds it was an infinitely small quantity of the fourth dimension. This peculiarity of the latter manifolds therefore might well be called flatness in the smallest parts. For present purposes, however, the most important peculiarity of these manifolds, on whose account alone they have been examined here, is this, that the metric relations of the doubly extended ones can be represented geometrically by surfaces and those of the multiply extended ones can be reduced to those of the surfaces contained within them, which still requires a brief discussion.

3

In the conception of surfaces the inner metric relations, which involve only the lengths of paths within them, are always bound up with the way the surfaces are situated with respect to points outside them. One can, however, abstract from external relations by considering deformations that leave the lengths of lines within the surfaces unaltered, that is, by considering arbitrary bendings – without stretching – of such surfaces, and by regarding all surfaces obtained from one another in this way as equivalent. For example, it follows that arbitrary cylindrical or conical surfaces are equivalent to a plane, since they can be formed from a plane by mere bending under which the inner metric relations remain the same, and all theorems about the plane – hence all of planimetry – retain their validity; on the other hand, they count as essentially different from the sphere, which cannot be transformed into the plane without stretching. According to the previous investigations, the inner metric relation at every point of a doubly extended quantity, if its line element can be expressed as the square root of a differential expression of the second degree, which is the case with surfaces, is characterized by the curvature. This quantity can be given a visual interpretation for surfaces as the product of the two curvatures of the surface at this point, or by the fact that its product with an infinitely small triangle formed from shortest lines is, in proportion to the radius, half the excess of the sum of its angles over two right angles. The first definition would presuppose the theorem that the product of the two radii of curvatures is unaltered by mere bendings of a surface, the second, that at each point the excess over two right angles of the sum of the angles of any infinitely small triangle is proportional to its area. To give a tangible meaning to the curvature of an n-fold extended manifold at a given point and in a given surface direction through it, one must proceed from the fact that a shortest line emanating from a point is completely determined if its initial direction is given. Consequently one obtains a certain surface if one prolongs all the initial directions from the given point that lie in the given surface element into shortest lines, and this surface has a definite curvature at the given point, which is equal to the curvature of the n-fold extended manifold at the given point in the given surface direction.

4

Before applying these results to Space, it is still necessary to make some general considerations about flat manifolds, that is, about manifolds in which the square of the line element can be represented as the sum of squares of complete differentials.

In a flat n-fold extended manifold the curvature in every direction, at every point, is zero; but according to the preceding investigation, in order to determine the metric relations it suffices to know that at each point the curvature is zero in $n(n-1)/2$ independent surface directions. The manifolds whose curvature is everywhere zero can be considered as a special case of those manifolds whose curvature is everywhere constant. The common character of those manifolds whose curvature is constant can be expressed as follows: Figures can be moved in them without stretching. For obviously figures could not be freely shifted and rotated in them if the curvature were not the same in all directions, at all points. On the other hand, the metric properties of the manifold are completely determined by the curvature; they are therefore exactly the same in all the directions around any one point as in the directions around any other, and thus the same constructions can be effected starting from either, and consequently, in the manifolds with constant curvature figures can be given any arbitrary position. The metric relations of these manifolds depend only on the value of the curvature, until it may be mentioned, as regards the analytic presentation, that if one denotes this value by α, then the expression for the line element can be put in the form

$$\frac{1}{1+\frac{\alpha}{4}\Sigma x^2}\sqrt{\Sigma(dx)^2}.$$

5

The consideration of *surfaces* with constant curvature may serve for a geometric illustration. It is easy to see that the surfaces whose curvature is positive can always be developed onto a sphere whose radius is the reciprocal of the curvature; but in order to survey the multiplicity of these surfaces, let one of them be given the shape of a sphere, and the others the shape of surfaces of rotation that touch it along the equator. The surfaces with greater curvature than the sphere will then touch the sphere from inside and take a form like the portion of the surface of a ring, which is situated away from the axis; they could be developed upon zones of spheres with smaller radii, but would go round more than once. Surfaces with smaller positive curvature are obtained from spheres of larger radii by cutting out a portion bounded by two great semicircles, and bringing together the cut-lines. The surface of curvature zero will be a cylinder standing on the equator; the surfaces with negative curvature will touch this cylinder from outside and be formed like the part of the surface of a ring which is situated near the axis. If one regards these surfaces as possible positions for pieces of surface moving in them, as Space is for bodies, then pieces of surface can be moved in all these surfaces without stretching. The surfaces with positive curvature can always be so formed that pieces of surface can even be moved arbitrarily without bending, namely on spherical surfaces, but those with negative curvature cannot. Aside from this independence of position for surface pieces, in surfaces with zero curvature an independence of position for directions also occurs, which does not hold in the other surfaces.

III. Applications to space

1

Following these investigations on the determination of the metric relations of an n-fold extended quantity, the conditions may be given which are sufficient and necessary for the determination of the metric relations of Space, if we assume the independence of lines from configuration and the representability of the line element as the square root of a second-order differential expression, that is, flatness in the smallest parts.

First, these conditions may be expressed by saying that the curvature at every point equals zero in three surface directions, and thus the metric relations of Space are implied if the sum of the angles of a triangle always equals two right angles. But secondly, if one assumes with Euclid not only the existence of lines independently of configuration, but also of bodies, then it follows that the curvature is everywhere constant, and the angle sum in all triangles determined if it is known in one.

In the third place, finally, instead of assuming the length of lines to be independent of place and direction, one might assume that their length and direction is independent of place. According to this conception, changes or differences in position are complex quantities expressible in three independent units.

2

In the course of the previous considerations, the relations of extension or regionality were first of all distinguished from the metric relations, and it was found that different metric relations were conceivable along with the same relations of extension; then systems of simple metric specifications were sought by means of which the metric relations of Space are completely determined, and from which all theorems about it are a necessary consequence; it remains now to discuss the question how, to what degree, and to what extent these assumptions are borne out by experience. In this connection there is an essential difference between mere relations of extension and metric relations, in that among the first, where the possible cases form a discrete manifold, the declarations of experience are surely never completely certain, but they are not inexact, while for the second, where the possible cases form a continuous manifold, every determination from experience always remains inexact be the probability ever so great that it is nearly exact. This circumstance becomes important when these empirical determinations are extended beyond the limits of observation into the immeasurably large and the immeasurably small; for the latter may obviously become ever more inexact beyond the boundary of observation, but not so the former.

When constructions in Space are extended into the immeasurably large, unboundedness and infinitude are to be distinguished; one belongs to relations of extension, the other to metric relations. That Space is an unbounded triply extended manifold is an assumption which is employed for every apprehension of the external world, by which at every moment the domain of actual perception is supplemented, and the possible locations of a sought after object are constructed; and in these applications it is continually confirmed. The unboundedness of space consequently has a greater empirical certainty than any experience of the external world. But its infinitude does not in any way follow from this; quite to the contrary, Space would necessarily be finite if one assumed independence of bodies from position, and thus ascribed to it a constant curvature, as long as this curvature had ever so small a positive value. If one prolonged the initial directions lying in a surface direction into shortest lines, one would obtain an unbounded surface with constant positive curvature, and thus a surface which in a flat triply extended manifold would take the form of a sphere, and consequently be finite.

3

Questions about the immeasurably large are idle questions for the explanation of Nature. But the situation is quite different with questions about the immeasurably small. Upon the exactness with which we pursue phenomenon into the infinitely small, does our knowledge of their causal connections essentially depend. The progress of recent centuries in understanding the mechanisms of Nature depends almost entirely on the exactness of construction that has become possible through the invention of the analysis of the infinite and through the simple principles discovered by Archimedes, Galileo, and Newton, which modem physics makes use of. By contrast, in the natural sciences where the simple principles for such constructions are still lacking, to discover causal connections one pursues phenomenon into the spatially small, as far as the microscope permits. Questions about the metric relations of Space in the immeasurably small are thus not idle ones.

If one assumes that bodies exist independently of position, then the curvature is everywhere constant, and it then follows from astronomical measurements that it cannot be different from zero; or at any rate its reciprocal must be an area in comparison with which the range of our telescopes can be neglected. But if such an independence of bodies from position does not exist, then one cannot draw conclusions about metric relations in the infinitely small from those in the large; at every point the curvature can have arbitrary values in three directions, provided only that the total curvature of every measurable portion of Space is not perceptibly different from zero; still more complicated relations can occur if the line element cannot be represented, as was presupposed, by the square root of a differential expression of the second degree. Now it seems that the empirical notions on which the metric determinations of Space are based, the concept of a solid body and that of a light ray, lose their validity in the infinitely small; it is therefore quite definitely conceivable that the metric relations of Space in the infinitely small do not conform to the hypotheses of geometry, and in fact one ought to assume this as soon as it permits a simpler way of explaining phenomena.

The question of the validity of the hypotheses of geometry in the infinitely small is connected with the question of the basis for the metric relations of space. In connection with this question, which may indeed still be reckoned as part of the study of Space, the above remark is applicable, that in a discrete manifold the principle of metric relations is already contained in the concept of the manifold, but in a continuous one it must come from something else. Therefore either the reality underlying Space must form a discrete manifold, or the basis for the metric relations must be sought outside it, in binding forces acting upon it.

An answer to these questions can be found only by proceeding from the conception of phenomena which has hitherto been proven by experience, for which

Newton laid the foundation, and gradually modifying it driven by facts that cannot be explained by it; investigations like the one just made, which begin from general concepts, can serve only to insure that this work is not hindered by the restriction of concepts, and that progress in comprehending the connection of things is not obstructed by traditional prejudices.

This leads us away into the domain of another science, the realm of physics, into which the nature of the present occasion does not allow us to enter.

Solutions to selected exercises

I entreat you, leave the doctrine of parallel lines alone; you should fear it like a sensual passion; it will deprive you of your health, leisure, and peace – it will destroy all joy in your life. These gloomy shadows can swallow up a thousand Newtonian towers and never will there be light on earth; never will the unhappy human race reach absolute truth – not even in geometry.

F. BOLYAI (IN A LETTER TO J. BOLYAI)

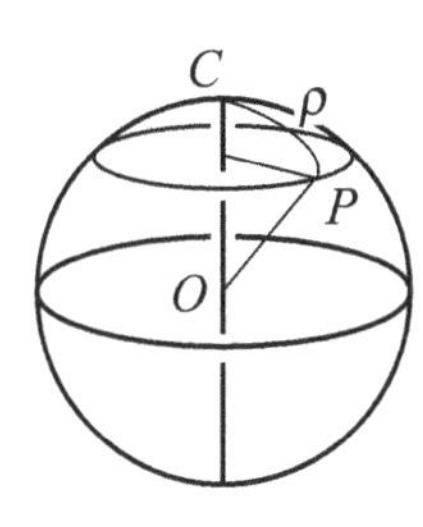

1.4. We can rotate the sphere so that the center C of the spherical circle of radius ρ is the North Pole. Suppose P is any point on the circle. If we denote the center of the sphere by O, then the angle $\angle POC$ is given by ρ/R radians. The circle lies in a plane, and the radius of this circle in the plane is given by $R\sin(\rho/R)$. Thus, its circumference is $2\pi R\sin(\rho/R)$. If we write the circumference as $\dfrac{2\pi\sin(\rho/R)}{1/R}$, then the limit as R goes to infinity is $2\pi\rho$.

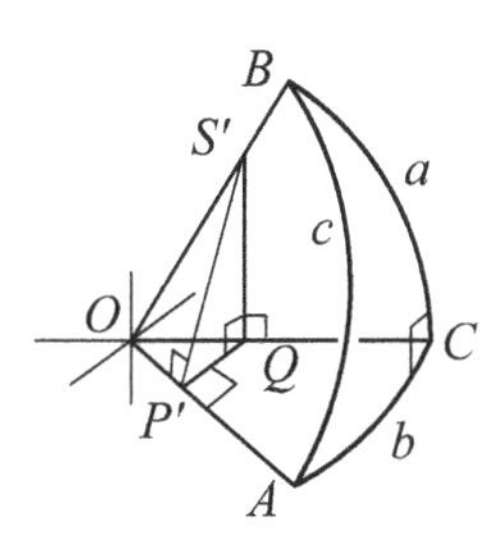

1.5. Many other relations derive from the diagram used to prove Theorem 1.3, the spherical law of sines. The diagram here is the analogous diagram for $\angle A$. Here we find:

$$\cos\angle A = \frac{P'Q}{P'S'} = \frac{P'Q}{OQ}\cdot\frac{OQ}{OS'}\cdot\frac{OS'}{P'S'} = \sin\frac{b}{R}\cos\frac{a}{R}\frac{1}{R\sin(c/R)}$$
$$= \sin\frac{b}{R}\cdot\cos\frac{a}{R}\cdot\frac{\sin\angle B}{\sin(b/R)} = \sin\angle B\cos(a/R).$$

1.10. Consider the plane determined by A, O, and B, denoted $\Pi(AOB)$. It meets the sphere in the great circle that contains the segment AB and the pole C' lies on it. The plane $\Pi(AOC)$ contains the great circle segment AC and the pole B'. The line segments $C'O$ and $B'O$ are perpendicular to the line $l = \Pi(AOB)\cap\Pi(AOC) = \overleftrightarrow{AO}$. The dihedral angle between $\Pi(AOB)$ and $\Pi(AOC)$ is the supplement of the angle $\angle C'OB'$ because the choice of pole puts B' on the other side of the plane

$\Pi(AOC)$. Thus $\angle A$, which is the dihedral angle between $\Pi(AOC)$ and $\Pi(AOB)$, is $\pi - \angle C'OB' = \pi - a'$.

1.11. The distance between points M and N may be given as R times the angle in radians between M and N. Since we are working with longitude and latitude, we can view the coordinates as describing points on the unit sphere. It follows that $\cos\angle MON = M \cdot N$. The rectangular coordinates corresponding to $M = (\lambda_1, \phi_1)$ are given by:

$$M = (\cos\phi_1 \cos\lambda_1, \sin\phi_1 \cos\lambda_1, \sin\lambda_1).$$

From the analogous formula for N we get:

$$M \cdot N = \cos(\phi_2 - \phi_1)\cos\lambda_1 \cos\lambda_2 + \sin\lambda_1 \sin\lambda_2.$$

So $\angle MON = \arccos(M \cdot N)$. To obtain the compass heading of the great circle segment MN, complete the triangle with the North Pole P. The angle PMN determines the compass heading. By the law of sines:

$$\frac{\sin\angle PMN}{\sin m} = \frac{\sin\angle MPN}{\sin\widehat{MN}}.$$

The angles m and $\angle MPN$ are found from the coordinates: $m = \pi/2 - \lambda_2$, and $\angle MPN = \phi_2 - \phi_1$. The segment $\widehat{MN}$ is is determined by the distance between M and N, and so one can just compute from the law of sines to determine $\angle PMN$.

2.4. From the assumptions of Proposition I.4, and part 6 of the axioms of congruence, we know that $\angle ABC \cong \angle A'B'C'$ and $\angle BCA \cong \angle B'C'A'$. Along $B'C'$ there is a point X with $B'X \cong BC$ (by part 1) of the axioms of congruence. Consider the triangles ΔABC and $\Delta A'B'X$. Then $AB \cong A'B'$, $BC \cong BX$, and $\angle ABC \cong \angle A'B'X$. Applying part 6 of the axioms of congruence we know that $\angle B'A'X \cong \angle BAC \cong \angle B'A'C'$. Since C' is on $B'C'$, $C' = X$ by 4. Then $BC \cong B'X \cong B'C'$.

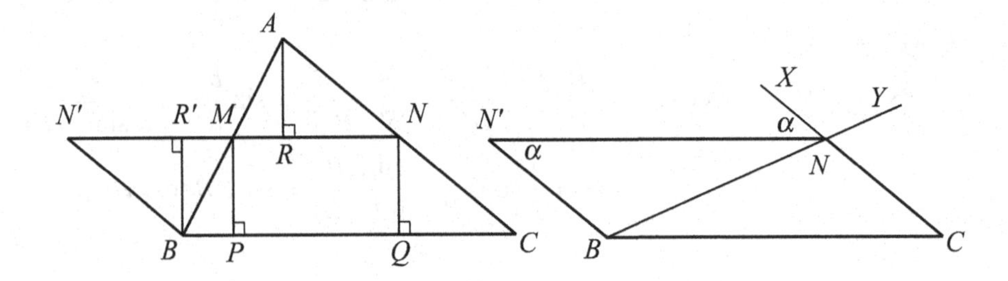

2.10. Let ΔABC be on base BC and let M denote the midpoint of AB and N the midpoint of AC. Join MN and extend it past M to N' where $MN \cong MN'$. Construct the perpendiculars AR to MN and MP and NQ to BC. If we choose the point R' on MN' with $MR \cong MR'$, then the perpendicular to NN' from R' meets B to form $\Delta BR'M \cong \Delta MRA$. Joining B to N' we obtain $\Delta MBN' \cong \Delta MAN$. It follows from the construction that $BN' \cong CN$. Consider the quadrilateral $BCNN'$. By the Saccheri–Legendre Theorem, the angle sum in the quadrilateral is less than or equal to 2π. Let α denote $\angle BN'N$. From the congruence $\Delta BR'M \cong \Delta MRA$, we have $\angle N'NC \cong \pi - \alpha$. Join the diagonal BN extended to BY, and extend CN to CX. By vertical angles we have $\angle BNC \cong \angle YNX$. Since $\angle N'NX$ is supplementary to $\angle N'NC$, we have $\angle XNN' \cong \alpha$. It follows from the Saccheri–Legendre Theorem that $\alpha + \angle NBN' \leq \angle YNX + \alpha$, and so $\angle NBN' \leq \angle CNB$. Since the greater angle subtends the greater side (Proposition I.19), $BC \geq NN' = 2MN$. This proof does not use Postulate V.

2.13. Let AB denote a diameter of a circle with center O and let C be a point on the circle not A or B. Join OC to form two triangles ΔAOC and ΔBOC. Since AO, BO, and CO are radii, they are congruent and the two triangles are isosceles triangles. Let $\angle OAC = \alpha = \angle OCA$ and $\angle OBC = \beta = \angle OCB$. The angle sum of ΔABC is expressed as $2\alpha + 2\beta$ and the Saccheri–Legendre Theorem implies $\angle ACB = \alpha + \beta \leq \pi/2$. In chapter 3 we see that we need Postulate V to deduce that the third vertex is a right angle.

2.14. We leave it to you to construct the sum. The product, quotient, and square root are given in the picture that follows. The arguments for product and quotient are based on Proposition I.43 (yes, we are assuming Postulate V), and the identification of a product with the area of a rectangle with the given lengths as sides. The argument for the square root follows by similar right triangles.

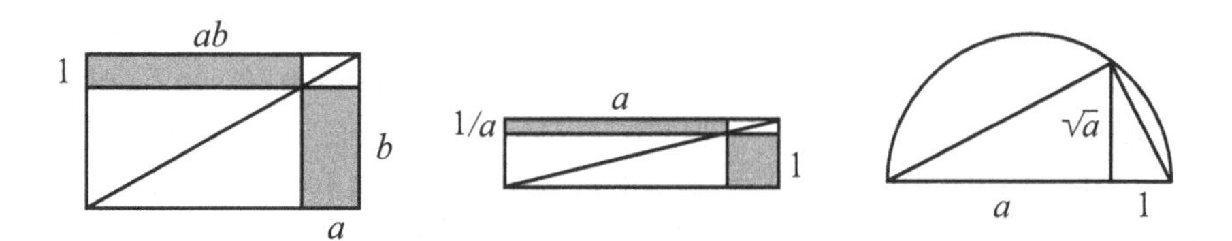

3.5. The error is subtle and depends on the equivalent of Postulate V given by Theorem 3.9, no. 10. There exists an acute angle such that every point in the interior angle is on a line intersecting both rays of the angle away from the vertex. In the drawing $\angle CAB$ would have to be such an angle to guarantee a line through D is on a line meeting $\overleftrightarrow{AC}$ at F and $\overleftrightarrow{AB}$ at E. Were Postulate V to fail, it would be possible to choose D in angle $\angle CAB$ so that any line through D meeting $\overleftrightarrow{AC}$ did not meet $\overleftrightarrow{AB}$ and any line through D meeting $\overleftrightarrow{AB}$ did not meet $\overleftrightarrow{AC}$.

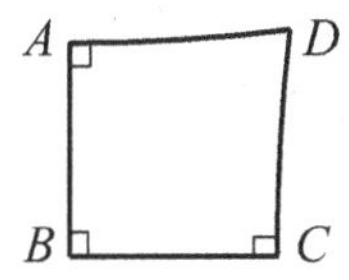

3.7. One can "double" a Lambert quadrilateral by reflecting it across side AB to form $CC'DD'$, a Saccheri quadrilateral. Then **HRA** implies the angles of the Saccheri quadrilateral are all right and hence it is so for the Lambert quadrilateral. Under

HAA, we know by Theorem 3.12, $DD' > CC'$ and so $AD > BC$ by halving. By reflecting across BC we see $CD > AB$.

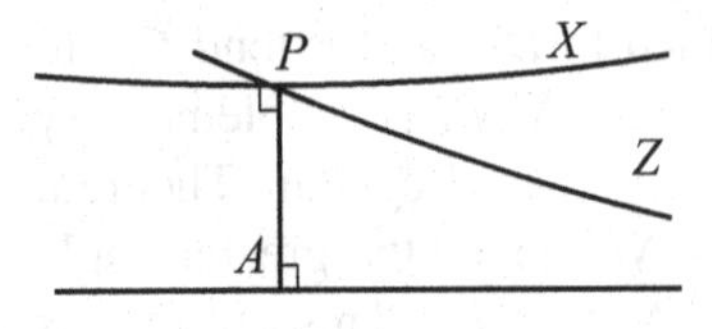

3.10. Suppose l and m are nonintersecting lines in a plane for which **HAA** holds. Suppose l and m have a common perpendicular line $\overleftrightarrow{AP}$ where A is on l, P is on m. Then there is a line $\overleftrightarrow{PZ}$ which is asymptotic to l and $\angle APZ$ is acute. If X lies on m in the direction of $\overrightarrow{PZ}$, then $\angle XPZ$ is acute and hence rays $\overrightarrow{PZ}$ and $\overrightarrow{PX}$ diverge. Since l lies on the other side of $\overrightarrow{PZ}$ from m, m and l diverge in this direction. By reflection across $\overleftrightarrow{AP}$, they diverge in the other direction as well.

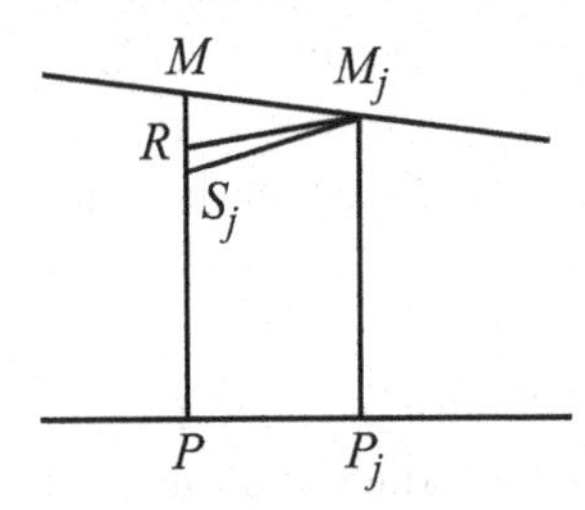

3.12. The failure of continuity means that there are points M_i converging to M for which length(P_iM_i) does not converge to length(PM). That is, there is a value $\epsilon_0 > 0$ such that for any $\delta > 0$ there is some M_j with length(MM_j) $< \delta$ and $|$length(P_iM_i) $-$ length(PM)$| > \epsilon_0$. Let R lie on PM with length(PR) $=$ ϵ_0. Then suppose length(P_jM_j) $<$ length(PM). Let $PS_j \cong P_jM_j$. Then $PP_jM_jS_j$ is a Saccheri quadrilateral and S_j lies between R and P. As M_j gets closer to M we find $\angle M_jRP$ becomes obtuse. However, $\angle M_jS_jP$ is acute and exterior to ΔM_jRS_j, a contradiction. Similarly, we get a contradiction when length(PM) $>$ length(P_jM_j).

4.3. We have shown earlier in the chapter that any acute angle is the angle of parallelism for some length. Suppose $\angle XAY$ is a given acute angle. Let $\overrightarrow{AZ}$ bisect $\angle XAY$. Then $\angle XAZ \cong \angle YAZ$ and each is the angle of parallelism for some length x. Let W lie on $\overrightarrow{AZ}$ with AW of length x. Then construct $\overleftrightarrow{WW'}$ perpendicular to $\overrightarrow{AZ}$. Then $\overleftrightarrow{WW'}$ is parallel to AX since $\angle XAW \cong \Pi(AW) = \Pi(x)$, and $\overleftrightarrow{WW'}$ is similarly parallel to AY.

4.10. Choose a point Q on n with $Q \neq P$. Choose points R on l and S on l' with $PR \cong PS$, and consider the right triangles ΔQPR and ΔQPS. These triangles are congruent by Side–Angle–Side. By Proposition I.7, however, $\Delta QPR = \Delta QPS$ and so $R = S$. Hence, $l = \overleftrightarrow{PR} = \overleftrightarrow{PS} = l'$.

4.17. To complete the proof of the hyperbolic law of sines, in an arbitrary triangle, construct two altitudes. Since $\sin(\pi - \alpha) = \sin(\alpha)$, whether the altitude meets the opposite side inside or outside the triangle is unimportant. Then, for each pair of right triangles sharing the altitude, the proven law of sines give the relations.

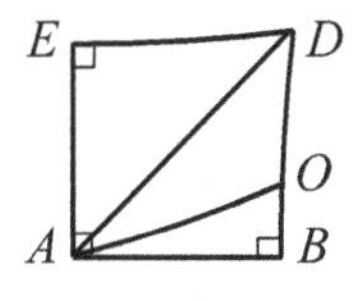

To prove $\angle OAE \cong \Pi(x)$ it is sufficient to prove $\sin(\angle OAE) = \sin(\Pi(x))$. By the relations developed to prove the Lobachevskiĭ–Bolyai Theorem we know $\sin(\Pi(x)) = \dfrac{1}{\cosh(x/k)}$. The Hyperbolic Pythagorean Theorem implies:

$$\frac{1}{\cosh(AE/k)} = \frac{\cosh(ED/k)}{\cosh(AD/k)} = \frac{\cosh(AB/k)\cosh(OB/k)}{\cosh(AB/k)\cosh(BD/k)} = \frac{\cosh(OB/k)}{\cosh(BD/k)}.$$

The relations developed for general right triangles imply that:

$$\cosh(OB/k) = \frac{\cos(\angle BAO)}{\sin(\angle BOA)}, \quad \cosh(BD/k) = \frac{\cos(\angle DAB)}{\sin(\angle BDA)}.$$

Therefore,

$$\frac{\cosh(OB/k)}{\cosh(BD/k)} = \frac{\cos(\angle BAO)\sin(\angle BDA)}{\sin(\angle BOA)\cos(\angle BDA)} = \cos(\angle BAO)\frac{\sin(\angle BDA)}{\sin(\angle BOA)\sin(\angle DAE)},$$

since $\angle DAB + \angle DAE =$ a right angle. By the cases of the law of sines for $\triangle DBA$, $\triangle AOB$ and $\triangle AED$, we find:

$$\sin(\angle DAE) = \frac{\sinh(ED/k)}{\sinh(AD/k)} = \frac{\sinh(AO/k)}{\sinh(AD/k)} = \frac{\sinh(AB/k)/\sin(\angle BOA)}{\sinh(AB/k)/\sin(\angle BDA)} = \frac{\sin(\angle BDA)}{\sin(\angle BOA)}.$$

Thus we have established that:

$$\sin(\Pi(x)) = \frac{1}{\cosh(AE/k)} = \frac{\cosh(OB/k)}{\cosh(BD/k)} = \cos(\angle BAO) = \sin(\angle OAE).$$

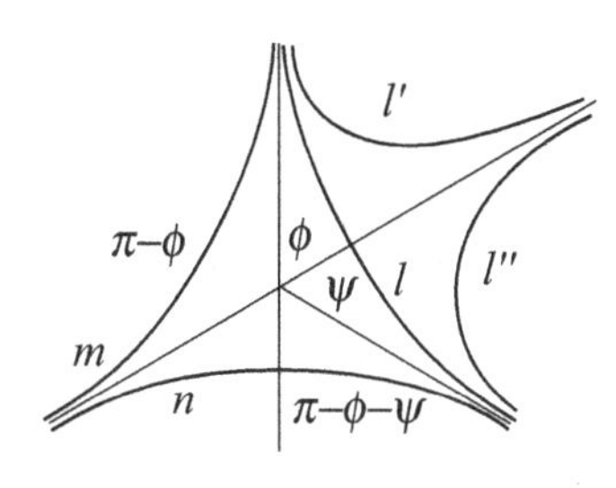

4.19. By adding lines l' and l'' parallel to the rays defining angle ϕ and ψ we see that $f(\pi - \phi - \psi) + A = f(\pi - \phi) + f(\pi - \psi)$ by considering the areas determined by l', m, and the line to which they are parallel that makes angle ϕ, and the similarly for l'', n, and the line that makes angle ψ. Since $f(\alpha) + f(\pi - \alpha) = A$, we get:

$$A - f(\phi + \psi) + A = A - f(\phi) + A - f(\psi).$$

Thus $f(\phi + \psi) = f(\phi) + f(\psi)$. This implies $f(\phi) = \lambda\phi + c$. Since $f(0) = 0$, we deduce $f(\alpha) = \lambda\alpha$. Taking $f(0) + f(\pi) = A$, we see $A = \lambda\pi$. Finally, in diagram D, we see that:

$$\text{area}\,\triangle + f(\alpha) + f(\beta) + f(\gamma) = A,$$

and this gives area $\triangle = A - \lambda\alpha - \lambda\beta - \lambda\gamma = \lambda(\pi - \alpha - \beta - \gamma)$.

5.5. The conversion between polar and rectangular coordinates is given by:

$$x = \rho(\theta)\cos\theta, \quad y = \rho(\theta)\sin\theta.$$

This transformation yields the rectangular curve as $(x(\theta), y(\theta))$ and we have:

$$x' = \rho' \cos\theta - \rho \sin\theta, \quad y' = \rho' \sin\theta + \rho \cos\theta.$$

Then $(x')^2 + (y')^2 = (\rho' \cos\theta - \rho \sin\theta)^2 + (\rho' \sin\theta + \rho \cos\theta)^2 = \rho^2 + (\rho')^2$ and the formula for arc length follows. Similarly, if one computes $x'y'' - x''y'$ in the new coordinates, one gets $2(\rho')^2 - \rho\rho'' + \rho^2$. The formula for oriented curvature follows.

5.8. The forces applied to the cable are due to gravity and to the tension of the chain keeping it in equilibrium. Since the chain is at rest, tension is constant along the chain and may be identified with a horizontal unit vector at the origin, the lowest point of the chain. Measuring from this point to a point along the chain, the force of gravity at a point is proportional to the arc length to that point since the weight is uniformly distributed along its length. Resolving the two forces we get $dy/dx = cs$ where c is a constant consisting of the gravitational constant, the mass per unit length, and the constant tension. It follows that:

$$\left(\frac{d^2y}{dx^2}\right)^2 = c^2\left(1 + \left(\frac{dy}{dx}\right)^2\right).$$

A solution to this equation comes from the relation $\cosh^2 x - \sinh^2 x = 1$. Taking the constant into consideration, consider $y(x) = (1/c)\cosh(cx) - (1/c)$. Then $y(0) = 0$ and $(y'')^2 = c^2(1 + (y')^2)$.

5.12. The simplest parametrization of $y^2 = ax$ is $\alpha(t) = (t^2/a, t)$. The tangent vector is given by $\alpha'(t) = (2t/a, 1)$, and so the unit tangent is given by $T(t) = \left(\frac{2t}{\sqrt{4t^2+a^2}}, \frac{a}{\sqrt{4t^2+a^2}}\right)$ and, thus, $N(t) = \left(-\frac{a}{\sqrt{4t^2+a^2}}, \frac{2t}{\sqrt{4t^2+a^2}}\right)$. The oriented curvature is given by:

$$\kappa_\alpha(t) = \frac{x'y'' - x''y'}{((x')^2 + (y')^2)^{3/2}} = \frac{-2a^2}{(4t^2+a^2)^{3/2}}.$$

To compute the involute and evolute, we use the direct definition for the involute and the curve of centers of curvature for the parabola $y^2 = ax$. The involute takes the form:

$$\beta(t) = \alpha(t) + \left(\int_0^t \|\alpha'(u)\|\, du\right)\alpha'(t).$$

We leave it to the reader to compute the arc length integral. The evolute is determined by the curve of centers of curvature and so is given by:

$$C_\alpha(t) = \alpha(t) + \frac{1}{\kappa_\alpha(t)} N(t) = (t^2/a, t) - \frac{(\sqrt{4t^2+a^2})^3}{2a^2}\left(-\frac{a}{\sqrt{4t^2+a^2}}, \frac{2t}{\sqrt{4t^2+a^2}}\right)$$

$$= \left(\frac{3t^2}{a} + \frac{a}{2}, -\frac{4t^3}{a^2} \right),$$

which is a semicubical parabola.

6.6. Since $T \times N = B$, $N \times B = T$ and $B \times T = N$, the Frenet–Serret formula suggests that the Darboux vector is a linear combination of vectors in the frame. To achieve the Frenet–Serret conditions, we need $\mathbf{w} \times T = \kappa N$, $\mathbf{w} \times N = -\kappa T + \tau B$, and $\mathbf{w} \times B = -\tau N$. The reader can check that $\mathbf{w} = \kappa B + \tau T$ works.

6.8. By choosing appropriate coordinates we can arrange, by rotations and translations, to have the plane on which the space curve is projected be the x-y-plane, and the osculating plane of the space curve at the point where it is orthogonal to the x-y-plane to be the y-z-plane. Assume that the space curve has unit speed. Then $T(s) = \alpha'(s)$ and $\kappa_\alpha(s)N(s) = \alpha''(s)$. Since both T and N lie in the osculating plane, at the special point we have $\alpha'(s) = (0, y', z')$ and $\alpha''(s) = (0, y'', z'')$. The orthogonal projecting in this case is given by $(x, y, z) \mapsto (x, y)$ and so $(\mathrm{Pr} \circ \alpha(s))' = (0, y')$ and $(\mathrm{Pr} \circ \alpha(s))'' = (0, y'')$. Computing the plane curvature at this point we see from $x'y'' - x''y'$ that $\kappa_\pm(\mathrm{Pr} \circ \alpha, s) = 0$ at the special point.

6.10. To lie entirely on a sphere, the unit speed curve $\alpha(s)$ must satisfy $(\alpha(s) - C) \cdot (\alpha(s) - C) = r^2$. This leads to the relations:

$$\begin{aligned} (\alpha(s) - C) \cdot \alpha'(s) &= 0 \\ (\alpha(s) - C) \cdot \alpha''(s) + \alpha'(s) \cdot \alpha'(s) &= 0 \\ (\alpha(s) - C) \cdot \alpha'''(s) &= 0. \end{aligned}$$

Thus $(\alpha(s) - C) \cdot (\kappa N) = -1$ or $(\alpha(s) - C) \cdot N = -1/\kappa$. Taking a derivative of this relation gives $(\alpha(s) - C) \cdot (-\kappa T + \tau B) = \kappa'/(\kappa^2)$, and since $T = \alpha'(s)$, we have $(\alpha(s) - C) \cdot B = \kappa'/(\tau\kappa^2)$. When we take the derivative of this relation we get $\alpha'(s) \cdot B + (\alpha(s) - C) \cdot (-\tau N) = -\tau(\alpha(s) - C) \cdot N = \tau/\kappa = (\kappa'/(\tau\kappa^2))'$, as desired.

6.14. Since $\beta(t) \in l^\perp_{\beta(t)}$, we can write $\beta(t) = \alpha(t) + r(t)N_\alpha(t)$. Taking the derivative we get:

$$\beta'(t) = \alpha'(t) + r'(t)N_\alpha(t) + r(t)(-\kappa_\alpha(t)T_\alpha(t) + \tau_\alpha(t)B_\alpha(t)).$$

However, since $l^\perp_{\alpha(t)} = l^\perp_{\beta(t)}$, it follows that $N_\beta(t) = \pm N_\alpha(t)$, and since $\beta'(t) \cdot N_\beta = 0 = \beta'(t) \cdot N_\alpha(t)$, we have $r'(t) = 0$ and $r(t) = r$, a constant. Also $(T_\alpha(t) \cdot T_\beta(t))' = (\kappa_\alpha N_\alpha) \cdot T_\beta + T_\alpha \cdot (\kappa_\beta N_\beta)$, but $N_\alpha = \pm N_\beta$ implies $(T_\alpha \cdot T_\beta)' = 0$ and so the angle between $\alpha'(t)$ and $\beta'(t)$ is constant along the curves. Denote this angle by θ. Let's assume that α is parametrized by arc length. Then:

$$\cos\theta = T_\alpha \cdot T_\beta = \alpha'(s) \cdot \frac{\beta'(s)}{\|\beta'(s)\|} = T_\alpha \cdot \frac{(1 - r\kappa_\alpha(s))T_\alpha + r\tau_\alpha(s)B_\alpha}{\sqrt{(1 - r\kappa_\alpha)^2 + (r\tau_\alpha)^2}}$$

$$= \frac{(1 - r\kappa_\alpha)}{\sqrt{(1 - r\kappa_\alpha)^2 + (r\tau_\alpha)^2}}.$$

It follows that $\cos^2\theta((1 - r\kappa_\alpha)^2 = (r\tau_\alpha)^2) = (1 - r\kappa_\alpha)^2$. Rearranging the terms we deduce $r\kappa_\alpha + (r\cot\theta)\tau_\alpha = 1$. Conversely, suppose $r\kappa_\alpha + c\tau_\alpha = 1$. Then $c = r\cot\theta$ for some angle θ. Let $\beta(s) = \alpha(s) + rN_\alpha(s)$. Then $\beta'(s) = (1 - r\kappa_\alpha)T_\alpha + r\tau_\alpha B_\alpha$. Computing $\beta'/\|\beta'\|$ we find:

$$\frac{\beta'(s)}{\|\beta'(s)\|} = \frac{(1 - r\kappa_\alpha)T_\alpha + r\tau_\alpha B_\alpha}{\sqrt{(1 - r\kappa_\alpha)^2 + (r\tau_\alpha)^2}} = \frac{(1 - r\kappa_\alpha)T_\alpha + r\tau_\alpha B_\alpha}{r\tau_\alpha \csc\theta} = (\cos\theta)T_\alpha + (\sin\theta)B_\alpha,$$

since $1 - r\kappa_\alpha = (r\cot\theta)\tau_\alpha$. Then:

$$\left(\frac{\beta'(s)}{\|\beta'(s)\|}\right)' = (\cos\theta)(\kappa_\alpha N_\alpha) + (\sin\theta)(-\tau_\alpha N_\alpha),$$

and so $N_\beta = \pm N_\alpha$ and $\alpha(s)$ and $\beta(s)$ are Bertrand mates.

7.4. Computing the coordinate tangent vectors we get:

$$x_u = \alpha'(u) + v\alpha''(u), \quad x_v = \alpha'(u).$$

Then $E = 1 + (\kappa_\alpha(u))^2 v^2$, $F = 1$, since $\alpha(u)$ is a unit speed curve, and $G = 1$ as well.

7.6. Suppose $\Phi\colon (U \subset \mathbb{R}^3) \to \mathbb{R}$ is a differentiable function with w a regular value of Φ. Write the partial derivatives of Φ by Φ_x, Φ_y, and Φ_z. By the Implicit Function Theorem, there is a function of two variables $f(u,v)$ and satisfying $\Phi(u,v,f(u,v)) = w$. Let $x(u,v) = (u,v,f(u,v))$ be a coordinate chart for the graph of f and hence part of $\Phi^{-1}(\{w\})$. The coordinate curves have tangents $x_u = (1,0,f_u)$ and $x_v = (0,1,f_v)$. To express E, F, and G in terms of Φ, apply implicit differentiation to $\Phi(u,v,f(u,v)) = w$. We get, assuming $\Phi_z \neq 0$,

$$0 = \frac{\partial\Phi}{\partial x} + \frac{\partial\Phi}{\partial z}\frac{\partial f}{\partial u}, \quad 0 = \frac{\partial\Phi}{\partial y} + \frac{\partial\Phi}{\partial z}\frac{\partial f}{\partial v},$$

and so $f_u = -\Phi_x/\Phi_z$ and $f_v = -\Phi_y/\Phi_z$. Hence,

$$E = 1 + \left(\frac{\Phi_x}{\Phi_z}\right)^2, \ F = \frac{\Phi_x\Phi_y}{(\Phi_z)^2}, \text{ and } G = 1 + \left(\frac{\Phi_y}{\Phi_z}\right)^2.$$

7.7. The tangent plane $T_p(S)$ is representable as the span of $\{x_u, x_v\}$, that is, the set of all linear combinations $sx_u + tx_v$ with s and t real numbers. The Jacobian of $x(u,v)$ is the matrix with columns given by x_u and x_v written as column vectors. It follows immediately that $J(x)(p)\begin{pmatrix} s \\ t \end{pmatrix} = sx_u + tx_v$. Hence $J(x)(p)(\mathbb{R}^2) = T_p(S)$.

7bis.5. The graph of a polar function is given by the curve $\alpha(\theta) = (\theta, \rho(\theta))$ in polar coordinates. The line element for the plane with polar coordinates is $ds^2 = \rho^2\,d\theta^2 + d\rho^2$. The reader can compute that $\alpha'(\theta)\cdot\alpha'(\theta) = (\rho')^2 + (\rho)^2$, and that $\alpha'(\theta)\cdot x_\theta = (\rho)^2$. Then, the condition of a constant angle with lines through the origin becomes:

$$\frac{\alpha'(\theta)\cdot x_\theta}{\|\alpha'(\theta)\|\cdot\|x_\theta\|} = \frac{\rho}{\sqrt{(\rho')^2+(\rho)^2}} = c, \text{ a constant.}$$

This equation leads to $\rho' = \dfrac{\sqrt{1-c^2}}{c}\rho$, which is solved by $\rho(\theta) = Ke^{(\sqrt{1-c^2}/c)\theta}$, a logarithmic spiral. The inverse of stereographic projection is given in polar coordinates by the geographic coordinates:

$$(\theta,\rho) \mapsto (\lambda(\theta,\rho), \phi(\theta,\rho)) = (\theta, \pm\left(\frac{\pi}{2} - 2\arctan(\rho)\right),$$

where the sign is positive for points with $\rho < 1$ and negative for $\rho \geq 1$. The formula for the loxodrome follows by composing these functions.

7bis.8. Let ψ denote the cone angle. The cone is a surface of revolution of the function $f(u) = u\tan\psi$. The coordinate chart for a surface of revolution gives $x(u,v) = (f(u)\cos v, f(u)\sin v, u)$. It is useful to change point of view to polar coordinates: Let (ρ,θ) map to:

$$y(\rho,\theta) = (\rho\sin\psi\cos\left(\frac{\theta}{\sin\psi}\right), \rho\sin\psi\cos\left(\frac{\theta}{\sin\psi}\right), \rho\cos\psi).$$

The reader can check that $y_\rho\cdot y_\rho = 1$, $y_\theta\cdot y_\rho = 0$, and $y_\theta\cdot y_\theta = \rho^2$, the line element for the polar plane. Divide through by $\cos\psi$ to compare this mapping with x to see that we have simply changed $u = \rho\cos\psi$ and $v = \theta/\sin\psi$. Thinking of a point on the cone as (ρ,θ) where θ is the angle along the circle measured from the x-axis at the distance ρ from the vertex, then $(\rho,\theta)\mapsto(\rho\cos\psi, \theta/\sin\psi)$ is the ideal map projection.

7bis.9. Put the radius of the sphere at one and suppose $0 < \phi_0 < \pi/2$ is a latitude in the Northern Hemisphere. Then the cone tangent to the sphere through this circle of latitude has cone angle ϕ_0 and vertex at $(0,0,1/\sin(\phi_0))$. For a point (λ,ϕ) on the sphere, the projection from the center to the cone goes to $(\rho,\theta) = (\cot(\phi_0) - \tan(\phi-\phi_0), \lambda)$. Apply the mapping in Exercise 7bis.8 to map these coordinates to the plane, giving the map projection.

8.8. Since trace(AB) = trace(BA), we know that trace(A) = trace($PP^{-1}A$) = trace($P^{-1}AP$). Thus, the trace of a matrix is independent of a choice of basis in which one expresses the matrix. Hence:

$$H = -\frac{1}{2}\text{trace}(dN_p) = -\frac{1}{2}(a_{11}+a_{22})$$
$$= -\frac{1}{2}\left(\frac{fF-eG}{EG-F^2}+\frac{fF-gE}{EG-F^2}\right) = \frac{Eg-2Ff+Ge}{2(EG-F^2)}.$$

8.10. Using the result of Exercise 8.8, $H = \dfrac{Eg-2fF+Ge}{EG-F^2}$. In the case $E=\phi=G$ and $F=0$, we have $H=\dfrac{1}{\phi}(g+e)$. Since $e=\langle N,x_{uu}\rangle$ and $g=\langle N,x_{vv}\rangle$, we see that $H=\dfrac{1}{\phi}\langle N,x_{uu}+x_{vv}\rangle$. If $\dfrac{\partial^2 f_i}{\partial u^2}+\dfrac{\partial^2 f_i}{\partial v^2}=0$, then $x_{uu}+x_{vv}=\mathbf{0}$ and $H=0$. To prove the converse we show that $x_{uu}+x_{vv}$ is a multiple of N. Then $H=0$ implies $x_{uu}+x_{vv}=\mathbf{0}$. The assumptions give us $x_u\cdot x_u=\phi=x_v\cdot x_v$ and $x_u\cdot x_v=0$. These conditions give the relations:

$$x_{uu}\cdot x_v + x_u\cdot x_{uv} = 0 = x_{uv}\cdot x_v + x_u\cdot x_{vv}, \quad x_{uu}\cdot x_u = x_{uv}\cdot x_v, \quad x_{uv}\cdot x_u = x_{vv}\cdot x_v.$$

It follows that:

$$x_{uu}\cdot x_v + x_u\cdot x_{uv} = x_{uu}\cdot x_v + x_{vv}\cdot x_v = (x_{uu}+x_{vv})\cdot x_v = 0.$$

Similarly, $(x_{uu}+x_{vv})\cdot x_u = 0$ and so $x_{uu}+x_{vv}=g(u,v)N$. Then $H=0$ implies $x_{uu}+x_{vv}=\mathbf{0}$, the harmonic condition.

8.12. The condition may be rewritten:

$$\frac{d}{dt}N(\alpha(t)) = dN_{\alpha(t)}(\alpha'(t)) = \lambda(t)\alpha'(t).$$

This reformulation shows that $\alpha'(t)$ is an eigenvector of $dN_{\alpha(t)}$, and so $\alpha'(t)$ is a principal direction. Conversely, if $\alpha(t)$ is a line of curvature, then $\alpha'(t)$ is a principal direction for all t and hence an eigenvector of $dN_{\alpha(t)}$. But that condition implies that $dN_{\alpha(t)}(\alpha'(t))=\lambda(t)\alpha'(t)$. The curvature of $\alpha(t)$ is computed using the second fundamental form which satisfies $\mathrm{II}_{\alpha(t)}(\alpha'(t)) = -\lambda(t)$.

9.4. We can write the expressions for the Christoffel symbols in this new notation as follows:

$$\begin{pmatrix}\Gamma_{11}^1\\ \Gamma_{11}^2\end{pmatrix} = \frac{1}{2}(g^{ij})\begin{pmatrix}(g_{11})_u\\ 2(g_{12})_u-(g_{11})_v\end{pmatrix} \quad \begin{pmatrix}\Gamma_{12}^1\\ \Gamma_{12}^2\end{pmatrix} = \frac{1}{2}(g^{ij})\begin{pmatrix}(g_{11})_v\\ (g_{22})_u\end{pmatrix}$$
$$\begin{pmatrix}\Gamma_{22}^1\\ \Gamma_{22}^2\end{pmatrix} = \frac{1}{2}(g^{ij})\begin{pmatrix}(2(g_{12})_v-(g_{22})_u\\ (g_{22})_v\end{pmatrix}.$$

The first equation for the Christoffel symbols follows by adding in the missing terms in the summation. The formula proving *Theorema Egregium* follows from the Gauss equations:

$$\begin{aligned}R_{1212} &= g_{11}R^1_{212} + g_{12}R^2_{212}\\ &= E[(\Gamma^1_{22})_u - (\Gamma^1_{12})_v + \Gamma^1_{22}\Gamma^1_{11} - \Gamma^1_{12}\Gamma^1_{12} + \Gamma^2_{22}\Gamma^1_{12} - \Gamma^2_{12}\Gamma^1_{22}]\\ &\quad + F[(\Gamma^2_{22})_u - (\Gamma^2_{12})_v + \Gamma^1_{22}\Gamma^2_{11} - \Gamma^1_{12}\Gamma^2_{12} + \Gamma^2_{22}\Gamma^2_{12} - \Gamma^2_{12}\Gamma^2_{22}]\\ &= EGK - F^2K = (EG - F^2)K.\end{aligned}$$

9.6. If we have two asymptotic directions, then we can take them to be the coordinate directions, and so $\mathrm{II}_p(x_u) = 0 = \mathrm{II}_p(x_v)$ from which it follows that $e = g = 0$. If x_u and x_v are perpendicular, then $F = 0$ and so $a_{11} = \dfrac{fF - eG}{EG - F^2} = 0$ and $a_{22} = \dfrac{fF - eE}{EG - F^2} = 0$ and hence the trace of dN_p is zero, that is, $H = 0$ and S is minimal. Conversely, $H = \dfrac{Eg - 2fF + eG}{2(EG - F^2)} = \dfrac{-fF}{EG - F^2} = 0$ implies that $fF = 0$. If $f = 0$, then $K = 0$ and the asymptotic directions are parts of lines, and in all directions $k_n = 0$. Hence we can choose perpendicular asymptotic directions. If $k \neq 0$, then $F = 0$ and the asymptotic directions x_u and x_v are perpendicular.

9.11. Since $H \neq 0$ there are distinct principal directions that we can take as coordinate curves. Since $K = 0$ on W, $k_1k_2 = 0$ and $k_1 + k_2 \neq 0$. We take W to be connected so that $k_1 = 0$ and $k_2 \neq 0$, but k_2 has the same sign throughout W. Since k_1 is a normal curvature, the coordinate curve for u is the asymptotic line. Suppose $\alpha(s) = x(u(s), v_0)$ is a unit speed parametrization of the asymptotic line. We show that $\dfrac{d}{ds}N(x(u(s), v_0)) = \mathbf{0}$. This implies that the $N(x(u(s)), v_0)$ is constant along $\alpha(s)$ and hence the tangent plane is constant along the asymptotic curve. Now $\dfrac{d}{ds}N(x(u(s), v_0)) = u'N_u$ and $N_v = 0$ along this curve because it is a coordinate curve. Since $N_u = a_{11}x_u + a_{22}x_v$, we see $N_u \cdot x_u = -\mathrm{II}_p(x_u) = 0$, so $a_{11} = 0$ along the curve. Thus $N(x(u(s)), v_0)$ is constant. At a point along the curve $\alpha(s_0)$, consider the plane through $\alpha(s_0)$ parallel to $\mathrm{Plane}(N, \alpha'(s_0))$. The curve of intersection of this plane with S has normal curvature $k_n = \mathrm{II}_{\alpha(s_0)}(\alpha'(s_0)) = 0$. The line in this plane in the direction of $\alpha'(s_0)$ satisfies the same differential equations as the asymptotic line, and so they are the same. Hence the asymptotic curve $\alpha(s)$ is a portion of a line in $\mathbb{R}^3$.

10.2. Suppose that $\rho\colon \mathbb{R}^3 \to \mathbb{R}^3$ is reflection across the plane Π. Then $\rho|_S\colon S \to S$ is an isometry. Let $\alpha(s)$ be a unit speed parametrization of $S \cap \Pi$. Then $\alpha(s)$ is planar and so $\alpha'(s)$ and $\alpha''(s)$ are in Π. Since reflection leaves $S \cap \Pi$ fixed, the normal to S along $\alpha(s)$ is left fixed by $\rho|_S$. It follows that $\alpha''(s) = k_n(T(s))N(\alpha(s)) + k_g(s)n_\alpha(s)$

has no component in the $n_\alpha(s)$, since $n_\alpha(s)$ is not in $\Pi = \text{Plane}(T(s), N(\alpha(s)))$. Hence $k_g(s) = 0$ and $\alpha(s)$ is a geodesic. Planes of symmetry for a sphere contain the origin and so give great circles. The obvious planes of symmetry of an ellipsoid give geodesics from the intersections of $\frac{x^2}{a^2} + \frac{y^2}{b^c} + \frac{z^2}{c^2} = 1$ with the x-y-, x-z-, and y-z-planes.

10.6. Let's suppose that $\alpha(s)$ is a unit speed curve on a surface of revolution. Then $\alpha(s) = x(u(s), v(s))$ and $\mathrm{I}_{\alpha(s)}(\alpha'(s), \alpha'(s)) = (u')^2E + (v')^2G = 1$. Taking a derivative with respect to t we get:

$$2u''u'E + (u')^3E_u + 2v''v'G + (v')2u'G_u = 0,$$

since $E_v = 0 = G_v$. This implies the relation:

$$u'' + (u')^2\frac{E_u}{2E} - (v')^2\frac{G_u}{2E} = -\frac{v'E}{2Eu'}\left(v'' + u'v'\frac{G_u}{G}\right).$$

Rewriting this relation in terms of Christoffel symbols for a surface of revolution, we get:

$$u'' + (u')^2\Gamma_{11}^1 + 2u'v'\Gamma_{12}^1 + (v')^2\Gamma_{22}^1 = -\frac{v'G}{2Eu'}((v'') + (u')^2\Gamma_{11}^2 + 2u'v'\Gamma_{12}^2 + (v')^2\Gamma_{22}^2).$$

The condition that the product of the radius at any point on the curve and the sine of the angle made by the curve with the meridian at that point is constant implies the vanishing of the right-hand side of the relation, and so the geodesic equations hold for $\alpha(s)$.

10.11. If a sequence of points $\{a_n\}$ in S is a Cauchy sequence in the metric determined by geodesics, then the sequence of lengths $\{d(a_0, a_n)\}$ is a Cauchy sequence in $\mathbb{R}$ and converges to some value ρ. Then, the circle of points a distance ρ from a_0 is compact and has the limit point of the sequence $\{a_n\}$. By compactness the limit exists and so S is closed in $\mathbb{R}^3$. The condition that $d(x,y) < r$ for all x, y in S means that a ball of radius $3r$ centered at any point in S contains all of S. This follows because a geodesic between points x and y in S has a length less than or equal to the distance $\|x - y\|$ in $\mathbb{R}^3$. Thus S is bounded and closed in $\mathbb{R}^3$, and hence compact.

11.4. A geodesic is closed if it can be parametrized $\gamma: \mathbb{R} \to S$ with some positive number p_0 for which $\gamma(s + p_0) = \gamma(s)$ for all $s \in \mathbb{R}$. Then $\gamma'(s) = \gamma'(s + p_0)$ and so taking one period $[0, p_0]$, we find $\gamma'(0) = \gamma'(p_0)$. With this in mind, we apply the Gauss–Bonnet formula to this case where there is only one piece of the curve to study, and the exterior angle is zero:

$$\iint_R K\,dA = 2\pi.$$

It follows that K must be nonnegative for some region. But K is everywhere nonpositive.

11.9. Since S is a compact, closed surface in $\mathbb{R}^3$, the integral $\iint_S K\,dA$ is finite and positive. However, by Proposition 11.9, the value of the integral is $2\pi\chi(S)$ and so $\chi(S) > 0$. It is now a topological theorem that a closed, orientable, compact surface in $\mathbb{R}^3$ is homeomorphic to a sphere whenever its Euler characteristic is positive. Hence, $S \cong S^2$. Now suppose $\mathbf{u}$ is any point in S^2. Consider the family of planes with equation

$$\mathbf{u}\cdot(x,y,z) = c, \text{ for } c \in \mathbb{R}.$$

The function $f(\mathbf{x}) = \mathbf{u}\cdot\mathbf{x}$ is continuous and since S is compact and closed, there are points on S that realize the maximum and minimum of $f(\mathbf{x})$. At such points $\mathbf{x}_0$ is a critical point of $f(\mathbf{x})$ and so $\mathbf{u}$ is perpendicular to the tangent space to S at $\mathbf{x}_0$. Thus, $\mathbf{u} = N(\mathbf{x}_0)$ and $N(\mathbf{x})$ is onto.

12.2. Let $x\colon (0,2\pi)\times(0,\epsilon_p)$ denote geodesic polar coordinates at p on S. It follows that $ds^2 = dr^2 + G(r,\theta)\,d\theta^2$, where $\sqrt{G} = r - K(p)\dfrac{r^3}{6} + \cdots$. The area of a geodesic circle of radius ρ is given by

$$\text{area}_p(\rho) = \int_0^{2\pi}\int_0^{\rho} \sqrt{G}\,dr\,d\theta.$$

Substituting and computing we get:

$$\text{area}_p(\rho) = \int_0^{2\pi}\int_0^{\rho} r - K(p)\frac{r^3}{6} + \cdots\,dr\,d\theta = \int_0^{2\pi} \frac{\rho^2}{2} - K(p)\frac{\rho^4}{24} + \rho^5(\text{stuff})\,d\theta$$

$$= \pi\rho^2 - K(p)\frac{\pi\rho^4}{12} + \rho^5\left(\int_0^{2\pi} \text{stuff}\,d\theta\right).$$

Rearranging and taking the limit we get:

$$\lim_{\rho\to 0+} \frac{12[\pi\rho^2 - \text{area}_p(\rho)]}{\pi\rho^4} = K(p).$$

12.5. From the equations $\lambda(u) = r\cos(u)$ and $(\mu'(u))^2 = 1 - (\lambda'(u))^2$, the function $\mu(u)$ must satisfy $\mu'(u) = \sqrt{1 - r^2\sin^2(u)}$ whose integral is an elliptic integral. Notice that the domain of u is restricted: If $0 < r < 1$, then we can take an entire branch of the sine, $0 \le u \le \pi$; however, the surface will have a sharp point where $u = 0$ and $u = \pi$ and look like a spindle. If $r = 1$, then the surface is part of a sphere. If $r > 1$, then $0 \le u \le \arcsin(1/r)$, which presents a bulge (*Wulst*) away from the axis, resulting in something like the outer half of a donut.

13.8. If a curve $\gamma(t) = x(u(t), v(t))$ is a geodesic, then it satisfies the geodesic equations. Since we are considering horizontal lines, we may assume $v' = 0 = v''$. The Christoffel symbols for the metric $ds^2 = du^2 + (f(u,v))^2\,dv^2$ satisfy $\Gamma^1_{11} = \Gamma^2_{11} = \Gamma^1_{22} = 0$.The equation $u'' + \Gamma^1_{11}(u')^2 + 2\Gamma^1_{12}u'v' + \Gamma^1_{22}(v')^2 = u'' = 0$, because $v' = 0 = \Gamma^1_{11}$. The second geodesic equation is $v'' + \Gamma^2_{11}(u')^2 + 2\Gamma^2_{12}u'v' + \Gamma^2_{22}(v')^2 = 0$, which holds when $v' = v'' = \Gamma^2_{11} = 0$. Thus, for $u(t) = at + b$, a solution to $u'' = 0$, we obtain that $\gamma(t)$ is a geodesic.

13.10. We show directly that $\mathbb{C}P^1$ is diffeomorphic to S^2. We think of S^2 as $\mathbb{C} \cup \{\infty\}$ where ∞ is identified with the North Pole, the point of projection for stereographic projection. Then a line in $\mathbb{C}^2$ is the set of all pairs $(z, \alpha z)$ for some fixed α in $\mathbb{C}$. We identify the line of these points with the inverse of stereographic projection applied to α. The point at infinity is identified with the line $(0,z)$. Notice that the lines $(z, \alpha z) \sim (z/\alpha, z) \mapsto (0,z)$ when $\alpha \mapsto \infty$. Similarly, as points in the plane go to infinity in norm, their stereographic projections go to the North Pole. Hence $\mathbb{C}P^1 \cong S^2$.

14.17. We know for transformational reasons that a non-Euclidean circle is a Euclidean circle in $\mathbb{H}$. It suffices to check that the radii and centers are related as described. The Euclidean circle is given by:

$$(x-a)^2 + (y - p\cosh(\rho))^2 = p^2\sinh^2(\rho).$$

The non-Euclidean radius is half the diameter and so is given by:

$$\ln\left(\frac{\sqrt{p^2\cosh^2(\rho) - p^2\sinh^2(\rho)}}{p\cosh(\rho) - p\sinh(\rho)}\right) = \ln\left(\frac{1}{\cosh(\rho) - \sinh(\rho)}\right) = \ln(e^{\rho}) = \rho.$$

Finally, the non-Euclidean circle has center $(a, \sqrt{p^2\cosh^2(\rho) - p^2\sinh^2(\rho)}) = (a, p)$, as required.

14.18. We present the case of a right triangle and leave the reader to fill in the general case. If we work in $\mathbb{D}_B$, then we can place a right triangle with angle $\angle A$ at the origin. The point $B = (x,y)$ and so the right angle is at $C = (x,0)$. In the Beltrami metric, the lengths of each side take a very nice form. Parametrize AC by $t \mapsto (t,0)$ for $0 \le t \le x$. Then $b = \displaystyle\int_0^x \frac{dt}{1-t^2}$. It follows immediately that $b = \dfrac{1}{2}\ln\left(\dfrac{1+x}{1-x}\right)$, and so $x = \tanh(b)$. The length a is obtained by parametrizing the BC by $t \mapsto (x,t)$ for $0 \le t \le y$. Since the metric determines the length by integrating

$\sqrt{Edx^2 + 2Fdx\,dy + Gdy^2}$, in this case we have:

$$a = \int_0^y \frac{\sqrt{1-x^2}}{1-x^2-t^2}\,dt = \int_0^y \frac{d(t/\sqrt{1-x^2})}{1-(t/\sqrt{1-x^2})^2}$$
$$= \int_0^{y/\sqrt{1-x^2}} \frac{dt}{1-t^2}.$$

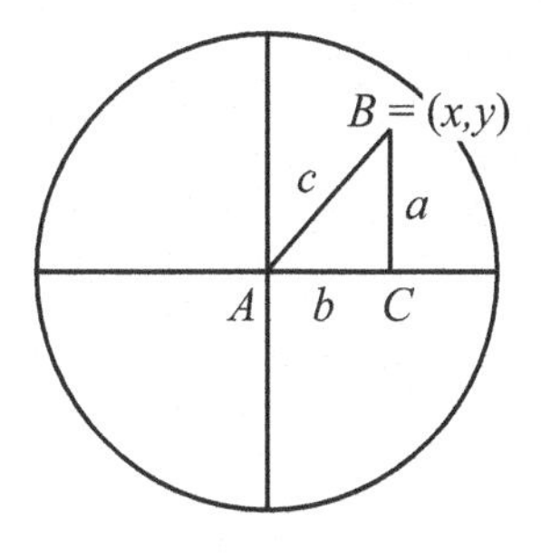

It follows that $\tanh(a) = y/\sqrt{1-x^2}$. Finally, parametrize AB by $t \mapsto (xt, yt)$ for $0 \le t \le 1$. Then:

$$c = \int_0^1 \frac{\sqrt{x^2+y^2}}{1-t^2(x^2+y^2)}\,dt = \int_0^{\sqrt{x^2+y^2}} \frac{dt}{1-t^2}.$$

Thus, $\tanh(c) = \sqrt{x^2+y^2}$. To prove the Hyperbolic Pythagorean Theorem we use the identity:

$$\frac{1}{\cosh^2(c)} = 1 - \tanh^2(c) = 1 - x^2 - y^2 = (1-x^2)(1 - \frac{y^2}{1-x^2})$$
$$= (1-\tanh^2(b))(1-\tanh^2(a)) = \frac{1}{\cosh^2(b)}\frac{1}{\cosh^2(a)},$$

and the relation $\cosh(c) = \cosh(a)\cosh(b)$ follows.

To prove the hyperbolic law of sines, observe that angles at the origin have the same measure in the Euclidean plane as in the hyperbolic plane because $ds^2 = du^2 + dv^2$ there. Working with the diagram, we see that $\sin\angle A = \frac{y}{\sqrt{x^2+y^2}}$. By our previous discussion we find:

$$\sin\angle A = \frac{y}{\sqrt{x^2+y^2}} = \frac{y}{\sqrt{1-x^2}}\frac{\sqrt{1-x^2}}{\sqrt{x^2+y^2}}$$
$$= \tanh(a)\frac{\sqrt{1-\tanh^2(b)}}{\tanh(c)} = \frac{\tanh(a)}{\cosh(b)\tanh(c)} = \frac{\sinh(a)}{\cosh(a)\cosh(b)\tanh(c)}$$
$$= \frac{\sinh(a)}{\cosh(c)\tanh(c)} = \frac{\sinh(a)}{\sinh(c)}.$$

Thus, $\frac{\sin\angle A}{\sinh(a)} = \frac{1}{\sinh(c)}$. Flipping the triangle over to interchange the roles of A and B, we can arrange that C lie along the x-axis again, and the length of c will not change, so the same argument gives $\frac{\sin\angle B}{\sinh(b)} = \frac{1}{\sinh(c)}$ and we have proved $\frac{\sin\angle A}{\sinh(a)} = \frac{\sin\angle B}{\sinh(b)}$.

15.8. At a point p in the manifold, we can choose the higher dimensional version of geodesic rectangular coordinates for which the Christoffel symbols vanish at p. Then, at p,

$$R^m_{ijk;l} = \frac{\partial^2 \Gamma^m_{ik}}{\partial x^l \partial x^j} - \frac{\partial^2 \Gamma^m_{ij}}{\partial x^l \partial x^k}.$$

It follows that:

$$R^m_{ijk;l} + R^m_{ikl;j} + R^m_{ilj;k} = \frac{\partial^2 \Gamma^m_{ik}}{\partial x^l \partial x^j} - \frac{\partial^2 \Gamma^m_{ij}}{\partial x^l \partial x^k} + \frac{\partial^2 \Gamma^m_{il}}{\partial x^j \partial x^k} - \frac{\partial^2 \Gamma^m_{ik}}{\partial x^j \partial x^l} + \frac{\partial^2 \Gamma^m_{ij}}{\partial x^k \partial x^l} - \frac{\partial^2 \Gamma^m_{il}}{\partial x^k \partial x^j} = 0.$$

Since everything involved is a tensor, this relation is obtained after a change of coordinate chart at p. Since p is arbitrary, the relation holds at every point.

15.12. If $Y = \sum_i a^i \frac{\partial}{\partial x^i}$, then let:

$$\nabla(\frac{\partial}{\partial x^j}, Y) = \sum_i b^i{}_{;j} \frac{\partial}{\partial x^i} = \sum_i \left(\frac{\partial b^i}{\partial x^j} + \sum_k b^k G^i_{jk} \right) \frac{\partial}{\partial x^i}.$$

Applied to $\nabla(\frac{\partial}{\partial x^j}, fY)$ we get:

$$\nabla(\frac{\partial}{\partial x^j}, fY) = \sum_i \left(\frac{\partial f}{\partial x^j} b^i + f \frac{\partial b^i}{\partial x^j} + f \sum_k b^k G^i_{jk} \right) \frac{\partial}{\partial x^i} = f \nabla(\frac{\partial}{\partial x^j}, Y) + \frac{\partial}{\partial x^j}(f) \cdot Y.$$

The other properties all follow by linearity. For a vector field Y along a curve $\alpha(t)$, we find:

$$\nabla(\alpha'(t), Y) = \nabla \left(\sum_i \frac{dx^i}{dt} \frac{\partial}{\partial x^i}, \sum_j a^j \frac{\partial}{\partial x^j} \right) = \sum_i \frac{dx^i}{dt} \nabla \left(\frac{\partial}{\partial x^i}, \sum_j a^j \frac{\partial}{\partial x^j} \right)$$

$$= \sum_i \frac{dx^i}{dt} \sum_k \left(\frac{\partial a^k}{\partial x^i} + \sum_l a^l \Gamma^k_{il} \right) \frac{\partial}{\partial x^k}$$

$$= \sum_k \sum_k \left(\frac{dx^i}{dt} \frac{\partial a^k}{\partial x^i} + \sum_l a^l \frac{dx^i}{dt} \Gamma^k_{il} \right) \frac{\partial}{\partial x^k}$$

$$= \sum_k \left(\frac{da^k}{dt} + \sum_{i,l} \Gamma^k_{il} a^i \frac{dx^l}{dt} \right) \frac{\partial}{\partial x^k} = \frac{DY}{dt}.$$

Bibliography

This text owes an enormous debt to the vast literature on geometry. I would like to call attention to the sources most relevant to each chapter for the reader who wishes to learn about a particular topic from a possibly different viewpoint. The whole project owes much to Gray (1979), Struik (1933, 1961), and Spivak (1970). Chapter 1 was inspired by an article by Busemann (1950). Chapters 2, 3, and 4 owe much to Euclid (trans. Heath, 1926), Gray (1979), Bonola (1955), Greenberg (2008), and Meschkowski (1964). Reading Saccheri (1986), Lobachevskiĭ in Bonola (1955), and Bolyai in Gray (2004) is great fun and highly recommended.

The middle section of the book in which classical differential geometry is developed follows the outline of the excellent treatments in Spivak (1970), do Carmo (1976), Struik (1961), Pogorelov (1950), and Hsiung (1981). I was lucky to have Yoder (1988) and Schröder (1988) when I was writing the first edition. These books inspired the digression on Huygens and chapter 7bis. Chapter 14 was inspired by reading Beltrami (1868) and Poincaré (1882), with some help from Gray (1989) and Coolidge (1940). Chapter 15 is my attempt to outline what Riemann said (inspired by Spivak (1970, Vol. 2)) based on hints found in Pinl (1981).

I have tried to include all the historical references made in the text as well as all the books I used in the writing. I have marked cited books and articles with a (c), recommended reading with an (r), and additional titles with and (a). The recent books by Berger (2006, 2010) offer much in the way of guiding an initiated reader into the modern subject of differential geometry. The history of higher dimensional manifolds is told well by Scholz (1999) and their origins in mechanics by Lützen. Happy reading.

Books

Ahlfors, L., *Complex Analysis*, McGraw-Hill, Columbus, OH, 3rd edition, 1979 (c).

Arnold, V.I., *Ordinary Differential Equations*, translated from the Russian by Richard A. Silverman, MIT Press, Cambridge, MA, 1973 (c).

Audin, M., *Geometry*, Berlin; New York: Springer, 2003 (a).

Bär, C., *Elementary Differential Geometry*, New York: Cambridge University Press, 2010 (a).

Berger, M., *A Panoramic View of Riemannian Geometry*, Berlin; New York: Springer, 2003 (r).

Berger, M., *Jacob's Ladder of Differential Geometry*; translated by Lester J. Senechal, New York; London: Springer, 2010 (r).

Bonola, R., *Non-Euclidean Geometry: A Critical and Historical Study of Its Development*, with writings of Lobachevskiĭ and Boylai (trans. by Halsted), translated by H. S. Carslaw, Dover Publications, Miniola, NY, 1955 (c,r).

Borsuk, K., Szmielew, W., *Foundations of Geometry*, North-Holland Pub. Co., Amsterdam, 1960 (c,r).

Bos, H., *Redefining Geometrical Exactness: Descartes' transformation of the early modern concept of construction*, Springer-Verlag, New York, 2001 (c,r).

Boyce, W.E., DiPrima, R.C., *Elementary Differential Equations*, Wiley, 9th edition, Hoboken, NJ, 2008 (c).

Brannan, D., Esplen, M.F., Gray, J.J., *Geometry*, Cambridge University Press, Cambridge, UK, 1999 (c,r).

Brentjes, S., Ahmad al-Karabisi's Commentary on Euclid's "Elements," in Sic Itur Ad Astra: Studien zur Geschichte der Mathematik und Naturwissenschaften, edited by M. Folkerts and R. Lorch, Harrassowitz Verlag, Wiesbaden, 2000 (c).

Buekenhout, F., *Handbook of Incidence Geometry: Buildings and Foundations*, North-Holland, Amsterdam New York, 1994 (c).

Butzer, P.L., Fehér, F., eds. E.B. Christoffel, the influence of his work on mathematics and the physical sciences. International Christoffel Symposium in Honour of Christoffel on the 150th Anniversary of His Birth (1979: Aachen, Germany, and Monschau, Germany), Birkhäuser Verlag, Basel; Boston, 1981 (a).

Carmo, M.P. do, *Differential Geometry of Curves and Surfaces*, Prentice-Hall, Englewood Cliffs, NJ, 1976 (c,r).

Carmo, M.P. do, *Riemannian Geometry*, Birkhäuser, Boston, 1992 (c).

Cartan, É., *Leçons sur al géometrie des espaces de Riemann*, Gauthier-Villar, Paris, 1928 (c).

Chern, S.-S., Chen, W.H., Lam, K.S., *Lectures on Differential Geometry*, World Scientific, Singapore; River Edge, NJ: 1999 (a,r).

Coddington, E., *A Brief Account of the Historical Development of Pseudospherical Surfaces from 1827 to 1887*, Columbia University thesis 1905 (c).

Conrad, B.P., *Differential Equations: A Systems Approach*, Prentice Hall, Upper Saddle River, NJ, 2002 (c).

Coolidge, J.L., *The Elements of Non-Euclidean Geometry*, Clarendon Press, Oxford, 1909 (c).

Coolidge, J.L., *A History of Geometrical Methods*, Oxford University Press, Oxford, 1940, reissued by Dover Publication, NY, 1963 (a,r).

Coxeter, H.S.M., *Non-Euclidean Geometry*, The University of Toronto Press, Toronto, 1947 (c).

Dacorogna, B., *Introduction to the Calculus of Variations*, Imperial College Press, London, 2004 (c).

Dombrowski, P., 150 years after Gauss' *Disquisitiones generales circa superficies curvas*, Astérique **62**, 1979 (c,r).

Dubrovin, B.A., Fomenko, A.T., Novikov, S.P., *Modern Geometry—Methods and Applications: Vol. 1*, The geometry of surfaces, transformation groups, and fields (1984); vol. 2, The geometry and topology of manifolds (1985); vol. 3, Introduction to homology theory (1990), trans. by S.P. Novikov and R.G. Burns, Springer-Verlag, New York (c,r).

Dugas, R., *A History of Mechanics*, Dover Publications, Mineola, NY, 1988 (c).

Einstein, A., *The Principle of Relativity*, selected papers of Einstein, Minkowski, and others, Dover Publications, NY, 1952 (c).

Encyklopädie der Mathematischen Wissenschaften, Geometrie, Vol. 3, especially section 3. Essays by H. von Mangoldt, R. von Lilienthal, G. Scheffers, A. Voss, H. Liebmann, E. Salkowski, R. Weitzenböck, L. Berwald. Edited by W.F. Mayer and H. Mohrmann, Teubner, Leipzig, 1902–27 (a).

Euclid, *The Thirteen Books of Euclid's Elements*, translated by Sir T. L. Heath, 3 vols., 2nd edition, Cambridge University Press, 1926. Reprinted by Dover Publications, New York, 1956 (c,r).

Fiala, F., *Mathematische Kartographie*, VEB Verlag Technik, Berlin, 1957 (c).

Gallot, S., Hulin, D., Lafontaine, J., *Riemannian Geometry*, 2nd ed., Springer, New York, 1990 (a).

Gauss, C.-F., Gesammelte Werke, hrsg. von der kgl. *Gesellschaft der Wissenschaften zu Göttingen*. Published 1870 by Dieterich in Göttingen. Band 1. Disquisitiones arithmeticae. Band 2. HŽhere Arithmetik. Band 3. Analysis. Band 4. Wahrscheinlichkeits-Rechnung und Geometrie. Band 5. Mathematische Physik. Band 6. Astronomische Abhandlungen (a,r).

Gauss, C.-F., Disquisitiones generales circa superficies curvas, Commentationes Societatis Regiae Scientiarum Göttingesis Recentiores. In the collected works, Volume VI, pp. 99–146. Reprinted in *150 years after Gauss' "Disquisitiones generales circa superficies curvas"* edited by P. Dombrowski, including an English translation from "General Investigations of Curved Surfaces" (published 1965) Raven Press, New York, translated by A.M.Hiltebeitel and J.C.Morehead, plus commentary. Société mathématique de France, 1979. Astérisque, 62 (c,r).

Gindikin, S. G., *Tales of Physicists and Mathematicians* (trans. A. Shuchat), Birkhäuser, Boston, 1988 (c).

Gray, A.; Abbena, E.; Salamon, S., *Modern Differential Geometry of Curves and Surfaces with Mathematica*, 3rd edition, Chapman and Hall, CRC, Boca Raton, FL, 2006 (a,r).

Gray, J.J., *Linear differential equations and group theory from Riemann to Poincaré*, Birkhaüser, Boston, 1985: 2nd edition, 2008 (c,r).

Gray, J.J., *Ideas of space: Euclidean, non-Euclidean, and relativistic*, Oxford University Press, Oxford, 2nd edition, 1989 (c,r).

Gray, J.J., Janos Bolyai, *Non-Euclidean Geometry and the Nature of Space*, Burndy Library Publications, MIT Press, Cambridge, MA, 2004 (c,r).

Gray, J.J., *Worlds out of Nothing; a course on the history of geometry in the 19th century*, Springer Undergraduate Mathematics Series. London: Springer, 2007 (a,r).

Greenberg, M.J., *Euclidean and Non-Euclidean Geometries*, W.H. Freeman and Co., New York, 4th edition, 2008 (c,r).

Guillemin, V., Pollack, A., *Differential Topology*, Prentice-Hall, Englewood Cliffs, NJ, 1974 (c,r).

Hicks, N., *Notes on Differential Geometry*, Van Nostrand Reinhold Co., London, 1971 (c,r).

Hilbert, D., *Grundlagen der Geometrie, Teubner, Leipzig, 1899*. Editions: 2nd, 1903; 3rd, 1909; 4th, 1913; 5th, 1922; 6th, 1923; 7th, 1930. English translation, Foundations of Geometry, translated by E.J. Townsend, Open Court Pub. Co., Chicago, 1902 (c,r).

Hilbert, D., *David Hilbert's lectures on the foundations of mathematics and physics, 1891–1933*. General editors, W. Ewald and M. Hallett, Springer, Berlin; New York, 2004 (a).

Hilbert, D; Cohn-Vóssen, S., *Geometry and the Imagination*; translated by P. Nemenyi, Chelsea Pub. Co., New York, 1952 (a).

Hopf, H., *Differential Geometry in the Large: Seminar lectures*, New York University, 1946 and Stanford University, 1956; with a preface by S.S. Chern, Berlin; Springer-Verlag, New York, 1983 (a).

Hsiung, C.-C., A *First Course in Differential Geometry*, Wiley, New York; 1981 (r).

Huygens, C., *Horologium oscillatorium sive de motu pendularium*, Muguet, Paris, 1673.

Jacobson, N., *Basic Algebra II*, Dover Publications, Mineola, NY, 2009 (c).

Katz, V., *A History of Mathematics*, Addison Wesley, Reading, MA. 3rd edition, 2008 (c,r).

Klein, F, *Vorlesungen §ber nicht-euklidische Geometrie*. Newly edited by W. Rosemann. Springer-Verlag, Berlin, 1928 (c).

Kobayashi, S., Nomizu, K., *Foundations of Differential Geometry*, in two volumes, John Wiley and Sons, Hoboken, NJ, 1963, 1969 (a,r).

Kreyszig, E., *Differential Geometry*, University of Toronto Press, Toronto, 1964 (a).

Kulczycki, S., *Non-Euclidean Geometry*. Translated from Polish by Stanislaw Knapowski, Pergamon Press, Oxford, New York, 1961 (a,r).

Kühnel, W., Differential Geometry : curves - surfaces - manifolds; trans. by Bruce Hunt. American Mathematical Society, Providence, RI, 2006 (a).

Lambert, J.H., *Beyträge zum Gebrauche der Mathematik und deren Anwendung*. Berlin, Dritte Theil, 1772 (c).

Lang, S. *Calculus of Several Variables*, UTM Series, Springer-Verlag, NY, 1987 (c).

Lang, S., *Linear Algebra*, UTM series, Springer-Verlag, NY, 2010 (c).

Lakatos, I., *Proofs and refutations: The logic of mathematical discovery*; edited by John Worrall and Elie Zahar, Cambridge, Cambridge University Press, New York, 1976 (a).

Laplace, P.S., *Mécanique Céleste*, vol. 1, 1829; vol. 2, 1832; vol. 3, 1834; and vol. 4, 1839, Hilliard, Gray, Little and Wilkins, Boston (c).

Laubenbacher, R., Pengelley, D., *Mathematical expeditions: Chronicles by the explorers.* UTM: Readings in Mathematics. Springer-Verlag, New York, 1999 (c, r).

Lee, J.M., *Introduction to smooth manifolds*, Springer, New York, 2003 (c).

Legendre, A.M., Éléments of Géometrie, Chez Firmin Didot, Paris, fifth edition, 1804 (a).

Lenz, H., *Nichteuklidische Geometrie*, Bibliographisches Institut, Mannheim, 1967 (a,r).

Levi-Civita, T., *The Absolute Differential Calculus.* Translated by M. Long, Blackie, London, 1929 (c,r).

Loria, G., *Spezielle Algebraische und Transscendente Ebene Kurven.* Theorie und Geschichte, trans. into German by Fritz Schütte, Teubner Verlag, Leipzig, 1902 (c,r).

Massey, W.S., *A Basic Course in Algebraic Topology, GTM* **127**, Springer-Verlag, New York, 1997, 3rd edition (c,r).

McCleary, J., *A First Course in Topology: Continuity and Dimension, STML/31*, American Mathematical Society, Providence, RI, 2006, (a).

McDonell, P.W., *Introduction to Map Projections*, Marcel Dekker, New York, 1979 (c).

Meschkowski, H., *Noneuclidean Geometry*, Academic Press, New York, 1964 (a,r).

Meyer, T.H., *Introduction to Geometrical and Physical Geodesy: Foundations of Geomatics*, ESRI Press, Redlands, CA, 2010 (c,r).

Millman, R.S., Parker, G.D., *Elements of Differential Geometry*, Englewood Cliffs, NJ, Prentice-Hall, 1977 (a).

Millman, R.S., Parker, G.D., *Geometry: A Metric Approach with Models*, Undergraduate Texts in Mathematics Series, Springer-Verlag, New York, 1981 (c).

Moise, E., *Geometric Topology in Dimensions 2 and 3*, Springer-Verlag, New York, 1977, (c).

Munkres, J., *Topology*, 2nd edition, Prentice-Hall, Upper Saddle River, NJ, 2000 (c,r).

Needham, T., *Visual Complex Analysis*, Oxford University Press, New York, 1999 (c,r).

Pais, A., *"Subtle is the Lord... " The Science and the Life of Albert Einstein*, Oxford University Press, New York, 1982 (c,r).

O'Neill, B., *Elementary Differential Geometry*, Academic Press, New York, 1966 (a,r).

O'Neill, B., *Semi-Riemannian Geometry: With Applications to Relativity*, Academic Press, Orlando, FL, 1983 (c,r).

Oprea, J., *The Mathematics of Soap Films: Explorations with Maple®*, American Mathematics Society, STML 10, Providence, RI, 2000 (c,r).

Playfair, J., Elements of Geometry; containing the first six books of Euclid, with two books on the geometry of solids. To which are added, elements of plane and spherical trigonometry, Bell and Bradfute, and G. G. and J. Robinson, London, 1795 (c,r).

Pogorelov, A.V., *Differential Geometry.* Trans. from the first Russian ed. by Leo F. Boron, P. Noordhoff, Groningen, ca. 1950 (c,r).

Poincaré, J.H., *Science and Méthode, Flammarion, Paris, 1908.* English translation, Science and Method, translated by Francis Maitland, with preface by Bertrand Russell, Thomas Nelson and Sons, London and New York, 1914 (c).

Prenowitz, W.; Jordan, M., *Basic Concepts of Geometry*, Ardsley House, New York, 1965 (a).

Pressley, A., *Elementary Differential Geometry*, London; New York : Springer, 2001 (a).

Proclus, A. *Commentary on the first Book of Euclid's Elements.* Translated, with introduction and Notes, by Glenn R. Morrow, Princeton University Press, Princeton, NJ, 1970 (c).

Richards, J., *Mathematical Visions: The Pursuit of Geometry in Victorian England*, Academic Press, New York, 1988 (a).

Riemann, B., *Gesammelte Mathematische Werke*, ed. R. Dedekind and H. Weber, Göttingen, 1892 with Supplement in 1902, Teubner, Leipzig. Reissued by Dover, New York, 1953 (c,r).

Robinson, A., *Elements of Cartography*, Wiley, New York, 1960 (a).

Rosenfeld, B., *A History of non-Euclidean Geometry* (trans. A. Shenitzer), 1st English edition, Springer-Verlag, New York, 1988 (c,r).

Ryan, P.J., *Euclidean and Non-Euclidean Geometry: An Analytic Approach*, Cambridge University Press, New York, 1986 (c,r).

Saccheri, Girolamo, *Euclides ab omni naevo Vindicatus* (Euclid vindicated of every flaw), Mediolani 1733. Translated from the Latin by George Bruce Halstead, AMS Chelsea Publishing, Providence, RI, 1986, (c,r).

Scholz, E., *Geschichte des Mannigfaltigkeitsbegriffs von Riemann bis Poincaré*. Birkhäuser, Basel-Boston-Stuttgart, 1980 (c,r).

Schouten, J.A., *Der Ricci-Kalkl*, Springer, Berlin, 1924; The Ricci Calculus, English translation, 1954 (c).

Schröder, E., *Kartenentwürfe der Erde*, Verlag Harri Deutsch, Thun, 1988 (c,r).

Sharpe, R., *Differential Geometry: Cartan's Generalization of Klein's Erlangen Program*, Springer-Verlag, GTM 166, New York, 1997 (c).

Shirokov, P.A., *A Sketch of the Fundamentals of Lobachevskian Geometry*. Prepared for publication by I.N. Bronshtein. Translated from the 1st Russian ed. by Leo F. Boron, with the assistance of Ward D. Bouwsma, P. Noordhoff, Groningen, 1964 (c,r).

Sommerville, D., *The Elements of Non-Euclidean Geometry*, Dover, New York, 1958 (a).

Sommerville, D., *Bibliography of Non-Euclidean Geometry*, Chelsea, House New York, 1970 (a).

Snyder, J.P., *Flattening the Earth: Two Thousand Years of Map Projections*. University of Chicago Press, Chicago and London, 1993 (c,r).

Sperry, P., *Short Course in Spherical Trigonometry*, Johnson Publ. Co., Richmond, VA, 1928 (a).

Spivak, M., *Calculus on Manifolds*, Westview Press, Boulder, CO, 1971 (c,r).

Spivak, M., *A Comprehension Introduction to Differential Geometry*. Vol. 1–2 1970, vol. 3–5, 1975, Publish or Perish Press, Boston, MA; 2nd editions, 1979 (c,a,r).

Stäckel, P., Engel, F., *Die Theorie der Parallellinien von Euklid bis auf Gauss*, Teubner, Leipzig, 1895 (c,r).

Stehney, A.K., Milnor, T.K., D'Atri, J.E., Banchoff, T.F., editors, Selected Papers on Geometry (The Raymond W. Brink selected mathematical papers; v. 4), Mathematical association of Amer; Washington, DC (1979) (a,r).

Stillwell, J.C., *Geometry of Surfaces*, Springer, New York, 1992 (a).

Stillwell, J.C., *Sources of Hyperbolic Geometry*, American Mathematics Society, Providence, RI, 1996 (c,r).

Stillwell, J.C., *Four Pillars of Geometry*, Springer, New York, 2005 (c, r).

Stoker, J.J., *Differential Geometry*, Wiley-Interscience, New York, 1969 (a).

Struik, D.J., *Lectures on Classical Differential Geometry*, Addison-Wesley Press, Cambridge, MA, 1950 (r).

Todhunter, I., *History of the Mathematical Theories of Attraction and Figure of the Earth from Newton to Laplace*, MacMillan and Co., London, 1873 (c).

Torretti, R., *Philosphy of Geometry from Riemann to Poincaré*, Reidel Publishing Co., Dordrecht, Holland, 1978 (c,r).

van Brummelen, G., *The Mathematics of the Heavens and the Earth: The Early History of Trigonometry*, Princeton University Press, Princeton, NJ, 2009 (c,r).

Vitale, G., *Euclide restituto, ovvero gli antichi elementi geometrici ristaurati e facilitati da Vitale Giordano da Bitonto*. Libri XV. ("Euclid Restored, or the ancient geometric elements rebuilt and facilitated by Giordano Vitale, 15 Books"), (1st edition 1680, Rome. 2nd edition with additions 1686, Rome) (c).

Warner, F., *Foundations of Differentiable Manifolds and Lie Groups*, GTM vol. 94, Springer-Verlag, New York, 1983 (c,r).

Weatherburn, C.E., *Differential Geometry of Three Dimensions*, Cambridge University Press, Cambridge. 1927 (a).

Weyl, H., *The Concept of a Riemann Surface*, translated by G.R. MacLane, Addison-Wesley, Reading, MA, 1955. Reissued by Dover Publications, NY, 2009 (c).

Willmore, T., *An Introduction to Differential Geometry*, Clarendon Press, Oxford, 1959 (a).

Yoder, J. G. *Unrolling Time: Christiaan Huygens and the Mathematization of Nature*, Cambridge University Press, New York, 1988 (c).

Zwikker, C., *Advanced Plane Geometry*, North-Holland Publ. Co., Amsterdam, 1950 (c).

Articles

Beltrami, E., Rizsoluztione del problema: "Riportare i punti di una superficie sopra un plano in modo che le linee geodetiche vengano rappresentate da linee rette," *Annali di Mathematiche pura ed applicata* (1)**7**(1865), 185–204 (c).

Beltrami, E., Saggio di interpretazione della Geometria non-Euclidea, *Giornale di Mat.*, **6**(1868), 284–312. Translated into English in (Stillwell 1996) (c,r).

Beltrami, E., Teoria fundamentale degli spazii di curvatura constante, *Annali. Di Mat.*, ser. II **2**(1869), 232–55. Translated into English in (Stillwell 1996) (c).

Bertrand, J., Démonstration d'un théorème de M. Gauss, *J. Math. Pure Appl.* **13**(1848), 80–6. Contains an account of Diguet's theorem (c).

Blanuša, D., Über die Einbettung hyperbolischer Räume in euklidische Räume, *Monatsh. Math.* 59 (1955), 217Ð229 (c).

Bonnet, P.O., Mémoire sur la théorie générale des surfaces, *Journal de l'École Polytechnique*, **32**(1848), 1–46 (c).

Bonnet, P.O., Mémoire sur la théorie des surfaces applicables sur une surface donnée, *J. École Poly.* **24**(1865), 209–30 (a).

Brooks, J., Push, S., The Cycloidal Pendulum, *The Amer. Math. Monthly*, **109**(2002), 463–465 (c).

Busemann, H., Non-Euclidean geometry, *Math. Mag.* **24**(1950), 19–34 (c).

Cayley, A., A sixth memoir upon quantics, *Phil. Trans. of the Royal Society of London*, **149**(1859), 61–90 (c).

Christoffel, E.B., Über dis Transformation der homogenen Differentialausdrücke zweiten Grades, *Crelle* **70**(1869), 46–70 (c).

Coddazi, D., Mémoire relatif à l'application des surfaces les unes sur les autres (envoyé au concours ouvert sur cette question en 1859 par l'Academie des Sciences), Mém. prés. div. sav. *Acad. Sci. Paris* (2)**27**(1883), 1–47 (c).

Doyle, P.H., Moran, D.A., A Short Proof that Compact 2-manifolds can be triangulated, *Inventiones Mathematiques* **5**(1968), 160–162 (c).

Einstein, A., Grossmann, M., Entwurf einer verallgemeinerten Relativitätstheorie und einer Theorie der Gravitation, B.G.Teubner (separatum), Leipzig (1913); with addendum by Einstein in *Zeitschrift für Mathematik und Physik*, 63(1914), pp. 225-61. (Papers, Vol. 4) (c).

Euler, L., De constructione aequationum ope motus tractorii aliisque ad methodum tangentium inversam pertinentibus, *Comm. acad. sci. Petroopl.* **8**(1736), 1741, 66–85. Opera Omnia, Series I, vol. XXXII83–107 (c).

Euler, L., Principes de la trigonomŐtrie sphŐrique, tirés de la méthode des plus grands et plus petits; *mémoires de l'Académie royale des sciences et belles-lettres* (Berlin) **9**(1753). [Berlin, 1755] (c).

Euler, L., Recherches sur la courbure des surfaces, E333, *M'emoires de l'academie des sciences de Berlin*, **16**(1760), 1767, 119–143. In Opera Omnia (1) 28, 1–22 (c).

Euler, L., De solidis quorum superficiem in planum explicare licet, *Novi Comm. Acad. Sci. Petropolitanae*, **16**(1771), 1772, 3–34. E419, in Opera Omnia (1) 28, 298–321 (c).

Euler, L., De mensura angulorum solidorum, Acta Academiae Sci. Imp. Petropolitinae, **2**(1781), 31–54. E514 in Opera Omnia (1) 26, 204–223 (c).

Euler, L., Methodus facilis omnia symptomata linearum curvarum non in eodem plano sitarum investigandi, *Acta Academiae Scientarum Imperialis Petropolitinae* 1782, 1786, 19–57. Opera Omnia: Series 1, **28**(1782), 348–38 (c).

Frenet, F., Sur les courbes à double courbure, extrait d'une thèse à la Faculté des Sciences de Toulouse, le 10 juillet 1847, *J. Math. Pure Appl.* **17**(1852), 437–447 (c).

Gauss, C.-F., Allgemeine Auflösung der Aufgabe: Die Theile einer andern gegebnen Fläche so abzubilden, dass die Abbildung dem Abgebildeten in den kleinsthen Theilen ähnlich wird (als Beantwortung der von der königlichen Societät der Wissenschaften in Copenhagen für 1822 aufgegebnen Preisfrage), *Astr. Abh.* (1825), 1–30 (a).

Gauss, C.-F., Beiträge zur Theorie der algebraischen Gleichungen, Juli 1849, *Gesammlte Werke* vol. 3 (1876) (c).

Gray, J.J., Non-Euclidean geometry – a re-interpretation, *Hist. Math.* **6**(1979), 236–58 (a,r).

Hazzidakis, J.N., Über einige Eigenschaften der Flächen mit constantem Krümmungsmass, *J. für reine und angew. Math.* **88**(1887), 68–73.

Hilbert, D., Über Flächen von konstanter Gausscher Krümmung, *TAMS* **1**(1901), 87–99 (c).

Hoffman, D., Meeks, W.H., Minimal surfaces based on the catenoid, *Amer. Math. Monthly* **97**(1990), 702–30 (c).

Holmgren, E., Sur les surfaces à courbure constant négative, *Comptes Rendus Acad. Sci. Paris*, Series A-B, **134**(1902), 740–43 (c).

Hopf, H.; Rinow, W., Über den Begriff der vollständigen differentialgeometrischen Fläche, *Comm. Math. Helv.* **3**(1931), 209–25 (c).

Hopf, H., Über die Drehung der Tangenten und Sehnen ebener Kurven, *Comp. Math.* **2**(1935), 50–62 (c).

Hopf, H., Zur Topologie der komplexen Mannigfaltigkeiten, Studies and Essays presented to R. Courant, Interscience Publishers Inc., New York, 1948, 167–185 (c).

Jacobi, C.G.J., Demonstration et amplificatio nova theorematis Gaussiani de quadrata integra triangula in data superficie e lineis brevissimis formati, *J. Math. Crelle* **16**(1837), 344–350 (c).

Klein, F., Über die sogenannte Nicht-Euclidische Geometrie, *Math. Ann.* **4**(1871), 573–625 (cf. *Ges. Math. Abh.* **1**, 244-350) (c).

Lagrange, J. L. "Sue les courbes tautochrones." Mém. de l'Acad. Roy. des Sci. et Belles-Lettres de Berlin 21, 1765. Reprinted in Oeuvres de Lagrange, tome 2, section deuxime: Mmoires extraits des recueils de l'Academie royale des sciences et Belles-Lettres de Berlin. Paris: Gauthier-Villars, pp. 317–332, 1868 (c).

Lambert, J.H., Theorie der Parallellinien, 1786. Excerpts in Stäckel, Engel (c).

Lambert, J. H., Observations trigonométriques. Mémoires de l'Académie royale des sciences de Berlin, année 1768/1770, 327–354.

Laubenbacher, R., Pengelley, D., *Mathematical expeditions: Chronicles by the explorers.* UTM: Readings in Mathematics. Springer-Verlag, New York, 1999 (c,r).

Lawlor, G., A new minimization proof for the brachistochrone, *Amer. Math. Monthly* **103**(1996), 242–249 (c).

Legendre, A.M., Éléments of Géometrie, Chez Firmin Didot, Paris, fifth edition, 1804 (a).

Levi-Civita, T., Nozione di parallelismo in una varietà qualunque, *Rend. Circ. Mat. Palermo* **42**(1917), 173–205 (c).

Liebmann, H., Über die Verbiegung der geschlossenen Flächen positiver Krümmung, *Math. Ann.* **53**(1900), 81–112 (c).

Lobachevskiĭ, N.I., O natschalach geometrii (Russian), Kasaner Bote 1829–30 (c).

Lobachevskiĭ, N.I., Imaginary geometry (Woobrashajemaja geometrija), Papers of the University of Kasan, 1835. Appeared in French in *J. für reine und angew. Math.* **17**(1837), 295–320 (c).

Lobachevskiĭ, N.I., New foundations of geometry with a complete theory of parallels (Nowja natschala geometrii s polnoj teorijij parallelnych), Papers of the University of Kasan, 1835–38 (c).

Lobachevskiĭ, N.I., Application of imaginary geometry to certain integrals (Primjenjenije woobrashajemoj geometrii k njekotorych integralach), Papers of the University of Kasan, 1836 (a).

Lumiste, Ü, Martin Bartels as researcher: his contribution to analytical methods in geometry. *Historia Math.* **24**(1997), 46–65 (c).

Lützen, J., Interactions between mechanics and differential geometry in the 19th century, *Arch. Hist. Exact Sci.* **49**(1995), 1–72 (c,r).

Mac Lane, S., Metric postulates for plane geometry, *Amer. Math. Monthly* **66**(1959), 543–55 (c,r).

Mainardi, G. Su la teoria generale delle superficie, *G. Ist. Lomb. Milano* (2)**9**(1857), 385–98 (c).

Malus, E. L., Traité d'Óptique, in Mémoires présentés à lInstitut des sciences par divers savants, **2**(1811), 214–302 (c).

McCleary, J., On Jacobi's remarkable curve theorem, *Historia Math.* **21**(1994), 377–85 (c).

McCleary, J., Trigonometries, *Amer. Math. Monthly*, **109**(2002), 623–38 (c,r).

Millman, R.S.; Stehney, A.K., The geometry of connections, *Amer. Math. Monthly*, **80**(1973), 475–500 (a,r).

Minding, F., Über die Curven des kürzesten Perimeters auf krummen Flächen, *J. Math. Crelle* **5**(1830), 297–304 (c).

Minding, F., Wie sich enscheiden läßt, ob zwei gegebene krumme Flächen auf einander abwickelbar sind oder nicht; nebst Bemerkungen über die Flächen von unveränderlichem Krümmungsmaße, *Crelle* **1**(1839), 370–87 (c).

Milnor, J.W., A problem in cartography, *Amer. Math. Monthly* **76**(1969), 1101–12 (a).

Milnor, J.W., Hyperbolic geometry: The first 150 years, *BAMS* **6**(1982), 9–24 (a,r).

Pinl, M., Christoffels Weg zum absoluten Differentialkalkül und sein Beitrag zur Theorie des Krümmungstensors, in *Butzer and Fehér* (1981), 474–79 (c,r).

Poincaré, J.H., Théorie des Groupes Fuchsiens, *Acta Mathematica* **1**(1882), 1–62. Translated into English in (Stillwell 1996) (c,r).

Poincaré, J.H., Analysis situs, *J. École Poly.* (2)**1**(1895), 1–123 (c).

Puiseux, V., Sur le même théorème, *J. Math. Pure Appl.* **13**(1848), 87–90. See (Bertrand) (c).

Radó, T., Über den Begriff der Riemannschen Fläche, *Acta Litt. Sci. Szeged* **2**(1925), 100–21 (c).

Reich, K., Die Geschichte der Differentialgeometrie von Gauss bis Riemann (1828–1868). *Arch. History Exact Sci.* **11**(1973/74), 273–382 (c,r).

Ricci, G., Levi-Civita, T., Méthodes de calcul différentiel absolu et leur applications, *Math. Annalen* **54**(1901), 125–201 (c).

Riemann, B., Über die Hypothesen, welche der Geometrie zu Grunde liegen, *Abhandlungen der Königlichen Gesellschaft der Wissenschaften zu Göttingen* **13**(1868) (c,r).

Riemann, B., Commentatio mathematica, qua respondere tentatur quaestioni ab Illma Academia Parisiensi propositae: "Trouver quel doit être l'état calorifique d'un corps solide homogène indéfeni pour qu'un système de courbes isothermes, à un instant donné, restent isothermes après un temps quelconque, de telle sorte que la température d'un point puisse s'exprimer en fonction du temps et de deux autres variables indépendantes." (1861), Gesammelte Mathematische Werke, 2nd ed., 391–404 (c).

Rodrigues, O., Recherches sur la théorie analytique des lignes et des rayons de courbure des surfaces, et sur la transformation d'une classe d'intégrales doubles, qui ont un rapport direct avec les formules de cette théorie, *École Poly. Corresp.* **3**(1814–16), 162–82 (c).

Rozendorn, È.R., A realization of the metric $ds^2 = du^2 + f_2(u)dv^2$ in a five-dimensional Euclidean space. (Russian) Akad. *Nauk Armjan. SSR Dokl.* **30**(1960), 197–99 (a).

Russell, B., Geometry, non-Euclidean, in Encyclopedia Britannica, Suppl. vol. 4, 1902. Cited in Heath (Euclid) (c).

Scholz, E., The concept of manifold, 1850–1950. In *History of topology*, edited by I.M. James, 25–64, North-Holland, Amsterdam, 1999 (c,r).

Scholz, E. Gauss und die Begründung der "höhere" Geodäsie, In S. S. Demidov; M. Folkerts; D. Rowe; C.-J. Scriba (Hrsg.): *Amphora. Festschrift für Hans Wuğing*. Basel: BirkhŁuser, 1992, 631–47 (c).

Scholz, E., C.F. Gauß Präzisionsmessungen terrestrischer Dreiecke und seine Überlegungen zur empirischen Fundierung der Geometrie in den 1820er Jahren. In: Folkerts, Menso; Hashagen, Ulf; Seising, Rudolf; (Hrsg.): Form, Zahl, Ordnung. Studien zur Wissenschafts- und Technikgeschichte. Ivo Schneider zum 65. Geburtstag. *Stuttgart: Franz Steiner Verlag*, 2004, 355Ð380 (c).

Serret, J.A., Sur quelques formules relatives à double courbure, *J. Math. Pure Appl.* **16**(1851), 193–207 (c).

Struik, D., Outline of a history of differential geometry. I, *Isis* **19**(1933), 92–120, II, **20**(1934), 161–91 (c,r).

Taurinus, F.A., Theorie der Parallellinien, published in 1825. In Engel and Stäckel (1895) (c).

Tchebychev, P.L., Sur la coupe des vêtements, OEuvres, vol. 2, 708 (c).

Weyl, H., Reine Infinitesimalgeometrie, *Math. Z.* **2**(1918), 384–411 (c).

Symbol index

Name index

Subject index

States

Printed in the United
By Bookmasters